Ivo Petráš (Ed.)

Handbook of Fractional Calculus with Applications

Handbook of Fractional Calculus with Applications

Edited by *J. A. Tenreiro Machado*

Volume 1: Theory
Anatoly Kochubei, Yuri Luchko (Eds.), 2019
ISBN 978-3-11-057081-6, e-ISBN (PDF) 978-3-11-057162-2,
e-ISBN (EPUB) 978-3-11-057063-2

Volume 2: Fractional Differential Equations
Anatoly Kochubei, Yuri Luchko (Eds.), 2019
ISBN 978-3-11-057082-3, e-ISBN (PDF) 978-3-11-057166-0,
e-ISBN (EPUB) 978-3-11-057105-9

Volume 3: Numerical Methods
George Em Karniadakis (Ed.), 2019
ISBN 978-3-11-057083-0, e-ISBN (PDF) 978-3-11-057168-4,
e-ISBN (EPUB) 978-3-11-057106-6

Volume 4: Applications in Physics, Part A
Vasily E. Tarasov (Ed.), 2019
ISBN 978-3-11-057088-5, e-ISBN (PDF) 978-3-11-057170-7,
e-ISBN (EPUB) 978-3-11-057100-4

Volume 5: Applications in Physics, Part B
Vasily E. Tarasov (Ed.), 2019
ISBN 978-3-11-057089-2, e-ISBN (PDF) 978-3-11-057172-1,
e-ISBN (EPUB) 978-3-11-057099-1

Volume 7: Applications in Engineering, Life and Social Sciences, Part A
Dumitru Băleanu, António Mendes Lopes (Eds.), 2019
ISBN 978-3-11-057091-5, e-ISBN (PDF) 978-3-11-057190-5,
e-ISBN (EPUB) 978-3-11-057096-0

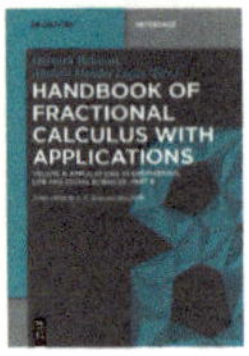

Volume 8: Applications in Engineering, Life and Social Sciences, Part B
Dumitru Băleanu, António Mendes Lopes (Eds.), 2019
ISBN 978-3-11-057092-2, e-ISBN (PDF) 978-3-11-057192-9,
e-ISBN (EPUB) 978-3-11-057107-3

Preface

Fractional Calculus (FC) originated in 1695, nearly at the same time as the conventional calculus. However, FC attracted limited attention and remained a pure mathematical exercise in spite of the contributions of important mathematicians, physicists, and engineers. FC had a rapid development during the last few decades, both in mathematics and applied sciences, being nowadays recognized as an excellent tool for describing complex systems, phenomena involving long range memory effects and nonlocality. A huge number of research papers and books devoted to this subject have been published, and presently several specialized conferences and workshops are organized each year. The FC popularity in all fields of science is due to its successful application in mathematical models, namely in the form of FC operators and fractional integral and differential equations. Presently, we are witnessing considerable progress both on theoretical aspects and applications of FC in areas such as physics, engineering, biology, medicine, economy, or finance.

The popularity of FC has attracted many researchers from all over the world and there is a demand for works covering all areas of science in a systematic and rigorous form. In fact, the literature devoted to FC and its applications is huge, but readers are confronted with a high heterogeneity and, in some cases, with misleading and inaccurate information. The Handbook of Fractional Calculus with Applications (HFCA) intends to fill that gap and provides the readers with a solid and systematic treatment of the main aspects and applications of FC. Motivated by these ideas, the editors of the volumes involved a team of internationally recognized experts for a joint publishing project offering a survey of their own and other important results in their fields of research. As a result of these joint efforts, a modern encyclopedia of FC and its applications, reflecting present day scientific knowledge, is now available with the HFCA. This work is distributed by several distinct volumes, each one developed under the supervision of its editors.

The aim of the sixth volume of the HFCA is a collection of various topics in the area of FC and its applications in control in order to highlight this very emerging research field so-called fractional-order control (FOC). The FOC is a field of control theory that uses the FC technique as the fractional-order integrator and fractional-order differentiator, respectively. It may be used in any control level, generally in basic control as well as advance control. The main advantage of the FOC is that the fractional-order operator (integral or derivative) weights history using a function that decays with a power law. In fact, such operator is a nonlocal operator and considers the whole history of signal (infinite memory). Volume six of the HFCA is structured into three groups devoted to control theory, design methods and toolboxes, and applications. In this volume, readers can find in 18 chapters a description of dynamical properties of fractional-order models (stability, observability, initial conditions problem, etc.) for continuous and discrete systems, identification methods, responses investigation in

https://doi.org/10.1515/9783110571745-201

various domains, and several toolboxes in Matlab/Simulink. Moreover, a number of control techniques as, for example, linear PID controller, CRONE controllers, nonlinear PID controller, various nonlinear control techniques as, for instance, sliding mode control and reset control, robust control (H_2, H_∞), optimal control, posicast control, adaptive control, lead-lag compensators, and the synchronization method together with their implementation techniques in analog and discrete forms and their applications in various areas are described as well.

My special thanks go to the authors of individual chapters that are excellent surveys of selected classical and new results in several important fields of FC. The editors believe that the HFCA will represent a valuable and reliable reference work for all scholars and professionals willing to develop research in the challenging and timely scientific area.

Ivo Petráš

Contents

Blas M. Vinagre, Inés Tejado, and S. Hassan HosseinNia

Nonlinear control methods

Abstract: This chapter explores the generalization to fractional order of some adaptive, sliding, and reset strategies as special types of nonlinear feedback control. Illustrative examples of application are given to demonstrate the benefits of fractional-order control in such classical nonlinear systems.

Keywords: Nonlinear, control, fractional, adaptive, sliding, reset

MSC 2010: 93B52, 93C10, 26A33, 93C40

1 Introduction

The interest for the introduction of fractional dynamics in nonlinear control is motivated, similar to other control fields, by the very good proven performances of fractional systems with respect to those of integer order. Many authors have proposed new fractional nonlinear control techniques, mostly inspired from classical control schemes. It is worth to notice that this effort toward the generalization of nonlinear control theory to fractional order has revealed important open problems, especially those related with stability analysis.

In this chapter, we will consider adaptive, sliding, and reset control as special types of nonlinear feedback control.

It is important to remark that, in the context of this chapter, fractional dynamics will be only considered in controllers in order to emphasize its benefits in control design; the plant models are assumed to be of integer order.

2 Adaptive control

Adaptive control covers a set of techniques which provide a systematic approach for automatic adjustment of controllers in order to achieve or maintain a desired performance in the control system in response to changes in the dynamics of the own system and external disturbances.

Blas M. Vinagre, Inés Tejado, Industrial Engineering School, University of Extremadura, Avenida de Elvas, s/n, 06006, Badajoz, Spain, e-mails: bvinagre@unex.es, itejbal@unex.es
S. Hassan HosseinNia, Faculty of Mechanical, Maritime and Materials Engineering, Technical University of Delft, Mekelweg 2, 2628 CD Delft, The Netherlands, e-mail: s.h.hosseinniakani@tudelft.nl

https://doi.org/10.1515/9783110571745-001

The way in which information is processed to tune the controller for achieving the desired performances will characterize the various adaptation techniques. We will now go on to present some basic schemes used in adaptive control, namely, gain scheduling (GS), auto-tuning, and model reference adaptive control (MRAC).

2.1 Gain scheduling

2.1.1 Generalities

GS is a special kind of open-loop adaptation or change of controller parameters. A block diagram of a system controlled with GS is shown in Figure 1. In many situations it is known, or can be measured or estimated, how the dynamics of a system changes with the operating conditions through auxiliary variables. With this information, it is then possible to reduce the effects of parameter variations simply by changing the parameters of the controller as functions of the auxiliary variables accordingly. Hence, the adaptation mechanism in this case is a simple look-up table stored in the hardware in which the controller is implemented, which gives the controller parameters for a given set of environment measurements. This can be done as quickly as the auxiliary variables respond to system changes. It should be said that this approach is an open-loop adaptive control system because the modifications of the system performance resulting from the change in controller parameters are not measured and feedback to a comparison-decision block in order to check the efficiency of the parameter adaptation.

It is difficult to give general rules for designing gain scheduled controllers. The key aspect is to identify the variables that reflect the system operation conditions, which will be used as scheduling variables, and find the relations or laws (ideally, simple expressions) between them and the controller parameters as a function of the operating conditions.

For example, consider a tank where the section A varies with height h as follows (Aström and Wittenmark [1]):

$$\frac{\mathrm{d}}{\mathrm{d}t}(A(h)h) = q_i - a\sqrt{2gh}, \tag{1}$$

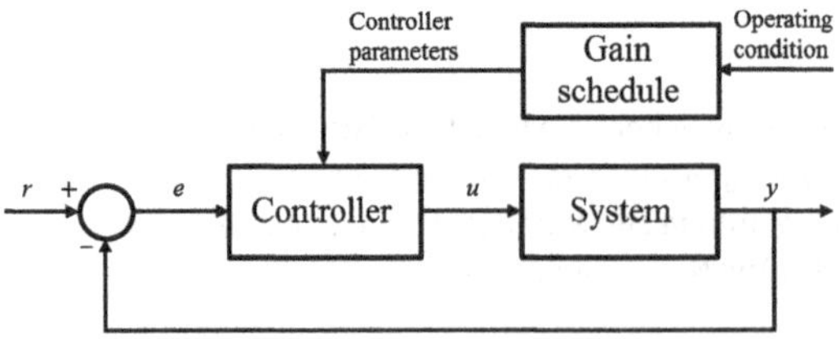

Figure 1: Block diagram of gain scheduling.

where q_i is the input flow, a the cross section of the outlet pipe, and g the gravity acceleration. Let q_i be the input and h the output of the system. Considering a PI controller and the linearized model at an operation point (q_i^0, h^0), doing some calculations it is possible to obtain the following expressions for the controller parameters:

$$k_p = 2\delta\omega_n A h^0, \quad T_i = \frac{2\delta}{\omega_n} \tag{2}$$

where δ and ω_n are the damping coefficient and the natural frequency of the closed-loop system, respectively. Hence, in this case it is sufficient to make the controller gains proportional to A.

However, when it is not possible to find the relations between auxiliary variables and controller parameters, another way of designing gain scheduled controllers is by minimizing a defined cost function over controller gains for the given operating conditions.

2.1.2 Gain and order scheduling

When a fractional PID controller replaces the conventional PID in the control scheme, gain and order scheduling (GOS) arises from the idea of improving the system performance obtained by solely using GS not only changing the controller gains, but also with the adjustment of its order (or orders), as shown in Figure 2.

For illustration purposes, consider the velocity servomotor in Tejado et al. [30] with transfer function

$$G(s) = \frac{0.92}{0.45s + 1} e^{-0.2s}, \tag{3}$$

to be controlled over the Internet by a fractional-order PI$^\lambda$ controller

$$C(s) = k_p + \frac{k_i}{s^\lambda}, \tag{4}$$

with $k_p = 1.05$, $k_i = 3.96$, and $\lambda = 0.9$ (referred to as nominal values of the controller parameters). Since the communication network is a multipurpose shared medium, a

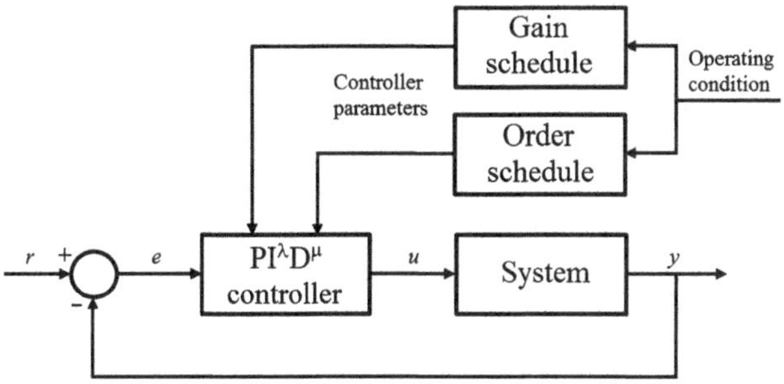

Figure 2: Block diagram of gain and order scheduling.

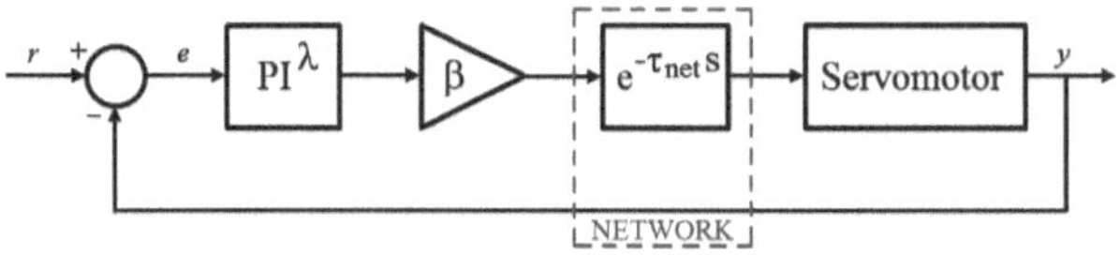

Figure 3: Block diagram for designing GOS for system (3).

delay appears in the control-loop, τ_{net}, which is time-varying, random, and dependent on the number of users connected to the network. That means that the actual delay of the system in closed-loop is also time-varying, that is, $L = L_0 + \tau_{\text{net}}$, being $L_0 = 0.2$ s the nominal delay of the system.

As operating conditions change with τ_{net}, a GOS approach is proposed to compensate its effects on the system performance. In particular, for the controller design the following cost function J is considered to be minimized:

$$J = \omega_1 J_1 + \omega_2 J_2, \tag{5}$$

where J_1 is the overshoot, J_2 is the integral of the squared error (ISE) of the closed-loop system, and ω_1 and ω_2 are the weights of each subfunction in J (in this case, $\omega_1 = \omega_2 = 0.5$).

Let β_{op} and λ_{op} denote the law of the gain scheduler and the order scheduler, respectively. In this case, simulations with the control scheme shown in Figure 3 are used to search for the optimal laws to adjust both the controller gains and the order as follows:

1. Subdivide the working range of the system for the network conditions into an adequate number of finite intervals.
2. Calculate the maximum value of β (referred to as $\beta_{\max}$) for each value of τ_{net} which guarantees the system's stability by applying the Nyquist stability criterion.
3. Simulate the closed-loop system changing β from 0 to $\beta_{\max}$ for each value of τ_{net} (as a constant delay) and evaluate J for each case. The order λ is fixed to its nominal value.
4. Obtain the law β_{op} as the value of β which minimizes J for each τ_{net} (see Figure 4(a)).
5. Repeat steps 3 and 4 changing the order λ for user-defined increments in the interval $(0, 2)$, while β takes the optimal value obtained previously. Similarly, λ_{op} will be the value of λ which minimizes J for each τ_{net}. The order scheduler law is represented in Figure 4(b).

For illustration purposes, experiments are divided into three scenarios, in which, firstly, gain scheduled controller will be compared with the (nonscheduled) nominal PI$^{\lambda}$ controller (scenario I) and, secondly, with the gain-order scheduled controller (scenario II), considering in both cases three network conditions: (a) $\tau_{\text{net}} \in [0.05, 0.12]$ s, (b) $\tau_{\text{net}} \in [0.12, 0.18]$ s, and (c) $\tau_{\text{net}} \in [0.18, 0.25]$ s.

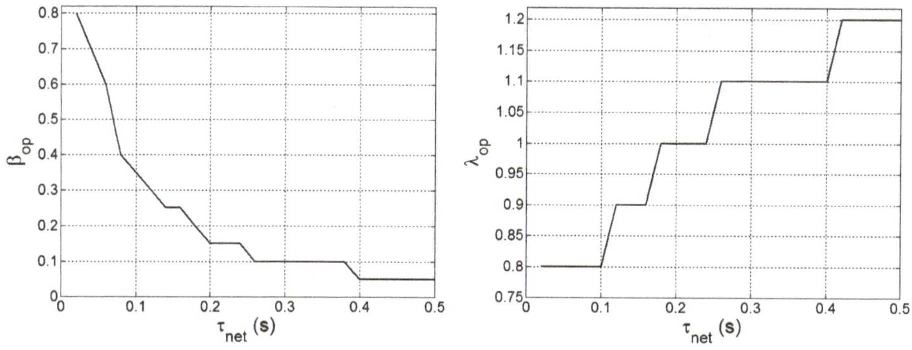

(a) (b)

Figure 4: Optimal laws of the: (a) Gain scheduler (b) Order scheduler.

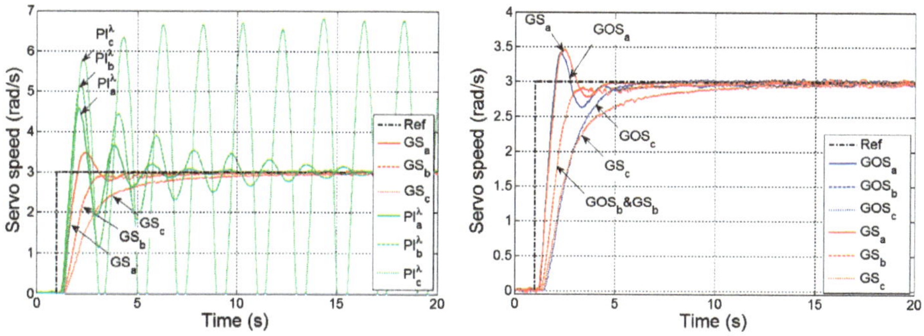

(a) (b)

Figure 5: Experimental responses of the servo when applying: (a) PI^{λ} controller and GS (scenario I) (b) GS and GOS (scenario II). (Notice that the scale of y-axis is not the same for the two scenarios).

Figure 5(a) shows the comparison of the responses obtained by using the nominal controller and the gain scheduled controller (denoted as GS). As can be observed, the servo performance is significantly better when using GS, obtaining an improvement with respect to the nonscheduled controller about 67 % in terms of the cost function J. Moreover, it is important to note that the improvement in terms of the overshoot is especially greater.

Figure 5(b) compares the responses when applying the two scheduled controllers. It can be seen that there exist only slight differences in the time responses of the servo for the cases (a) and (c), being a bit faster when using GOS, but definitely GOS improves the servo performance in comparison with GS in terms of J. However, relative stability of the system with this controller seems to be poorer, which may be caused by the order jumps in the optimal law.

2.1.3 Stability analysis

For simplicity, consider a first-order system with time-varying delay controlled by GOS together with a fractional PID of the form

$$C(s) = k_p + \frac{k_i}{s^{\lambda}} + k_d s^{\mu}, \tag{6}$$

where $\lambda, \mu \in \mathbb{R}^+$ are the orders of its integral and derivative parts, respectively. This control scheme leads to the consideration that the stability problem of the controlled system can be analyzed on the basis of switching systems using the frequency domain theory proposed in HosseinNia et al. [15].

Thus, the system and the controller can be written, respectively, as

$$G_j(s) = \frac{K}{Ts+1} e^{-(L_j)s}, \tag{7}$$

$$C_j(s) = \beta_j\left(k_p + \frac{k_i}{s^{\lambda_j}} + k_d s^{\mu_j}\right), \quad j = 1, 2, \dots, m, \tag{8}$$

where L_j refers to the time-varying system delay, β_j is the law of the gain scheduler, λ_j and μ_j are the laws corresponding to the order schedulers, and m is the number of switching subsystems in which the control problem is split.

Consider a switching system given by

$$\dot{x} = Ax, \quad A \in \operatorname{co}\{A_1, \dots, A_m\}, \tag{9}$$

where co denotes a convex combination and A_i, $i = 1, \dots, m$, is the switching subsystem, and its characteristic polynomial of order n as $c(s) = s^n + c_{n-1}s^{(n-1)} + \dots + c_1 s + c_0$ with

$$A = \begin{bmatrix} -c_{n-1} & -c_{n-2} & \cdots & -c_1 & -c_0 \\ 1 & 0 & \cdots & 0 & 0 \\ 0 & 1 & \cdots & 0 & 0 \\ \vdots & \vdots & \ddots & \vdots & \vdots \\ 0 & 0 & \cdots & 1 & 0 \end{bmatrix}.$$

Consider $c_1(s), c_2(s), \dots$, and $c_m(s)$, m stable polynomials of order n corresponding to the switching subsystems $\dot{x} = A_1 x, \dot{x} = A_2 x, \dots, \dot{x} = A_m x$, respectively, of system (9). Then system (9) is quadratically stable if and only if:

$$\left|\arg(c_{m+1}(j\omega)) - \arg(c_{m+2}(j\omega))\right| < \frac{\pi}{2}, \quad \forall\omega, \tag{10}$$

$$\left|\arg(c_{m+2}(j\omega)) - \arg(c_{m+3}(j\omega))\right| < \frac{\pi}{2}, \quad \forall\omega, \tag{11}$$

$$\vdots$$

$$|\arg(c_{2m-1}(j\omega)) - \arg(c_{2m}(j\omega))| < \frac{\pi}{2}, \quad \forall\omega, \tag{12}$$

where $c_{m+1}(s) = \frac{c_1(s)+c_2(s)+\cdots+c_{m-1}(s)}{m-1}$, $c_{m+2}(s) = \frac{c_1(s)+c_2(s)+\cdots+c_m(s)}{m-1}$, ..., $c_{2m}(s) = \frac{c_2(s)+c_3(s)+\cdots+c_m(s)}{m-1}$. Therefore, the quadratic stability of the controlled system will be proved by fulfilling the set of conditions (10)–(12).

2.2 Auto-tuning

Automatic tuning of conventional PID control has been of interest among control engineers over the past several decades since those controllers are the most widely used in industrial applications. This method must have the capability of setting controller parameters with minimal human intervention, but for the sake of robustness.

There is a wide variety of auto-tuning methods for integer controllers. Some of them aim in someway at the robustness of the controlled system (Tan et al. [29]). Among them, the relay test is recognized as the most effective instrument for auto-tuning (see, e. g., Boiko [3], Hang et al. [10]).

2.2.1 Classical relay test

There are three key reasons for advocating the use of relay test for auto-tuning purposes. The first is that this test will automatically generate an excitation input signal to ensure successful identification of the process information that is relevant for the PID controller design, which is achieved without a priori information about the system dynamics. The second is that the system is under closed-loop control with the nonlinear relay controller such that it is maintained at the operating condition chosen by the user. The third is that the sustained periodic oscillation generated by the relay feedback control produces an excellent signal-to-noise ratio for the estimation of critical process frequency information.

One of the variation of the standard relay test is shown in Figure 6, where a delay θ_a is introduced after the relay function. With this scheme, the following relations are given (Chen and Moore [5]):

$$\arg(G(j\omega_c)) = -\pi + \omega_c\theta_a, \tag{13}$$

$$|G(j\omega_c)| = \frac{\pi a}{4d} = \frac{1}{N(a)}, \tag{14}$$

where $G(j\omega_c)$ is the transfer function of the plant at the frequency ω_c, which is the frequency of the output signal y corresponding to the delay θ_a, d is the relay output, a is the amplitude of the output signal y, and $N(a)$ is the equivalent relay gain. This

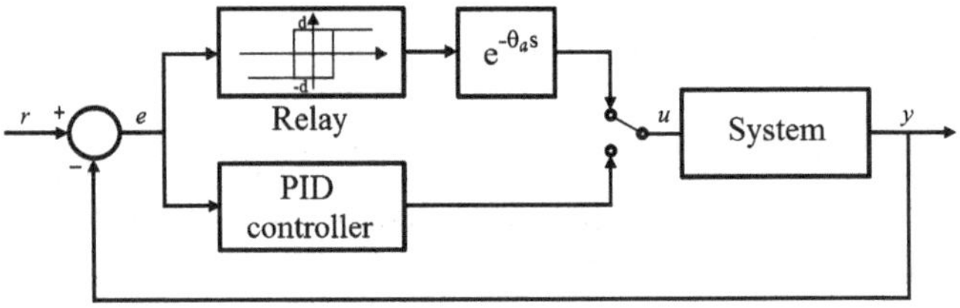

Figure 6: Block diagram of relay auto-tuning scheme with delay.

way, for each value of θ_a, a different point on the Nyquist plot of the plant is obtained. Therefore, a point on the Nyquist plot of the plant at a particular desired frequency ω_c can be identified, for example, at the gain crossover frequency required for the controlled system ($\omega_c = \omega_{cg}$).

An iterative method can be used to select the right value of θ_a which corresponds to a specific frequency ω_c (Chen and Moore [5]). Thus, the artificial time delay parameter can be updated using the following simple interpolation/extrapolation method:

$$\theta_n = \frac{\omega_c - \omega_{n-1}}{\omega_{n-1} - \omega_{n-2}}(\theta_{n-1} - \theta_{n-2}) + \theta_{n-1}, \tag{15}$$

where n represents the current iteration number. With the new θ_n, after the relay test, the corresponding frequency ω_n can be recorded and compared with the frequency ω_c so that the iteration can continue or stop. Two initial values of the delay (θ_{-1} and θ_0) and their corresponding frequencies (ω_{-1} and ω_0) are needed to start the iteration process. With these two pairs $(\theta_{-1}, \omega_{-1})$ and (θ_0, ω_0), the next value of θ_n is automatically obtained by using the interpolation/extrapolation method given by (15).

2.2.2 Relay test for fractional PID controllers

Following Monje et al. [21], let us consider a fractional-order lead-lag PID controller formulated as

$$C(s) = K_c x^{\mu}\left(\frac{\lambda_1 s + 1}{s}\right)^{\lambda}\left(\frac{\lambda_2 s + 1}{x\lambda_2 s + 1}\right)^{\mu}, \tag{16}$$

which can be divided into a fractional-order PI controller

$$\mathrm{PI}^{\lambda}(s) = \left(\frac{\lambda_1 s + 1}{s}\right)^{\lambda}, \tag{17}$$

and a fractional-order lead compensator plus a noise filter as

$$\mathrm{PD}^{\mu}(s) = K_c x^{\mu}\left(\frac{\lambda_2 s + 1}{x\lambda_2 s + 1}\right)^{\mu}. \tag{18}$$

It should be remarked that the fractional PI controller is used to cancel the slope of the phase of the plant at the gain crossover frequency ω_{cg}, which ensures a flat phase around the frequency of interest. On its part, the fractional-order lead compensator is designed to fulfill the design specifications of gain crossover frequency ω_{cg} and phase margin φ_m, following a robustness criterion based on the flatness of the phase curve of this compensator. This way, the resulting phase of the open-loop system will be the flattest possible, ensuring the maximum robustness to plant gain variations.

The auto-tuning method involves an iterative process with the following steps:

1. For a specified phase margin φ_m and gain crossover frequency ω_{cg}, the relay test is applied to the system and the resulting pairs (θ_n, ω_n) obtained from the n iterations of the test are saved and used for the calculation of the phase and magnitude of the plant at each frequency ω_n (following (13) and (14)). These values are used to obtain the slope of the phase curve of the system by

$$\upsilon = \frac{\phi_u - \phi_{n-1}}{\omega_u - \omega_{n-1}} \text{ sec,} \tag{19}$$

where ω_{n-1} is the frequency $n-1$ experimented with the relay test and ϕ_{n-1} its corresponding phase, and ϕ_u is the system phase corresponding to the frequency of interest $\omega_u = \omega_{cg}$. With the value of the slope, the parameters λ and λ_1 of the PI^{λ} controller can be directly obtained from

$$\lambda \frac{\lambda_1}{1 + (\lambda_1 \omega_{cg})^2} = -\upsilon, \tag{20}$$

$$(\lambda_1 \omega_{cg})^2 - 1 = 0 \quad \Rightarrow \quad \lambda_1 = \frac{1}{\omega_{cg}}. \tag{21}$$

2. Once the system $G_{flat}(s) = G(s)PI^{\lambda}(s)$ is defined, the controller $PD^{\mu}(s)$ will be designed so that the open-loop system $F(s) = G_{flat}(s)PD^{\mu}(s)$ satisfies the above mentioned specifications of gain crossover frequency ω_{cg} and phase margin φ_m. So, the following relations for the open-loop system have to be taken into account:

$$\begin{aligned} & G_{flat}(j\omega_{cg})k'\left(\frac{j\lambda_2\omega_{cg}+1}{jx\lambda_2\omega_{cg}+1}\right)^{\mu} = e^{j(-\pi+\varphi_m)} \\ \Rightarrow \quad & C'(j\omega_{cg}) = \left(\frac{j\lambda_2\omega_{cg}+1}{jx\lambda_2\omega_{cg}+1}\right)^{\mu} = \frac{e^{j(-\pi+\varphi_m)}}{G_{flat}(jw_{cg})k'} = a_1 + jb_1 \\ \Rightarrow \quad & \left(\frac{j\lambda_2\omega_{cg}+1}{jx\lambda_2\omega_{cg}+1}\right) = (a_1 + jb_1)^{1/\mu} = a + jb, \end{aligned} \tag{22}$$

$$x = \frac{a-1}{a(a-1)+b^2}, \tag{23}$$

where $k' = K_c x^{\mu} = 1$ in this case, and (a_1, b_1) is referred to as the "design point."

3. Select a very small initial value of μ (e. g., $\mu = 0.05$) and calculate the value of x and λ_2 using the relations in (22) and (23).
4. If the value of x obtained is negative, then the value of μ is increased a fixed step, and step 2 is repeated again. The smaller the fixed increase of μ, the more accurate the selection of parameter $\mu_{\min}$. Repeat step 2 until the value of x obtained is positive.
5. Once a positive value of x is obtained, the corresponding value of μ must be recorded as $\mu_{\min}$. Likewise, the value of λ_2 corresponding to $\mu_{\min}$ is also recorded. It is important to remark that the value of x will be close to zero and will ensure the maximum flatness of the phase curve of the compensator (iso-damping constraint).

Once the parameters of fractional PID controller (16) are obtained by the design method explained above, they can be related to those of the standard fractional PID controller of the form

$$C_{\mathrm{std}}(s) = K_{\mathrm{p}}\left(1 + \frac{1}{T_{\mathrm{i}}s}\right)^{\lambda}\left(1 + \frac{T_{\mathrm{d}}s}{1 + T_{\mathrm{d}}s/N}\right)^{\mu}. \tag{24}$$

Doing some manipulations in (24), the following transfer function for the controller can be obtained:

$$C_{\mathrm{std}}(s) = \frac{K_{\mathrm{p}}}{T_{\mathrm{i}}^{\lambda}}\left(\frac{T_{\mathrm{i}}s + 1}{s}\right)^{\lambda}\left(\frac{T_{\mathrm{d}}(1 + 1/N)s + 1}{1 + T_{\mathrm{d}}s/N}\right)^{\mu}. \tag{25}$$

Thus, comparing (16) and (25), the relations between parameters are: $T_{\mathrm{i}} = \lambda_1$, $K_{\mathrm{p}} = K_{\mathrm{c}}x^{\mu}\lambda_1^{\lambda}$, $N = (1 - x)/x$, and $T_{\mathrm{d}} = \lambda_2(1 - x)$.

2.3 Model reference adaptive control

2.3.1 Generalities

MRAC is one of the main approaches in adaptive control, in which the desired performance is expressed in terms of a reference model that establishes how the system output should respond ideally to the desired value. A block diagram of the MRAC approach is illustrated in Figure 7. As can be seen, MRAC consists of an inner-loop, which provides the ordinary control feedback, and an outer-loop, which adjusts the parameters in the inner-loop in such a way that the error e between the system output y and the model output y_m becomes small.

The key point is to determine the adjustment mechanism so that the closed-loop system is stable and the error, zero. Although in the literature several mechanisms have been proposed for that purpose, next the basis of the classical MIT rule is recalled

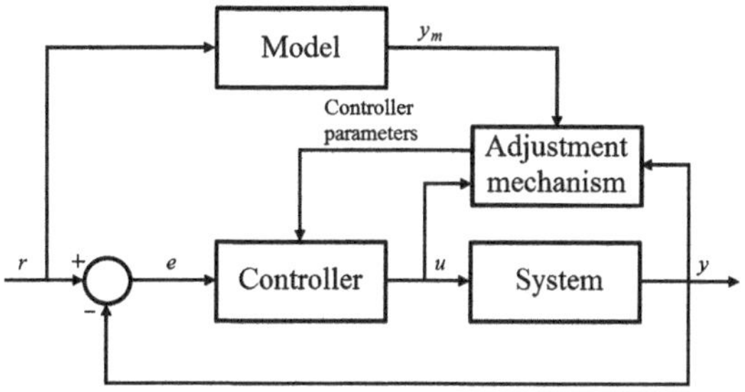

Figure 7: Block diagram of model reference adaptive control.

for a better understanding of the fractional versions proposed. In particular, two modifications of the conventional MRAC are described: the first one is based on the use of an adjustment rule of the fractional order, whereas the second uses a fractional-order system as reference model (see Vinagre et al. [35] for more details).

The MIT rule, also known as the gradient approach of MRAC, is based on the assumption that the parameters change more slowly than the other variables in the system. This assumes a quasi-stationary treatment, which is essential for the computation of the sensitivity derivatives that are needed in the adaptation.

Let θ denote the controller parameters to be updated. By using the criterion $J(\theta) = \frac{1}{2}e^2$, the adjustment rule for changing the parameters in the direction of the negative gradient of J is given by

$$\frac{\mathrm{d}\theta}{\mathrm{d}t} = -\gamma\frac{\partial J}{\partial\theta} = -\gamma e\frac{\partial e}{\partial\theta}, \tag{26}$$

where γ is the adaptation gain (a small positive number).

Taking into account the above mentioned assumption, the sensitivity derivative of the system, that is, the derivative $\frac{\partial e}{\partial\theta}$, can be evaluated under the consideration that θ is constant.

2.3.2 MRAC with fractional adjustment rule

Equation (26) of the MIT rule can be extended to fractional order as follows:

$$\frac{\mathrm{d}^\alpha\theta}{\mathrm{d}t^\alpha} = -\gamma\frac{\partial J}{\partial\theta} = -\gamma e\frac{\partial e}{\partial\theta}, \tag{27}$$

where $\alpha \in \mathbb{R}^+$ denotes the differentiation order. In this way, the rate of change of the parameters will depend not only on the adaptation gain γ, but also on the differentiation order α. Hence, the parameter updating rule can be expressed as

$$\theta = -\gamma\mathcal{I}^\alpha\left[\frac{\partial J}{\partial\theta}\right] = -\gamma\mathcal{I}^\alpha\left[e\frac{\partial e}{\partial\theta}\right], \tag{28}$$

with $\mathcal{I}^\alpha \equiv \mathcal{D}^{-\alpha}$.

Consider the following first-order system to be controlled (Aström and Wittenmark [1]):

$$\frac{dy}{dt} + ay = bu, \tag{29}$$

where y is the output, u is the input, and a and b are unknown constants or unknown slowly time-varying parameters. Assume that the corresponding reference model is given by

$$\frac{dy_m}{dt} + a_m y_m = b_m u_c, \tag{30}$$

where u_c is the reference input signal for the reference model, y_m is the output of the reference model, and a_m and b_m are known constants. Perfect model-following can be achieved with a control law defined as

$$u(t) = \theta_1 u_c(t) - \theta_2 y(t), \tag{31}$$

with $\theta_1 = \frac{b_m}{b}$ and $\theta_2 = \frac{a_m - a}{b}$.

From (27) and (29), assuming that $a + b\theta_2 \approx a_m$, and taking into account that b can be absorbed in γ, the equations for updating the controller parameters can be designed as (Monje et al. [20]):

$$\frac{d^\alpha \theta_1}{dt^\alpha} = -\gamma\left(\frac{1}{p + a_m} u_c\right)e, \tag{32}$$

$$\frac{d^\alpha \theta_2}{dt^\alpha} = \gamma\left(\frac{1}{p + a_m} y\right)e \tag{33}$$

where $p = \frac{d}{dt}$. Equivalently, in frequency domain, (32) and (33) can be written as

$$\theta_1 = -\frac{\gamma}{s^\alpha}\left(\frac{1}{s + a_m} u_c\right)e, \tag{34}$$

$$\theta_2 = \frac{\gamma}{s^\alpha}\left(\frac{1}{s + a_m} y\right)e. \tag{35}$$

Clearly, the conventional MRAC is the case when $\alpha = 1$. A block diagram of the above MRAC scheme for adjusting the unknown parameters θ_1 and θ_2 is shown in Figure 8.

Figure 9 shows simulation results of system (29) controlled by a MRAC with the fractional adjustment rule for two values of α, that is, $\alpha = 1$ (integer case) in Figure 9(a), and $\alpha = 1.25$ in Figure 9(b), with $a = 1$, $b = 0.5$, $a_m = b_m = 2$ and $\gamma = 3$. As can be observed from Figure 9(b), under the same conditions, the updating of the unknown parameters is faster when $\alpha = 1.25$, which demonstrates that the use of a

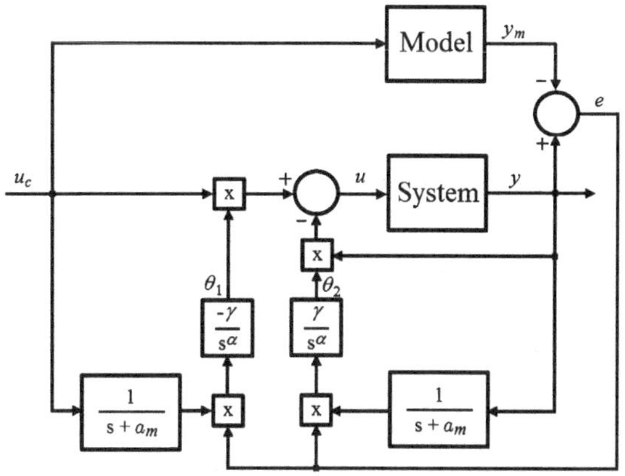

Figure 8: Block diagram of MRAC with fractional adjustment rule.

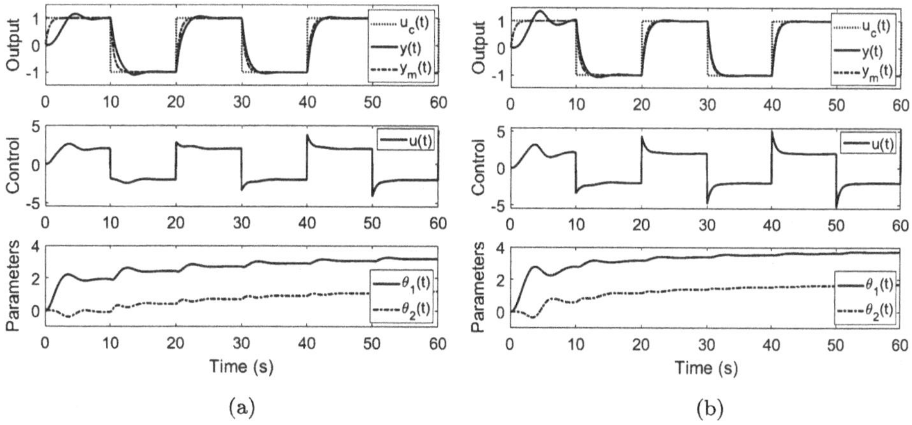

Figure 9: Simulation results for system (29) when applying a MRAC with fractional adjustment rule for: (a) $\alpha = 1$ (b) $\alpha = 1.25$.

slightly higher order in the fractional adjustment rule can improve the system performance in comparison with the conventional MRAC.

2.3.3 MRAC with fractional reference model

In the simplest MRAC problems, reference models are usually first- or second-order dynamic systems. Clearly, they can be replaced by those of fractional order so that transient response of the closed-loop system can be improved by choosing an appropriate value of the system order. Hence, in this second approach a fractional-order system is used as reference model in the MRAC scheme.

3 Sliding control

3.1 Generalities

Sliding mode control (SMC) is well known for its robustness to disturbances and parameter variations. Recently, similarly to other classical strategies, SMC has been extended to fractional order so as to allow more flexible system performances to be attained. The works reported in the literature about fractional sliding mode control (FSMC) mainly deal with applications to fractional systems. Jakovljevic et al. [17], Pashaei and Badamchizadeh [26], Majidabad et al. [19], Mujumdar et al. [22], Si-Ammour et al. [27], Valério and Sá da Costa [32] are a few examples of fractional systems which include, for example, nonlinear dynamics, uncertainty, external disturbances, delays and noncommensurability. Although receiving less attention from the community, FSMC has been also applied to integer order systems, including real systems, such as Buck converters in Calderón et al. [4], HosseinNia et al. [16] or a permanent magnet motor in Zhang et al. [36], as well as in others from a theoretical point of view (Corradini et al. [7]).

The basic idea of SMC is to use the input to make the state vector converge to a subspace, called sliding surface ($\mathcal{L}_S$), of the space where it may otherwise evolve, and stay there. The key points in designing SMC are as follows. First, a switching function (S) is chosen so that the control objective or design specifications are fulfilled on the surface $\mathcal{L}_S$. Deviations from the sliding surface are corrected because all trajectories converge back to the sliding surface. The dynamical behavior of the system when confined to such a surface is described as ideal sliding motion (equivalently, it is achieved when $S = 0$). Second, a control law (u) is selected to enforce the sliding mode, that is, such that the trajectories of the closed-loop motion are directed toward the surface and, therefore, existence and reachability conditions are satisfied. Likewise, an equivalent control law (u_{eq}) can be obtained when $\dot{S} = 0$ (refer to, e. g., Edwards and Spurgeon [8]).

3.2 Fractional SMC

There are several ways for introducing fractional dynamics in SMC (see Vinagre and Calderón [33] for details). One of them is the use of a *fractional-order switching function* to enforce fractional dynamics.

Consider the double integrator given by

$$\ddot{y}(t) = u(t). \tag{36}$$

For minimizing the time taken to induce sliding, maximizing the region in which the sliding takes place, to ensure that once the trajectories reach the sliding surface they are forced to remain there, and to reduce the amplitude of the high-frequency switching for limiting wear and tear on the actuators, a control law candidate is (Edwards and Spurgeon [8])

$$u(t) = -(m + \Phi)\dot{y}(t) - \Phi m y(t) - \rho\,\mathrm{sgn}(S(t)), \tag{37}$$

with $S(t) = my(t) + \dot{y}(t)$, and where m, Φ, and ρ are positive scalars corresponding to, respectively, the rate at which the sliding surface is attained, the dynamics of the sliding motion, and the reduction of the amplitude of the high-frequency switching.

By introducing the notation $\mathcal{D}^{\lambda} y \triangleq y^{(\lambda)}$, a generalization of the former classical PD switching function can be expressed of the form

$$S(y, y^{(\alpha)}) = my + y^{(1+\alpha)}, \quad 0 < \alpha < 1. \tag{38}$$

By doing so, the motion when confined to $\mathcal{L}_s$ satisfies the differential equation obtained from rearranging $S(y, y^{(\alpha)}) = 0$, namely,

$$y^{(1+\alpha)} = -my \tag{39}$$

being the corresponding characteristic equation

$$s^{1+\alpha} + m = 0, \tag{40}$$

which corresponds to a fractional integrator in closed loop. This dynamics exhibits an overshoot governed by the parameter α and a rate governed by the couple (m, α), namely, in the frequency domain, a gain crossover frequency $\omega_c = m^{1/(1+\alpha)}$.

The equivalent control can be obtained as follows:

$$\begin{aligned} S &= my + y^{(1+\alpha)} \\ \Longrightarrow \quad \dot{S} &= m\dot{y} + y^{(2+\alpha)} = m\dot{y} + u^{(\alpha)} = 0 \\ \Longrightarrow \quad u &= -my^{(1-\alpha)}. \end{aligned} \tag{41}$$

So, a control law can be proposed as

$$u(t) = -my^{(1-\alpha)} - \rho \mathcal{I}^{\beta}\,\mathrm{sgn}(S(t)), \quad 0 < \alpha, \beta < 1 \tag{42}$$

where the linear term is equal to the equivalent control, $\mathcal{I}^{\beta}(\cdot)$ means the fractional integral. The last term of (42) can be seen as a low-pass filter or an attenuation selective in frequency, and allows to reduce the amplitude of the high-frequency switching. In simulations, similar results can be obtained with (37) and (42).

3.3 Fractional SMC of integer-order systems by system augmentation

According to Theorem 1 in Sierociuk and Vinagre [28], the integer-order state space system

$$\dot{x}(t) = \boldsymbol{A}x(t) + \boldsymbol{B}u(t) + \boldsymbol{E}d(t), \qquad (43)$$
$$y(t) = \boldsymbol{C}x(t)$$

with initial conditions $x(0) = x_0$ and matrices $\boldsymbol{A} \in \mathbb{R}^{n\times n}$, $\boldsymbol{B} \in \mathbb{R}^{n\times m}$, $\boldsymbol{E} \in \mathbb{R}^{n\times m}$, $\boldsymbol{C} \in \mathbb{R}^{r\times n}$, can be rewritten in the form of fractional (rational) order system as

$$\mathcal{D}^{\alpha}\mathbf{x}(t) = \tilde{\boldsymbol{A}}\mathbf{x}(t) + \tilde{\boldsymbol{B}}u(t) + \tilde{\boldsymbol{E}}d(t), \qquad (44)$$
$$y(t) = \tilde{\boldsymbol{C}}\mathbf{x}(t)$$

where $\alpha = 1/q$ is the fractional order ($q \in \mathbb{Z}^{+}$), and

$$\tilde{\boldsymbol{A}} = \left[\begin{array}{c:c} \mathbf{0}_{p\times n} & \boldsymbol{I}_{p\times p} \\ \hdashline \boldsymbol{A} & \mathbf{0}_{n\times p} \end{array}\right], \quad \tilde{\boldsymbol{B}} = \left[\begin{array}{c} \mathbf{0}_{p\times m} \\ \hdashline \boldsymbol{B} \end{array}\right], \quad \tilde{\boldsymbol{E}} = \left[\begin{array}{c} \mathbf{0}_{p\times m} \\ \hdashline \boldsymbol{E} \end{array}\right],$$

$$\tilde{\boldsymbol{C}} = \left[\begin{array}{c:c} \boldsymbol{C} & \mathbf{0}_{p\times r} \end{array}\right], \quad \mathbf{x}(t) = \begin{bmatrix} x(t) \\ x_{a,1}(t) \\ \vdots \\ x_{a,q-2}(t) \\ x_{a,q-1}(t) \end{bmatrix},$$

being $x_{a,1}(t) = \mathcal{D}^{\alpha}x(t)$, $x_{a,i}(t) = \mathcal{D}^{\alpha}x_{a,i-1}(t)$ for $i = 2,\dots,q-1$, the components of the augmented state vector (i. e., the pseudo-states) with initial conditions $x(0) = \begin{bmatrix} x_0^T & x_{a,1}^T(0) & \cdots & x_{a,q-1}^T(0) \end{bmatrix}^T$ ($(-)^T$ denotes the transpose operator). Taking into account the system augmentation in such a way, a robust controller via SMC is going to be designed similarly to HosseinNia et al. [12].

For illustration purposes, let consider the position of a servo system in state space form given by

$$\dot{x}(t) = \begin{bmatrix} 0 & 1 \\ 0 & -1/\tau \end{bmatrix} x(t) + \begin{bmatrix} 0 \\ K/\tau \end{bmatrix} u(t) \qquad (45)$$

where τ and K are the time constant and the gain of the system, respectively. Thus, the augmented system of order $\alpha = 0.5$ is given by the following matrices:

$$\tilde{\boldsymbol{A}} = \begin{bmatrix} 0 & 1 & 0 & 0 \\ 0 & 0 & 1 & 0 \\ 0 & 0 & 0 & 1 \\ 0 & 0 & -1/\tau & 0 \end{bmatrix}, \quad \tilde{\boldsymbol{B}} = \begin{bmatrix} 0 \\ 0 \\ 0 \\ K/\tau \end{bmatrix}, \quad \tilde{\boldsymbol{C}} = \begin{bmatrix} 1 & 0 & 0 & 0 \end{bmatrix}, \quad \tilde{\boldsymbol{E}} = \begin{bmatrix} 0 \\ 0 \\ 0 \\ 1 \end{bmatrix}.$$

Let consider the switching function as

$$S = \lambda_1(x_1 - r) + \lambda_2 x_2 + \lambda_3 x_3 + \lambda_4 x_4, \tag{46}$$

where the parameters λ_i have to be tuned. Thus, taking $S = 0$, the ideal sliding motion corresponds to the dynamics of the following fractional-order system:

$$\frac{X_1(s)}{R(s)} = \frac{1}{\frac{\lambda_4}{\lambda_1}s^{1.5} + \frac{\lambda_3}{\lambda_1}s + \frac{\lambda_2}{\lambda_1}s^{0.5} + 1}. \tag{47}$$

Notice that: (i) the poles of fractional system (47) are the roots of the polynomial $\frac{\lambda_4}{\lambda_1}\sigma^3 + \frac{\lambda_3}{\lambda_1}\sigma^2 + \frac{\lambda_2}{\lambda_1}\sigma + 1 = 0$; and (ii) S and ideal sliding motion for the integer-order case can be obtained by taking $\lambda_2 = \lambda_4 = 0$ in (46) and (47), respectively.

Two design strategies will be considered here: particle swarm optimization (PSO), and pole placement. The design and implementation procedure of each strategy for the FSMC will be as follows (further details can be found in Tejado et al. [31]):

1. Determine the optimal values of the parameters λ_i and the closed-loop poles of the fractional-order system which correspond to the ideal sliding motion when λ_i take such optimal values.
2. Determine the values of the vector λ for ideal sliding motion considering the previously calculated closed-loop poles as design specifications. (This is to obtain similar dynamics with both strategies.)

The optimal parameters obtained for both the integer and fractional controllers are given in Table 1 (the values of the maximum control law $U_{\max}$ and the minimum energy $G_{\min}$ are also included).

Figure 10 shows the comparison of the system performance when using both the FSMC and the SMC for an unit step response (in Figure 10(a)) and ideal sliding motion (in Figure 10(b)). As can be observed, the response is faster when applying the FSMC than the obtained with the SMC (both designed by PSO). Only slight differences can be found in the control laws: SMC requires higher control efforts, as can be stated from the values of $U_{\max}$ and $G_{\min}$ in Table 1. Notice that the phenomenon of chattering with SMC for position control is nearly negligible. To show the additional control possibilities with the FSMC derived by the pseudo-states x_2 and x_4, their gains, namely, λ_2 and λ_4, are going to be changed. The system responses for two situations with λ_2 and λ_4

Table 1: FSMC and SMC parameters for the position control of system (45).

	λ_1	λ_2	λ_3	λ_4	$U_{\max}$	$G_{\min}$
FSMC	3.04	0.04	0.18	0.01	3.73	90.07
SMC	4.16	–	0.36	–	5.16	105.15

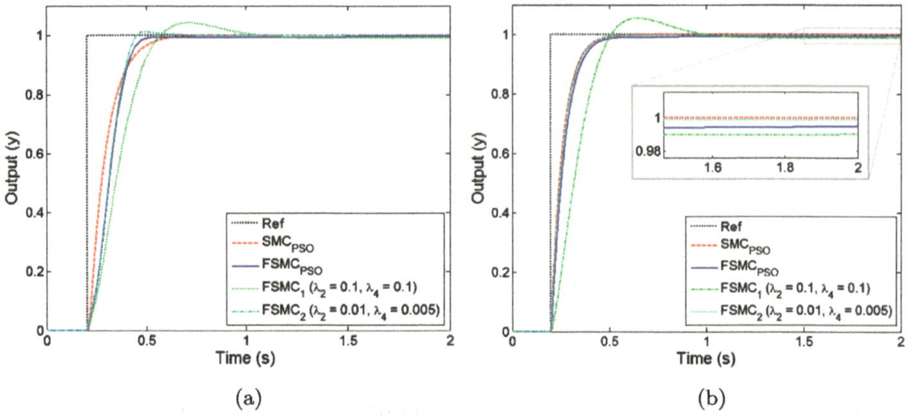

Figure 10: Results for position control of system (45) when applying SMC and FSMC: (a) Simulated outputs (b) Ideal sliding motion.

higher and smaller than the optimal case are included in the figure: (1) $\lambda_2 = \lambda_4 = 0.1$, and (2) $\lambda_2 = 0.01$, $\lambda_4 = 0.005$. As can be seen, the lower the values of λ_2 and λ_4, the faster the response of the servo and the closer to the integer case.

Figure 10(b) shows the position responses of the servo for ideal sliding motion choosing different placements of the poles: case 0) $p_1 = -18.66$, $p_{2,3} = 0.33 \pm j4.02$; case 1) $p_1 = -3.72$, $p_{2,3} = 0.96 \pm j2.69$; and case 2) $p_1 = -36.40$, $p_{2,3} = 0.2 \pm j4.08$ (again, the values of λ_i will be the optimal ones previously obtained). Notice that $|\arg(p_i)| \leq \alpha\pi/2 = \pi/4$, so these poles are stable. Although faster, these responses are similar to those shown in Figure 10(a). Indeed, the differences arise from the approximation used for the fractional operator in the simulations when applying the controllers designed by PSO.

In what concerns the simulations, the following considerations should be remarked: (1) fifth-order Oustaloup's approximation was used in the frequency range [0.01, 100] rad/s for fractional-order operators for the FSMC case (see, e. g., Oustaloup [24]); (2) the control law was implemented with the pseudo-sign function $W(\phi)$ as $u = -S - W(\phi)$, being $W(\phi) = \frac{S}{|S|+\phi}$ with $\phi = 0.01$ in accordance with Edwards and Spurgeon [8]; and (3) the function `fode_sol()` was taken from Monje et al. [20] to obtain the step responses for ideal sliding motion.

4 Reset control

Reset control is a nonlinear technique which has gained popularity over the years and has the advantage of fitting within the framework of PID for improved performance. Reset involves the resetting of a subset of controller states when a reset condition is met. Reset was first introduced by Clegg [6] for integrators to improve performance.

Significant work can be found in literature showing the advantages of reset control in reducing the overshoot, increasing the precision and bandwidth of the motion which is impossible using linear controller due its limitation like waterbed effects and Bode gain and phase relation. This section presents the properties of general fractional-order reset strategies which reset controller states to fixed or variable nonzero values and are able to eliminate or reduce the overshoot of the first and higher-order systems.

4.1 Generalities

4.1.1 Reset elements

The most popular reset controller can be defined using the following differential inclusions:

$$\begin{aligned} \mathcal{D}^{\alpha} x_r(t) &= A_r x_r(t) + B_r e(t) \quad \text{if } e(t) \neq 0, \\ x_r(t^+) &= A_\rho x_r(t) \quad \text{if } e(t) = 0, \\ u(t) &= C_r x_r(t) + D_r e(t) \end{aligned} \tag{48}$$

where A_r, B_r, C_r, D_r are the state space matrices of the base linear system, A_ρ is the reset matrix which determines the state after reset values (it is defined to reset only the appropriate states of controller), α is the differentiation order, $e(t)$ is the error signal, and $u(t)$ is the control signal. Clearly, when $\alpha = 1$ the reset controller will be equivalent to that of integer order. Reset controller (48) generally consists of both linear and nonlinear reset parts.

In what the reset part of controller defined by (48) is concerned, it has been presented as different reset elements in the literature as follows:

1. *Clegg integrator*: Clegg or reset integrator (CI) is the first introduction of reset technique in the literature (Clegg [6]). The action of resetting the integrator output to zero when input crosses zero results in favored behavior of reducing phase lag from 90° to 38.1°. CI is the most extensively studied and applied reset element due to its advantages in reducing the overshoot and increasing the phase margin of the system. The matrices of CI for (48) are: $A_r = 0$, $B_r = 1$, $C_r = 1$, $D_r = 0$, $A_\rho = 0$. Likewise, fractional-order Clegg integrator (FCI) can be defined for the given system parameters, and assuming $0 < \alpha \leq 1$ (see, e. g., Vinagre et al. [34]).
2. *First-order reset element*: CI was extended to a first-order reset element (FORE) by Horowitz and Rosenbaum [11]. FORE provides the advantage of filtering frequency placement unlike CI, and has been used for narrow-band phase compensation in Li et al. [18]. The matrices of FORE for (48) where the base linear filter has corner frequency ω_r are: $A_r = -\omega_r$, $B_r = \omega_r$, $C_r = 1$, $D_r = 0$, $A_\rho = 0$. Similarly to CI, fractional-order FORE can be defined if the order of the system is $0 < \alpha \leq 1$.

3. *Generalized first-order reset element*: FORE was generalized in Guo et al. [9] to obtain the generalized FORE (GFORE) to provide an additional degree of freedom: it has a nonzero resetting parameter γ (such that $A_\rho = \gamma$), which allows to control the level of reset. Notice that, when $\gamma = 1$, the controller results in a linear filter. The parameter γ is used to influence the amount of nonlinearity, and hence, phase lag.

FCI and FORE can be used to replace one part of fractional-order PID controller. Therefore, in order to be able to tune such controller, these nonlinear elements should be analyzed in the frequency domain by means of the describing function.

4.1.2 Describing function

The nonlinearity of reset elements creates the problem of designing controllers in frequency domain, especially using industry popular loop shaping technique which uses Bode, Nyquist, and Nichols plots. In the literature, sinusoidal input describing function analysis has been used to analyze reset elements in frequency domain. In fact, the phase lag reduction advantage was seen by Clegg in 1958 using this technique. Although the describing function does not accurately capture all the frequency domain aspects of reset, it is useful in providing necessary information for both the design and the analysis.

Describing function for a given fractional-order reset controller can be calculated by

$$N(A,\omega) = \frac{2j\omega}{\pi A} \int_0^{\pi} y(t) e^{-j\omega t}\, dt. \tag{49}$$

The describing function of generic integer reset systems as simplified by (48) is provided in Guo et al. [9] and this is used to obtain understanding of the system in frequency domain. The sinusoidal input describing function is obtained as

$$G(j\omega) = C_r^T (j\omega - A_r)^{-1} (I + j\Theta_\rho(\omega)) B_r + D_r, \tag{50}$$

where

$$\Theta_\rho = \frac{2}{\pi} (I + e^{\frac{\pi A_r}{\omega}}) \left(\frac{I - A_\rho}{I + A_\rho e^{\frac{\pi A_r}{\omega}}} \right) \left(\left(\frac{A_r}{\omega} \right)^2 + I \right)^{-1},$$

being I the identity matrix of proper dimension.

Table 2 shows the describing functions of all defined elements. From this table, it is clear that CI gives a phase lead of almost 52° with respect to a classical integrator (also increases the gain with a factor of about 1.62). Figure 11(a) shows the fundamental property of CI and FORE by means of the Nichols chart.

Table 2: Describing functions of reset elements (Vinagre et al. [34]).

Name	Base filter	Describing function
CI	$\frac{1}{s}$	$\frac{4}{\pi\omega}(1 - \mathrm{j}\frac{\pi}{4})$
FCI	$\frac{1}{s^\alpha}$	$\frac{4}{\pi\omega^\alpha}(\sin(\alpha\frac{\pi}{2}) + \frac{\pi}{4}e^{-\mathrm{j}\alpha\frac{\pi}{2}})$
FORE	$\frac{K}{s+b}$	$\frac{K}{b+\mathrm{j}\omega}(1 + \mathrm{j}\frac{2\omega^2(1+e^{-b\frac{\pi}{\omega}})}{\pi(b^2+\omega^2)})$
Fractional FORE	$\frac{K}{s^\alpha+b^\alpha}$	$\frac{K}{b^\alpha+(\mathrm{j}\omega)^\alpha} - \mathrm{j}\frac{2B\omega(1+e^{-b^\alpha\frac{\pi}{\omega}})}{\pi(b^\alpha+\mathrm{j}\omega))}$, where $B = \frac{-K\omega^\alpha\sin(\alpha\frac{\pi}{2})}{(\omega^\alpha\cos(\alpha\frac{\pi}{2})+b^\alpha)^2+(\omega^\alpha\sin(\alpha\frac{\pi}{2}))^2}$

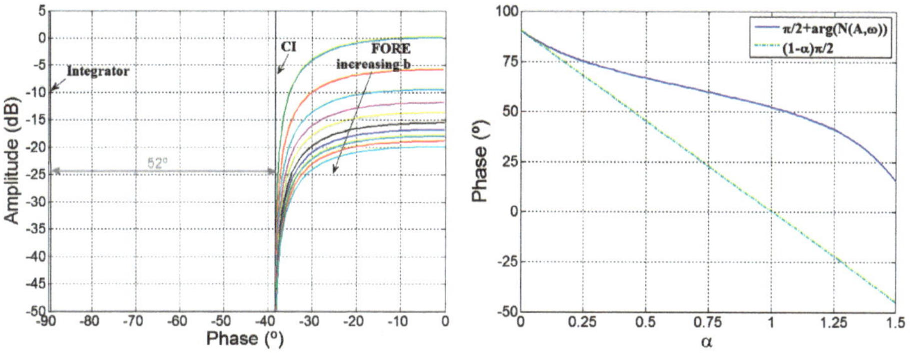

(a) (b)

Figure 11: Describing function analysis: (a) Nichols chart of FORE and CI with respect to the classical integrator (b) Phase difference between CI and FCI.

This concept has been generalized to fractional-order integrators by Vinagre et al. [34], HosseinNia et al. [14]. It was showed that FCI has a tunable phase lag. Figure 11(b) compares the phase difference between both the fractional-order CI and the fractional-order integrator in comparison with the classical (integer-order) linear integrator. As can be observed, the phase lag depends on the value of α for both cases, but is always higher when using the FCI for $\alpha < 1$. Note that this phase difference can be considered as the phase margin to be added to the system. As an example, CI cannot compensate 60° but this specification can be achieved by a FCI with order $\alpha = 0.5$. Similar response is expected for fractional-order FORE at frequencies higher than the cutoff frequency, that is, b.

4.2 General forms for reset control

As commented, linear controllers like PIDs are well accepted in industry due to their simple structures and wide applicability. However, they are limited by the Bode gain and phase relation. In other words, increasing the gain at low-frequency and decreasing the gain at high-frequency decreases the phase margin and stability. The solution lies on the application of nonlinear control. However, most of nonlinear controllers are complicated to be applied in practice. On one hand, reset controllers have great performance in reducing the overshoot due to their phase advantage and, on the other, they have simple structure as their base linear controllers. Thus, these controllers are the best candidate to overcome limitation of linear controllers (refer to Baños and Barreiro [2] for more information).

Despite all advantages of reset controllers, there is one limitation which makes reset controller (48) infeasible in practice. This is called the limit cycle, which may occur if the states reset to zero. Therefore, it is clear that reset controllers need new ways to reset so as to avoid the occurrence of such problems; some of them are described in Nesic et al. [23], Panni et al. [25], Zheng et al. [37, 38]. Next, two general single-input and single-output (SISO) reset controllers will be introduced, including fractional-order dynamics. In these strategies, in order to avoid limit cycle, the system resets to a nonzero value (HosseinNia et al. [13]).

4.2.1 Reset when error crosses zero

Let consider a general fractional-order system with the following state space representation:

$$\begin{aligned} \mathcal{D}^{\alpha_p} x_p(t) &= A_p x_p(t) + B_p u_r(t), \\ y(t) &= C_p x_p(t), \end{aligned} \tag{51}$$

where $\alpha_p \in \mathbb{R}^+$ is the basic order of the plant, $A_p \in \mathbb{R}^{n_p \times n_p}$, $B_p \in \mathbb{R}^{n_p \times 1}$, $C_p \in \mathbb{R}^{1 \times n_p}$ are its state, input and output matrices, respectively.

Taking into account the properties of reset strategies, the following general fractional-order reset PI controller can be derived, where its state is reset to $K \cdot r$ when error crosses zero. It can be represented as

$$\begin{aligned} \mathcal{D}^{\alpha} x_r(t) &= A_r x_r(t) + B_r e(t), \quad && e(t) \neq 0, \\ x_r(t^+) &= A_\rho x_r(t) + B_\rho \frac{K}{C_r} r, \quad && e(t) = 0, \\ u_r(t) &= C_r x_r(t) + D_r e(t), \end{aligned} \tag{52}$$

where $K = \frac{1}{P(0)}$, the inverse of the DC gain of the system, is given by

$$K = \begin{cases} -\frac{1}{C_p A_p^{-1} B_p}, & \text{if } A_p \text{ is invertible} \\ 0, & \text{otherwise} \end{cases} \tag{53}$$

where $A_r = A_\rho = 0$, $B_r = B_\rho = 1$, $C_r = \frac{k_p}{\tau_i}$, and $D_r = k_p$. Although (52) is valid for a PI, it can be extended easily to any controller by means of the FCI.

4.2.2 Reset periodically at fixed instants

Likewise, the controller described above can be reshaped to reset periodically when $t = t_k$, which results in a new general reset controller as follows:

$$\begin{aligned} \mathcal{D}^{\alpha}x_r(t) &= A_r x_r(t) + B_r e(t), \quad t \neq t_k, \\ x_r(t^+) &= A_\rho x_r(t) + B_\rho\left(\frac{Kr - D_r e(t_k)}{C_r}\right), \quad t = t_k, \\ u_r(t) &= C_r x_r(t) + D_r e(t). \end{aligned} \tag{54}$$

Due to the fact that reset happens periodically, and not necessarily when error is zero, it will take place at a variable nonzero value, which is function of the DC gain of the system.

Next, two examples of application of the aforementioned general reset controller for first- and second-order systems are given. The first example compares different strategies with error zero crossing, whereas the second one shows controllers with periodic reset for second-order systems.

Consider the system in Baños and Barreiro [2]

$$\begin{aligned} \dot{x}_p &= -0.5x_p + 1.5u, \\ y &= x_p, \end{aligned} \tag{55}$$

whose transfer function is $P(s) = \frac{1.5}{s+0.5}$, controlled by a PI with reset integrator with $k_p = 2$ and $\tau_i = 0.15$ (the base PI controller was tuned to set the rising time to 0.31 s).

Four strategies are designed for system (55): a classical PI controller, a PI with reset (proportional-CI, PCI), a reset controller given by (48) with a PCI, and a general reset controller given by (52). In what the selection of α is concerned, there is a trade off to overcome limit cycle and settling time: the lower its value, the higher the ability to avoid the limit cycle but the larger the settling time. For this reason, α was set to 0.9 in the fractional controllers.

Simulation results (step responses and control signals) for all these cases are plotted in Figure 12. As can be observed, PI and PCI cause an undesirable overshoot and limit cycle, respectively. Taking into account the control signals, the system output when applying the PI tends to its equilibrium, that is, $K = 0.33$ when $t \to \infty$. By contrast, the PCI always resets to zero (when error is zero) which causes the limit cycle. However, the fractional-order PCI and general reset do not have this problem and are able to avoid the occurrence of limit cycle by resetting to the equilibrium point.

Now, consider the dynamics of a micro-actuator system described by Zheng et al. [37, 38]:

$$\begin{aligned} \dot{x}_{p_1}(t) &= x_{p_2}(t), \\ \dot{x}_{p_2}(t) &= -a_1 x_{p_1}(t) - a_2 x_{p_2}(t) + bu(t), \\ y(t) &= x_{p_1}(t), \end{aligned} \tag{56}$$

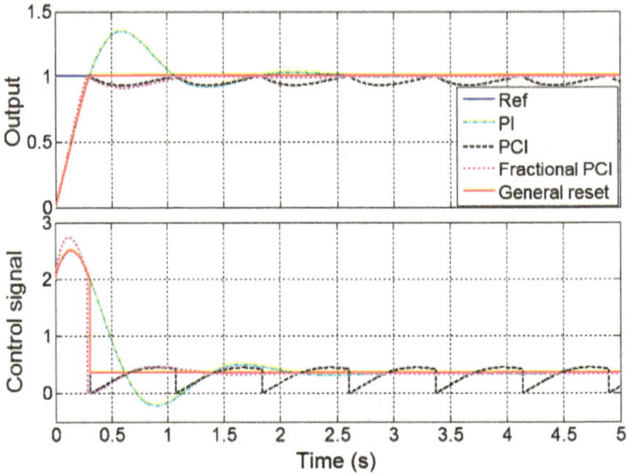

Figure 12: Responses of system (55) when applying different integer and fractional-order reset controllers.

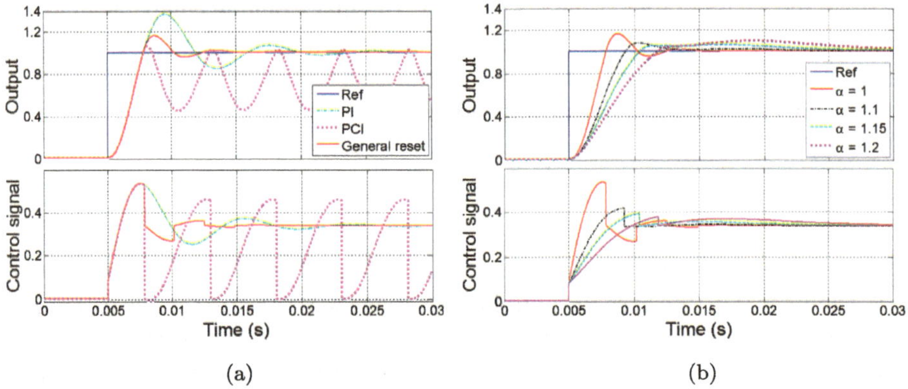

Figure 13: Responses of system (56) when applying controllers with zero crossing reset of: (a) Integer order (b) Fractional order.

where x_{p_1} and x_{p_2} are position and velocity of the moving stage, with $a_1 = 10^6$, $a_2 = 1810$, and $b = 3 \times 10^6$. This system can be also given by its transfer function $P(s) = \frac{b}{s^2+a_2 s+a_1}$.

Consider fractional reset controllers given by (52) and (54) with a PI as base linear controller, and with $C_r = \frac{k_p}{\tau_i} = 300$, $D_r = k_p = 0.08$. Also consider a PI and PCI to control system (56).

The step responses and control signals when applying all reset controllers of integer order (with $\alpha = 1$) are shown in Figure 13(a). The step response obtained with the PCI shows a limit cycle which is eliminated using general reset controller by resetting the control states to a nonzero value, that is, $\frac{K}{D_r}$. Figure 13(b) compares general reset control for different values of α. It can be seen that the higher the value of α, the lower the overshoot but the slower the response. Thus, a trade off between an integer and

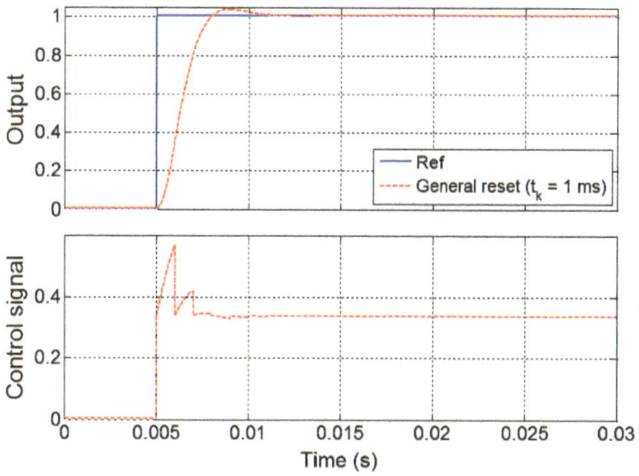

Figure 14: Responses of system (56) when applying controller with periodic reset.

a fractional-order general reset controller (in this case $\alpha = 1.1$) may be a good way to eliminate limit cycle and reduce overshoot at the same time.

Simulation results when applying general reset control with periodic resetting are depicted in Figure 14 for $t_k = 1$ ms. As can be seen, the system has almost no overshoot. It should be remarked that the value of t_k has to be chosen so that both overshoot and rise time are minimized. In particular, the lower the value of t_k, the lower the overshoot. However, overdoing it may reduce settling time.

Bibliography

[1] K. J. Aström and B. Wittenmark, *Adaptive Control*, Addison-Wesley Publishing Company, 1989.

[2] A. Baños and A. Barreiro, *Reset Control Systems*, Springer Verlag, 2012.

[3] I. Boiko, *Non-parametric Tuning of PID Controllers. Advances in Industrial Control*, Springer-Verlag, London, 2013.

[4] A. J. Calderón, B. M. Vinagre, and V. Feliu, Fractional order control strategies for power electronic Buck converters, *Signal Processing*, **86** (2006), 2803–2819.

[5] Y. Q. Chen and K. L. Moore, Relay feedback tuning of robust PID controllers with iso-damping property, *IEEE Transactions on Systems, Man, and Cybernetics, Part B (Cybernetics)*, **35**(1) (2005), 23–31.

[6] J. C. Clegg, A nonlinear integrator for servomechanisms, *Transactions of the American Institute of Electrical Engineers, Part II: Applications and Industry*, **77**(1) (1958), 41–42.

[7] M. L. Corradini, R. Giambo, and S. Pettinari, On the adoption of a fractional-order sliding surface for the robust control of integer-order LTI plants, *Automatica*, **51**(11) (2015), 364–3671.

[8] C. Edwards and S. K. Spurgeon, *Sliding Mode Control. Theory and Applications*, Taylor & Francis Ltd, 1998.

[9] Y. Guo, Y. Wang, and L. Xie, Frequency-domain properties of reset systems with application in hard-disk-drive systems, *IEEE Transactions on Control Systems Technology*, **17**(6) (2009), 1446–1453.

[10] C. C. Hang, K. J. Aström, and Q. C. Wang, Relay feedback auto-tuning of process controllers—A tutorial review, *Journal of Process Control*, **12**(1) (2002), 143–162.

[11] I. Horowitz and P. Rosenbaum, Non-linear design for cost of feedback reduction in systems with large parameter uncertainty, *International Journal of Control*, **21**(6) (1975), 977–1001.

[12] S. H. HosseinNia, D. Sierociuk, A. J. Calderón, and B. M. Vinagre, Augmented system approach for fractional order SMC of a DC-DC Buck converter, in *Proceedings of the 4th IFAC Workshop Fractional Differentiation and Its Applications (FDA'10)*, 2010.

[13] S. H. HosseinNia, I. Tejado, D. Torres, B. M. Vinagre, and V. Feliu, A general form for reset control including fractional order dynamics, in *Proceedings of the 19th IFAC World Congress*, pp. 2028–2033, 2014.

[14] S. H. HosseinNia, I. Tejado, and B. M. Vinagre, Fractional-order reset control: Application to a servomotor, *Mechatronics*, **23**(7) (2013), 781–788.

[15] S. H. HosseinNia, I. Tejado, and B. M. Vinagre, Stability of fractional order switching systems, *Computer & Mathematics with Applications*, **66**(5) (2013), 585–596.

[16] S. H. HosseinNia, I. Tejado, B. M. Vinagre, and D. Sierociuk, Boolean-based fractional order SMC for switching systems: Application to a DC-DC Buck converter, *Journal of Signal Image and Video Processing*, **6**(3) (2012), 445–451.

[17] B. Jakovljevic, A. Pisano, M. R. Rapaic, and E. Usai, On the sliding-mode control of fractional-order nonlinear uncertain dynamics, *International Journal of Robust Nonlinear Control*, **26**(4) (2016), 782–798.

[18] Y. Li, G. Guo, and Y. Wang, Nonlinear mid-frequency disturbance compensation in hard disk drives, in *Proceedings of the 16th IFAC Triennial World Congress*, pp. 31–36, 2005.

[19] S. Majidabad, H. Shandiz, A. Hajizadeh, and H. Tohidi, Robust block control of fractional-order systems via nonlinear sliding mode techniques, *Journal of Control Engineering and Applied Informatics*, **17**(1) (2015), 31–40.

[20] C. A. Monje, Y. Q. Chen, B. M. Vinagre, D. Xue, and V. Feliu, *Fractional-Order Systems and Controls. Fundamentals and Applications*, Springer-Verlag, London, 2010.

[21] C. A. Monje, B. M. Vinagre, V. Feliu, and Y. Q. Chen, Tuning and auto–tuning of fractional order controllers for industry applications, *Control Engineering Practice*, **16**(7) (2008), 789–812.

[22] A. Mujumdar, B. Tamhane, and S. Kurode, Observer-based sliding mode control for a class of noncommensurate fractional-order systems, *IEEE-ASME Transactions on Mechatronics*, **20**(15) (2015), 2504–2512.

[23] D. Nesic, A. R. Teel, and L. Zaccarian, Stability and performance of SISO control systems with first-order reset elements, *IEEE Transactions on Automatic Control*, **56**(11) (2011), 2567–2582.

[24] A. Oustaloup, *La Commande CRONE: Commande Robuste dOrdre Non Entier*, Hermes, 1991.

[25] F. S. Panni, D. Alberer, and L. Zaccarian, Set point regulation of an EGR valve using a FORE with hybrid input bias estimation, in *Proceedings of the 2012 American Control Conference (ACC'12)*, pp. 4221–4226, 2012.

[26] S. Pashaei and M. Badamchizadeh, A new fractional-order sliding mode controller via a nonlinear disturbance observer for a class of dynamical systems with mismatched disturbances, *ISA Transactions*, **63** (2016), 39–48.

[27] A. Si-Ammour S. Djennoune, and M. Bettayeb, A sliding mode control for linear fractional systems with input and state delays, *Communications in Nonlinear Science and Numerical Simulation*, **14** (2009), 2310–2318.

[28] D. Sierociuk and B. M. Vinagre, State and output feedback fractional control by system augmentation, in *Proceedings of the 4th IFAC Workshop on Fractional Differentiation and Its Applications (FDA'10)*, 2010.

[29] K. K. Tan, S. Huang, and R. Ferdous, Robust self-tuning PID controller for nonlinear systems, *Journal of Process Control*, **12**(7) (2002), 753–761.

[30] I. Tejado, S. H. HosseinNia, and B. M. Vinagre, Adaptive gain-order fractional control for network-based applications, *Fractional Calculus & Applied Analysis*, **17**(2) (2014), 462–482.

[31] I. Tejado, A. Djari, and B. M. Vinagre, Two strategies for fractional sliding mode control of integer order systems by system augmentation: Application to a servomotor, in *Proceedings of the 20th IFAC World Congress*, pp. 8103–8108, 2017.
[32] D. Valério and J. Sá da Costa, Fractional sliding mode control of MIMO nonlinear noncommensurable plants, *Journal of Vibration and Control*, **20**(7) (2014), 1052–1065.
[33] B. M. Vinagre and A. J. Calderón, On fractional sliding mode control, in *Proceedings of the 7th Portuguese Conference on Automatic Control*, 2006.
[34] B. M. Vinagre, C. A. Monje, and I. Tejado, Reset and fractional integrators in control applications, in *Proceedings of the 8th International Carpathian Control Conference (ICCC'07)*, pp. 754–757, 2007.
[35] B. M. Vinagre, I. Petras, I. Podlubny, and Y. Q. Chen, Using fractional order adjustement rules and fractional order reference models in model–reference adaptive control, *Nonlinear Dynamics*, **29**(1–4) (2002), 269–279.
[36] B. Zhang, Y. Pi, and Y. Luo, Fractional order sliding-mode control based on parameters auto-tuning for velocity control of permanent magnet synchronous motor, *ISA Transactions*, **51**(5) (2012), 649–656.
[37] J. Zheng, Y. Guo, M. Fu, Y. Wang, and L. Xie, Improved reset control design for a PZT positioning stage, in *Proceedings of the IEEE International Conference on Control Applications*, pp. 1272–1277, 2007.
[38] J. Zheng, Y. Guo, M. Fu, Y. Wang, and L. Xie, Development of an extended reset controller and its experimental demonstration, *IET Control Theory & Applications*, **2**(10) (2008), 866–874.

Jocelyn Sabatier and Christophe Farges

Dynamical properties of fractional models

Abstract: This chapter proposes a presentation of several fractional models commonly encountered in the literature. It highlights that these models are of infinite dimension. In order to compute the model output, and depending on the used modeling approach, that implies to take into account all the model past or to consider infinite spatial dimensions. Stability results dedicated to commensurate models are then given, a special attention being paid to LMI tools that allow stability analysis but also the control laws design. The chapter also presents results on fractional models initialization and observability, demonstrating that these notions are complex in the case of fractional models and lead to deadlock since they require the knowledge of the model past since an infinite time.

Keywords: Fractional models, stability, observability, initial conditions

MSC 2010: 26A33, 93A30, 93C05

1 Introduction

At least within a frequency interval, numerous physical systems exhibit a behavior that can be captured by a fractional operator. These systems are found in various application domains such as:

- electrochemistry in which charges diffusion in batteries can be described by *Randles' models* [38, 39];
- thermal conduction where the exact solution of the heat equation in a semi-infinite media link the heat rate to the surface temperature by a fractional differentiation of order 0.5 [3, 7];
- biology for modeling of complex dynamics in biological tissues [21];
- mechanics with the dynamical property of viscoelastic materials and for the wave propagation problems in these materials [22];
- acoustics where the fractional differentiation is used to model visco-thermal losses in wind instruments [27];
- robotics through environmental modeling [32];
- electrical distribution network [10];
- explosive materials modeling [14].

It therefore seems natural to use this operator to build models that capture this type of input-output behavior or for the design of control laws [35, 37].

Jocelyn Sabatier, Christophe Farges, Univ. Bordeaux, Bordeaux INP, CNRS, IMS, UMR 5218, 33405 Talence, France, e-mail: jocelyn.sabatier@u-bordeaux.fr

https://doi.org/10.1515/9783110571745-002

For fractional models defined in the frequency range, two major classes can be cited: explicit transfer functions (commensurable or not) and implicit transfer functions. In order to benefit from the framework of the state space description as it exists for integer models, the pseudo-state space description, associated with commensurable explicit transfer functions, was subsequently introduced. If the latter is similar to a state space description by its writing, it remains only a model of the input-output type and does not allow to take into account the internal behavior or to correctly define initial conditions. The diffusive representation, presenting a distributed state of infinite dimension, responds to the definition of the state space description as it allows the initial conditions to be correctly defined. It remains nonetheless a representation dedicated to input-output modeling for reasons that will be developed later. The next section is thus dedicated to an overview of the definition of these input–output models. Then, some elementary properties will be analyzed in the next sections.

2 Input–output fractional models

2.1 Dynamical models: general case

The general case of a Multi-Input, Multi-Output (MIMO) Linear Time Invariant (LTI) models is considered here where the inputs are gathered into the vector $u(t) \in \mathbb{R}^m$, and where the outputs are gathered into the vector $y(t) \in \mathbb{R}^p$. Only dynamical models are considered here. These models are supposed causal: their behavior at a time t, characterized by the vector $y(t)$, and depends on present and past inputs $u(t)$. Causality principle can be associated with the state concept.

The model state, denoted $\eta(t)$, may consist of a finite or an infinite number of variables and contains the necessary information to describe the effect of the past inputs $u(t)$ on the future behavior of the model outputs $y(t)$.

Definition 1 (State of a dynamical model [16]). The state $\eta(t)$ is the minimal amount of variable (information) such that, if $\eta(t_0)$ is known at $t = t_0$, then $y(t_1)$ and $\eta(t_1)$ can be determined in a unique way for any $t_1 \geq t_0$ if $u(t)$ is known within the interval $[t_0, t_1]$.

Variables $u(t)$, $y(t)$ and $\eta(t)$ describing the dynamic behavior of the model are shown in Figure 1.

The behavior of the systems studied in this chapter will be captured by models having linearity and time invariance property.

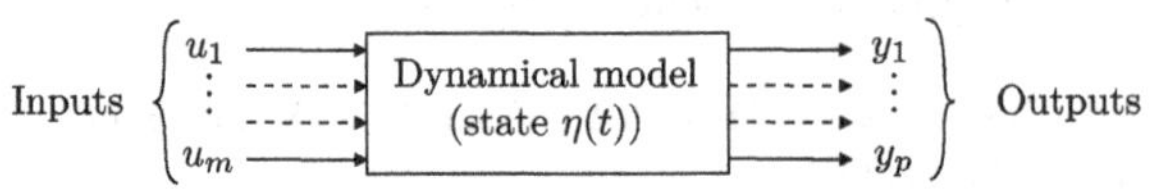

Figure 1: Inputs, outputs, and states of a dynamical model.

Property 1 (Linearity). Let $y_i(t)$ the outputs of a model whose inputs are $u_i(t)$. The model is said linear if for a linear combination of inputs:

$$u(t) = \sum_i \alpha_i u_i(t), \quad \alpha_i \in \mathbb{R} \tag{1}$$

output $y(t)$ is a linear combination of the elementary responses to each input applied separately:

$$y(t) = \sum_i \alpha_i y_i(t). \tag{2}$$

The majority of models describing physical systems are nonlinear. Their behavior in a domain can however be considered linear in first approximation.

Property 2 (Time invariance). Let $y(t)$ the output of a model in response to an input $u(t)$. The model is time invariant if, for any $T \geq 0$, the same input shifted by a time T:

$$u_d(t) = u(t+T) \tag{3}$$

leads to the same output shifted by a time T:

$$y_d(t) = y(t+T). \tag{4}$$

Stating from these general considerations about dynamical models, various dynamical, linear and time-invariant (LTI) fractional models are defined in the following sections.

2.2 Fractional transfer matrices

A transfer matrix is a modeling tool often used in automatic control. It permits to characterize the frequency input–output behavior of a system at rest at $t = 0$ ($u(t) = 0$ $\forall\, t < 0$). It is defined as the operator that links the Laplace transform of the output signal $Y(s) = \mathcal{L}\{y(t)\}$ to the Laplace transform of the input signal $U(s) = \mathcal{L}\{u(t)\}$:

$$Y(s) = M(s)U(s) \tag{5}$$

where $M(s) \in \mathbb{C}^{p \times m}$ is the system transfer matrix. Each element of this matrix is a transfer function.

In the case of integer LTI models, transfer function $G(s)$ is a rational function of the Laplace variable s:

$$G(s) = \frac{N(s)}{D(s)} \tag{6}$$

where $N(s)$ and $D(s)$ are polynomials with integer powers of s.

In the case of fractional LTI models, functions $N(s)$ and $D(s)$ involve fractional powers. They are denoted *explicit transfer function* if fractional powers are directly on the Laplace variable s (e. g., $s^{0.5}$) and *implicit transfer function* if fractional powers are on polynomials of the Laplace variable s (e. g., $(s^2+s+1)^{0.5}$).

2.2.1 Explicit fractional transfer functions

General form of an explicit transfer function is

$$G(s) = \frac{\sum_{k=1}^{N_n} \alpha_k s^{\nu_{n_k}}}{\sum_{l=1}^{N_d} \beta_l s^{\nu_{d_l}}} \tag{7}$$

where $\{\alpha_k, \beta_l\} \in \mathbb{R}$, $\{\nu_{n_k}, \nu_{d_l}\} \in \mathbb{R}_+$, $\forall\, k \in \{1 \cdots N_n\}$ and $\forall\, l \in \{1 \cdots N_d\}$. If all the orders are multiple of a same order, the model is said to be commensurate.

Definition 2 (Commensurate fractional transfer function). Transfer function (7) is said *commensurate* if and only if all the differentiation orders in this function are multiple of a same order $\nu = \gcd(\nu_{n_1}, \ldots, \nu_{n_{N_n}}, \nu_{d_1}, \ldots, \nu_{d_{N_d}})$ denoted model *commensurate order.*

Thus transfer function (7) becomes

$$G(s) = \frac{N(s)}{D(s)} = \frac{\sum_{k=0}^{N'_n} \alpha'_k (s^\nu)^k}{\sum_{l=0}^{N'_d} \beta'_l (s^\nu)^l} = \frac{C_0 \prod_{k=1}^{N_\mathfrak{z}} ((s^\nu)^k - \mathfrak{z}_k)}{\prod_{l=1}^{N_\mathfrak{p}} ((s^\nu)^l - \mathfrak{p}_l)}. \tag{8}$$

Using change of variable $\mathfrak{s} = s^\nu$, functions $N(s)$ and $D(s)$ in (8) becomes polynomials. Roots of these polynomials, denoted $\mathfrak{z}_k$ and $\mathfrak{p}_l$, are respectively denoted transfer s^ν-zeros and s^ν-poles. Such a change of variables allows the extension of various analysis and synthesis methods dedicated to the integer transfer functions to explicit commensurate transfer functions.

2.2.2 Implicit fractional transfer functions

In the most general case, an implicit fractional transfer function involves fractional powers of polynomials of the Laplace variable. In most studies, this is a first-order polynomial and an implicit fractional transfer function is defined by

$$G(s) = C_0 \frac{\prod_{k=1}^{N_z} (s - z_k)^{\nu_{z_k}}}{\prod_{l=1}^{N_p} (s - p_l)^{\nu_{p_l}}} \tag{9}$$

where $C_0 \in \mathbb{R}$ and $\{\nu_{z_k}, \nu_{p_l}\} \in \mathbb{R}_+$, $\forall\, k \in \{1 \cdots N_z\}$ and $\forall\, l \in \{1 \cdots N_p\}$. The zeros and poles of the considered transfer function are real or complex: $\{z_k, p_l\} \in \mathbb{C}$.

Implicit transfer functions were first introduced by Davidson and Cole to model dielectric relaxation in glycerol, propylene glycol, and n-propanol [9]. The implicit transfer function used in their work and involving a single pole is known in the literature as the Davidson–Cole transfer function.

Remark 1. Fractional transfer functions are presented here with respect to their ability to capture particular frequency behaviors (e. g., long memory behaviors). They are thus often obtained by identification from input–output experimental data. However, in direct physical phenomena modeling approach, differential equations involving fractional derivatives of signals can be derived. In this case, an explicit transfer function is obtained by taking the Laplace transform of the input–output differential equation. Transfer functions with implicit terms can be derived from the solving of partial differential equations.

2.3 Pseudo-state space description of fractional models

For integer models, state space description concept permitted to derive powerful analysis and synthesis tools. To take benefit from such a situation, the pseudo- state space description, associated with commensurable explicit transfer functions, has been introduced for fractional models. If the latter is similar to a state space description by its writing, it remains only a model of input–output type and does not allow to take into account the model internal behavior.

2.3.1 Pseudo-state space description

An implicit fractional model defined by (8) can be rewritten as the pseudo-state space description:

$$\begin{cases} \mathbf{D}_0^\nu x(t) = Ax(t) + Bu(t), \\ y(t) = Cx(t) + Du(t), \end{cases} \qquad \{x(t), u(t)\} = 0 \quad \forall\, t < 0 \tag{10}$$

where ν is the commensurate order, $x(t) \in \mathbb{R}^n$ is the *pseudo-state vector*, $A \in \mathbb{R}^{n\times n}$, $B \in \mathbb{R}^{n\times m}$, $C \in \mathbb{R}^{p\times n}$ and $D \in \mathbb{R}^{p\times m}$ are constant matrices. To obtain such a description, the variable change $\mathfrak{s} = s^\nu$ is required. It permits to rewrite (8) as a rational function of $\mathfrak{s}$. Matrices A, B, C, and D are then deduced as for integer models and meet the following relations:

$$G(\mathfrak{s}) = C(\mathfrak{s}I - A)^{-1}B + D \quad \Leftrightarrow \quad G(s) = C(s^\nu I - A)^{-1}B + D. \tag{11}$$

Block diagram associated to description (10) is given by Figure 2. Contrary to what relation (10) might suggest, this figure highlights that the pseudo-state space description does not explicitly involve the fractional differentiation operator but the fractional-order integration operator. Thus, it is not necessary to specify which particular definition is used for $\mathbf{D}_0^\nu$ in equation (10). Moreover, even if the system is assumed to have zero initial conditions at $t = 0$, namely if the system is supposed at rest

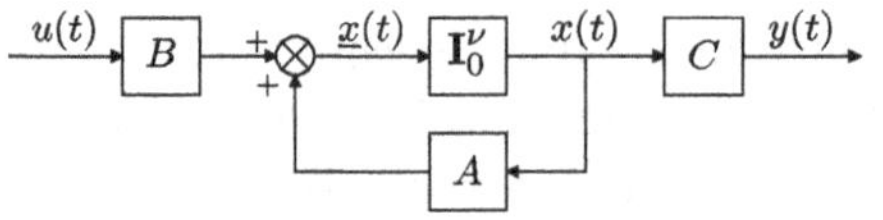

Figure 2: Block diagram of description (10).

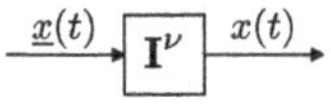

Figure 3: Block diagram of an order v fractional integrator pseudo state space description.

$(u(t) = y(t) = x(t) = 0,\ \forall t < 0)$, it is important to note that $x(t)$ cannot be considered as a state for the fractional model.

To better illustrate such a concept, a simple fractional model is used: a fractional integrator supposed at rest at $t = 0$. This integrator can also be represented by the description (10) where $A = 0$, $B = 1$, $C = 1$, and $D = 0$. The corresponding block diagram is shown in Figure 3.

For an integer integrator, $v = 1$ and (10) is really a state space description. At $t_1 > 0$. State x can be computed if the inputs between 0 et t_1 are known:

$$x(t_1) = \int_0^{t_1} \underline{x}(\tau)d\tau = x_1 = \text{cst}. \tag{12}$$

Values of $x(t)$ at later times t_1 are given by

$$x(t) = \int_0^{t} \underline{x}(\tau)d\tau = \underbrace{\int_0^{t_1} \underline{x}(\tau)d\tau}_{x_1=\text{cst}} + \int_{t_1}^{t} \underline{x}(\tau)d\tau, \quad t > t_1. \tag{13}$$

Thus, $x(t)$ can be computed if $\underline{x}$ is known within t_1 and t. Integrator output at time t thus summarizes the whole model past. Variable $x(t)$ is really the state of the dynamic model, in agreement with the Definition 1.

Let us apply the same reasoning to the fractional integrator case of order v. From the definition of fractional integration, value of x at $t_1 > 0$ can be computed if the inputs between $t = 0$ and t_1 are known:

$$x(t_1) = \frac{1}{\Gamma(v)} \int_0^{t_1} (t_1 - \tau)^{v-1} \underline{x}(\tau)d\tau = x_1 = \text{cst}. \tag{14}$$

Variable $x(t)$, $\forall t > t_1$, is thus given by

$$\begin{aligned} x(t) &= \frac{1}{\Gamma(v)} \int_0^{t} (t - \tau)^{v-1} \underline{x}(\tau)d\tau \\ &= \underbrace{\frac{1}{\Gamma(v)} \int_0^{t_1} (t - \tau)^{v-1} \underline{x}(\tau)d\tau}_{\alpha(t)\neq x_1} + \frac{1}{\Gamma(v)} \int_{t_1}^{t} (t - \tau)^{v-1} \underline{x}(\tau)d\tau. \end{aligned} \tag{15}$$

Two notable differences can be highlighted with respect to the integer case. First, term $\alpha(t)$ in equation (15) is not a constant but depends on the considered time t. Moreover, even if $x_1 = x(t_1)$ is known, it is not possible to compute $\alpha(t)$. Output x of the integrator is thus not a state for the fractional model.

The same analysis can be held for the general case of a pseudo-state description given by (10). Outputs $x(t)$ of the integrators in the block diagram of Figure 2 thus cannot be considered as the model state variables.

Remark 2. For computing $\alpha(t)\ \forall t$, it is thus necessary to know $\underline{x}(t)\ \forall t \in [0; t_1]$, thus all the model past. Fractional models are thus infinite dimensional because an infinite amount of information is necessary to summarize their past.

To illustrate this remark, the following fractional-order model is considered:

$$\frac{d^\nu}{dt^\nu} x(t) = u(t), \quad y(t) = x(t), \tag{16}$$

which corresponds to a fractional integrator of order ν. Let us also consider two different input signals $u_1(t)$ and $u_2(t)$ creating model history:

$$\begin{aligned} u(t) &= u_i(t) = A_i(t + t_i) \quad \text{for } -t_i < t < 0, \\ u(t) &= u_i(t) = 0 \quad \text{else}, \end{aligned} \qquad i = \{1, 2\}. \tag{17}$$

A constraint is also imposed to these two inputs so that at $t = 0$, the two resulting outputs $y_1(t)$ and $y_2(t)$ are equal:

$$y_1(0) = y_2(0). \tag{18}$$

As the response of a fractional integrator of order ν to a ramp of slope A_i is

$$A_i \frac{t^{1+\nu}}{\Gamma(2+\nu)} \tag{19}$$

the constraint (18) thus becomes

$$A_2 = A_1 \frac{t_1^{1+\nu}}{t_2^{1+\nu}}. \tag{20}$$

To obtain the response for $t > 0$, it must be noticed that signals $u_i(t)$ can be split into three elementary signals as shown in Figure 4.

For $t > 0$, the response $y_i(t)$ of the model to the input $u_i(t)$ is thus:

$$y_i(t) = \frac{A_i}{\Gamma(2+\nu)} (t + t_i)^{1+\nu} - \frac{A_i}{\Gamma(2+\nu)} t^{1+\nu} - \frac{A_i}{\Gamma(1+\nu)} t^\nu. \tag{21}$$

Responses $y_1(t)$ and $y_2(t)$ are plotted on Figure 5 for $A_1 = 1$, $t_1 = 1$, $t_2 = 10$ and, according to (20), $A_2 = 0.0398$. This figure highlights that it is not enough to consider

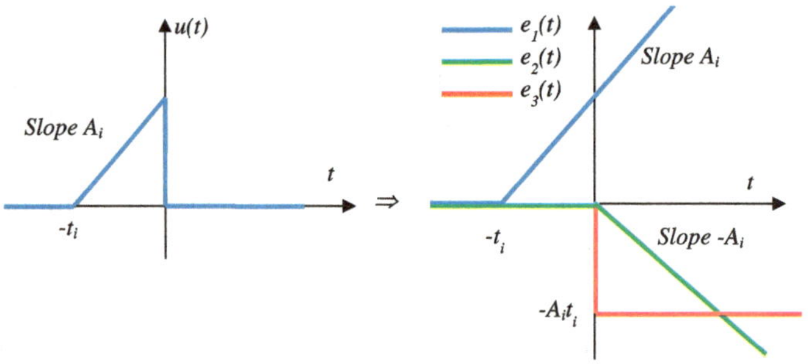

Figure 4: Signal $u(t)$ split into 3 elementary signals.

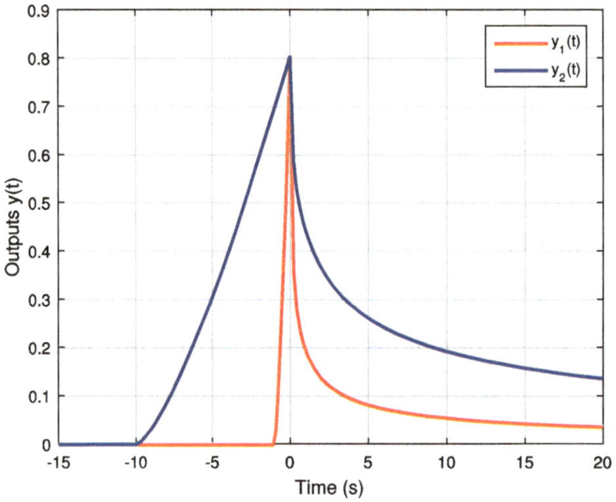

Figure 5: Comparison of signals $y(t)$.

only the output value at time $t = 0$ to compute the model future as two equal values of $y(0)$ produce different behaviors of $y(t)$ for $t > 0$.

For all these reasons, $x(t)$ in relation (10) does not represent the real state of the fractional model (which is of infinite dimension). It is thus denoted pseudo state. In the sequel, description (10) will therefore only be used with null initial conditions and to represent an input-output behavior of the modeled system.

2.4 Modal decomposition based representation

The representation proposed in this section was developed in [46, 47] for SISO models and extended to MIMO models in [43]. This representation, also known as diffusive representation [29], can be directly derived from the impulse response of a fractional order model as reminded in the sequel (see [43] for details).

2.4.1 Modal decomposition of a fractional order model

As for integer models, modal decomposition of model (10) is given by

$$\begin{cases} \frac{d^\nu}{dt^\nu} x_J(t) = J\, x_J(t) + B_J\, u(t), \\ y(t) = C_J\, x_J(t) + D\, u(t) \end{cases} \tag{22}$$

where J is a Jordan matrix whose diagonal is constituted of matrix A eigenvalues denoted λ_l, $l \in \{1, \ldots, r\}$, of multiplicity n_l.

If the model is supposed at rest and introducing inverse Laplace transform, output $y(t)$ is then given by

$$y(t) = \mathcal{L}^{-1}\{C_J(s^\nu I - J)^{-1} B_J\} * u(t) + Du(t) \tag{23}$$

where matrix $(s^\nu I - J)^{-1}$ is

$$\begin{aligned} (s^\nu I - J)^{-1} = \operatorname{diag}(&(s^\nu I - J_{n_1}(\lambda_1))^{-1}, \ldots, (s^\nu I - J_{n_l}(\lambda_l))^{-1}, \\ &\ldots, (s^\nu I - J_{n_r}(\lambda_r))^{-1}) \end{aligned} \tag{24}$$

where Laplace variable s is supposed to belong to a subset of $\mathbb{C}$ such that $(s^\nu I - J)^{-1}$ exists and where each Jordan block is given by

$$(s^\nu I - J_{n_l}(\lambda_l))^{-1} = \begin{bmatrix} \frac{1}{s^\nu - \lambda_l} & (\frac{1}{s^\nu - \lambda_l})^2 & \cdots & (\frac{1}{s^\nu - \lambda_l})^{n_l} \\ & \frac{1}{s^\nu - \lambda_l} & & \vdots \\ & & \ddots & (\frac{1}{s^\nu - \lambda_l})^2 \\ & & & \frac{1}{s^\nu - \lambda_l} \end{bmatrix}; \quad \sum_{l=1}^{r} n_l = n. \tag{25}$$

Output $y(t)$ thus stems from a linear combination of n terms called modes:

$$h^{\lambda_l,q}(t) = \mathcal{L}^{-1}\left\{\left(\frac{1}{s^\nu - \lambda_l}\right)^q\right\}, \quad q \in \{1, \ldots, n_l\}, \quad l \in \{1, \ldots, r\}. \tag{26}$$

2.4.2 Computation of modes

Each mode given by (26) can be decomposed into integer and diffusive parts [33]:

$$h^{\lambda_l,q}(t) = h_i^{\lambda_l,q}(t) + h_d^{\lambda_l,q}(t). \tag{27}$$

The integer part $h_i^{\lambda_l,q}(t)$ results from the computation of the residues of the poles associated to the mode $h^{\lambda_l,q}(t)$. Note that, unlike integer order models, poles (which are solutions of the equation $s^\nu - \lambda_l = 0$) are not equal to matrix A eigenvalues. In order to extend the notion of poles to fractional-order models, let us associate to each eigenvalue λ_l the set of integers $\mathbb{K}_{\lambda_l}$ defined as

$$\mathbb{K}_{\lambda_l} = \left\{k \in \mathbb{Z} \,:\, -\frac{\nu}{2} - \frac{\arg(\lambda_l)}{2\pi} < k < \frac{\nu}{2} - \frac{\arg(\lambda_l)}{2\pi}\right\}. \tag{28}$$

If the set $\mathbb{K}_{\lambda_l}$ is not empty, then $n_{\mathbb{K}_{\lambda_l}} = \mathrm{card}(\mathbb{K}_{\lambda_l})$ poles denoted $p_j^{\lambda_l}, j \in \{1, \ldots, n_{\mathbb{K}_{\lambda_l}}\}$, can be associated to the eigenvalue λ_l. Those poles are defined by $p_j^{\lambda_l} = |p_j^{\lambda_l}| e^{j\theta_j^{\lambda_l}}$ where

$$|p_j^{\lambda_l}| = |\lambda_l|^{\frac{1}{\nu}}, \quad \theta_j^{\lambda_l} = \frac{1}{\nu}\arg(\lambda_l) + \frac{2k\pi}{\nu} \tag{29}$$

and $j = k - k_{min} + 1$, $k \in \mathbb{K}_{\lambda_l}$, $k_{min} = \min_{k\in\mathbb{K}_{\lambda_l}} k$. $h_i^{\lambda_l,q}(t)$ is thus defined by

$$h_i^{\lambda_l,q}(t) = \sum_{j=1}^{n_{\mathbb{K}_{\lambda_l}}} \frac{p_j^{\lambda_l}}{\lambda_l^q} Q_{q-1}\left(\frac{1}{\nu}, tp_j^{\lambda_l}\right) e^{tp_j^{\lambda_l}} \tag{30}$$

where $Q_\kappa(\alpha, \beta)$ is a two variable polynomial defined by [33]

$$\begin{cases} Q_0(\alpha,\beta) = \alpha, \\ \kappa Q_\kappa(\alpha,\beta) = (\alpha\beta + \alpha - \beta)Q_{\kappa-1}(\alpha,\beta) + \alpha\beta\frac{\partial}{\partial\beta}Q_{\kappa-1}(\alpha,\beta). \end{cases} \tag{31}$$

As demonstrated in [24, 33], diffusive part of the mode $h_d^{\lambda_l,q}(t)$ is given by

$$h_d^{\lambda_l,q}(t) = \int_0^{+\infty} \mu^{\lambda_l,q}(\chi) e^{-t\chi} d\chi, \tag{32}$$

where

$$\mu^{\lambda_l,q}(\chi) = \frac{1}{\pi}\frac{\sum_{j=0}^{q-1}(-1)^j \binom{q}{j} (\lambda_l)^j \chi^{\nu(q-j)} \sin(\nu\pi(q-j))}{(\chi^{2\nu} - 2\lambda_l\chi^\nu\cos(\nu\pi) + \lambda_l^2)^q}, \quad \binom{q}{j} = \frac{q!}{j!(q-j)!}. \tag{33}$$

2.4.3 Fractional-order model as the association of integer and diffusive submodels

As for modes, model (10) time response $y(t)$ with $D = 0$, can be defined as

$$y(t) = y_i(t) + y_d(t), \quad y_i(t) \in \mathbb{R}^p, \quad y_d(t) \in \mathbb{R}^p, \tag{34}$$

where $y_i(t)$ and $y_d(t)$ are given by

$$y_{i,d}(t) = \int_0^t C_J \begin{bmatrix} h_{i,d}^{\lambda_1}(\tau) & & \\ & \ddots & \\ & & h_{i,d}^{\lambda_r}(\tau) \end{bmatrix} B_J u(t-\tau)d\tau \tag{35}$$

where

$$h_{i,d}^{\lambda_l}(\tau) = \begin{bmatrix} h_{i,d}^{\lambda_l,1}(\tau) & h_{i,d}^{\lambda_l,2}(\tau) & \cdots & h_{i,d}^{\lambda_l,n_l}(\tau) \\ & h_{i,d}^{\lambda_l,1}(\tau) & & \vdots \\ & & \ddots & h_{i,d}^{\lambda_l,2}(\tau) \\ & & & h_{i,d}^{\lambda_l,1}(\tau) \end{bmatrix}. \tag{36}$$

Considering the diffusive part, using relation (32), relation (35) becomes

$$y_d(t) = C_J \int_0^{+\infty} \begin{bmatrix} \mu^{\lambda_1}(\chi) & & \\ & \ddots & \\ & & \mu^{\lambda_r}(\chi) \end{bmatrix} w(\chi, t) d\chi = C_J \int_0^{+\infty} \mu(\chi) w(\chi, t) d\chi \tag{37}$$

with

$$\mu^{\lambda_l}(\chi) = \begin{bmatrix} \mu^{\lambda_l,1}(\chi) & \mu^{\lambda_l,2}(\chi) & \cdots & \mu^{\lambda_l,n_l}(\chi) \\ & \mu^{\lambda_l,1}(\chi) & & \vdots \\ & & \ddots & \mu^{\lambda_l,2}(\chi) \\ & & & \mu^{\lambda_l,1}(\chi) \end{bmatrix} \quad \text{and} \quad w(\chi, t) = \int_0^t e^{-\tau\chi} B_J u(t-\tau) d\tau \tag{38}$$

and thus

$$\dot{w}(\chi, t) = -\chi w(\chi, t) + B_J u(t), \quad w(\chi, t) \in \mathbb{R}^m, \quad \chi \in \mathbb{R}^+. \tag{39}$$

Relations (37) and (39) define the differential system

$$\begin{cases} \dot{w}(\chi, t) = -\chi w(\chi, t) + B_J u(t), \\ y_d(t) = C_J \int_0^{+\infty} \mu(\chi) w(\chi, t) d\chi \end{cases} \tag{40}$$

where initial conditions are defined by $w(\chi, 0) = \rho(\chi)$. Model (40) also writes as

$$\begin{cases} \dot{w}(\chi, t) = -\chi w(\chi, t) + B_J u(t), \\ y_d(t) = \frac{1}{2} C_J \int_0^{+\infty} \mu(\chi) w(\chi, t) d\chi + \frac{1}{2} C_J \int_0^{+\infty} \mu(\chi) w(\chi, t) d\chi. \end{cases} \tag{41}$$

Using the changes of variable $z = \frac{\sqrt{\chi}}{2\pi}$ and $z = -\frac{\sqrt{\chi}}{2\pi}$ applied respectively to the first and second integral of (41) and if $\Psi(z, t) = w(4\pi^2 z^2, t)$ denotes spatial Fourier transform of $\Phi(\zeta, t)$ with $\Psi(z, 0) = \rho(4\pi^2 z^2)$, $z \in \mathbb{R}$, it is demonstrated in [29], and in [46] for the commensurate case, that relation (41) becomes

$$\begin{cases} \frac{\partial \Psi(z,t)}{\partial t} = (2i\pi z)^2 \Psi(z, t) + B_J \mathcal{F}\{u(t)\}, \\ y_d(t) = C_J \int_{-\infty}^{+\infty} \left(\int_{-\infty}^{+\infty} 4\pi^2 z \mu(4\pi^2 z^2) e^{-2i\pi z\zeta} dz \right) \Phi(\zeta, t) d\zeta. \end{cases} \tag{42}$$

Using change of variables $\tilde{z} = -z$ and applying inverse Fourier transform to last equation leads to

$$\begin{cases} \{\frac{\partial\Phi(\zeta,t)}{\partial t}\} = \frac{\partial^2\Phi(\zeta,t)}{\partial\zeta^2} + B_J u(t)\delta(\zeta), \\ y_d(t) = C_J \int_{-\infty}^{+\infty} m(\zeta)\Phi(\zeta,t)d\zeta \end{cases} \tag{43}$$

with $m(\zeta) = \mathcal{F}^{-1}\{-4\pi^2\tilde{z}\mu(4\pi^2\tilde{z}^2)\}$, $\Phi(\zeta,0) = \mathcal{F}^{-1}\{\rho(4\pi^2\tilde{z}^2)\}$, $\zeta \in \mathbb{R}$.

If now relation (35) is taken into account, then a LTI fractional-order model can be described by

$$\dot{\eta}(t) = A_P\eta(t) + B_P u(t), \quad y_i(t) = C_P\eta(t), \tag{44}$$

$$\frac{\partial\Phi(\zeta,t)}{\partial t} = \frac{\partial^2\Phi(\zeta,t)}{\partial\zeta^2} + B_J u(t)\delta(\zeta), \quad y_d(t) = C_J \int_{-\infty}^{+\infty} m(\zeta)\Phi(\zeta,t)d\zeta, \tag{45}$$

$$y(t) = y_i(t) + y_d(t) \tag{46}$$

with initial conditions $\eta(0)$ and $\Phi(\zeta,0)$. For matrix A distinct eigenvalues ($r = n$), matrices A_P, B_P, and C_P in (44) are defined in [43]. Relations (44) define a state-space description whose time response is given by the convolution product (35).

Any fractional model can thus be seen as an infinite dimensional model (defined on an infinite space domain) described by a set of diffusion equations associated to a classical integer model.

Figure 6 is a representation of the interpretation previously given for the submodel whose output is $y_d(t)$. $\Phi(\zeta,t)$ is the real submodel state vector with n components (of infinite dimension with respect to ζ) and $u(t)$ is both the input of the integer submodel and of the diffusive submodel, applied for the latter at $\zeta = 0$. To conclude, it can be said that a fractional-order model is at the boundary of two classes of models: classical integer models and distributed parameters models.

Remark 3. The need to know all the model past evoked previously to obtain its future evolution, is translated in this representation by the need to know its state on an infinite dimension space.

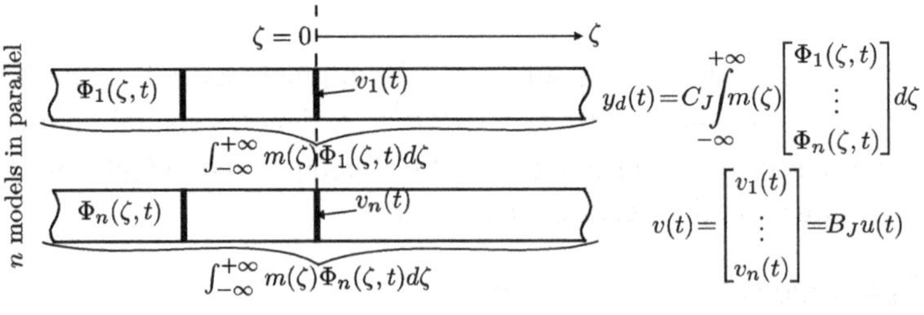

Figure 6: A representation of sub-model whose output is $y_d(t)$.

3 Fractional models stability

Stability is a notion whose importance for the control of dynamical systems is paramount and that can be defined in different ways. The definition adopted in this chapter is first reminded as well as its characterization from the values of the transfer function poles. Stability analysis methods specific to commensurate models formulated in terms of Linear Matrix Inequalities (LMI) are then presented but results for noncommensurate models was also proposed by the authors in [44]. The focus is on one of them, based on the formalism of generalized LMI regions, and providing a stability necessary and sufficient condition.

3.1 Input–output stability of a fractional model

The fractional models considered in this chapter being only able to account for the input–output behavior of the modeled systems, only the input–output stability, also denoted external stability, is studied. A system is said to be stable from the input–output point of view if any bounded input signal in the $\mathfrak{L}_q$ norm sense generates an output signal itself bounded.

Definition 3 ($\mathfrak{L}_q$ stability [18]). A model Σ of input u and output y, described by the transfer matrix $G(s)$, is $\mathfrak{L}_q$-stable, $1 < q < \infty$, if and only if

$$\sup_{u \in \mathfrak{L}_q, u \neq 0} \frac{\|g * u\|_p}{\|u\|_p} < \infty, \tag{47}$$

where $g(t)$ denotes the model impulse response.

When the $\mathfrak{L}_\infty$ norm is used in the previous definition, such a stability is denoted Bounded Input Bounded Output (BIBO) stability.

Definition 4 (Fractional model BIBO stability [23]). A fractional LTI model whose impulse response is defined by $g(t) = \mathcal{L}^{-1}\{G(s)\}$ is BIBO stable if and only if $\forall\, u(t) \in \mathfrak{L}_\infty(\mathbb{R}_+, \mathbb{R})$, $y(t) = g(t) * u(t) \in \mathfrak{L}_\infty(\mathbb{R}_+, \mathbb{R})$.

Various methods make it possible to evaluate the input-output stability of a dynamical model according to the definition used ($\mathfrak{L}_q$ stability or BIBO stability).

If the Definition 4 is used, it is well known that an integer model is BIBO stable if and only if all its pole real parts are negative. This result has been extended to fractional models as reminded in [5].

Theorem 1 (BIBO stability using poles analysis [5]). *Let a strictly proper fractional model defined by the transfer function:*

$$G(s) = \frac{N(s)}{D(s)}, \tag{48}$$

where $N(s)$ and $D(s)$ have no common roots. Then $G(s)$ is BIBO stable if the poles of G are all in the stability domain:

$$\mathcal{D}_s = \{s \in \mathbb{C} : \mathrm{Re}(s) < 0\}, \tag{49}$$

or equivalently if $\mathrm{Re}\,(p_i) < 0\ \forall i$ *where p_i denotes the ith root of $D(s)$.*

Unlike integer models, D is not a polynomial of the s variable. For an implicit transfer function given by relation (9), $D(s) = \prod_{l=1}^{N_p} (s - p_l)^{\nu_{p_l}}$ makes the poles p_l appear directly and it is easy to conclude on the stability. However, for an explicit transfer function given by relation (7), $D_s(s) = \sum_{l=1}^{N_d} \beta_l s^{\nu_{d_l}}$ does not make the poles appear directly and their computation is more complex.

3.2 Stability of explicit commensurate fractional models

As stated in the previous section, the stability of an explicit model described by a transfer function of the form (7) depends on the sign of the real part of N_p poles denoted p_l that is of the roots of

$$D(s) = \sum_{l=1}^{N_d} \beta_l s^{\nu_{d_l}}. \tag{50}$$

For commensurate models, differentiation orders are all multiple of the same order ν, and as shown by relation (8) of Section 2.2.1, $D(s)$ can be written as

$$D(s) = \sum_{l=0}^{N'_d} \beta'_l (s^\nu)^l. \tag{51}$$

Change of variable $\mathfrak{s} = s^\nu$ permits to write D as a polynomial of $\mathfrak{s}$:

$$D(\mathfrak{s}) = \sum_{l=0}^{N'_d} \beta'_l \mathfrak{s}^l. \tag{52}$$

The roots of $D(\mathfrak{s})$, denoted s^ν-poles and noted $\mathfrak{p}_l$, can be computed using the tools commonly used for integer models. To conclude on the stability of fractional models, it remains to characterize the link between the sign of the N_p poles p_l and the values of the N'_p s^ν-poles $\mathfrak{p}_l$. Clearly, it is necessary to study the transformation of the stability domain for the poles $\mathcal{D}_s$ into stability domain for s^ν-poles $\mathcal{D}_\mathfrak{s}$ via variable change $\mathfrak{s} = s^\nu$.

Stability domain $\mathcal{D}_s$ given by (49) is defined by

$$\mathcal{D}_s = \{s \in \mathbb{C} : \mathrm{Re}(s) < 0\}. \tag{53}$$

Using $s = \rho e^{i\theta}$, thus

$$\mathcal{D}_S = \left\{ s = \rho e^{i\theta} \in \mathbb{C} : |\theta| > \frac{\pi}{2} \right\}. \tag{54}$$

Change of variable $\mathfrak{s} = s^\nu$ thus leads to

$$\mathcal{D}_\mathfrak{s} = \{\mathfrak{s} = s^\nu \in \mathbb{C} : s \in \mathcal{D}_S\}. \tag{55}$$

or:

$$\mathcal{D}_\mathfrak{s} = \left\{ \mathfrak{s} = (\rho e^{i\theta})^\nu = \rho^\nu e^{i\nu\theta} \in \mathbb{C} : |\theta| > \frac{\pi}{2} \right\}. \tag{56}$$

Using $\mathfrak{s} = \rho' e^{i\theta'}$, argument of the s^ν-poles is given by $\theta' = \nu\theta$. Condition $|\theta| > \frac{\pi}{2}$ thus implies $|\theta'| > \nu\frac{\pi}{2}$ and the stability domain becomes

$$\mathcal{D}_\mathfrak{s} = \left\{ \mathfrak{s} = \rho' e^{i\theta'} \in \mathbb{C} : |\theta'| > \nu\frac{\pi}{2} \right\}. \tag{57}$$

BIBO stability of an explicit commensurate transfer function can thus be characterized by the belonging of s^ν-poles to the stability domain $\mathcal{D}_\mathfrak{s}$. Moreover, as detailed in the Section 2.3, a pseudo-state representation associated with the transfer function $G(s)$ is a realization of the rational transfer $G(\mathfrak{s})$. The eigenvalues of the A matrix of the pseudo-state space description are therefore the poles of $G(\mathfrak{s})$, namely s^ν-poles of $G(s)$. The following stability criterion is thus given.

Theorem 2 (BIBO stability analysis on a fractional transfer function or on its pseudo state space description). *An explicit commensurate transfer function defined by*

$$G(s) = \frac{N(s)}{D(s)} = \frac{\sum_{k=0}^{N_n'} \alpha_k' (s^\nu)^k}{\sum_{l=0}^{N_d'} \beta_l' (s^\nu)^l}, \tag{58}$$

whose pseudo-state space description is given by (10) *is BIBO stable if and only if the s^ν-poles of $G(s)$, which are also the eigenvalues of the matrix A, belong to the stability domain:*

$$\mathcal{D}_\mathfrak{s} = \left\{ \mathfrak{s} = \rho' e^{i\theta'} \in \mathbb{C} : |\theta'| > \nu\frac{\pi}{2} \right\}. \tag{59}$$

Stability domain $\mathcal{D}_\mathfrak{s}$ described by relation (59) is illustrated by Figure 7.

Remark 4. Stability Theorem 2 results from Denis Matignon's work [23] dedicated to fractional pseudo-state space description. It is also a direct consequence of the explicit calculation of the poles of the transfer function as detailed [34]. Another proof based on Cauchy's theorem is also proposed in [40, 42].

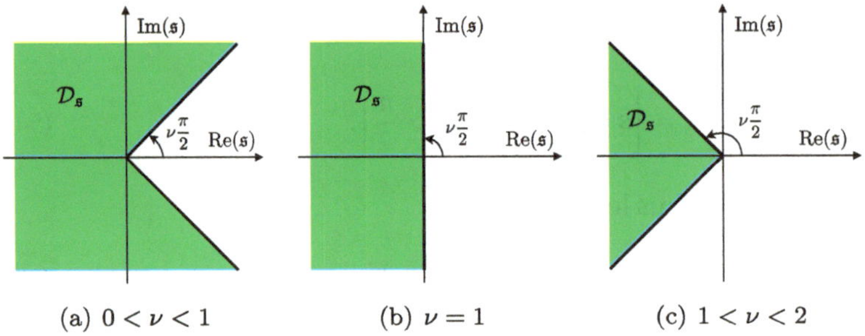

Figure 7: Stability domain $\mathcal{D}_s$ as a function of the commensurate order v.

3.3 LMI results for commensurate model stability analysis

If the explicit calculation of the s^{ν}-poles of $G(s)$ or the eigenvalues of the A matrix makes it possible to conclude on the stability, various works have been devoted to the characterization of the $\mathcal{D}_s$ domain membership, leading criteria formulated using LMIs. These LMI criteria fall into two categories according to the commensurable order ν.

If $1 < \nu < 2$, $\mathcal{D}_s$ domain shown in Figure 7(c) is convex and the characterization of the matrix A eigenvalues belonging to this domain can be solved using the LMI region formalism introduced in [6], and generalized in [36] through the $\mathbb{D}_R$ stability concept dedicated to integer models.

Theorem 3 ([30]). *A fractional commensurate model with* $1 < \nu < 2$ *of pseudo- state space description* (10) *is stable if and only if* $\exists X \in \mathbb{R}^{n\times n} > 0$ *s.t.*

$$\begin{bmatrix} (XA^T + AX)\sin(\nu\frac{\pi}{2}) & (XA^T - AX)\cos(\nu\frac{\pi}{2}) \\ (AX - XA^T)\cos(\nu\frac{\pi}{2}) & (XA^T + AX)\sin(\nu\frac{\pi}{2}) \end{bmatrix} < 0. \tag{60}$$

If $0 < \nu < 1$, stability domain depicted by Figure 7(a) is not convex. However, various LMI conditions were proposed [48, 49]. These results are now reminded, and commented on their respective pessimism as well as their ability to be used in the context of control law design.

The first method is based on an algebraic transformation of the model considering that the fractional order ν is rational, leading to the following theorem.

Theorem 4 ([28]). *A commensurate fractional model of order* $0 < \nu < 1$ *and pseudo state space description* (10) *is BIBO stable if* $\exists\, P \in \mathbb{R}^{n\times n} > 0$ *s.t.*

$$(A^{\frac{1}{\nu}})^T P + P(A^{\frac{1}{\nu}}) < 0. \tag{61}$$

As shown in [30], this method is however pessimistic because the stability domain is not fully characterized by LMI (61).

A necessary and sufficient condition was proposed in [30] using a geometric transformation of the stability domain for the case $0 < \nu < 1$.

Theorem 5 ([30]). *A commensurate fractional model of order* $0 < \nu < 1$ *and pseudo state space description* (10) *is BIBO stable iff* $\exists\, P \in \mathbb{R}^{n\times n} > 0$ *s.t.*

$$\left(-(-A)^{\frac{1}{2-\nu}}\right)^T P + P\left(-(-A)^{\frac{1}{2-\nu}}\right) < 0. \tag{62}$$

It is shown in [30] that the stability domain of a fractional model of order $0 < \nu < 1$ is fully characterized by Theorem 5, which confirms that it is a necessary and sufficient condition. Although this result is interesting for analysis, its use to derive a LMI design condition is difficult because of the nonlinearity with respect to the matrix A. If the pseudo-state feedback control law $u(t) = r(t) - Kx(t)$ is considered where $r(t)$ represents the reference signal and K is a constant matrix gain, the state matrix of the closed loop is $(A - BK)$. To analyze the closed-loop stability, Theorem 5 can be used, leading to matrix inequality:

$$\left(-(-A+BK)^{\frac{1}{2-\nu}}\right)^T P + P\left(-(-A+BK)^{\frac{1}{2-\nu}}\right) < 0. \tag{63}$$

Closed-loop state matrix raised to the power $\frac{1}{2-\nu}$ makes it difficult to obtain a linearising variable change, allowing the synthesis of K gain.

Another stability condition based on the instability domain characterization is also presented in [30]. Indeed, for $0 < \nu < 1$, the instability domain of a commensurable fractional model is convex.

Theorem 6 ([30]). *A commensurate fractional model of order* $0 < \nu < 1$ *and pseudo state space description* (10) *is BIBO stable* $\nexists Q \geq 0$ *s.t.*

$$rAQ + \bar{r}QA^T < 0, \quad r = e^{j(1-\nu)\frac{\pi}{2}}. \tag{64}$$

Although Theorem 6 provides a necessary and sufficient stability condition, the algorithm used for the LMI (64) resolution must be chosen carefully. Indeed, the nonconvergence of this LMI toward a matrix Q implies the stability of the model. A fault in the algorithm can therefore easily lead to bad conclusions. In addition, no synthesis LMI can be deduced from Theorem 6 because the stability LMI is based on the nonexistence of a matrix.

Results provided by Theorems 5 and 6 are therefore either pessimistic or unsuited to control laws design. New LMI stability criteria were thus developed.

In order to avoid the nonconvexity problem of the stability domain of a commensurate fractional model of order $0 < \nu < 1$, Generalized LMI (GLMI) concept, introduced in [6] and developed in [1, 2] was used.

Definition 5 (GLMI region [6]). A region $\mathcal{D}$ in the complex plane is an order l region if $\exists\, \theta_k \in \mathbb{C}^{l\times l}$, $\psi_k \in \mathbb{C}^{l\times l}$, $H_k \in \mathbb{C}^{l\times l}$ and $J_k \in \mathbb{C}^{l\times l}$, $\forall\, k \in \{1, \dots, m\}$ s.t.

$$\mathcal{D} = \left\{z \in \mathbb{C} \,:\, \exists\, w = [w_1 \cdots w_m]^T \in \mathbb{C}^m \text{ t.q. } f_{\mathcal{D}}(z,w) < 0,\ g_{\mathcal{D}}(w) = 0\right\}, \tag{65}$$

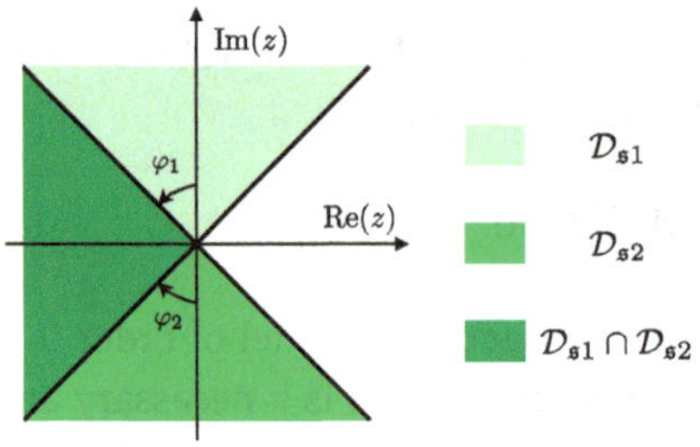

Figure 8: Stability domains $\mathcal{D}_{\mathfrak{s}}$ as the union of two half-planes.

where

$$f_{\mathcal{D}}(z, w) = \sum_{k=1}^{m} (\theta_k w_k + \theta_k^* \bar{w_k} + \psi_k z w_k + \psi_k^* \bar{w_k} \bar{z}) \tag{66}$$

and

$$g_{\mathcal{D}}(w) = \sum_{k=1}^{m} (H_k w_k + J_k \bar{w_k}). \tag{67}$$

Stability domain $\mathcal{D}_{\mathfrak{s}}$ depicted in Figure 7(a) can be seen as the union of two half-planes:

$$\mathcal{D}_{\mathfrak{s}} = \mathcal{D}_{\mathfrak{s}1} \cup \mathcal{D}_{\mathfrak{s}2}. \tag{68}$$

As shown in Figure 8, domains $\mathcal{D}_{\mathfrak{s}1}$ and $\mathcal{D}_{\mathfrak{s}2}$ result from rotations of the left complex half-plane of angles $\varphi_1 = \varphi$ and $\varphi_2 = -\varphi$, respectively, where $\varphi = (1 - \nu)\frac{\pi}{2}$, or

$$\mathcal{D}_{\mathfrak{s}i} = \{z \in \mathbb{C} : \mathrm{Re}(z\,\mathrm{e}^{\mathrm{j}\varphi_i}) < 0\}, \quad \forall\, i \in \{1, 2\}, \tag{69}$$

that can also be written:

$$\mathcal{D}_{si} = \{z \in \mathbb{C} : \exists w_i \in \mathbb{R}^+ \text{ t.q. } e^{j\varphi_i z w_i} + e^{-j\varphi_i \bar{z} \bar{w}_i} < 0\}. \tag{70}$$

It is shown in [1] that the union of m convex regions of the form:

$$\mathcal{D}_k = \{z \in \mathbb{C} : f_k(z) = \alpha_k + \beta_k z + \beta_k^* \bar{z} < 0\}, \quad \forall k \in \{1 \cdots m\}, \tag{71}$$

is a GLMI region of the form (65) of order $l = m + 1$ with

$$\theta_k = \frac{1}{2}\begin{bmatrix} \Theta_k & \mathbb{O}_{1\times m} \\ \mathbb{O}_{m\times 1} & -\varepsilon_k^m \end{bmatrix}, \quad \psi_k = \begin{bmatrix} \Psi_k & \mathbb{O}_{1\times m} \\ \mathbb{O}_{m\times 1} & \mathbb{O}_m \end{bmatrix}, \quad \Theta_k = \begin{bmatrix} \alpha_k & \cdots & 0 \\ \vdots & \ddots & \vdots \\ 0 & \cdots & 0 \end{bmatrix},$$
$$\Psi_k = \begin{bmatrix} \beta_k & \cdots & 0 \\ \vdots & \ddots & \vdots \\ 0 & \cdots & 0 \end{bmatrix}, \quad H_k = -J_k = \varepsilon_{k+1}^{m+1}, \tag{72}$$

where $\epsilon_p^q \in \mathbb{R}^{q\times q}$ is given by: $\varepsilon_p^q(\rho,\sigma) = 1$ if $\rho = \sigma = p$ or $\varepsilon_p^q(\rho,\sigma) = 0$ else. Thus regions $\mathcal{D}_{si}$ given by (70) being of the form (71)–(72) with $\alpha_k = 0$ and $\beta_k = e^{j\varphi_k}$, stability domain $\mathcal{D}_s$ is a GLMI region of the form (65) of order $l = 3$ with:

$$\theta_1 = \frac{1}{2}\begin{bmatrix} 0 & 0 & 0 \\ 0 & -1 & 0 \\ 0 & 0 & 0 \end{bmatrix}, \quad \theta_2 = \frac{1}{2}\begin{bmatrix} 0 & 0 & 0 \\ 0 & 0 & 0 \\ 0 & 0 & -1 \end{bmatrix},$$
$$\psi_1 = \begin{bmatrix} e^{j(1-\nu)\frac{\pi}{2}} & 0 & 0 \\ 0 & 0 & 0 \\ 0 & 0 & 0 \end{bmatrix}, \quad \psi_2 = \begin{bmatrix} e^{-j(1-\nu)\frac{\pi}{2}} & 0 & 0 \\ 0 & 0 & 0 \\ 0 & 0 & 0 \end{bmatrix}, \tag{73}$$

$$H_1 = \begin{bmatrix} 0 & 0 & 0 \\ 0 & 1 & 0 \\ 0 & 0 & 0 \end{bmatrix}, \quad H_2 = \begin{bmatrix} 0 & 0 & 0 \\ 0 & 0 & 0 \\ 0 & 0 & 1 \end{bmatrix}, \quad J_1 = \begin{bmatrix} 0 & 0 & 0 \\ 0 & -1 & 0 \\ 0 & 0 & 0 \end{bmatrix}, \quad J_2 = \begin{bmatrix} 0 & 0 & 0 \\ 0 & 0 & 0 \\ 0 & 0 & -1 \end{bmatrix}. \tag{74}$$

To evaluate if the eigenvalues of a matrix belong to a GLMI region of the complex plane, the following lemma is used.

Lemma 1 ([6]). *Let $A \in \mathbb{C}^{n\times n}$ and $\mathcal{D}$ a GLMI region. A is $\mathcal{D}$-stable, that is, A matrix eigenvalues belong to $\mathcal{D}$, iff $\exists$ m matrices $X_k \in \mathbb{C}^{n\times n}$ s.t.*

$$\sum_{k=1}^{m}(\theta_k \otimes X_k + \theta_k^* \otimes X_k^* + \psi_k \otimes (AX_k) + \psi_k^* \otimes (AX_k)^*) < 0 \tag{75}$$

and

$$\sum_{k=1}^{m}(H_k \otimes X_k + J_k \otimes X_k^*) = \mathbb{O}_{nl\times nl}. \tag{76}$$

Using Lemma 1 on the GLMI region $\mathcal{D}_s$ of the complex plane defined by (65)–(67) and (73)–(74), the following necessary and sufficient stability LMI condition of a fractional model of order $0 < \nu < 1$ can be obtained.

Theorem 7 ([11–13]). *A commensurate fractional model of order $0 < \nu < 1$ and pseudo state space description (10) is stable iff $\exists X = X^* \in \mathbb{C}^{n\times n} > 0$ s.t.*

$$(rX + \bar{r}\bar{X})^T A^T + A(rX + \bar{r}\bar{X}) < 0 \tag{77}$$

where $r = e^{j(1-\nu)\frac{\pi}{2}}$.

Unlike other results in the literature, the LMI (77) fully characterizes the stability domain of a fractional model of order $0 < \nu < 1$ and allows, as shown in the next section, the synthesis of stabilizing controllers.

3.4 LMI results for pseudo-state feedback stabilization of fractional commensurate models

The design problem of a pseudo-state feedback control law is considered. This law is defined by the equation:

$$u(t) = y_r(t) + Kx(t), \tag{78}$$

where $K \in \mathbb{R}^{m\times n}$ is a constant matrix gain and $y_r(t) \in \mathbb{R}^m$ is the reference signal. Control law (78) applied to (10) leads to the closed-loop model:

$$\begin{cases} \mathbf{D}_0^{\nu}x(t) = (A + BK)\,x(t) + B\,y_r(t), \\ y(t) = (C + DK)\,x(t) + D\,y_r(t). \end{cases} \tag{79}$$

Using Theorem 2, such a control problem can be formulated as follows.

Problem 1. For a given pair of matrices ($A \in \mathbb{R}^{n\times n}, B \in \mathbb{R}^{n\times m}$), find a real matrix $K \in \mathbb{R}^{m\times n}$ such that: $|\operatorname{Arg}(\lambda(A + BK))| > \nu\frac{\pi}{2}$.

If $1 < \nu < 2$, Theorem 3 applied to closed-loop model (79) leads to conclude that the closed loop is stable if there exists $K \in \mathbb{R}^{m\times n}$ and $X \in \mathbb{R}^{n\times n} > 0$ such that:

$$\begin{bmatrix} (X(A + BK)^T + (A + BK)X)\sin(\nu\frac{\pi}{2}) & (X(A + BK)^T - (A + BK)X)\cos(\nu\frac{\pi}{2}) \\ ((A + BK)X - X(A + BK)^T)\cos(\nu\frac{\pi}{2}) & (X(A + BK)^T + (A + BK)X)\sin(\nu\frac{\pi}{2}) \end{bmatrix} < 0. \tag{80}$$

However, inequality (80) is no more linear. To obtain a LMI formulation, the change of variable $Y = KX$ classically used for integer models can also be used and leads to the following theorem.

Theorem 8 ([30]). *A fractional commensurate model of order $1 < \nu < 2$ and pseudo state space description* (10) *can be stabilized using a pseudo-state feedback iff $\exists X \in \mathbb{R}^{n\times n} > 0$ and $Y \in \mathbb{R}^{m\times n}$ s.t.*

$$\begin{bmatrix} (XA^T + Y^TB^T + AX + BY)\sin(\nu\frac{\pi}{2}) & (XA^T + Y^TB^T - AX - BY)\cos(\nu\frac{\pi}{2}) \\ (AX + BY - XA^T - Y^TB^T)\cos(\nu\frac{\pi}{2}) & (XA^T + Y^TB^T + AX + BY)\sin(\nu\frac{\pi}{2}) \end{bmatrix} < 0. \tag{81}$$

The stabilizing controller gain is then given by

$$K = YX^{-1}. \tag{82}$$

Finding such a change of variables in the case $0 < \nu < 1$ can appear more difficult because the variable X appearing in the LMI (77) is complex when the controller gain K to be determined is real. However, the LMI (77) of Theorem 7 has been formulated to make the term $(rX + \bar{r}\overline{X})$ appears, which is real. Using the change of variables $Y = K(rX + \bar{r}\overline{X})$, the following theorem is obtained.

Theorem 9. *A commensurate fractional model of order* $0 < \nu < 1$ *and pseudo- state space description* (10) *can be stabilized using a pseudo-state feedback iff* $\exists X = X^* \in \mathbb{C}^{n \times n} > 0$ *and* $Y \in \mathbb{R}^{m \times n}$ *s.t.*

$$(rX + \bar{r}\overline{X})^T A^T + A(rX + \bar{r}\overline{X}) + Y^T B^T + BY < 0 \tag{83}$$

where $r = e^{j(1-\nu)\frac{\pi}{2}}$. *The stabilizing controller gain is then given by*

$$K = Y(rX + \bar{r}\overline{X})^{-1}. \tag{84}$$

4 Fractional models and initial conditions

In a series of papers, Lorenzo and Hartley [15, 19, 20], have demonstrated that initial conditions are not taken into account in the same way whether Riemann–Liouville or Caputo definitions are considered. The goal of this section is to go further and to demonstrate that neither Riemann–Liouville nor Caputo definitions permit to take into account initial conditions in a consistent way in relation to a physical system behavior. The demonstration is done on a counterexample based on the following model which corresponds to a fractional integrator:

$$\frac{d^\nu x(t)}{dt^\nu} = u(t), \quad 0 < \nu < 1. \tag{85}$$

As in the Lorenzo and Hartley work, the demonstration is based on the shift of the time origin. Model (85) is supposed to be at rest when time $t < 0$. Input $u(t)$ is also supposed to be defined by

$$u(t) = H(t) - H(t - t_0) \tag{86}$$

where $H(\cdot)$ denotes the Heaviside function. Time response of model (85) is thus defined by

$$x(t) = \frac{t^\nu}{\Gamma(\nu+1)} H(t) - \frac{(t - t_0)^\nu}{\Gamma(\nu+1)} H(t - t_0). \tag{87}$$

The change of variable

$$\tau = t - 2t_0 \tag{88}$$

is now used so that at time $\tau = 0$, the model is not at rest.

Using Riemann–Liouville's definition, Laplace transform of the fractional derivative of $x(\tau)$ is defined by

$$\mathcal{L}\{D^\nu x(\tau)\} = s^\nu \overline{x}(s) - \sum_{k=0}^{n-1} s^k [D^{\nu-k-1} x(\tau)]_{\tau=0} \tag{89}$$

with $n-1 < \nu < n$ and where $\overline{x}(s)$ denotes the Laplace transform of $x(\tau)$ with zero initial conditions. Laplace transform applied to equation (85), leads to

$$s^{\nu}\overline{x}(s) - \sum_{k=0}^{n-1} s^{k}\left[D^{\nu-k-1}x(\tau)\right]_{tau=0} = 0, \tag{90}$$

(here $n = 1$) and the corresponding time response is defined by

$$x(\tau) = \mathcal{L}^{-1}\left\{\frac{1}{s^{\nu}}\right\}\left[I^{1-\nu}x(\tau)\right]_{\tau=0} \tag{91}$$

where

$$\left[I^{1-\nu}x(\tau)\right]_{\tau=0} = \mathcal{L}^{-1}\left\{\frac{1}{s^{1-\nu}}\left(\frac{e^{2t_0 s}}{s^{\nu+1}} - \frac{e^{t_0 s}}{s^{\nu+1}}\right)\right\}_{\tau=0} = t_0. \tag{92}$$

The time response of model (85), is thus given by

$$x(\tau) = t_0\frac{\tau^{\nu-1}}{\Gamma(\nu)}, \quad \tau \geq 0. \tag{93}$$

Using Caputo's definition, Laplace transform of the fractional derivative of $x(\tau)$ is defined by

$$\mathcal{L}\{D^{\nu}x(\tau)\} = s^{\nu}\overline{x}(s) - \sum_{k=0}^{n-1} s^{\nu-k-1}\left[D^{k}x(\tau)\right]_{\tau=0}. \tag{94}$$

Laplace transform applied to equation (85), leads to

$$s^{\nu}\overline{x}(s) - \sum_{k=0}^{n-1} s^{\nu-k-1}\left[D^{k}x(\tau)\right]_{\tau=0} = 0, \tag{95}$$

(here $n = 1$) and the corresponding time response is given by

$$x(\tau) = L^{-1}\left\{\frac{s^{\nu-1}}{s^{\nu}}\right\}\left[x(\tau)\right]_{\tau=0} \tag{96}$$

where, using (87)

$$\left[x(\tau)\right]_{\tau=0} = \frac{(2t_0)^{\nu}}{\Gamma(\nu+1)} - \frac{(2t_0 - t_0)^{\nu}}{\Gamma(\nu+1)}. \tag{97}$$

Time response of model (85) is thus given by

$$x(\tau) = \frac{(2t_0)^{\nu} - t_0^{\nu}}{\Gamma(\nu+1)}H(\tau), \quad \tau \geq 0. \tag{98}$$

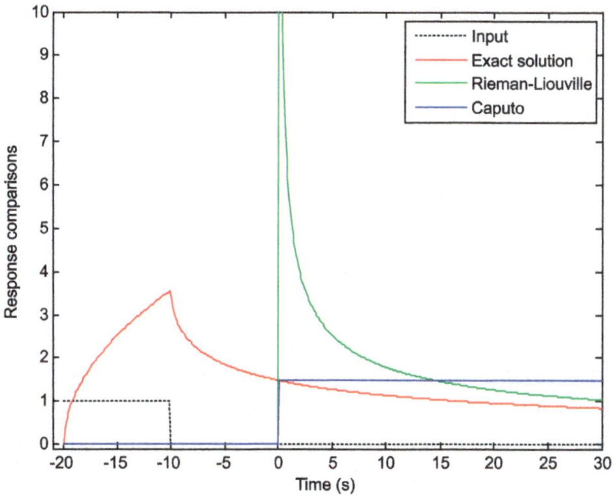

Figure 9: Comparison of the exact response of model (85) with the responses obtained with Riemann–Liouville and Caputo definitions ($t_0 = 10s$).

In Figure 9, model (85) response to the input given by relation (86) is compared to the response obtained using Riemann–Liouville's and Caputo's definitions, initial conditions being taken into account. This figure highlights that the three time responses are completely different. Using a the counterexample, it has been demonstrated in this section that neither the Riemann–Liouville's nor the Caputo's definitions are compatible with the real behavior of a fractional model. Using these definitions, model response to $u(t)$ does not match model response to nonzero initial conditions. To solve this problem, representation (44), (45), and (46) can be used though the initialization of functions $\eta(t)$ and $\Phi(\zeta, t)$, but initial conditions $\eta(0)$ and $\Phi(\zeta, 0)$ are defined on an infinite dimensional space domain.

This counterexample was extend to partial differential equations case in [41]. To avoid these initialization problems, another solution presented in [45] consists in using a description based on fractional integration as suggested in Section 2.3, but it requires the knowledge of all the model past.

5 Fractional models observability

As described in the previous section, representation (44), (45), and (46) permits to take into account initial conditions correctly in relation to the system physical behavior [46]. It can also be used to obtain results on the observability of fractional models. Observability of fractional models has been addressed in several papers [4, 17, 25, 26, 31]. However, most of the criteria presented in these papers:

- suppose that the system is represented by a pseudo-state space description as (10), and the observability of vector $x(t)$ is evaluated;

- Caputo's definition is used for fractional derivative and initial conditions are taken into account in the demonstrations.

These results are thus questionable because, according to the analysis done in the previous sections:
- representation (10) is not a state space description and vector $x(t)$ is not the system state;
- Caputo's definition does not permit to take into account initial conditions in a convenient way [46].

In representation (44), (45), and (46), $\eta(t)$ and $\Phi(\zeta,t)$ really met the definition of state variables (see Definition 1). According to Definition 1, $\eta(t_0)$ and $\Phi(\zeta,t_0)$ are the minimal amount of information required to compute the future behavior of the model. The fractional model state notion being now clarified, it is possible to discuss the state observability notion.

Response of model (45) to initial condition $\Phi(\zeta,0)$ is defined by

$$y_d(t) = C_J \int_{-\infty}^{+\infty} m(\zeta)\Phi(\zeta,t)d\zeta \tag{99}$$

with

$$\Phi(\zeta,t) = \frac{1}{2\sqrt{\pi t}} \int_{-\infty}^{+\infty} e^{-\frac{\zeta-z}{4t}}\Phi(\zeta,0)dz. \tag{100}$$

Formally, equations (99) and (100) can be written

$$y_d(t) = C_J C\Phi(\zeta,t) \tag{101}$$

and

$$\Phi(\zeta,t) = T(t,\zeta)\Phi(\zeta,0)dz \tag{102}$$

where C and $T(t,\zeta)$ are two linear operators from $\mathbb{R}^m$ to $\mathbb{R}^m$. According to [8], page 154, two observability notions relate to infinite models, namely the exact and the approximate observability. For these two notions, exact observability is ensured if the knowledge of the output on $\mathcal{L}_2[0,\tau]$ determines the state uniquely, that is, if function

$$G = C_J C T(t,\zeta)\Phi(\zeta,0) \tag{103}$$

that defines the observability map is injective. Moreover, the model is only approximately observable on $[0,\tau)$ if

$$\ker G = 0. \tag{104}$$

Remark 5. If the model is exactly observable, operator is invertible and its in-verse is continuous and depends continuously on $y_d(t)$. If the model is approximately observable, it is invertible and its inverse is not necessarily bounded. Thus, an error on the output $y(t)$ can lead to a wrong estimation of the initial state.

Theorem 10 ([43]). *Assuming that the state of the integer order submodel of* (44) *is observable, the state of fractional model* (45) *is not exactly observable but approximately observable.*

This theorem highlights that the need to know all the model past restricts the properties of a fractional models such as (10).

6 Conclusions

This chapter proposes an overview of existing fractional models for the linear time invariant case. It also proposes stability results for these models, mainly under the form of LMI conditions. Some of these LMI can be used for the design of stabilizing control laws. Important results on fractional models initialization and observability are also presented.

But through these results, this chapter is above all an opportunity to raise essential questions in the field of fractional modeling. The physical consistency of the fractional models described by equations of the form (7), (10), or (44) to (46), and mainly their state remain questionable. Representation (44) to (46) introduces the signal $\Phi(\zeta, t)$ (or equivalently $w(\chi, t)$ in (40)) that can be viewed as a possible state as it makes it possible to solve properly the model initialization. However, such an initialization involves the state $\Phi(\zeta, t)$ (or $w(\chi, t)$) which is also not related to the modeled system. Moreover, this model also exhibits infinitely small and high time constants (or infinite memory), which is not physically consistent and usually stems from making ζ (or χ) artificially infinite (thus introducing infinitely small time constants). Representation (44) to (46) and the associated signal $\Phi(\zeta, t)$ (or $w(\chi, t)$) are thus not eligible to represent the internal behavior of a modeled physical system. The signal $\Phi(\zeta, t)$ (or $w(\chi, t)$) has no physical link with the modeled system state. Fractional models described by equations of the form (7), (10), or (44) to (46) are thus adapted to study the input–output behavior of a system, but not its internal properties (initialization, internal stability, controllability, and observability among others). This comes from the fact that these representations are obtained based on an input–output modeling approach. For example, they do not result from an infinitesimal modeling approach, as can be the case for my integer models. The results obtained from the analysis of these models are thus very restrictive: infinite amount of information for model initialization [46], observability [43], controllability, flatness, etc.

The authors therefore question the usefulness of studying the internal properties of models whose states have no link with physical variables of the modeled system and do not depict its internal behavior. The authors are thus currently developing different tools in order to be able to describe the internal behavior of systems whose input–output behavior is well captured by fractional models.

Bibliography

[1] O. Bachelier, *Commande des systemes lineaires incertains: placement de poles robuste en d-stabilite*, PhD thesis, INSA, Toulouse, France, 1998. 1998ISAT0028.

[2] O. Bachelier and B. Pradin, Bounds for uncertain matrix root-clustering in a union of subregions, *International Journal of Robust and Nonlinear Control*, **9**(6) (1999), 333–359.

[3] J.-L. Battaglia, O. Cois, L. Puigsegur, and A. Oustaloup, Solving an inverse heat conduction problem using a non-integer identified model, *International Journal of Heat and Mass Transfer*, **44**(14) (2001), 2671–2680.

[4] M. Bettayeb and S. Djennoune, A note on the controllability and the observability of fractional dynamical systems, in *IFAC-FDA Workshop*, 2006.

[5] C. Bonnet and J. R. Partington, Coprime factorizations and stability of fractional differential systems, *Systems & Control Letters*, **41**(3) (2000), 167–174.

[6] M. Chilali, *Méthodes LMI Pour L'analyse Et la Synthèse Multi-critère*, PhD thesis, Paris IX—Dauphine, Paris, France, 1996.

[7] O. Cois, *Systèmes linéaires non entiers et identification par modèle non entier: application en thermique*, PhD thesis, University of Bordeaux, Bordeaux, France, 7 2002.

[8] R. F. Curtain and H. Zwart, *An Introduction to Infinite-Dimensional Linear Systems Theory*, Springer-Verlag, New York, USA, 1995.

[9] D. W. Davidson and R. H. Cole, Dielectric relaxation in glycerol, propylene glycol, and propanol, *The Journal of Chemical Physics*, **19**(12) (1951), 1484–1490.

[10] O. Enacheanu, *Modélisation fractale des réseaux électriques*, PhD thesis, Université Joseph Fourier—Grenoble I, 2008.

[11] C. Farges, M. Moze, and J. Sabatier, Pseudo-state feedback stabilization of commensurate fractional order systems, *Automatica*, **46**(10) (2010), 1730–1734.

[12] C. Farges, M. Moze, and J. Sabatier, Pseudo state feedback stabilization of commensurate fractional order systems, in *European Control Conference (ECC)*, Budapest, Hungary, Aug. 2009, pp. 3395–3400.

[13] C. Farges, J. Sabatier, and M. Moze, Stabilisation par retour de pseudo-état des systèmes non entiers commensurables, in *Conférence Internationale Francophone d'Automatique (CIFA)*, Nancy, France, Jun. 2010.

[14] C. Gruau, D. Picart, R. Belmas, E. Bouton, F. Delmaire-Size, J. Sabatier, and H. Trumel, Ignition of a confined high explosive under low velocity impact, *International Journal of Impact Engineering*, **36**(4) (2009), 537–550.

[15] T. T. Hartley and C. F. Lorenzo, Dynamics and control of initialized fractional-order systems, *Nonlinear Dynamics*, **29**(1–4) (2018), 201–233.

[16] D. Hinrichsen and A. J. Pritchard, *Mathematical Systems Theory I: Modelling, State Space Analysis, Stability and Robustness*, vol. 48, 1st edition, Springer-Verlag, 2005.

[17] R. Hotzel and M. Fliess, Gain estimation for fractional linear systems, *IFAC Proceedings Volumes*, **31**(18) (1998), 221–226.

[18] H. K. Khalil, *Nonlinear Systems*, Pearson Education, 3rd edition, Prentice Education International, New Jersey, USA, 2000.

[19] C. F. Lorenzo and T. T. Hartley, Initialization in fractional order systems, in *Proceedings of the European Control Conference*, pp. 1471–1476, 2001.

[20] C. F. Lorenzo and T. T. Hartley, Initialization of fractional differential equations: theory and application, in *Proceedings of the ASME 2007 International Design Engineering Technical Conferences*, DETC2007–34814, 2007.

[21] R. L. Magin, *Fractional Calculus in Bioengineering*, Begell House Publishers Inc., USA, 2006.

[22] F. Mainardi, *Fractional Calculus and Waves in Linear Viscoelasticity*, Imperial College Press, London, UK, 2010.

[23] D. Matignon, Stability results for fractional differential equations with applications to control processing, in *Computational Engineering in Systems Applications*, pp. 963–968, 1996.

[24] D. Matignon, Stability properties for generalized fractional differential systems, *ESAIM. Proceedings*, **5**, (1998), 145–158.

[25] D. Matignon and B. D'Andréa-Novel, Observer-based controllers for fractional differential systems, in *Conference on Decision and Control*, pp. 4967–4972, 1996.

[26] D. Matignon and B. D'Andréa-Novel, Some results on controllability and observability of finite-dimensional fractional differential systems, in *Computational Engineering in Systems Applications*, pp. 952–956, 1996.

[27] D. Matignon, B. D'Andréa-Novel, P. Depalle, and A. Oustaloup, Viscothermal losses in wind intruments: a non integer model, in *International Symposium on the Mathematical Theory of Networks and Systems (MTNS)*, Regensburg, Germany, August 1993.

[28] S. M. Momani and R. El-Khazali, Stability analysis of composite fractional systems, in *Intelligent Systems and Control*, Tampa, Florida, 2001.

[29] G. Montseny, Diffusive representation of pseudo-differential time-operators, *ESAIM. Proceedings*, **5** (1998), 159–175.

[30] M. Moze and J. Sabatier, LMI tools for stability analysis of fractional order systems, in *ASME International Design Engineering Technical Conferences & Computers and Information in Engineering Conference*, Long Beach, CA, USA, Sep. 2005, pp. 1–9.

[31] F. Nanot, *Derivateur generalise et representation generalisees des systemes lineaires*, PhD thesis, Bordeaux 1 University, 3 1996.

[32] B. Orsoni, *Dérivée généralisée en planification de trajectoire et génération de mouvement*, PhD thesis, LAPS, Université de Bordeaux, 2002.

[33] A. Oustaloup, *Systemes asservis lineaires d ordre fractionnaire: theorie et pratique*, Masson, Paris, France, 1983.

[34] A. Oustaloup, *La dérivation non entière: théorie, synthèse et applications*, Traité des nouvelles technologies. Série Automatique, Hermès, France, Paris, 1995.

[35] A. Oustaloup, O. Cois, P. Lanusse, P. Melchior, X. Moreau, and J. Sabatier, The crone approach: theoretical developments and major applications, in *Proceedings of the 2nd IFAC Workshop on Fractional Differentiation and Its Applications*, pp. 324–354, 2006.

[36] D. Peaucelle, *Formulation générique de problèmes en analyse et commande robuste par fonctions de lyapunov dépendant des paramètres*, PhD thesis, Université Toulouse III, Paul Sabatier, Toulouse, France, Jul. 2000.

[37] S. Pritesh and A. Sudhir, Review of fractional pid controller, *Mechatronics*, **38**(1) (2016), 29–41.

[38] S. Rodrigues, N. Munichandraiah, and A. K. Shukla, A review of state-of-charge indication of batteries by means of a.c. impedance measurements, *Journal of Power Sources*, **87**(1–2) (2000), 12–20.

[39] J. Sabatier, M. Aoun, A. Oustaloup, G. Grégoire, F. Ragot, and P. Roy, Fractional system identification for lead acid battery sate charge estimation, *Signal Processing*, **86**(10) (2006), 2645–2657.

[40] J. Sabatier and C. Farges, On stability of commensurate fractional order systems, *International Journal of Bifurcation and Chaos*, **22**(4) (2012), 1–8.

[41] J. Sabatier and C. Farges, Comments on the description and initialization of fractional partial differential equations using Riemann–Liouville's and Caputo's definitions, *Journal of Computational and Applied Mathematics*, **339** (2018), 30–39.

[42] J. Sabatier and C. Farges, On stability of commensurate fractional systems, in *IFAC Workshop on Fractional Differentiation and Its Applications (FDA)*, Badajoz, Spain, Oct. 2010.

[43] J. Sabatier, C. Farges, M. Merveillaut, and L. Feneteau, On observability and pseudo state estimation of fractional order systems, *European Journal of Control*, **18**(3) (2012), 260–271.

[44] J. Sabatier, C. Farges, and J.-C. Trigeassou, A stability test for non-commensurate fractional order systems, *Systems & Control Letters*, **62** (2013), 739–746.

[45] J. Sabatier, C. Farges, and J.-C. Trigeassou, Fractional systems state space description: some wrong ideas and proposed solutions, *Journal of Vibration and Control*, **20**(7) (2014), 1076–1084.

[46] J. Sabatier, M. Merveillaut, R. Malti, and A. Oustaloup, How to impose physically coherent initial conditions to a fractional system?, *Communications in Nonlinear Science and Numerical Simulation*, **15**(5) (2010), 1318–1326.

[47] J. Sabatier, M. Merveillaut, R. Malti, and A. Oustaloup, On a representation of fractional order systems: interests for the initial condition problem, in *3rd IFAC Workshop on Fractional Differentiation and Its Applications (FDA)*, Ankara, Turkey, Nov. 2008.

[48] J. Sabatier, M. Moze, and C. Farges, Lmi stability conditions for fractional order systems, *Computers & Mathematics with Applications*, **59**(5) (2010), 1594–1609.

[49] J. Sabatier, M. Moze, and C. Farges, On stability of fractional order systems, in *3rd IFAC Workshop on Fractional Differentiation and Its Applications (FDA)*, Ankara, Turkey, Nov. 2008.

Ivo Petráš

Modified versions of the fractional-order PID controller

Abstract: In this chapter, a brief survey of the modified versions of the fractional-constant-order PID controller is presented. Both linear and nonlinear versions of such controllers are presented together with illustrative examples and some notes for their practical implementation in control systems. For simulation purposes, the Matlab functions are mentioned as well as the links for their free download.

Keywords: Fractional control, fractional-order PID controller, implementation

MSC 2010: 26A33, 34A08, 93A30, 93C95, 93D15

1 Introduction

There are many compensators and controllers in practice. However, PID controller is the most commonly used series controller. In many process control literatures, the PID controller is often described by different equations. Such modifications have many reasons—as for example, to avoid pure derivative action, remove steady-state error, reduce settling time, and so on.

Moreover, one of the possible modifications of the classical PID controller is to use a fractional calculus technique in integral and derivative actions, respectively. In this manner, we obtain a fractional-order PID controller. There linear and nonlinear fractional-order PID controllers and their various modifications are presented.

It is necessary to mention, that except fractional-order controllers in PID structure, there exist many other fractional-order controllers/compensators as, for example, [4, 9, 13, 29]: CRONE controller, TID compensator, lead-lag compensator, etc.

In this chapter, the standard notation for denoting the left-sided fractional-order integrodifferential operator of a function $f(t)$ defined in the interval $[a, t]$ (bounds of the operation) as ${}_aD_t^{\alpha}f(t)$ and for $a = 0$ as $D_t^{\alpha}f(t)$ with $\alpha \in \mathbb{R}$ is used.

There exist three main definitions of the fractional-order operator: Riemann–Liouville (RL), Caputo (C), and Grünwald–Letnikov definition (GL). For zero initial conditions and lower terminal $a = 0$, the Laplace transform of fractional-order (deriva-

Acknowledgement: The results described in this chapter were presented by the author at the following conferences: IEEE International Joint Conference on Neural Networks (2016), and IEEE International Carpathian Control Conference (2014, 2015, 2016). Thus an acknowledgement is given to the Institute of Electrical and Electronics Engineers (IEEE).

Ivo Petráš, Technical University of Košice, BERG Faculty, B. Němcovej 3, 04200 Košice, Slovakia, e-mail: ivo.petras@tuke.sk

https://doi.org/10.1515/9783110571745-003

tive/integral) operator D_t^α defined by RL, C, or GL definition, reduces to [27]:

$$\mathscr{L}\{D_t^\alpha f(t)\} = s^\alpha F(s). \tag{1}$$

2 Linear fractional-order controllers

The fractional-order $PI^\lambda D^\delta$ (a.k.a. $PI^\lambda D^\mu$) controller (FOC) was proposed in [28] as a generalization of the PID controller with integrator of real order λ and differentiator of real order δ. The transfer function of such parallel controller in the Laplace domain has the form:

$$C(s) = \frac{U(s)}{E(s)} = K_p + T_i s^{-\lambda} + T_d s^\delta, \quad (\lambda, \delta > 0), \tag{2}$$

where K_p is the proportional constant, T_i is the integration constant and T_d is the differentiation constant.

The internal structure (see Figure 1) of the fractional-order controller consists of the parallel connection of the proportional, integration, and derivative part. The transfer function (2) corresponds in time domain to the fractional differential equation of the form:

$$u(t) = K_p e(t) + T_i D_t^{-\lambda} e(t) + T_d D_t^\delta e(t). \tag{3}$$

Taking $\lambda = 1$ and $\delta = 1$, we obtain a classical PID controller. If $\delta = 0$ and $T_d = 0$, we obtain a PI^λ controller, etc. All of these types of controllers are particular cases of the fractional-order controller, which is more flexible and gives an opportunity to better adjust the dynamical properties of the fractional-order control system.

It can also be mentioned that there are many other modifications of the fractional $PI^\lambda D^\delta$ controller [8–10, 17, 22, 32].

For Matlab implementation of the FOC in discrete form, the function `DFOC()` can be used [15].

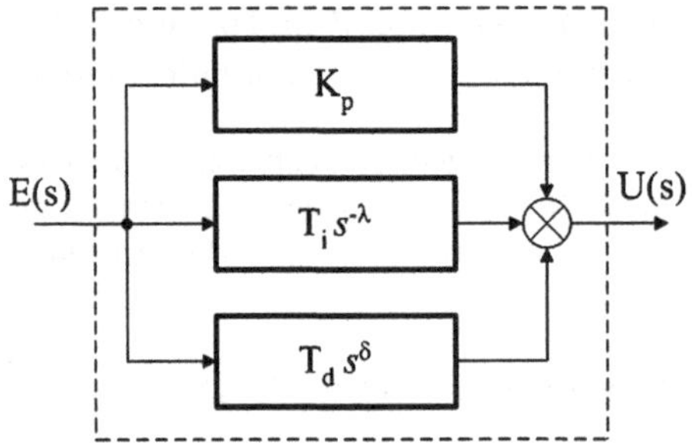

Figure 1: General structure of a fractional-order $PI^\lambda D^\delta$ controller.

3 Nonlinear fractional-order controllers

In this section, a nonlinear fractional-order controller is defined. Following the traditional structure of the nonlinear PID controller, which is well known in literature, we can write a new formula for nonlinear fractional-order $PI^{\lambda}D^{\delta}$ controller (NFOC) of the form [20, 21, 24]:

$$u(t) = f(e)[K_p\,e(t) + T_i\,D_t^{-\lambda}e(t) + T_d\,D_t^{\delta}e(t)], \tag{4}$$

where $f(e)$ is nonlinear function of variable e.

Various definitions of the non-linear function $f(e)$ can be used, for instance:

1. A widely used nonlinear function can have the form [2]:

$$f(e) = K_0 + (1 - K_0)|e(t)|, \quad K_0 \in \langle 0, 1\rangle. \tag{5}$$

   When $K_0 = 1$ in (5), we obtain a classical form of the linear fractional-order controller (3). For $K_0 \neq 1$, we have a 6 degrees of freedom (6DOF) controller.
   In Figure 2, defined by (5), function $f(e)$ for various values of parameter K_0 depended on variable e is depicted.
2. For the aforementioned nonlinearity $f(e)$, the other various piece-wise linear functions of nonlinear gain can be also used, as it is depicted in Figure 3.

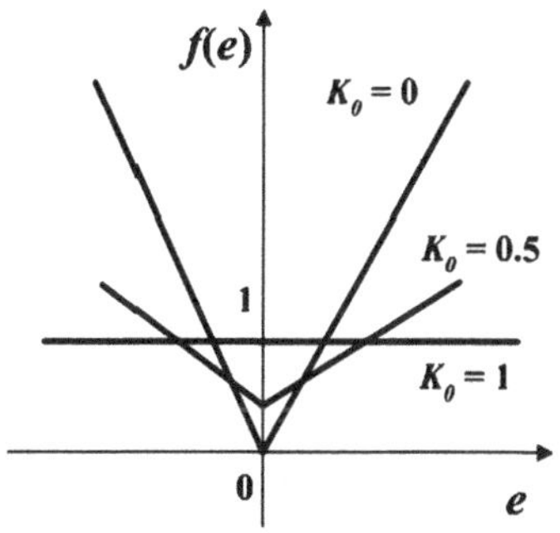

Figure 2: Behavior of the function (5) for various values of K_0.

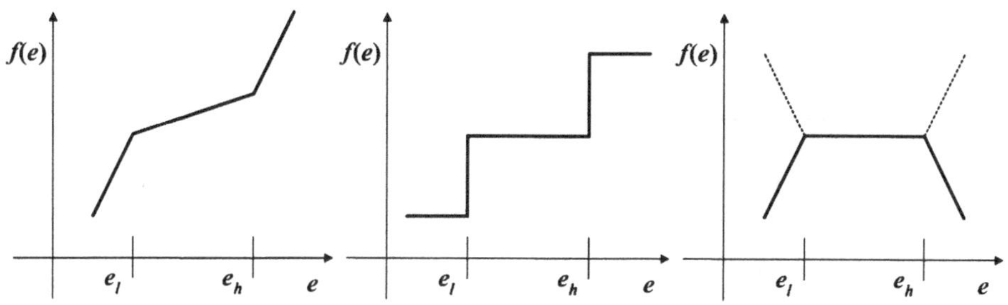

Figure 3: Various nonlinear gains $f(e)$.

3. For desired low e_l and high e_h control error bounds, we obtain a controller with variable gain and function $f(e)$ is defined as [2]:

$$f(e) = 1 \quad \text{for } e > e_h,\\ f(e) = 1 \quad \text{for } e < e_l,\\ f(e) = K_0 \quad \text{for } e \in \langle e_l, e_h \rangle,\ K_0 \geq 0. \tag{6}$$

When K_0 is 0 within interval $\langle e_l, e_h \rangle$, the output of the controller does not change and, therefore, the actuator behavior is much smoother.
4. When we consider a scaled error function of form $f(e) = k(e).e(t)$, where the nonlinear gain $k(e)$ represents any general nonlinear function of the error $e(t)$, which is bounded in a sector, then we can write a general form of the nonlinear fractional-order $PI^\lambda D^\delta$ controller:

$$\begin{aligned} u(t) &= [K_p + T_i\, D_t^{-\lambda} + T_d\, D_t^{\delta}] f(e) \\ &= K_p[k(e).e(t)] + T_i\, D_t^{-\lambda}[k(e).e(t)] + T_d\, D_t^{\delta}[k(e).e(t)] \\ &= K_p(\cdot)\, e(t) + T_i(\cdot)\, D_t^{-\lambda} e(t) + T_d(\cdot)\, D_t^{\delta} e(t), \end{aligned} \tag{7}$$

where $K_p(\cdot)$, $T_i(\cdot)$, and $T_d(\cdot)$ are time-varying controller parameters, which may depend on system state, input, control error, or other variables [30].
In Figure 4, the internal structure of a general nonlinear fractional-order $PI^\lambda D^\delta$ controller (7) is depicted. We should take into account that by using the nonlinear controller parameters as in equation (7), we obtain a controller with variable structure. Controllers of such kind are often called super-twisting controllers [26].
A nonlinear fractional-order $PI^\lambda D^\delta$ controller can be also tuned as combination of the linear and nonlinear parameters—for instance, pure gain for proportional part and nonlinear time constant for integration and derivation constants, and so on.
5. Other general modifications of the nonlinear fractional-order PID controller (4) is a controller with piece-wise linear gain which is depending on signal $v(t)$ and then the proportional controller gain is $K_p = f(v)$, where piece-wise linear function $f(v)$ can be defined, for instance, as namely in Figure 3. Similarly, we could

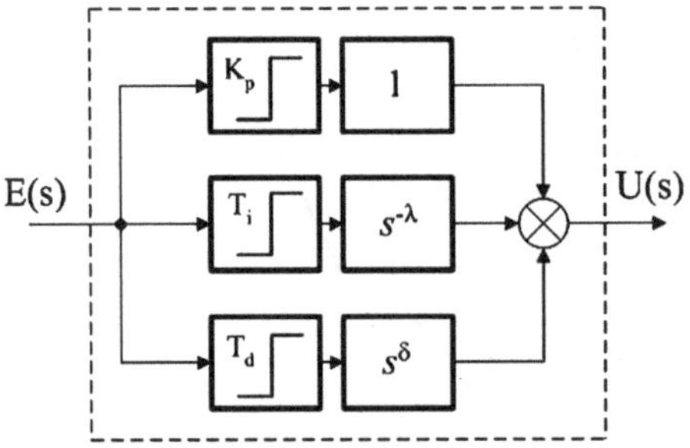

Figure 4: General structure of a nonlinear fractional-order $PI^\lambda D^\delta$ controller.

modify other controller constants, T_i and T_d. As for the external signal $v(t)$, we can consider: control value $u(t)$, controlled value $y(t)$, control error $e(t)$, or other additional external value. Such modification is a particular case of controller (7).

When we take into account all possible nonlinear functions and all possible particular cases of the fractional-order controllers, we deal with a new wide class of controllers. This opens a new area of research and many questions are still open. For instance, parameters design and tuning of nonlinear fractional-order controllers, selection of proper nonlinear functions, and so on. On the other hand, it is possible to turn to the experimental approach, employ well-known tuning methods proposed for linear fractional-order controllers described in already published works [3, 8–10, 16, 32], and then experimentally select nonlinearity and find its necessary parameters. However, we still have to check the performance of a control loop via simulation before applying the proposed controller to real object in order to verify and validate requirements on the control system.

For Matlab implementation of the NFOC with nonlinearity (5) in discrete form, the function `NFOC()` can be used [19].

4 Modification of the control actions in the fractional-order PID controllers

Several modifications of the fractional-order control can be used in order to avoid such problems as wind-up effect, actuator saturation, and derivative action limitation. Among most used ones belong to the following [14, 16]:

- *Filtering the desired value* $r(t)$: filtering the desired value $r(t)$, known as setpoint tracking, by the first- or second-order filter is a very frequently used trick to avoid problems with derivative action. Instead of a step change of the desired value, which could be a problem especially for the derivative part of the controller, the control algorithm executes slow change of the desired value, and thus changes of the control signal are not that extreme. For most applications, the first-order filter is satisfactory. We recommend the first-order prefilter in the discrete form:

  $$H_p(z) = \frac{k_f}{1 - k_f z^{-1}},$$

  where k_f is the prefilter constant.
- *Using a controlled value in proportional and derivative parts of controller*: the above mentioned problem related to step changes of the control signal due to step changes of the desired value $r(t)$ can also be solved via replacing the control error $e(t) = r(t) - y(t)$ by controlled value $y(t)$. This modification can help a lot,

especially when desired value has changed rapidly and, therefore, the actuator becomes saturated.

- *Filtering the derivative action*: due to noisy signal on measured controlled value, the differentiation of noise can involve inappropriate changes of control signal. Derivative action is more sensitive to higher-frequency terms in the inputs. Because of this, the derivative part of the controller can be filtered by the first- and second-order high frequency filter. For the first-order filter in the derivative part and with a genuine integral action, we can write the transfer function of the fractional-order controller in the form:

$$C(s) = \frac{U(s)}{E(s)} = K_p + T_i \frac{s^{1-\lambda}}{s} + \frac{T_d s^{\delta}}{T_f s + 1}, \quad (\lambda, \delta > 0), \tag{8}$$

where $T_f = N/T_d$ is the filter constant. For $N = 0$, we obtain the usual FOC described by relation (2).

- *Limitation of integral action*: this limitation is also known as wind-up of the controller. It is due to the fact that actuator has also limitations and, for instance, if the actuator is at the end position and the control error is not zero, the integral part of the controller rapidly grows, the controller calculates unreal value of the control signal and, therefore, the actuator stays at the end position until the sign of control error is changed. This problem is known as wind-up or integral saturation, and it can be solved via limitation of integral part in the controller. Another possibility for avoiding wind-up is to introduce limiters of the desired values, so that the controller output will never reach the actuator bounds. There are also some other standard anti-wind-up schemes, which can be also used for the fractional-order control, as for instance [12]:
  - conditional integration,
  - automatic reset configuration,
  - back-calculation.

  Filtering can be also obtained automatically if the derivative is implemented by taking difference between the reference signal and a its filtered version as it was done in [14]. Instead of filtering just the derivative, it is also possible to use a usual controller and *filter the measured signal*. The transfer function of the fractional-order controller with the filter is then

$$C(s) = \frac{U(s)}{E(s)} = (K_p + T_i s^{-\lambda} + T_d s^{\delta}) \frac{1}{(T_f s + 1)^{\sigma}}, \tag{9}$$

  where σth order filter is used.

Obviously, there are many other modifications of the control algorithms, which help us implement fractional-order controller in practice. For instance, we can mention initial conditions for a nonimpact controller connection to control loop, analog and digital filtering of measured values, and so on.

5 Some other possible modifications of the fractional-order PID controllers

Besides the linear and nonlinear fractional-order PID controllers presented in previous sections, there are many additional modifications. Here are listed for illustration just a few of them:

- Fractional-order PID controllers presented in [31]:

$$C(s) = K_p\left(1 + \frac{1}{T_i s}\right)^{\lambda}(1 + T_d s)^{\delta}$$

or

$$C(s) = K_p\left(1 + \frac{1}{T_i s^{\lambda}}\right)(1 + T_d s^{\delta}).$$

- Fractional-order [PI] and [PD] controllers, respectively, introduced in [8]:

$$C(s) = \left(K_p + \frac{T_i}{s}\right)^{\lambda} \quad \text{and} \quad C(s) = (K_p + T_d s)^{\delta}.$$

- Fractional-order PID controllers suggested in [11]:

$$C(s) = K_p \frac{T_i s^{\lambda} + 1}{T_i s^{\lambda}}(T_d s^{\delta} + 1)$$

and

$$C(s) = K_p\left(1 + \frac{1}{T_i s^{\lambda}} + T_d s^{\delta}\right)\frac{1}{T_f s + 1}$$

or

$$C(s) = K_p \frac{T_i s^{\lambda} + 1}{T_i s^{\lambda}} \frac{T_d s^{\delta} + 1}{(T_d/N)s + 1}.$$

6 Implementation technique

Implementation techniques for the FOC have been described in several works. An analogue implementation was proposed in works [5, 25], and a digital implementation was suggested in works [3, 32]. In this chapter, we will focus only on the digital implementation technique which we adopt also for the slightly modified NFOC.

However, the pure derivative action cannot be used in the controller in practice. For this reason, a first-order lag filter can be used to approximate the derivative action. Taking this into account, we can write a new formula for the modified fractional-order nonlinear controller with the first-order filter on derivative action (see (8)), in the following form [24]:

$$T_f \frac{du(t)}{dt} + u(t) = f(e)\left[K_p\left[T_f \frac{de(t)}{dt} + e(t)\right]\right.$$
$$\left. + T_i[T_f D_t^{1-\lambda} e(t) + D_t^{-\lambda} e(t)] + T_d D_t^{\delta} e(t)\right], \tag{10}$$

where $f(e)$ is a nonlinear function of variable e. Various definitions of the non-linear function $f(e)$ can be used. When $K_0 = 1$ in (5), we obtain the form of a linear fractional-order controller. For $K_0 \neq 1$, we get a 7 degrees of freedom (7DOF) controller with the following set of the parameters: K_p, T_i, λ, T_d, δ, K_0, N. Obviously, for $N = 0$, that is, $T_f = 0$ in (10), we obtain the usual NFOC described by relation (4).

Having tuned the controllers, to implement them we have to take into account other considerations, such as memory size and computational cost required by the algorithm.

There are several possible approximation techniques for fractional-order differentiators/integrators. In general, we may use a direct and indirect implementation techniques for numerical approximation of the fractional operator arising from various definitions of the fractional-order derivatives or integrals. This approach is based on the fact that for a wide class of functions, three definitions—RL, and C, and GL definitions—are equivalent.

The relation for the explicit numerical approximation of αth derivative or integral (direct form) in discrete time domain at the points kh, ($k = 1, 2, \ldots$), is based on the GL definition and has the following form [6]:

$${}_{(k-L_m/h)}D_{t_k}^{\alpha} f(t) \approx h^{-\alpha} \sum_{j=v}^{k} (-1)^j \binom{\alpha}{j} f(t_{k-j}) = h^{-\alpha} \sum_{j=v}^{k} c_j^{(\alpha)} f(t_{k-j}), \tag{11}$$

where L_m is the "memory length," $t_k = kh$, h is the time step of calculation and $c_j^{(\alpha)}$ are binomial coefficients. For their calculation, we can use the following expression [6]:

$$c_0^{(\alpha)} = 1, \quad c_j^{(\alpha)} = \left(1 - \frac{1+\alpha}{j}\right) c_{j-1}^{(\alpha)}. \tag{12}$$

When we consider the time step of calculation h in relation (11) as a sampling period T_s and the approximation of continuous time t as a set of points with the sampling interval T_s, where $t \approx kT_s$ and k is a position of point in discrete time, we can write the discrete version of the fractional-order nonlinear controller.

Finally, the discrete version of the nonlinear fractional-order controller (10) has the following form [24]:

$$\begin{aligned} u(k) = f(e)\Bigg[& K_p\{T_f T_s^{-1}[e(k) - e(k-1)] + e(k)\} \\ & + T_i\left\{ T_f T_s^{\lambda-1} \sum_{j=v}^{k} c_j^{(1-\lambda)} e(k-j) + T_s^{\lambda} \sum_{j=v}^{k} c_j^{(-\lambda)} e(k-j) \right\} \\ & + T_d T_s^{-\delta} \sum_{j=v}^{k} c_j^{(\delta)} e(k-j) \Bigg] \frac{T_s}{T_s + T_f} + T_f T_s^{-1} u(k-1), \end{aligned} \tag{13}$$

where $c_j^{(1-\lambda)}$, $c_j^{(-\lambda)}$ and $c_j^{(\delta)}$ are binomial coefficients calculated according to relation (12) and where

$$\begin{aligned} v &= 0 \quad \text{for } k < (L_m/T_s), \\ v &= k - (L_m/T_s) \quad \text{for } k > (L_m/T_s). \end{aligned} \tag{14}$$

For calculating the control error e, we can use a usual formula $e = r - y$, where r is the desired (reference) value and y is the controlled (measured) value.

Using the discrete version of the fractional-order controller in the form (13)–FIR filter, we can apply the short memory principle and use just L_m/T_s number of the previous values of control error to calculate a new control output. It is very useful and effective for capacity of the memory and load of the processor on microprocessor devices as, for instance, PLC [7].

When we use the indirect implementation technique in the form of approximation as rational function (IIR filter), the number of coefficients of the controller is less then in the case of direct approximation (FIR filter), but the accuracy of approximation is lower. On the other hand, the implementation on microprocessor devices requires less memory capacity and less processor time. The choice of fractional-order operator approximation is a very important task and it should be considered very carefully.

We have to take into account that a real controller should have the limitations for control actions due to elimination of wind-up effect and actuator saturation. Then the output value from the controller (control value—$u(k)$) is limited in range $(u_{\min}; u_{\max})$ or $\pm u_{\lim}$, where we define d as $d = |u_{\lim}|$. In fact, using this limitation, which is standard in practice, we obtain in general a nonlinear controller with the piece-wise nonlinearity defined as: $a(k) = 0.5 \times (|u(k) + d| - |u(k) - d|)$, where $a(k)$ is the final control action to the actuator (see also Figure 8 in Section 7.3).

7 Illustrative examples

7.1 Effect of filtering a derivative action

In this section, an effect of a filtered derivative action is demonstrated [18].

Let us consider filtered derivative action in the FOC (2) as it was demonstrated in equation (8). Thus we obtain system:

$$D(s) = \frac{T_d s^\delta}{T_f s + 1}, \quad (\delta > 0), \tag{15}$$

where $T_f = N/T_d$ is the filter constant and $N \geq 0$. Following the well-known technique of the order compensation in the classical PID controller, which is

$$T_n \frac{d^n u(t)}{dt^n} + \cdots + T_2 \frac{d^2 u(t)}{dt^2} + T_1 \frac{du(t)}{dt} + u(t) = K_p\, e(t) + T_i \int_0^t e(\tau) d\tau + T_d \frac{de(t)}{dt},$$

we can rewrite equation (15) to the following general form:

$$D(s) = \frac{T_d s^\delta}{T_f s^\mu + 1}, \quad (\delta, \mu > 0). \tag{16}$$

In Figure 5 and Figure 6, the step responses of the system (16) for various values of the parameters μ and N and fixed parameters $T_d = 3$ and $\delta = 0.5$ are shown.

For approximating fractional operator s^r, $r \in \mathbb{R}$, the CFE method applied on the Tustin rule for sampling period $T_s = 0.1$ sec and $n = 5$ was used.

This illustrative example demonstrated the effect of derivative action filtering in order to avoid the problem of actuator saturation and oscillations that often occur in applications. Such problematic behavior may destroy the actuator.

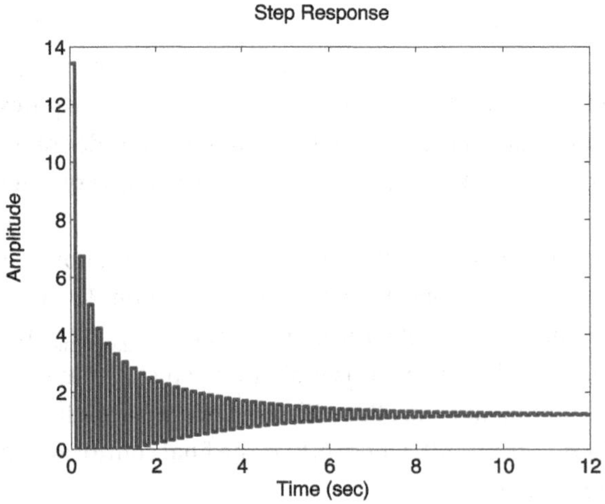

Figure 5: Step response of system (16) for the parameters $T_d = 3$, $\delta = 0.5$, $\mu = 0$, and $N = 0$ (without filtering derivative action).

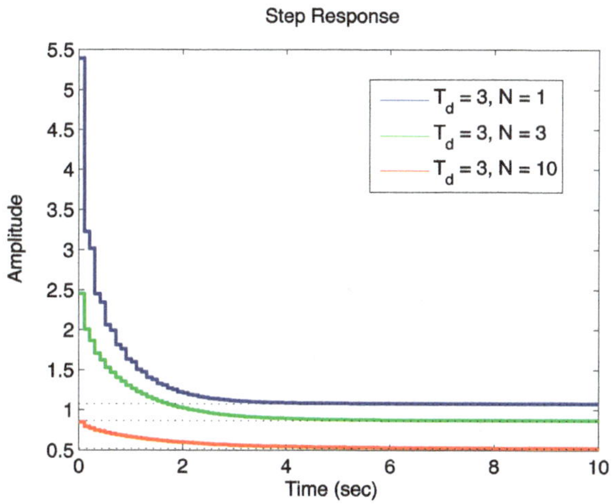

Figure 6: Step response of system (16) for the parameters $T_d = 3$, $\delta = \mu = 0.5$, and various values of the filter parameter N.

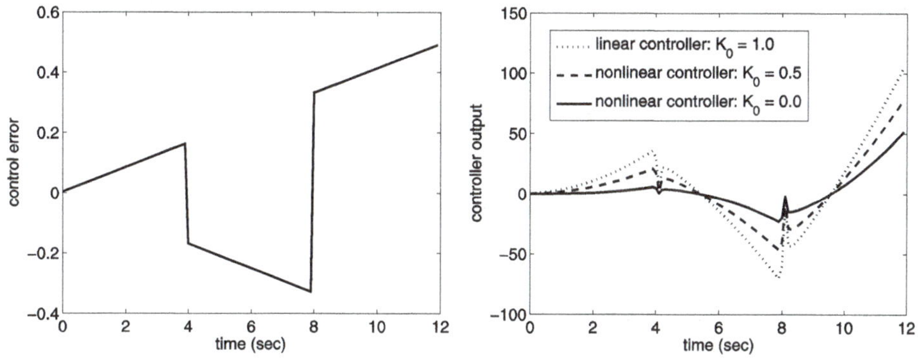

Figure 7: Controller response to control error for the various values of parameter K_0 in controller (4) with nonlinearity (5), where controller parameters are: $K_p = 20.5$, $T_i = 100$, $T_d = 11.31$, $\lambda = 1$, $\delta = 0.82$, sampling period $T_s = 0.01$ sec.

7.2 Effect of nonlinear part in controller

In this section, the effect of the nonlinear part in the nonlinear fractional-order controller represented by equation (17) is presented.

Let us consider that controller has the form [23]:

$$u(t) = f(e)\left[20.5\,e(t) + 100\int_0^t e(\tau)d\tau + 11.31\,D_t^{0.82}e(t)\right], \tag{17}$$

where for the nonlinear part $f(e)$ the function (5) is used for various values of the parameter K_0.

In Figure 7, the control error in the form of the ramp function and the controller (17) response to the control error for the sampling period $T_s = 0.01$ sec are depicted. As we can observe, very small changes of control error can produce a large change of controller output in the case of linear fractional-order controller. In the case of the nonlinear fractional-order controller, we see a smooth response of the controller which is more appropriate for the actuator (especially mechanical). This shows the main advantages of the proposed non-linear fractional-order controller.

7.3 Effect of limitation for actuator saturation

For verification and validation of the proposed control algorithm, an electrical heater as a controlled object was chosen. The mathematical model of this laboratory object (controlled plant) has the following form [3]:

$$P(s) = \frac{0.57}{75s + 1}, \tag{18}$$

which is a simple first-order system. The maximal temperature of the blowing hot air is 55°C. In order to compare the results, the linear integer-order and fractional-order controllers parameters were designed by using a simple parameters tuner `fpidtune()` in Matlab, which has been proposed in the book [33].

For the controlled system parameters (18) and the criterion based on minimization of the integral weighted time square error (ITSE), the following controller parameters for the classical linear PID controller were tuned:

$$C_{\mathrm{PID}}(s) = 20 + \frac{0.37}{s} + 0.1s \tag{19}$$

and for the fractional-order linear $\mathrm{PI}^{\lambda}\mathrm{D}^{\delta}$ controller:

$$C_{\mathrm{FOC}}(s) = 17.90 + \frac{18.35}{s^{0.37}} + 9.24s^{0.10}. \tag{20}$$

There is a lack of methods for the design of nonlinear fractional-order controllers and, therefore, for the demonstration purposes linear fractional-order controllers are used.

Generally, in practice, the output of the controller is followed by the nonlinearity of actuator saturation. Figure 8 shows the control loop structure with feedback connection of a nonlinear fractional-order controller and a real plant (electrical heater).

In Figure 9, the simulation results for the plant (18) and controllers (19) and (20) for the simulation time 60 sec without limitation for controller output (actuator input) for zero initial conditions are depicted.

In Figure 10, the simulation results for the plant (18) and controllers (19) and (20) for the simulation time 300 sec with limitation for controller output (actuator input) within the interval (0–5 V) for zero initial conditions are depicted.

We observe that limitation of the controller output influences the control performance. In this case, the settling time and the overshoot are much larger then in the case without limitation. Such observations are very typical in practical implementations.

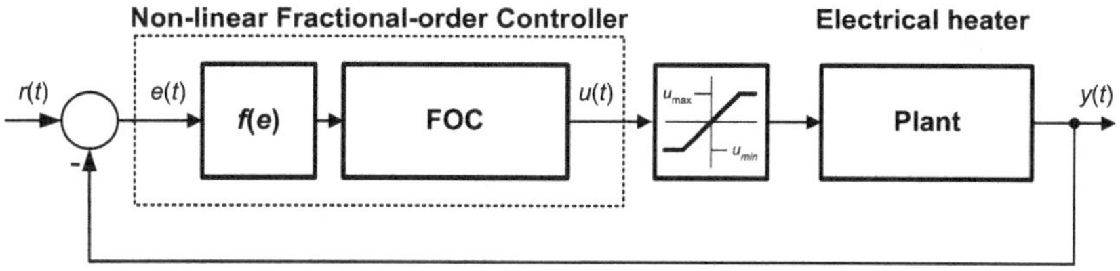

Figure 8: Structure of the control loop.

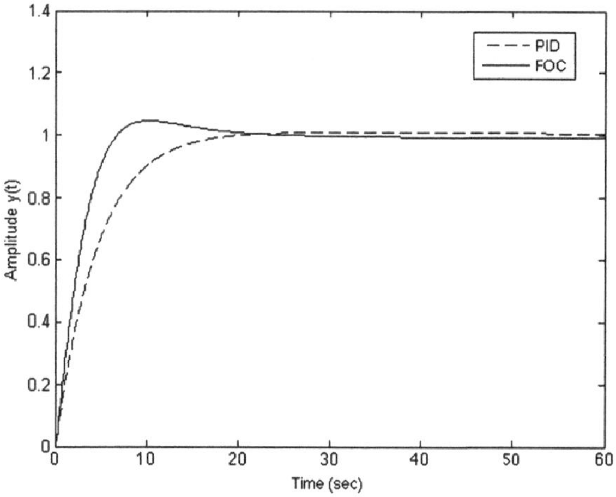

Figure 9: Simulation results for the plant (18) and controllers (19) and (20) without limitation for controller output.

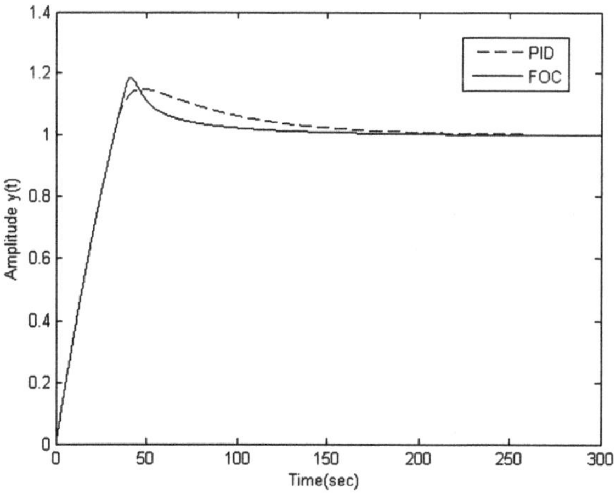

Figure 10: Simulation results for the plant (18) and the controllers (19) and (20) with limitation for the controller output.

8 Conclusions

In this chapter, various modifications of fractional-order PID controllers are presented. The nonlinear fractional-order controller (NFOC) with filtered derivative action is presented. Since linear and nonlinear PID controllers have long history and tradition (more than 95 % of the control loops are of PI/PID type [1]), the NFOC opens a new area of research. Classical nonlinear PID controllers have two basic classes of the applications: (1) nonlinear systems where a nonlinear PID controller is used to accommodate the nonlinearity in order to achieve consistent response across a range of conditions; and (2) linear systems, where a nonlinear PID controller is used in order to achieve performance not achievable by linear compensation. The main reason why the nonlinear controller performs better behavior is that it provides higher gain when error is small and lower gain when error is large. Due to this reason, we focused on the nonlinear fractional-order PID controller.

When a nonlinear fractional-order controller is used, it is not necessary to use the nonlinear control techniques, such as for instance reset control or sliding mode control, where a chattering problem could be occurred. It is important to note that any standard design techniques known from the literature can be used for the fractional-order controller parameters design [3, 9, 13]. For the design of the controller's nonlinear part, the recommendations described in this chapter can be used. For practical implementation of the NFOC both methods, direct and indirect approximation techniques, can be used. These methods are applicable in industry and were already described in many works; see, for example, [3, 4, 8, 10, 16, 32].

Appendix. A list of Matlab functions

For simulation purposes, the Matlab functions have been created. A linear fractional-order PID controller has been implemented as function `DFOC()` and a nonlinear fractional-order PID controller has been implemented as function `NFOC()`. They are published at the Matlab Central File Exchange, MathWorks, Inc., as freely downloadable functions. Here is the description of both functions, respectively:

- *Linear Fractional-Order PID Controller* (provides a discrete transfer function of the linear fractional-order PID controller for given parameters) [15]:

  ```
  function [controller] = DFOC(K, Ti, Td, m, d, Ts, n, method)
  ```

  where `K`, `Ti`, `Td`, `m`, `d` are controller parameters, `Ts` is the sampling period, `n` is the order of approximation and `method` is the used approximation method:
  - CFE of Euler rule: "CFE_Euler"
  - CFE of Tustin rule: "CFE_Tustin"

   - PSE of Euler rule: "PSE_Euler"
   - PSE of Tustin rule: "PSE_Tustin"

  For the above approximation techniques the additional Matlab functions have been created:
   - Digital Fractional Order Differentiator/integrator—IIR type www.mathworks.com/matlabcentral/fileexchange/3672
   - Digital Fractional Order Differentiator/integrator—FIR type www.mathworks.com/matlabcentral/fileexchange/3673
   - Digital Fractional-Order Differentiator and Integrator—new IIR type www.mathworks.com/matlabcentral/fileexchange/31358
- *Nonlinear Fractional-Order PID Controller* (provides a model of the nonlinear fractional-order PID controller for given parameters) [19]:

```
function uc = NFOC(e, K0, Ts, Kp, Ti, Td, lambda, delta, k)
```

  where `K0,Kp,Ti,Td,lambda,delta` are the controller parameters, `Ts` is the sampling period, `e` is the vector of control errors. Parameter `K0` is for the non-linear function `f(e(t)) = K0 + (1 - K0) * |e(t)|` and `k` is an actual position in the time sequence. The output parameter `uc` is the control action (single value) applicable for the actuator.

Bibliography

[1] K. J. Astrom and T. Hagglund, *Advanced PID Control*, ISA, USA, 2006.

[2] V. Bobal, J. Bohm, R. Prokop, and J. Fessl, *Practical Aspects of Self-Tuning Controllers: Algorithms and Implementation*, VUT, Brno, 1999 (in Czech).

[3] R. Caponetto, G. Dongola, L. Fortuna, and I. Petráš, *Fractional Order Systems: Modeling and Control Applications*, World Scientific, Singapore, 2010.

[4] Y. Q. Chen, I. Petráš, and D. Xue, Fractional order control—a tutorial, in *Proc. of the American Control Conference*, St. Louis, MO, USA, June 10–12, 2009.

[5] I. Dimeas, I. Petráš, and C. Psychalinos, New analog implementation technique for fractional-order controller: a DC motor control, *AEÜ. International Journal of Electronics and Communications*, **78** (2017), 192–200.

[6] Ľ. Dorčák, Numerical models for the simulation of the fractional-order control systems, in *UEF-04-94, The Academy of Sciences, Institute of Experimental Physics*, Košice, Slovakia, 1994.

[7] M. Köver-Dorčo and I. Petráš, Creation of fractional-order PID controller function block in automation studio environment, in *Proc. of the IEEE ICCC 2015*, Szilvásvárad, Hungary, May 27–30, 2015, pp. 251–254.

[8] Y. Luo and Y. Q. Chen, *Fractional Order Motion Control*, John Wiley & Sons, UK, 2013.

[9] C. A. Monje, Y. Q. Chen, B. M. Vinagre, D. Xue, and V. Feliu, *Fractional Order Systems and Control*, Springer, London, 2010.

[10] C. A. Monje, B. M. Vinagre, V. Feliu, and Y. Q. Chen, Tuning and auto-tuning of fractional order controllers for industry applications, *Control Engineering Practice*, **16** (2008), 798–812.

[11] F. Padula and A. Visioli, *Advances in Robust Fractional Control*, Springer, Berlin, 2015.

[12] F. Padula, A. Visioli, and M. Pagnoni, On the anti-windup schemes for fractional-order PID controllers, in *Proc. of the IEEE ETFA2012*, Krakow, Poland, 17–21 September, 2012, pp. 1–4.
[13] I. Petráš, *Fractional-Order Nonlinear Systems*, Springer, Berlin, 2011.
[14] I. Petráš, Practical aspects of tuning and implementation of fractional-order controllers, in *Proc. of the ASME/IEEE DETC2011*, Washington, USA, August 28–31, 2011, pp. 1–10.
[15] I. Petráš, Discrete fractional-order PID controller, Matlab Central File Exchange, MathWorks, Inc., 2011, url: http://www.mathworks.com/matlabcentral/fileexchange/33761.
[16] I. Petráš, Tuning and implementation methods for fractional-order controllers, *Fractional Calculus & Applied Analysis. An International Journal for Theory and Applications*, **15** (2012), 282–303.
[17] I. Petráš, An adaptive fractional-order controller, in *Proc. of the IEEE ICCC2013*, Rytro, Poland, 26–29 May 2013, pp. 297–301.
[18] I. Petráš, Practical aspects for implementation of fractional-order controllers, in *Proc. of the IEEE ICCC2014*, Velke Karlovice, Czech Republic, 28–30 May 2014, pp. 428–431.
[19] I. Petráš, Non-linear fractional-order PID controller, Matlab Central File Exchange, MathWorks, Inc., 2015, url: http://www.mathworks.com/matlabcentral/fileexchange/51190.
[20] I. Petráš, A note on fractional-order non-linear controller: possible neural network approach to design, in *Proc. of the IEEE IJCNN2016*, Vancouver, Canada, 24–29 July, 2016, pp. 603–608.
[21] I. Petráš, Fractional-order nonlinear controllers: design and implementation notes, in *Proc. of the IEEE ICCC2016*, High Tatras, Slovakia, May 29–June 1, 2016, pp. 579–583.
[22] I. Petráš, An introduction to class of fractional-order extremal control: first results, in *Proc. of the IEEE ICCC2018*, Szilvásvárad, Hungary, May 28–31, 2018, pp. 542–547.
[23] I. Petráš and Ľ. Dorčák, Fractional-order control systems: modelling and simulation, *Fractional Calculus & Applied Analysis. An International Journal for Theory and Applications*, **6**(2) (2003), 205–232.
[24] I. Petráš and M. Köver-Dorčo, An effective algorithm for implementation of non-linear fractional-order controller on PLC, in *Proc. of the IEEE ICCC2016*, High Tatras, Slovakia, May 29–June 1, 2016, pp. 584–589.
[25] I. Petráš, I. Podlubny, P. O'Leary, Ľ. Dorčák, and B. M. Vinagre, *Analogue Realization of Fractional Order Controllers*, FBERG, Košice, 2002.
[26] A. Pisano, M. Rapaic, Z. Jelicic, and E. Usai, Nonlinear fractional PI control of a class of fractional-order systems', in *Proc. of the IFAC Conference on Advances in PID Control-PID'12*, Brescia, Italy, March 28–30, 2012.
[27] I. Podlubny, *Fractional Differential Equations*, Academic Press, San Diego, USA, 1999.
[28] I. Podlubny, Fractional-order systems and $PI^{\lambda}D^{\mu}$-controllers, *IEEE Transactions on Automatic Control*, **44**(1) (1999), 208–214.
[29] P. Shah and S. Agashe, Review of fractional PID controller, *Mechatronics*, **38** (2016), 29–41.
[30] Y. X. Su, D. Sun, and B. Y. Duan, Design of an enhanced nonlinear PID controller, *Mechatronics*, **15**(8) (2005), 1005–1024.
[31] D. Valério and J. S. da Costa, *An Introduction to Fractional Control*, IET, London, 2012.
[32] B. M. Vinagre, C. A. Monje, A. J. Calderon, and J. I. Suarez, Fractional PID controllers for industry application. A brief introduction, *Journal of Vibration and Control*, **13**(9–10) (2007), 1419–1429.
[33] D. Xue and Y. Q. Chen, *Modeling, Analysis and Design of Control Systems in Matlab and Simulink*, World Scientific, Singapore, 2015.

Christophe Farges, Jocelyn Sabatier, and Mathieu Chevrié

$\mathcal{H}_\infty$ and $\mathcal{H}_2$ control of fractional models

Abstract: This chapter is dedicated to the design of control laws for systems described by commensurate fractional-order models. Control specifications are formulated in terms of $\mathcal{H}_\infty$ and $\mathcal{H}_2$ constraints on performance transfers of the closed-loop model. Two main approaches are proposed in order to design the control law. The first one consists in approximating the fractional-order model by an uncertain integer order model. The control law is then obtained using robust control design techniques. The second approach is to develop control design tools that are specific to fractional order models.

Keywords: $\mathcal{H}_\infty$ norm, $\mathcal{H}_2$ norm, control, linear matrix inequalities

MSC 2010: 26A33, 93B36, 93B51, 93B52

1 Introduction and problem statement

Fractional-order models allow to describe accurately and with a limited number of parameters the dynamical behavior of a wide range of physical phenomena arising in various applicative fields such as electrochemistry [32, 33], heat science [2, 7], biology [8, 23], mechanics [24], acoustics [27], robotics [28], electronics [12], image processing [42], and economics [36]. The idea is then to use these compact models for the design of control laws [30, 35].

In this chapter, the general control problem formulation, introduced by Doyle in [10] and relying on the control design scheme of Figure 1, is considered. It stems from the reformulation of the different control specifications as an optimization problem of the form:

$$\begin{aligned} \min_{K_f} \quad & \left\| T_{zw}^{\Sigma_f \star K_f}(s) \right\|_q \\ \text{subjected to} \quad & \Sigma_f \star K_f \text{ is stable,} \end{aligned} \tag{1}$$

where:
- Σ_f is the fractional-order model of the generalized plant that contains the process model along with the weighting functions;
- K_f is the control law to be designed (fractional-order control law in the more general case);
- $\Sigma_f \star K_f$ denotes the closed-loop model resulting from the interconnection of Σ_f and K_f;

Christophe Farges, Jocelyn Sabatier, Mathieu Chevrié, Univ. Bordeaux, Bordeaux INP, CNRS, IMS, UMR 5218, 33405 Talence, France, e-mail: jocelyn.sabatier@u-bordeaux.fr

https://doi.org/10.1515/9783110571745-004

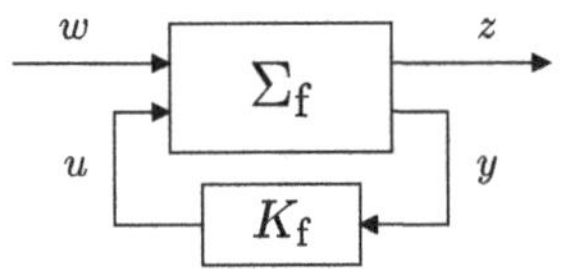

Figure 1: General control design scheme for the $\mathcal{H}_2$ and $\mathcal{H}_\infty$ control problems.

- $T_{zw}^{\Sigma_f \star K_f}(s)$ denotes the transfer matrix of system $\Sigma_f \star K_f$ between performance input vector w and performance output vector z;
- $\|\cdot\|_q$ denotes the q-norm of a given system.

Two systems norms are considered for the performance transfer of the closed-loop model $\Sigma_f \star K_f$, namely $\mathcal{H}_\infty$ and $\mathcal{H}_2$ norms that are defined as follows.

Definition 1. $\mathcal{H}_\infty$ norm of the transfer $T_{zw}^{\Sigma_f \star K_f}(s)$ of $\Sigma_f \star K_f$ is defined by

$$\|T_{zw}^{\Sigma_f \star K_f}(s)\|_\infty = \sup_\omega \overline{\sigma}(T_{zw}^{\Sigma_f \star K_f}(j\omega)) \tag{2}$$

where $\overline{\sigma}(T_{zw}^{\Sigma_f \star K_f}(j\omega))$ is the largest singular value of $T_{zw}^{\Sigma_f \star K_f}(j\omega)$ at frequency ω.

Definition 2. $\mathcal{H}_2$ norm of the transfer $T_{zw}^{\Sigma_f \star K_f}(s)$ of $\Sigma_f \star K_f$ is defined by

$$\|T_{zw}^{\Sigma_f \star K_f}(s)\|_2 = \sqrt{\frac{1}{\pi}\int_0^{+\infty} |T_{zw}^{\Sigma_f \star K_f}(j\omega)|^2 d\omega}. \tag{3}$$

If the $\mathcal{H}_\infty$ norm is considered, the generalized plant Σ_f usually stems from a loop-shaping approach and the optimization problem (1) becomes a feasibility problem, that is, the cost function of (1) is replaced by $\|T_{zw}^{\Sigma_f \star K_f}(s)\|_\infty < 1$. If the $\mathcal{H}_2$ norm is considered, Σ_f can stem for instance from the reformulation of the linear quadratic Gaussian problem as a $\mathcal{H}_2$ optimization problem. The interested reader can refer to [38] for an overview of the numerous control problems that can be formulated using the general control problem formulation (1).

As detailed in Chapter 2, fractional-order models description and properties depend on the fact that they are commensurate or not, explicit or implicit, etc. In this chapter, explicit commensurate fractional-order models are considered. Indeed, if some analytical formulae to compute the $\mathcal{H}_2$ norm of implicit fractional models are available in the literature [4, 5, 26], their extension to synthesis appear to be tedious. Concerning $\mathcal{H}_\infty$ norm, its computation for implicit models has not been considered yet in the literature. Considered models being commensurate and explicit, they admit a pseudo state space description of the form:

$$\Sigma_f : \begin{cases} \mathfrak{D}_0^\nu x(t) = Ax(t) + B_w w(t) + Bu(t), \\ z(t) = C_z x(t) + D_{zw} w(t) + D_{zu} u(t), \quad \{x(t), u(t), w(t)\} = 0 \; \forall\, t < 0 \\ y(t) = Cx(t) + D_{yw} w(t), \end{cases} \tag{4}$$

where v is the commensurate order, $x(t) \in \mathbb{R}^n$ is the pseudo state, $u(t) \in \mathbb{R}^m$ is the control input, $y(t) \in \mathbb{R}^p$ is the measured output, $w(t) \in \mathbb{R}^{m_w}$ is the performance input and $z(t) \in \mathbb{R}^{p_z}$ is the performance output.

Two main approaches to tackle problem (1) are considered in this chapter. The first approach, detailed in Section 2, relies on the construction of an uncertain integer order model whose possible realizations include Σ_f. Although artificially adding uncertainty, and thus leading to suboptimal solutions, this approach allows to use efficient and reliable robust control techniques developed for integer order models. In Section 3 are presented control design techniques that are specifically developed for fractional models. All the proposed design methods are formulated in terms of Linear Matrix Inequalities (LMI).

2 Methods based on an uncertain integer order model reformulation

As a first attempt to tackle a control problem involving a fractional-order model, a simple approach is to approximate it by an integer order model using one of the numerous methods available in the literature [9, 21, 29, 40, 41]. Control design techniques developed in the integer order case can then be applied.

However, if approximation error is not taken into account, obtained results cannot be guaranteed with respect to the original fractional-order model. Hence, as announced in the Introduction of this chapter and as illustrated in Figure 2, the idea developed in this section is to describe the fractional-order model Σ_f as an uncertain model Σ_u corresponding to the interconnexion of an integer order model Σ_i with a normalized uncertainty Δ such that $\|\Delta(s)\|_\infty < 1$. Fractional- order model Σ_f is then seen as a possible realization of the uncertain integer order model $\Sigma_u = \Delta \star \Sigma_i$.

Once Σ_u is obtained, the original optimization problem:

$$\begin{aligned} \gamma_q = \min_{K_f} \quad & \|T_{zw}^{\Sigma_f \star K_f}(s)\|_q \\ \text{subjected to} \quad & \Sigma_f \star K_f \text{ is stable} \end{aligned} \tag{5}$$

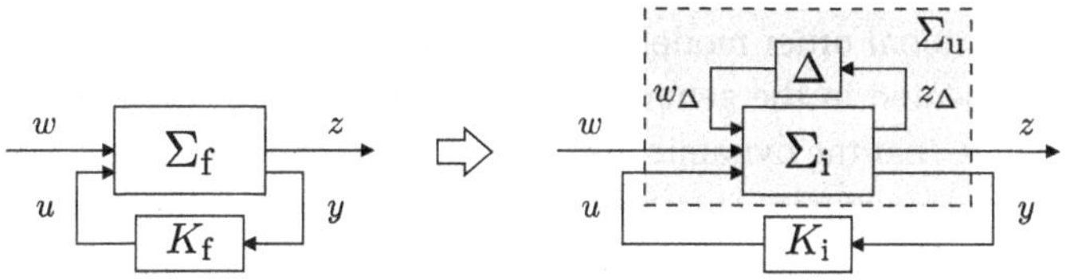

Figure 2: General control design scheme based on a fractional-order model Σ_f and its reformulation involving an uncertain integer order model Σ_u.

is replaced by:

$$\begin{aligned} \tilde{\gamma}_q = \min_{K_\mathrm{i}} \quad & \|T_{zw}^{\Sigma_\mathrm{i}\star K_\mathrm{i}}(s)\|_q \\ \text{subjected to} \quad & \Sigma_\mathrm{i} \star K_\mathrm{i} \text{ is stable} \\ & \|T_{z_\Delta w_\Delta}^{\Sigma_\mathrm{i}\star K_\mathrm{i}}(s)\|_\infty < 1, \end{aligned} \tag{6}$$

where K_i is an integer order control law, as its design is based on an integer order model. New $\mathcal{H}_\infty$ constraint $\|T_{z_\Delta w_\Delta}^{\Sigma_\mathrm{i}\star K_\mathrm{i}}(s)\|_\infty < 1$ in the optimization problem (6) allows to ensure that the closed-loop system is stable whatever the normalized uncertainty Δ and that the achieved level of performance by the optimal control law K_i^* is guaranteed whatever the normalized uncertainty Δ, and thus that it is guaranteed with respect to the original fractional-order model, that is,

$$\|T_{zw}^{\Sigma_\mathrm{f}\star K_\mathrm{i}^*}(s)\|_q < \tilde{\gamma}_q. \tag{7}$$

Of course, controller K_i^* solution of optimization problem (6) is a suboptimal solution with respect to optimization problem (5), and thus

$$\tilde{\gamma}_q \geq \gamma_q. \tag{8}$$

In Section 2.1, a methodology to obtain the uncertain integer order model Σ_u is detailed. Based on this approximation, robust control techniques are proposed in Section 2.2 to address the $\mathcal{H}_\infty$ and $\mathcal{H}_2$ control problems.

2.1 Construction of the uncertain integer order model

The methodology proposed in this section to obtain the uncertain integer order model Σ_u approximating the fractional-order model relies on the decomposition of Σ_f into an integer order submodel and a diffusive submodel. Indeed, such a representation offers the advantage of already exhibiting an integer order submodel.

2.1.1 Decomposition of the fractional-order model intro an integer order submodel and a diffusive submodel

The decomposition of a fractional-order model into an integer order submodel and a diffusive submodel is presented in the general case in Chapter 2. For the sake of simplicity, let us assume here that the dynamic matrix A of Σ_f pseudo state space description (4) has n distinct eigenvalues. This assumption simplifies the derivations presented in the remaining of this section (please note that the general case could nevertheless be addressed using the same methodology). Let us also denote the augmented input vector $\bar{u}(t) = \begin{bmatrix} w^T(t) & u^T(t) \end{bmatrix}^T \in \mathbb{R}^{\bar{m}}$, $\bar{m} = m_w + m$, and the augmented output vector $\bar{y}(t) = \begin{bmatrix} z^T(t) & y^T(t) \end{bmatrix}^T \in \mathbb{R}^{\bar{p}}$, $\bar{p} = p_z + p$. Representation (4) then admits the following diagonal form:

$$\Sigma_{\mathrm{f}} : \begin{cases} \mathfrak{D}_0^\nu \, \bar{x}(t) = \bar{A}\bar{x}(t) + \bar{B}\bar{u}(t), \\ \bar{y}(t) = \bar{C}\bar{x}(t) + \bar{D}\bar{u}(t), \end{cases} \tag{9}$$

where $\bar{A} = \mathrm{diag}(\lambda_1, \dots, \lambda_n)$ and λ_l, $l \in \{1, \dots, n\}$, denote the eigenvalues of A.

The output of model (9) can be decomposed into integer and diffusive parts:

$$\bar{y}(t) = \bar{y}_{\mathrm{i}}(t) + \bar{y}_{\mathrm{d}}(t). \tag{10}$$

The integer part can be seen as the response of the following integer order submodel of state $\eta_{\mathrm{i}}(t) \in \mathbb{R}^N$, $N = \sum_{l=1}^n n_l$:

$$\Sigma_{\mathrm{i}}^{\mathrm{sub}} : \begin{cases} \dot{\eta}_{\mathrm{i}}(t) = A_{\mathrm{p}}\eta_{\mathrm{i}}(t) + B_{\mathrm{p}}\bar{u}(t), \\ \bar{y}_{\mathrm{i}}(t) = C_{\mathrm{p}}\eta_{\mathrm{i}}(t), \end{cases} \tag{11}$$

with

$$A_{\mathrm{p}} = \left[\begin{array}{ccc|c|ccc} p_{\lambda_1,1} & & & & & & \\ & \ddots & & & & & \\ & & p_{\lambda_1,n_1} & & & & \\ \hline & & & \ddots & & & \\ \hline & & & & p_{\lambda_n,1} & & \\ & & & & & \ddots & \\ & & & & & & p_{\lambda_n,n_n} \end{array}\right], \quad B_{\mathrm{p}} = \left[\begin{array}{c} \frac{p_{\lambda_1,1}}{\nu\lambda_1}\bar{B}_1 \\ \vdots \\ \frac{p_{\lambda_1,n_1}}{\nu\lambda_1}\bar{B}_1 \\ \hline \vdots \\ \hline \frac{p_{\lambda_n,1}}{\nu\lambda_n}\bar{B}_n \\ \vdots \\ \frac{p_{\lambda_n,n_n}}{\nu\lambda_n}\bar{B}_n \end{array}\right], \tag{12}$$

$$C_{\mathrm{p}} = \left[\begin{array}{ccc|c|ccc} \bar{C}_1 & \cdots & \bar{C}_1 & \cdots & \bar{C}_n & \cdots & \bar{C}_n \end{array}\right]. \tag{13}$$

where $\bar{B}_l$, $l \in \{1, \dots, n\}$, denote the lines of matrix $\bar{B}$ and $\bar{C}_l$, $l \in \{1, \dots, n\}$, denote the columns of matrix $\bar{C}$. $p_{\lambda_l,q}$, $q \in \{1, \dots, n_l\}$, are the n_l poles associated to λ_l that can be determined according to

$$p_{\lambda_l,q} = |p_{\lambda_l,q}| \mathrm{e}^{\mathrm{j}\theta_{\lambda_l,q}}, \quad \text{with} \begin{cases} |p_{\lambda_l,q}| = |\lambda_l|^{\frac{1}{\nu}}, \\ \theta_{\lambda_l,q} = \frac{1}{\nu}\arg(\lambda_l) + \frac{2k\pi}{\nu}, \\ k \in \mathbb{K}_{\lambda_l}, \end{cases} \tag{14}$$

and

$$\mathbb{K}_{\lambda_l} = \left\{ k \in \mathbb{Z} : -\frac{\nu}{2} - \frac{\arg(\lambda_l)}{2\pi} < k < \frac{\nu}{2} - \frac{\arg(\lambda_l)}{2\pi} \right\}, \quad n_l = \mathrm{card}(\mathbb{K}_{\lambda_l}). \tag{15}$$

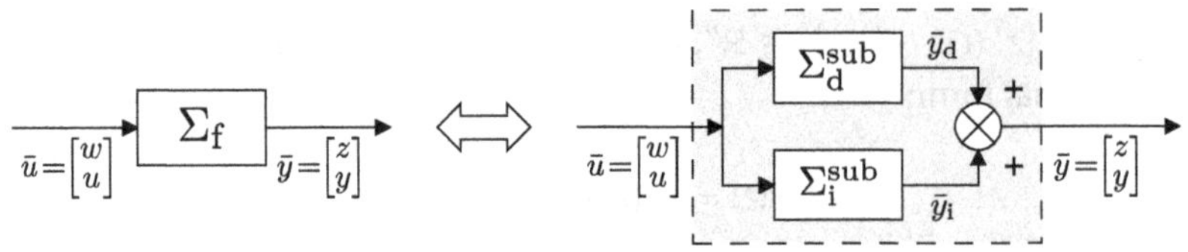

Figure 3: Decomposition of Σ_f into an integer order submodel and a diffusive submodel.

The diffusive part can be seen as the response of the following diffusive model of distributed state $w(\chi, t) \in \mathbb{R}^n$:

$$\Sigma_d^{sub} : \begin{cases} \dot{w}(\chi, t) = -\chi \, w(\chi, t) + \bar{B}\,\bar{u}(t), \\ \bar{y}_d(t) = \bar{C} \int_0^{+\infty} \mu(\chi) w(\chi, t) d\chi, \end{cases} \quad \chi \in \mathbb{R}_+ \tag{16}$$

where $\mu(\chi) = \mathrm{diag}(\mu_{\lambda_1}(\chi), \dots, \mu_{\lambda_n}(\chi))$ and

$$\mu_{\lambda_l}(\chi) = \frac{1}{\pi} \frac{\chi^\nu \sin(\nu\pi)}{\chi^{2\nu} - 2\chi^\nu \lambda_l \cos(\nu\pi) + \lambda_l^2}. \tag{17}$$

As illustrated in Figure 3, the fractional-order model Σ_f can thus be seen as the association of the integer order submodel Σ_i^{sub} of state $\eta_i(t)$ given by (11) and the diffusive submodel Σ_d^{sub} of distributed state $w(\chi, t)$ given by (16).

2.1.2 Uncertain model obtained by considering the whole diffusive submodel as an uncertainty

Given the block diagram of Figure 3, it is possible to obtain an integer order approximation of Σ_f by considering the diffusive submodel Σ_d^{sub} as a dynamic uncertainty bounded by the $\mathcal{H}_\infty$ norm γ_d of Σ_d^{sub}:

$$\gamma_d = \left\| T_{\bar{y}_d \bar{u}}^{\Sigma_d^{sub}}(s) \right\|_\infty, \tag{18}$$

where transfer matrix $T_{\bar{y}_d \bar{u}}^{\Sigma_d^{sub}}(s)$ is obtained by taking Laplace transform of (16):

$$T_{\bar{y}_d \bar{u}}^{\Sigma_d^{sub}}(s) = \bar{C} \int_0^{+\infty} \frac{1}{s + \chi} \mu(\chi) d\chi \bar{B}. \tag{19}$$

This approach is illustrated in Figure 4 and finally leads to an uncertain model resulting from the interconnection of the following integer order model Σ_i^0 with a normalized uncertainty Δ:

$$\Sigma_i^0 : \begin{cases} \dot{\eta}_i(t) = A_p \, \eta_i(t) + B_p \, \bar{u}(t), \\ z_\Delta(t) = \bar{u}(t), \\ \bar{y}(t) = C_p \, \eta_i(t) + \gamma_d \, w_\Delta(t) \end{cases} \tag{20}$$

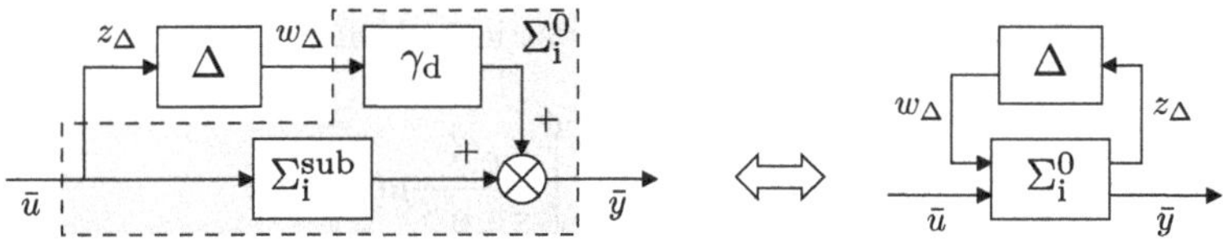

Figure 4: Uncertain model obtained by considering the diffusive submodel as an uncertainty.

where $\eta(t) \in \mathbb{R}^N$ is the state of dimension N equal to the total number of poles of the integer order submodel, matrices A_p, B_p, and C_p are given by equations (12)–(13) and scalar γ_d is given by (18)–(19).

Such an approach, considered in the single-input single-output (SISO) case in [19], has the following drawback. Depending on the considered fractional-order model, the diffusive part of the model response $\bar{y}_d(t)$ can be dominating with respect to the integer part $\bar{y}_i(t)$. In some cases, the integer part can even be null. This is the case when no pole $p_{\lambda_l,q}$ is generated by any of the eigenvalues λ_l, that is, when the set $\mathbb{K}_{\lambda_l}$ given by (15) is empty $\forall\, l \in \{1,\dots,n\}$.

In order to overcome this drawback, a methodology to include a part of the diffusive submodel behavior into the nominal model is proposed in the next section.

2.1.3 Uncertain model obtained by including a part of the diffusive submodel behavior in the nominal model

The objective of this section is to reduce the uncertainty by including a part of the diffusive submodel behavior given by transfer matrix (19) into the nominal model.

2.1.3.1 Rewriting the diffusive submodel as an uncertain integer model

Considering the change of variables $\chi = e^{-\zeta}$, and thus $d\chi = -e^{-\zeta}d\zeta$, relation (19) becomes

$$T_{\bar{y}_d\bar{u}}^{\Sigma_d^{sub}}(s) = \bar{C}\left[\int_{-\infty}^{+\infty} \frac{e^{-\zeta}}{s+e^{-\zeta}}\mu(e^{-\zeta})d\zeta\right]\bar{B} = \bar{C}H_d(s)\bar{B}. \tag{21}$$

The integral in the expression of diagonal transfer matrix $H_d(s)$ of equation (21) can be separated into three parts, leading to

$$H_d(s) = \operatorname{diag}(H_{d,1}(s),\dots,H_{d,n}(s)) = H_1(s) + H_{12}(s) + H_2(s), \tag{22}$$

where $H_1(s)$, $H_{12}(s)$, and $H_2(s)$ are diagonal transfer matrices given by

$$\begin{aligned} H_1(s) &= \int_{-\infty}^{\zeta_1} \frac{e^{-\zeta}}{s+e^{-\zeta}}\mu(e^{-\zeta})d\zeta; \quad H_{12}(s) = \int_{\zeta_1}^{\zeta_2} \frac{e^{-\zeta}}{s+e^{-\zeta}}\mu(e^{-\zeta})d\zeta; \\ H_2(s) &= \int_{\zeta_2}^{+\infty} \frac{e^{-\zeta}}{s+e^{-\zeta}}\mu(e^{-\zeta})d\zeta, \end{aligned} \tag{23}$$

with $\mu(e^{-\zeta}) = \mathrm{diag}(\mu_{\lambda_1}(e^{-\zeta}), \dots, \mu_{\lambda_n}(e^{-\zeta}))$ and, according to (17)

$$\mu_{\lambda_l}(e^{-\zeta}) = \frac{1}{\pi}\frac{e^{-\zeta v}\sin(v\pi)}{e^{-2\zeta v} - 2e^{-\zeta v}\lambda_l\cos(v\pi) + \lambda_l^2}. \tag{24}$$

In order to determine an uncertain model associated to the diffusive submodel $T_{\bar{y}_d\bar{u}}^{\Sigma_d^{sub}}(s)$, upper bounds, in terms of gain, for transfers $H_1(s)$, $H_{12}(s)$ and $H_2(s)$ are now proposed (see reference [34] for detailed computations in the SISO case).

Keeping dominant terms when $z \to -\infty$ in expressions of $\mu_{\lambda_l}(e^{-\zeta})$ given by (24), transfer $H_1(s)$ can be upperbounded in terms of gain according to

$$|H_1(j\omega)| \le \overline{H_1}, \quad \forall\omega \in \mathbb{R}_+, \tag{25}$$

where $\overline{H_1}$ is a constant diagonal matrix given by

$$\overline{H_1} = \mathrm{diag}(m_1(\lambda_1, \zeta_1), \dots, m_1(\lambda_n, \zeta_1))\frac{|\sin(v\pi)|}{v\pi}e^{v\zeta_1}, \tag{26}$$

$$m_1(\lambda_l, \zeta_1) = \begin{cases} \frac{|\sin(v\pi)|}{v\pi}e^{v\zeta_1}\frac{1}{|\mu_{\lambda_l}(e^{-\zeta_1})|} & \text{if } \frac{|\sin(v\pi)|}{v\pi}e^{v\zeta_1} > |\mu_{\lambda_l}(e^{-\zeta_1})|, \\ 1 & \text{else.} \end{cases} \tag{27}$$

Using the same approach when $z \to +\infty$, transfer $H_2(s)$ can be upper-bounded in terms of gain according to

$$|H_2(j\omega)| \le |\overline{H_2}(j\omega)|, \quad \forall\omega \in \mathbb{R}_+, \tag{28}$$

where $\overline{H_2}(s)$ is an integer order diagonal transfer matrix given by

$$\overline{H_2}(s) = \mathrm{diag}(m_2(\lambda_1, \zeta_1), \dots, m_2(\lambda_n, \zeta_1))\frac{|\sin(v\pi)|}{v\pi|\lambda_l^2|}\frac{e^{-v\zeta_2}}{se^{\zeta_2}+1}, \tag{29}$$

$$m_2(\lambda_l, \zeta_2) = \begin{cases} \frac{|\sin(v\pi)|}{v\pi|\lambda_l^2|}\frac{e^{-v\zeta_2}}{se^{\zeta_2}+1}\frac{1}{|\mu_{\lambda_l}(e^{-\zeta_2})|} & \text{if } \frac{|\sin(v\pi)|}{v\pi|\lambda_l^2|}\frac{e^{-v\zeta_2}}{se^{\zeta_2}+1} > |\mu_{\lambda_l}(e^{-\zeta_2})|, \\ 1 & \text{else.} \end{cases} \tag{30}$$

Upperbounds on $H_1(s)$ and $H_2(s)$ were obtained by taking the limit of their expressions when z tends toward $-\infty$ and $+\infty$, respectively. Such an approach is not possible for $H_{12}(s)$. It is thus proposed here to use the trapezoidal rule with α intervals in order to approximate the integrals in the expression of $H_{12}(s)$. Hence, transfer $H_{12}(s)$ can be approximated as follows:

$$|H_{12}(\mathrm{j}\omega)| = |\widetilde{H}_{\mathrm{d}}(\mathrm{j}\omega)| + E_{12}(\omega), \quad \forall \omega \in \mathbb{R}_+ \tag{31}$$

where $\widetilde{H}_{\mathrm{d}}(s)$ is an integer order diagonal transfer matrix given by

$$\widetilde{H}_{\mathrm{d}}(s) = \mathrm{diag}(\widetilde{H}_{\mathrm{d},\lambda_1}(s), \dots, \widetilde{H}_{\mathrm{d},\lambda_n}(s)), \tag{32}$$

$$\begin{aligned}\widetilde{H}_{\mathrm{d},\lambda_l}(s) = {} & \frac{\mathrm{e}^{-\zeta_1}}{s + \mathrm{e}^{-\zeta_1}} \int_{\zeta_1}^{\zeta_1+h} \mu_{\lambda_l}(\mathrm{e}^{-\zeta}) \left(\frac{\zeta_1 + h - \zeta}{h} \right) d\zeta \\ & + \sum_{k=1}^{\alpha-1} \frac{\mathrm{e}^{-\zeta_1 - kh}}{s + \mathrm{e}^{-\zeta_1 - kh}} \left(\int_{\zeta_1+(k-1)h}^{\zeta_1+kh} \mu_{\lambda_l}(\mathrm{e}^{-\zeta}) \left(\frac{\zeta - \zeta_1 - (k-1)h}{h} \right) d\zeta \right. \\ & \left. + \int_{\zeta_1+kh}^{\zeta_1+(k+1)h} \mu_{\lambda_l}(\mathrm{e}^{-\zeta}) \left(\frac{\zeta_1 + (k+1)h - \zeta}{h} \right) d\zeta \right) \\ & + \frac{\mathrm{e}^{-\zeta_2}}{s + \mathrm{e}^{-\zeta_2}} \int_{\zeta_1+(\alpha-1)h}^{\zeta_2} \mu_{\lambda_l}(\mathrm{e}^{-\zeta}) \left(\frac{\zeta - \zeta_1 + (\alpha-1)h}{\zeta_2 - \zeta_1 - (\alpha-1)h} \right) d\zeta, \end{aligned} \tag{33}$$

where $h = \frac{\zeta_2 - \zeta_1}{\alpha}$. An upperbound in terms of gain on $E_{12}(\omega)$ is given by

$$\overline{E_{12}}(\omega) = \frac{1}{2}\sqrt{\frac{1}{30}} h^{\frac{5}{2}} \sum_{k=0}^{\alpha-1} I_{\mu,k} \max_{z \in [\zeta_1+kh, \zeta_1+(k+1)h]} \left| \frac{\omega\, \mathrm{e}^{\zeta}}{(\omega^2 \mathrm{e}^{2\zeta} + 1)} \right|, \quad \forall \omega \in \mathbb{R}_+ \tag{34}$$

where $I_{\mu,k} = \|\mu(\mathrm{e}^{-\zeta})\|_{\mathcal{L}_2}$ for $\zeta \in [\zeta_1 + kh, \zeta_1 + (k+1)h]$ can be computed with high precision since the discretization step used for the evaluation of this integral has no impact on the order of the obtained approximation. Based on the above results, the diagonal transfer $H_{\mathrm{d}}(s)$ corresponding to the diffusive submodel can be upperbounded in terms of gain as follows:

$$|H_{\mathrm{d}}(\mathrm{j}\omega)| \le |\widetilde{H}_{\mathrm{d}}(\mathrm{j}\omega)| + \overline{E_{\mathrm{d}}}(\omega), \tag{35}$$

where $\widetilde{H}_{\mathrm{d}}(s)$ is the diagonal integer order transfer given by (32) and where

$$\overline{E_{\mathrm{d}}}(\omega) = \mathrm{diag}(\overline{E_{\mathrm{d},\lambda_1}}(\omega), \dots, \overline{E_{\mathrm{d},\lambda_n}}(\omega)) = \overline{H_1} + \overline{E_{12}}(\omega) + \overline{H_2}(\mathrm{j}\omega), \tag{36}$$

$\overline{H_1}$ being the constant matrix given by (26), $\overline{E_{12}}(\omega)$ the function of ω given by (34), and $\overline{H_2}(s)$ the diagonal integer order transfer given by (29).

Upperbound (35) is used in the next paragraphs to rewrite the diffusive submodel and the whole fractional-order model as integer order models affected by additive or multiplicative uncertainties. These representations involve a state space realization of transfer $\widetilde{H}_{\mathrm{d}}(s)$ given by (32) that is denoted as follows:

$$\widetilde{H}_{\mathrm{d}} : \begin{cases} \dot{\widetilde{\eta}}_{\mathrm{d}}(t) = \widetilde{A}_{\mathrm{d}}\, \widetilde{\eta}_{\mathrm{d}}(t) + \widetilde{B}_{\mathrm{d}}\, \bar{u}(t), \\ \widetilde{y}_{\mathrm{d}}(t) = \widetilde{C}_{\mathrm{d}}\, \widetilde{\eta}_{\mathrm{d}}(t) + \widetilde{D}_{\mathrm{d}}\, \bar{u}(t). \end{cases} \tag{37}$$

2.1.3.2 Rewriting the whole fractional-order model as an uncertain integer order model affected by additive uncertainty

Considering relation (35), the diagonal part of the diffusive submodel can also be seen as an uncertain model corresponding to the integer model $\widetilde{H}_{12}$ affected by a nonrational uncertainty $\overline{E}_{\mathrm{d}}(\omega)$.

Let $W_{\mathrm{a},\lambda_i}(s)$, $i \in \{1, \ldots, n\}$, rational transfer functions upperbounding $\overline{E_{\mathrm{d},\lambda_i}}(\omega)$ given by (36) in terms of gain for all ω, that is,

$$\overline{E_{\mathrm{d},\lambda_i}}(\omega) \leq |W_{\mathrm{a},\lambda_i}(\mathrm{j}\omega)|, \quad \forall \omega \in \mathbb{R}_+, \quad \forall i \in \{1, \ldots, n\}. \tag{38}$$

Let the following state space realization associated to the diagonal transfer matrix $W_{\mathrm{a}}(s) = \mathrm{diag}(W_{\mathrm{a},\lambda_1}(s), \ldots, W_{\mathrm{a},\lambda_n}(s))$:

$$W_{\mathrm{a}} : \begin{cases} \dot{\eta}_{\mathrm{a}}(t) = A_{\mathrm{a}}\, \eta_{\mathrm{a}}(t) + B_{\mathrm{a}}\, w_{\Delta}(t), \\ y_{\mathrm{a}}(t) = C_{\mathrm{a}}\, \eta_{\mathrm{a}}(t) + D_{\mathrm{a}}\, w_{\Delta}(t). \end{cases} \tag{39}$$

Then, as illustrated in Figure 5, the diffusive submodel, and thus the whole fractional model can be seen as uncertain models. The uncertain model associated to Σ_{f} corre-

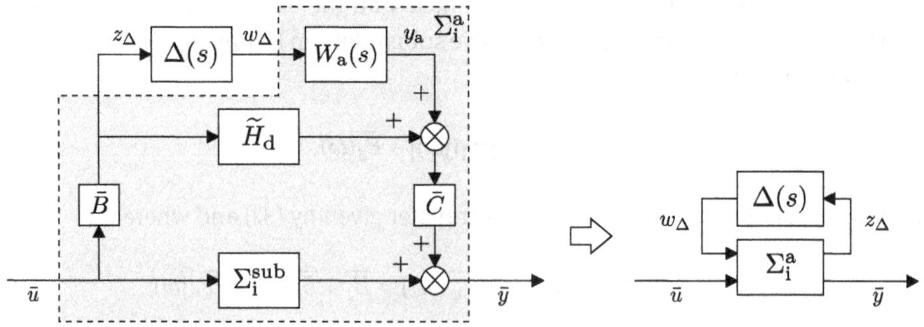

Figure 5: Uncertain model obtained by including a part of the diffusive submodel into the nominal model (additive form).

sponds to the interconnection of the integer model $\Sigma_{\mathrm{i}}^{\mathrm{a}}$ given by

$$\Sigma_{\mathrm{i}}^{\mathrm{a}}: \begin{cases} \begin{bmatrix} \dot{\eta}_{\mathrm{i}}(t) \\ \dot{\tilde{\eta}}_{\mathrm{d}}(t) \\ \dot{\eta}_{\mathrm{a}}(t) \end{bmatrix} = \begin{bmatrix} A_{\mathrm{p}} & \mathbb{0} & \mathbb{0} \\ \mathbb{0} & \tilde{A}_{\mathrm{d}} & \mathbb{0} \\ \mathbb{0} & \mathbb{0} & A_{\mathrm{a}} \end{bmatrix} \begin{bmatrix} \eta_{\mathrm{i}}(t) \\ \tilde{\eta}_{\mathrm{d}}(t) \\ \eta_{\mathrm{a}}(t) \end{bmatrix} + \begin{bmatrix} B_{\mathrm{p}} \\ \tilde{B}_{\mathrm{d}}\bar{B} \\ \mathbb{0} \end{bmatrix} \bar{u}(t) + \begin{bmatrix} \mathbb{0} \\ \mathbb{0} \\ B_{\mathrm{a}} \end{bmatrix} w_{\Delta}(t), \\ z_{\Delta}(t) = \bar{B}\bar{u}(t), \\ \bar{y}(t) = \begin{bmatrix} C_{\mathrm{p}} & \bar{C}\tilde{C}_{\mathrm{d}} & C_{\mathrm{a}} \end{bmatrix} \begin{bmatrix} \eta_{\mathrm{i}}(t) \\ \tilde{\eta}_{\mathrm{d}}(t) \\ \eta_{\mathrm{a}}(t) \end{bmatrix} + \bar{C}D_{\mathrm{a}}w_{\Delta}(t). \end{cases} \tag{40}$$

with a diagonal uncertainty $\Delta(s) = \mathrm{diag}(\delta_1(s), \ldots, \delta_n(s))$, $\|\delta_i(s)\|_\infty = 1$.

Approximation error $\overline{E_{\mathrm{d}}}(\omega)$ being constant in high frequency, some of the possible realizations of the uncertain model $\Sigma_{\mathrm{u}}^{\mathrm{a}} = \Delta \star \Sigma_{\mathrm{i}}^{\mathrm{a}}$ will exhibit constant gain in high frequency. A noticeable drawback of using an additive uncertainty is thus that the $\mathcal{H}_2$ norm of $\Sigma_{\mathrm{u}}^{\mathrm{a}}$ would be infinite.

2.1.3.3 Rewriting the whole fractional-order model as an uncertain integer order model affected by multiplicative uncertainty

In order to overcome the problem of constant gain in high frequency, a possible solution is to consider multiplicative uncertainty. Let us consider the following relative errors between the fractional order transfers $H_{\mathrm{d},\lambda_i}(s)$ given by (22) and the integer transfers $\widetilde{H}_{\mathrm{d},\lambda_i}(s)$ given by (32):

$$l_{\lambda_i}(s) = \frac{H_{\mathrm{d},\lambda_i}(s)}{\widetilde{H}_{\mathrm{d},\lambda_i}(s)} - 1, \quad \forall i \in \{1, \ldots, n\}. \tag{41}$$

Transfers $l_{\lambda_i}(s)$ being fractional, rational filters $W_{\mathrm{m},\lambda_i}(s)$ such that

$$|l_{\lambda_i}(\mathrm{j}w)| \le |W_{\mathrm{m},\lambda_i}(\mathrm{j}\omega)|, \quad \forall \omega \in \mathbb{R}, \quad \forall i \in \{1, \ldots, n\}, \tag{42}$$

are introduced. Let the following state space realization associated to the diagonal transfer matrix $W_{\mathrm{m}}(s) = \mathrm{diag}(W_{\mathrm{m},\lambda_1}(s), \ldots, W_{\mathrm{m},\lambda_n}(s))$:

$$W_{\mathrm{m}}: \begin{cases} \dot{\eta}_{\mathrm{m}}(t) = A_{\mathrm{m}}\, \eta_{\mathrm{m}}(t) + B_{\mathrm{m}}\, w_{\Delta}(t), \\ y_{\mathrm{m}}(t) = C_{\mathrm{m}}\, \eta_{\mathrm{m}}(t) + D_{\mathrm{m}}\, w_{\Delta}(t). \end{cases} \tag{43}$$

Then, as illustrated in Figure 6, the uncertain model associated to Σ_{f} corresponds to the interconnection of the integer model $\Sigma_{\mathrm{i}}^{\mathrm{m}}$ given by

$$\Sigma_{\mathrm{i}}^{\mathrm{m}}: \begin{cases} \begin{bmatrix} \dot{\eta}_{\mathrm{i}}(t) \\ \dot{\tilde{\eta}}_{\mathrm{d}}(t) \\ \dot{\eta}_{\mathrm{m}}(t) \end{bmatrix} = \begin{bmatrix} A_{\mathrm{p}} & \mathbb{0} & \mathbb{0} \\ \mathbb{0} & \tilde{A}_{\mathrm{d}} & \tilde{B}_{\mathrm{d}}C_{\mathrm{m}} \\ \mathbb{0} & \mathbb{0} & A_{\mathrm{m}} \end{bmatrix} \begin{bmatrix} \eta_{\mathrm{i}}(t) \\ \tilde{\eta}_{\mathrm{d}}(t) \\ \eta_{\mathrm{m}}(t) \end{bmatrix} + \begin{bmatrix} B_{\mathrm{p}} \\ \tilde{B}_{\mathrm{d}}\bar{B} \\ \mathbb{0} \end{bmatrix} \bar{u}(t) + \begin{bmatrix} \mathbb{0} \\ \tilde{B}_{\mathrm{d}}D_{\mathrm{m}} \\ B_{\mathrm{m}} \end{bmatrix} w_{\Delta}(t), \\ z_{\Delta}(t) = \bar{B}\bar{u}(t), \\ \bar{y}(t) = \begin{bmatrix} C_{\mathrm{p}} & \bar{C}\tilde{C}_{\mathrm{d}} & \bar{C}C_{\mathrm{m}} \end{bmatrix} \begin{bmatrix} \eta_{\mathrm{i}}(t) \\ \tilde{\eta}_{\mathrm{d}}(t) \\ \eta_{\mathrm{m}}(t) \end{bmatrix} + \bar{C}\widetilde{D}_{\mathrm{d}}\bar{u}(t) + \bar{C}D_{\mathrm{m}}w_{\Delta}(t) \end{cases} \tag{44}$$

with a diagonal uncertainty $\Delta(s) = \mathrm{diag}(\delta_1(s), \ldots, \delta_n(s))$, $\|\delta_i(s)\|_\infty = 1$.

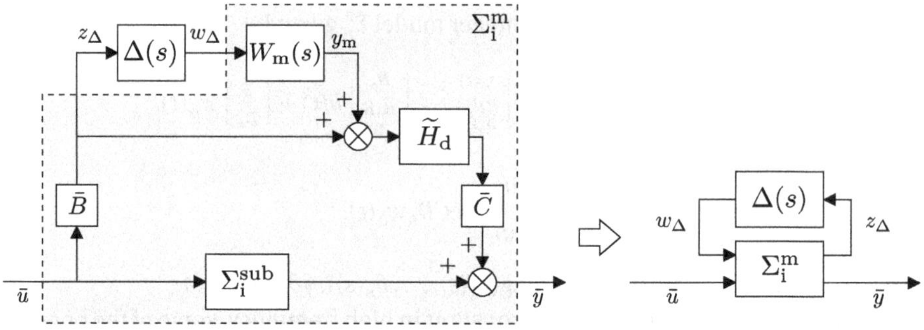

Figure 6: Uncertain model obtained by including a part of the diffusive submodel in the nominal model (multiplicative form).

2.1.3.4 Conclusion

In this section, three different uncertain integer order models whose realizations contain the original fractional-order models have been proposed. The first one, depicted in Figure 4 and involving Σ_i^0 given by (20), was obtained by considering the whole diffusive submodel as an uncertainty. The second one, depicted in Figure 5 and involving Σ_i^a given by (40), was obtained by including a part of the diffusive submodel in the nominal model and considering an additive uncertainty. The last one, depicted in Figure 6 and involving Σ_i^m given by (44), was obtained by including a part of the diffusive submodel in the nominal model and considering a multiplicative uncertainty.

For the last two cases, and as detailed in [34], the nominal model relies on the choice of boundaries ζ_1 and ζ_2 and on the number of intervals α used for the discretization of the interval. This choice results from a compromise between the size of the uncertainty (size impacting the conservatism associated to the uncertain model) and the order of the integer model Σ_i.

By separating the augmented input vector $\bar{u}(t)$ into $w(t)$ and $u(t)$ on one hand and the augmented output vector $\bar{y}(t)$ into $z(t)$ and $y(t)$ on the other hand, each of the three integer models (Σ_i^0, Σ_i^a or Σ_i^m) can be rewritten in the generic form:

$$\Sigma_i : \begin{cases} \dot{\tilde{x}}(t) = \tilde{A}\tilde{x}(t) + \tilde{B}_\Delta w_\Delta(t) + \tilde{B}_w w(t) + \tilde{B}u(t), \\ z_\Delta(t) = \tilde{C}_\Delta \tilde{x}(t) + \tilde{D}_{z_\Delta w_\Delta} w_\Delta(t) + \tilde{D}_{z_\Delta u} u(t), \\ z(t) = \tilde{C}_z \tilde{x}(t) + \tilde{D}_{zw} w(t) + \tilde{D}_{zu} u(t), \\ y(t) = \tilde{C}\tilde{x}(t) + \tilde{D}_{yw_\Delta} w_\Delta(t) + \tilde{D}_{yw} w(t). \end{cases} \tag{45}$$

The fractional model Σ_f given by (4) is thus seen as a possible realization of an uncertain model that corresponds to the interconnexion of the integer order model Σ_i given by (45) with a normalized uncertainty Δ such that $\|\Delta(s)\|_\infty < 1$.

2.2 $\mathcal{H}_\infty$ and $\mathcal{H}_2$ control based on the uncertain integer order model

Applying methodology described in Section 2.1 allows to rewrite the fractional model Σ_f given by (4) as an uncertain model Σ_u corresponding to the interconnexion of the integer order model Σ_i given by (45) with a normalized uncertainty Δ.

Based on the uncertain model Σ_u, the optimization problem (5) is replaced by the optimization problem (6) which corresponds to a classical robust control problem for an uncertain integer order model. Methodology to solve such a problem however differs if $\mathcal{H}_\infty$ or $\mathcal{H}_2$ norm is considered for the cost function. These two cases are detailed in the next two subsections. In each case, a LMI based design methodology borrowed for the literature dedicated to integer order systems is proposed. In both cases, full order dynamic output feedback control laws of the form:

$$K_i : \begin{cases} \dot{\tilde{x}}_K(t) = \tilde{A}_K\,\tilde{x}_K(t) + \tilde{B}_K\,y(t), \\ u(t) = \tilde{C}_K\,\tilde{x}_K(t) + \tilde{D}_K\,y(t), \end{cases} \tag{46}$$

where $\tilde{x}_K(t)$ denotes the state of the control law of same dimension as $\tilde{x}(t)$, are considered.

2.2.1 $\mathcal{H}_\infty$ control based on the uncertain integer order model

As mentioned in Section 1, when $\mathcal{H}_\infty$ norm is considered, Σ_f usually stems from a loop-shaping approach and the optimization problem (5) becomes a feasibility problem, that is, cost function of (5) is replaced by $\|T_{zw}^{\Sigma_f\star K_f}(s)\|_\infty < 1$. Optimization problem (6) then writes as the feasibility problem:

$$\begin{aligned} \text{Find}\quad & K_i \\ \text{subjected to}\quad & \Sigma_i \star K_i \text{ is stable} \\ & \|T_{zw}^{\Sigma_i\star K_i}(s)\|_\infty < 1 \\ & \|T_{z_\Delta w_\Delta}^{\Sigma_i\star K_i}(s)\|_\infty < 1, \end{aligned} \tag{47}$$

where Σ_i is the integer order model given by (45) and K_i is the full order dynamic output feedback control law given by (46). Denoting $\bar{z}(t) = \begin{bmatrix} z_\Delta^T(t) & z^T(t) \end{bmatrix}^T$ and $\bar{w}(t) = \begin{bmatrix} w_\Delta^T(t) & w^T(t) \end{bmatrix}^T$, system (45) writes

$$\Sigma_i : \begin{cases} \dot{\tilde{x}}(t) = \tilde{A}\tilde{x}(t) + \tilde{B}_{\bar{w}}\bar{w}(t) + \tilde{B}u(t), \\ \bar{z}(t) = \tilde{C}_{\bar{z}}\tilde{x}(t) + \tilde{D}_{\bar{z}\bar{w}}\bar{w}(t) + \tilde{D}_{\bar{z}u}u(t), \\ y(t) = \tilde{C}\tilde{x}(t) + \tilde{D}_{y\bar{w}}\bar{w}(t), \end{cases} \tag{48}$$

with

$$\tilde{B}_{\bar{w}} = \begin{bmatrix} \tilde{B}_\Delta & \tilde{B}_w \end{bmatrix}, \quad \tilde{D}_{y\bar{w}} = \begin{bmatrix} \tilde{D}_{yw_\Delta} & \tilde{D}_{yw} \end{bmatrix},$$
$$\tilde{C}_{\bar{z}} = \begin{bmatrix} \tilde{C}_\Delta \\ \tilde{C}_z \end{bmatrix}, \quad \tilde{D}_{\bar{z}\bar{w}} \begin{bmatrix} \tilde{D}_{z_\Delta w_\Delta} & \mathbb{0} \\ \mathbb{0} & \tilde{D}_{zw} \end{bmatrix}, \quad \tilde{D}_{\bar{z}u} = \begin{bmatrix} \tilde{D}_{z_\Delta u} \\ \tilde{D}_{zu} \end{bmatrix}, \tag{49}$$

and problem (47) can be recast into

$$\begin{aligned} \text{Find} \quad & K_\mathrm{i} \\ \text{subjected to} \quad & \Sigma_\mathrm{i} \star K_\mathrm{i} \text{ is stable} \\ & \| T_{\bar{z}\bar{w}}^{\Sigma_\mathrm{i} \star K_\mathrm{i}}(s) \|_\infty < 1. \end{aligned} \tag{50}$$

Problem (50) is a standard $\mathcal{H}_\infty$ control problem that can be solved using LMI as described for instance in [37], leading to the following theorem.

Theorem 1. *Problem (50) admits a solution if and only if there exist matrices $X = X^T > 0$, $Y = Y^T > 0$, $\hat{A}$, $\hat{B}$, $\hat{C}$, and $\hat{D}$ such that*

$$\begin{bmatrix} \mathrm{Sym}\{\tilde{A}X + \tilde{B}\hat{C}\} & \star & \star & \star \\ \hat{A} + (\tilde{A} + \tilde{B}\hat{D}\tilde{C})^T & \mathrm{Sym}\{Y\tilde{A} + \hat{B}\tilde{C}\} & \star & \star \\ (\tilde{B}_{\bar{w}} + \tilde{B}\hat{D}\tilde{D}_{y\bar{w}})^T & (Y\tilde{B}_{\bar{w}} + \hat{B}\tilde{D}_{y\bar{w}})^T & -I & \star \\ \tilde{C}_{\bar{z}}X + \tilde{D}_{\bar{z}u}\hat{C} & \tilde{C}_{\bar{z}} + \tilde{D}_{\bar{z}u}\hat{D}\tilde{C} & \tilde{D}_{\bar{z}\bar{w}}\hat{D}\tilde{D}_{y\bar{w}} & -I \end{bmatrix} < 0, \quad \begin{bmatrix} X & I \\ I & Y \end{bmatrix} > 0. \tag{51}$$

Let X, Y, $\hat{A}$, $\hat{B}$, $\hat{C}$, and $\hat{D}$ the solutions of LMI feasibility problem (51). Then state space matrices of the control law K_i can be computed as follows:

1. *Choose M and N such that $MN^T = I - XY$ (choose for instance $M = (XY - I)^{\frac{1}{2}}$ and $N = -M^T$);*
2. *Find $\tilde{A}_K$, $\tilde{B}_K$, $\tilde{C}_K$, and $\tilde{D}_K$ according to:*

$$\tilde{D}_K = \hat{D}, \tag{52}$$
$$\tilde{C}_K = (\hat{C} - \tilde{D}_K\tilde{C}X)M^{-T}, \tag{53}$$
$$\tilde{B}_K = N^{-1}(\hat{B} - Y\tilde{B}\tilde{D}_K), \tag{54}$$
$$\tilde{A}_K = N^{-1}(\hat{A} - N\tilde{B}_K\tilde{C}X - Y\tilde{B}\tilde{C}_K M^T - Y(\tilde{A} + \tilde{B}\tilde{D}_K\tilde{C})X)M^{-T}. \tag{55}$$

Please note that considering problem (50) instead of problem (47) introduces additional conservatism as structure of the uncertainty is ignored. Nevertheless, if LMI constraints (51) are feasible, then the integer order control law retrieved using Theorem 1 is also a solution for problem (50), and thus a guaranteed control law with respect to the original fractional-order model Σ_f.

2.2.2 $\mathcal{H}_2$ control based on the uncertain integer order model

When $\mathcal{H}_2$ norm is considered, problem (6) writes

$$\begin{aligned} \tilde{\gamma}_2 = \min_{K_\mathrm{i}} \quad & \|T_{zw}^{\Sigma_\mathrm{i}\star K_\mathrm{i}}(s)\|_2 \\ \text{subjected to} \quad & \Sigma_\mathrm{i} \star K_\mathrm{i} \text{ is stable} \\ & \|T_{z_\Delta w_\Delta}^{\Sigma_\mathrm{i}\star K_\mathrm{i}}(s)\|_\infty < 1. \end{aligned} \tag{56}$$

Problem (56) is a $\mathcal{H}_2$ synthesis problem under the $\mathcal{H}_\infty$ constraint, which is a particular case of multiobjective synthesis. If $\mathcal{H}_2$ or $\mathcal{H}_\infty$ synthesis problems considered separately admit necessary and sufficient LMI solutions, the multiobjective problem cannot be solved without conservatism. A simple solution, known in the literature as quadratic synthesis, consists in using common Lyapunov matrices for the $\mathcal{H}_2$ and $\mathcal{H}_\infty$ problems. This approach, borrowed from [37], is reminded in the next theorem.

Theorem 2. *Let the optimization problem*

$$\Gamma_2 = \min_{X=X^T>0, Y=Y^T>0, \hat{A}, \hat{B}, \hat{C}, \hat{D}, Q} \quad \mathrm{trace}(Q) \tag{57}$$

under LMI constraints (58)–(60) *and equality constraint* (61) *defined as*

$$\begin{bmatrix} \mathrm{Sym}\{\tilde{A}X + \tilde{B}\hat{C}\} & \star & \star & \star \\ \hat{A} + (\tilde{A} + \tilde{B}\hat{D}C)^T & \mathrm{Sym}\{Y\tilde{A} + \hat{B}\tilde{C}\} & \star & \star \\ (\tilde{B}_\Delta + \tilde{B}\hat{D}\tilde{D}_{yw_\Delta})^T & (Y\tilde{B}_\Delta + \hat{B}\tilde{D}_{yw_\Delta})^T & -I & \star \\ \tilde{C}_{z_\Delta}X + \tilde{D}_{z\Delta u}\hat{C} & \tilde{C}_{z_\Delta} + \tilde{D}_{z_\Delta u}\hat{D}\tilde{C} & \tilde{D}_{z_\Delta w_\Delta}\hat{D}\tilde{D}_{yw_\Delta} & -I \end{bmatrix} < 0, \tag{58}$$

$$\begin{bmatrix} \tilde{A}X + X\tilde{A}^T + \tilde{B}\hat{C} + (\tilde{B}\hat{C})^T & \star & \star \\ \hat{A} + (\tilde{A} + \tilde{B}\hat{D}\tilde{C})^T & \tilde{A}^TY + Y\tilde{A} + \hat{B}\tilde{C} + (\hat{B}\tilde{C})^T & \star \\ (\tilde{B}_w + \tilde{B}\hat{D}\tilde{D}_{yw_2})^T & (Y\tilde{B}_w + \hat{B}\tilde{D}_{yw_2})^T & -I \end{bmatrix} < 0, \tag{59}$$

$$\begin{bmatrix} X & \star & \star \\ I & Y & \star \\ \tilde{C}_2X + \tilde{D}_{z_2u}\hat{C} & \tilde{C}_z + \tilde{D}_{z_2u}\hat{D}\tilde{C} & Q \end{bmatrix} > 0, \tag{60}$$

$$\tilde{D}_{z_2w_2} + \tilde{D}_{z_2u}\hat{D}\tilde{D}_{yw_2} = 0. \tag{61}$$

Let $\Gamma_2, X, Y, \hat{A}, \hat{B}, \hat{C}, \hat{D}$, *and* Q *be the solutions of optimization problems* (57)–(61). *State space matrices* $\tilde{A}_K$, $\tilde{B}_K$, $\tilde{C}_K$, *and* $\tilde{D}_K$ *of the control law can be determined as in Theorem 1. Integer order control law* K_i *whose matrices are given by* (52)–(55) *is a suboptimal controller for optimization problem* (56).

Theorem 2 provides a LMI formulation for the design of an integer order control law K_i that guarantees an $\mathcal{H}_2$ cost for the uncertain integer order closed-loop system, and thus for the original fractional-order model Σ_f. Control law K_i remains however suboptimal for two main reasons: the uncertainty that is artificially introduced when reformulating the fractional model as an uncertain integer order model; the use of a quadratic approach in order to obtain a LMI formulation for the multiobjective synthesis problem. Hence, according to notation introduced previously, achieved $\mathcal{H}_2$ cost is such that

$$\gamma_2 < \tilde{\gamma}_2 < \sqrt{\Gamma_2}. \tag{62}$$

As detailed in Section 2.1, the size of the uncertainty can be reduced by augmenting the order of the nominal model. The value of $\tilde{\gamma}_2$ will be closer to γ_2 when this order is high. However, such an augmentation of the order will directly impact the order of control law K_i. Difference between $\sqrt{\Gamma_2}$ and $\tilde{\gamma}_2$ can be reduced by introducing additional Lyapunov variables dedicated to the $\mathcal{H}_2$ and $\mathcal{H}_\infty$ constraints as described for instance in [11].

3 Methods dedicated to commensurate fractional-order models

Control law design results of Section 2 rely on the reformulation of the fractional- order model as an uncertain integer order model. The optimization problem (5) is then recast as problem (6) that can be solved using efficient and reliable robust control techniques developed for integer order models.

A drawback of this approach is that solving problem (6) only allows to obtain suboptimal control laws because of the uncertainty that is artificially introduced in the model. In this section are thus proposed control law design techniques specifically dedicated to systems described by commensurate fractional-order models and that allow to directly address optimization problem (5).

This section is organized as follows. The $\mathcal{H}_\infty$ control problem is first considered. LMI methods for the computation of the $\mathcal{H}_\infty$ norm are proposed in Section 3.1 and are used in Section 3.2 for the design of control laws. Finally, some remarks on the $\mathcal{H}_2$ control problem are given in Section 3.3.

3.1 LMI formulations for the computation of the $\mathcal{H}_\infty$ norm

In order to address problem (5) with $\mathcal{H}_\infty$ norm in the optimization criterion:

$$\begin{aligned} \gamma_\infty = \min_{K_\mathrm{f}} \quad & \|T_{zw}^{\Sigma_\mathrm{f}\star K_\mathrm{f}}(s)\|_\infty \\ \text{subjected to} \quad & \Sigma_\mathrm{f} \star K_\mathrm{f} \text{ is stable,} \end{aligned} \tag{63}$$

the first step is to obtain a LMI formulation for the analysis problem, that is, for a given control law K_f, an LMI formulation to evaluate $\|T_{zw}^{\Sigma_f\star K_f}(s)\|_\infty$. To this purpose, let a pseudo state space representation associated to $T_{zw}^{\Sigma_f\star K_f}(s)$:

$$\Sigma_f \star K_f : \begin{cases} \mathfrak{D}_0^\nu \hat{x}(t) = \mathcal{A}\hat{x}(t) + \mathcal{B}_w w(t), \\ z(t) = \mathcal{C}_z\hat{x}(t) + \mathcal{D}_{zw} w(t), \end{cases} \quad \{x(t), w(t)\} = 0 \; \forall\, t < 0. \tag{64}$$

3.1.1 Computation of the $\mathcal{H}_\infty$ norm based on a pseudo-Hamiltonian matrix

The $\mathcal{H}_\infty$ norm of an integer order model can be computed iteratively by evaluating, at each step of the algorithm, the eigenvalues localization of a particular Hamiltonian matrix. This method is known as γ-iteration [3]. This problem can also be formulated in terms of LMI, leading to the bounded real lemma [18]. The extension of this approach to fractional models is discussed in this section.

According to Definition 1, the $\mathcal{H}_\infty$ norm of the transfer $T_{zw}^{\Sigma_f\star K_f}(s)$ of stable model $\Sigma_f \star K_f$ defined by (64) is less than γ if and only if

$$\forall\, \omega \in \mathbb{R}, \quad \max_{i=\{1\cdots \min(m_w,p_z)\}} \sqrt{\lambda_i(T_{zw}^{\Sigma_f\star K_f}(j\omega)^* T_{zw}^{\Sigma_f\star K_f}(j\omega))} < \gamma \tag{65}$$

which can be rewritten as a positive definiteness problem:

$$\forall\, \omega \in \mathbb{R}, \quad \Phi(j\omega) = (\gamma^2 I - T_{zw}^{\Sigma_f\star K_f}(j\omega)^* T_{zw}^{\Sigma_f\star K_f}(j\omega)) > 0. \tag{66}$$

Inequality (66) is of infinite dimension as it depends on $\omega \in \mathbb{R}$. However, noticing that function $\Phi(j\omega)$ is continuous and positive when $\omega \to \infty$, inequality (66) is equivalent to the nonsingularity of $\Phi(j\omega)$. Hence, the problem is equivalent to verify if $\Phi(s) = \gamma^2 I - T_{zw}^{\Sigma_f\star K_f}(s)^* T_{zw}^{\Sigma_f\star K_f}(s)$ has no pure imaginary zero, that is, $\Phi(s)^{-1}$ has no pure imaginary pole, or equivalently that $\Phi(s)^{-1}$ has no s^ν-pole on $\mathbb{C}_{\nu 0} = \{(j\omega)^\nu = |\omega|^\nu e^{\mathrm{sign}(\omega)\nu j\frac{\pi}{2}}, \omega \in \mathbb{R}\}$.

Let H_γ be the dynamic matrix associated to a pseudo state space description of $\Phi(s)^{-1}$ when $\omega > 0$ and H_γ^- the one obtained when $\omega < 0$. $\Phi(s)^{-1}$ has no s^ν-pole on $\mathbb{C}_{\nu 0}$ if H_γ has no eigenvalue on $\mathbb{C}_{\nu 0}^+ = \{(j\omega)^\nu, \omega \in \mathbb{R}^+\}$ and if H_γ^- has no eigenvalue on $\mathbb{C}_{\nu 0}^- = \{(j\omega)^\nu, \omega \in \mathbb{R}^-\}$. Since $T_{zw}^{\Sigma_f\star K_f}(j\omega)$ is an even function of ω, it is possible to show that the eigenvalues of H_γ and H_γ^- are conjugate. H_γ has thus an eigenvalue on segment $\mathbb{C}_{\nu 0}^+$ if and only if H_γ^- has an eigenvalue on segment $\mathbb{C}_{\nu 0}^-$. Hence, it is only required to check if H_γ has eigenvalues on $\mathbb{C}_{\nu 0}^+$, as indicated in the following theorem.

Theorem 3. *Let the stable pseudo state space description* (64), *a real positive number* $\gamma > \overline{\sigma}(\mathcal{D}_{zw})$ *et* $R = (\gamma^2 I - \mathcal{D}_{zw}^T\mathcal{D}_{zw})^{-1}$. *The inequality* $\|T_{zw}^{\Sigma_f\star K_f}(s)\|_\infty < \gamma$ *holds if and only if the pseudo-Hamiltonian matrix:*

$$H_\gamma \triangleq \begin{bmatrix} \mathcal{A} + \mathcal{B}_w R\mathcal{D}_{zw}^T\mathcal{C}_z & \mathcal{B}_w R\mathcal{B}_w^T \\ e^{\nu j\pi}\mathcal{C}_z^T(I + \mathcal{D}_{zw}R\mathcal{D}_{zw}^T)\mathcal{C}_z & e^{\nu j\pi}(\mathcal{A}^T + \mathcal{C}_z^T\mathcal{D}_{zw}R\mathcal{B}_w^T) \end{bmatrix} \tag{67}$$

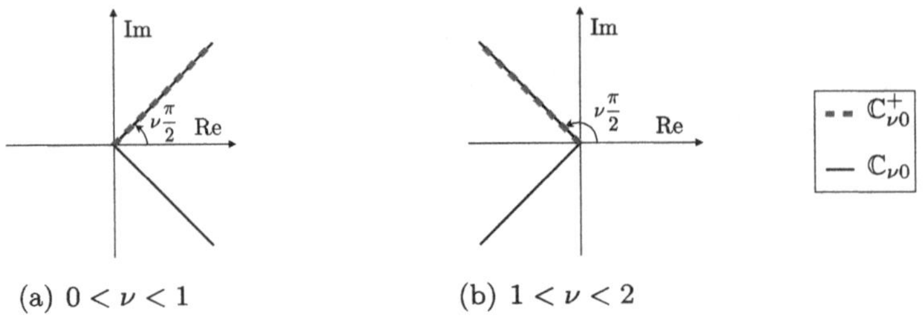

Figure 7: Set $\mathbb{C}_{\nu 0}$ and half-line $\mathbb{C}^+_{\nu 0}$ according to commensurate order ν.

has no eigenvalue on half-line $\mathbb{C}^+_{\nu 0}$ represented in Figure 7 and defined as follows:

$$\mathbb{C}^+_{\nu 0} = \{(j\omega)^\nu = |\omega|^\nu e^{\nu j\frac{\pi}{2}}, \omega \in \mathbb{R}^+\}. \tag{68}$$

A dichotomy on γ then allows to retrieve the $\mathcal{H}_\infty$ norm of the fractional-order model by the iterative application of Theorem 3.

In the case of integer order models, there exists a result similar to the one of Theorem 3 for which the pseudo-Hamiltonian matrix H_γ is replaced by a Hamiltonian matrix H and the half-line $\mathbb{C}^+_{\nu 0}$ by the imaginary axis. In that case, the following theorem is used.

Theorem 4. *Let $\tilde{A} = \tilde{A}^* \in \mathbb{C}^{n\times n}, S = S^T > 0 \in \mathbb{R}^{n\times n}$ and $Q = Q^T \in \mathbb{R}^{n\times n}$. The Hamiltonian matrix $H = \begin{bmatrix} \tilde{A} & S \\ -Q & -\tilde{A}^* \end{bmatrix}$ has no eigenvalue on the imaginary axis iff $\exists\, X = X^* \in \mathbb{C}^{n\times n}$ satisfying the Riccati inequality:*

$$\tilde{A}^* X + X\tilde{A} + XSX + Q < 0. \tag{69}$$

In the case of fractional-order models, the matrix H_γ is complex and does not have the properties of a Hamiltonian matrix. Moreover, when $\|T^{\Sigma_f \star K_f}_{zw}(s)\|_\infty > \gamma$, the eigenvalues of H_γ are on $\mathbb{C}^+_{\nu 0}$ and not on the imaginary axis. Two transformations are thus applied to H_γ. First, matrix $H'_\gamma = e^{j\theta} H_\gamma$ is obtained by rotating H_γ with an angle of $\theta = (1-\nu)\frac{\pi}{2}$. Hence, if H_γ has no eigenvalue on $\mathbb{C}^+_{\nu 0}$, then H'_γ has no eigenvalue on half-line $\mathbb{C}^+_0$ of pure imaginary numbers with a positive imaginary part. A similarity transformation with $U = \mathrm{diag}(I, rI)$, $r = e^{j\theta}$, then allows to obtain a Hamiltonian matrix $H = UH'_\gamma U^{-1}$. Theorem 4 can thus be applied with

$$\tilde{A} = r(\mathcal{A} + \mathcal{B}_w R \mathcal{D}^T_{zw} \mathcal{C}_z), \quad S = \mathcal{B}_w R \mathcal{B}^T_w, \quad Q = \mathcal{C}^T_z (I + \mathcal{D}_{zw} R \mathcal{D}^T_{zw}) \mathcal{C}_z. \tag{70}$$

Applying twice the Schur complement leads to Theorem 5.

Theorem 5. *Let the stable pseudo state space description (64), a real positive number $\gamma > \overline{\sigma}(\mathcal{D}_{zw})$ and the following inequality with $r = e^{(1-\nu)j\frac{\pi}{2}}$:*

$$\begin{bmatrix} \bar{r}\mathcal{A}^T X + Xr\mathcal{A} & \star & \star \\ \mathcal{B}^T_w X & -\gamma^2 I & \star \\ r\mathcal{C}_z & \mathcal{D}_{zw} & -I \end{bmatrix} < 0. \tag{71}$$

Case 1 $(1 < \nu < 2)$: $\|T_{zw}^{\Sigma_f \star K_f}(s)\|_\infty < \gamma$ *iff* $\exists X = X^* \in \mathbb{C}^{n\times n}$ *s.t.* (71).
Case 2 $(0 < \nu < 1)$: *if* $\exists X = X^* \in \mathbb{C}^{n\times n}$ *s.t.* (71), *then* $\|T_{zw}^{\Sigma_f \star K_f}(s)\|_\infty < \gamma$.

As shown in [13], the fact that Theorem 4 tests the belonging of H eigenvalues to an imaginary axis and not only to $\mathbb{C}_0^+$ introduces some conservatism in Theorem 5 in the case $0 < \nu < 1$. Moreover, if constraint $X > 0$ is added to Theorem 5, stability of the model is also attested. When $1 < \nu < 2$, this does not add any conservatism to the approach. When $0 < \nu < 1$, the additional conservatism is important.

Finally, the term γ^2 appearing linearly in LMI (71), Theorem 5 can be recast into an optimization problem leading to $\|T_{zw}^{\Sigma_f \star K_f}(s)\|_\infty$.

Remark. Another LMI formulation based on the pseudo-Hamiltonian matrix H_γ was proposed in [14]. This formulation, obtained using the generalized LMI theory [1, 6], is however difficult to extend to synthesis as it involves several complex variables and does not ensure the stability of the system.

3.1.2 LMI methods based on the generalized KYP lemma

Methods proposed in this section follow the works of [22] and use the generalized Kalman–Yakubovič–Popov (KYP) lemma to characterize the domain $\mathbb{C}_{\nu 0}$.

Lemma 1 ([20]). *Let the matrices* $A \in \mathbb{R}^{n\times n}$, $B \in \mathbb{R}^{n\times m}$, $\Theta = \Theta^* \in \mathbb{C}^{m\times m}$, $\Phi = \Phi^* \in \mathbb{C}^{2\times 2}$, $\Psi = \Psi^* \in \mathbb{C}^{2\times 2}$ *and the set* Λ *defined as*

$$\Lambda(\Phi, \Psi) := \left\{\lambda \in \mathbb{C} : \begin{bmatrix}\lambda \\ 1\end{bmatrix}^* \Phi \begin{bmatrix}\lambda \\ 1\end{bmatrix} = 0, \begin{bmatrix}\lambda \\ 1\end{bmatrix}^* \Psi \begin{bmatrix}\lambda \\ 1\end{bmatrix} \geq 0\right\}. \tag{72}$$

Consider the following hypothesis:

1. *if* $H(\lambda) = (\lambda I_n - A)^{-1}B$, *then*

$$\begin{bmatrix}H(\lambda) \\ I_m\end{bmatrix}^* \Theta \begin{bmatrix}H(\lambda) \\ I_m\end{bmatrix} < 0, \quad \forall \lambda \in \Lambda; \tag{73}$$

2. *there exist* $P, Q \in \mathbb{H}_n$, $Q > 0$ *such that*

$$\begin{bmatrix}A & B \\ I_n & 0\end{bmatrix}^* (\Phi \otimes P + \Psi \otimes Q) \begin{bmatrix}A & B \\ I_n & 0\end{bmatrix} + \Theta < 0. \tag{74}$$

Then 2. $\Rightarrow$ 1. *is always true. Moreover, if* Λ *is a curve in* $\mathbb{C}$, *then* 2. $\Leftrightarrow$ 1.

According to Definition 2, $\mathcal{H}_\infty$ norm of model (64) is given by

$$\|T_{zw}^{\Sigma_f \star K_f}(s)\|_\infty = \sup_{\omega \geq 0} \overline{\sigma}(T_{zw}^{\Sigma_f \star K_f}(j\omega)) = \sup_{\omega \geq 0} \left[\mathcal{C}_z(\Lambda(\omega)I - \mathcal{A})^{-1}\mathcal{B}_w + \mathcal{D}_{zw}\right] \tag{75}$$

where $\lambda(\omega) = \mathrm{e}^{\mathrm{j}\nu\frac{\pi}{2}}\omega^{\nu}$, $\omega \geq 0$, is a half-line of the complex plane that can be represented by the set Λ defined according to relation (72) with

$$\Phi = \begin{bmatrix} 0 & \mathrm{e}^{\mathrm{j}\theta} \\ \mathrm{e}^{-j\theta} & 0 \end{bmatrix}, \quad \Psi = \begin{bmatrix} 0 & 1-\nu \\ 1-\nu & 0 \end{bmatrix}. \tag{76}$$

Moreover, according to equation (66), $\|T_{zw}^{\Sigma_f \star K_f}(s)\|_\infty < \gamma$ also writes

$$T_{zw}(j\omega)T_{zw}^{*}(j\omega) - \gamma^2 I < 0, \quad \forall\omega \geq 0 \quad \Leftrightarrow \quad \begin{bmatrix} H(\lambda) \\ I_m \end{bmatrix}^{*} \Theta \begin{bmatrix} H(\lambda) \\ I_m \end{bmatrix} < 0, \quad \forall\lambda \in \Lambda \tag{77}$$

with

$$H(\lambda) = (\lambda I_n - \mathcal{A}^T)^{-1}\mathcal{C}_z^T, \quad \Theta = \begin{bmatrix} \mathcal{B}_w\mathcal{B}_w^T & \mathcal{B}_w\mathcal{D}_{zw}^T \\ \mathcal{D}_{zw}\mathcal{B}_w^T & \mathcal{D}_{zw}\mathcal{D}_{zw}^T - \gamma^2 I_n \end{bmatrix}. \tag{78}$$

Hence, according to Lemma 1, relation (77) is equivalent to the existence of $P = P^* \in \mathbb{C}^{n \times n}$ and $Q = Q^* \in \mathbb{C}^{n \times n}$, $Q > 0$, such that

$$\begin{aligned} &\begin{bmatrix} \mathcal{A} & I_n \\ \mathcal{C}_z & 0 \end{bmatrix} \left(\begin{bmatrix} 0 & \mathrm{e}^{\mathrm{j}\theta} \\ \mathrm{e}^{-\mathrm{j}\theta} & 0 \end{bmatrix} \otimes P + \begin{bmatrix} 0 & 1-\nu \\ 1-\nu & 0 \end{bmatrix} \otimes Q \right) \begin{bmatrix} \mathcal{A}^T & \mathcal{C}_z^T \\ I_n & 0 \end{bmatrix} \\ &+ \begin{bmatrix} \mathcal{B}_w\mathcal{B}_w^T & \mathcal{B}_w\mathcal{D}_{zw}^T \\ \mathcal{D}_{zw}\mathcal{B}_w^T & \mathcal{D}_{zw}\mathcal{D}_{zw}^T - \gamma^2 I_n \end{bmatrix} < 0, \end{aligned} \tag{79}$$

where $\theta = (1-\nu)\frac{\pi}{2}$. Applying the Schur complement, this inequality becomes

$$\begin{bmatrix} \mathrm{Sym}\{\mathcal{A}(\mathrm{e}^{\mathrm{j}\theta}P + (1-\nu)Q)\} & * & * \\ \mathcal{C}_z(\mathrm{e}^{\mathrm{j}\theta}P + (1-\nu)Q) & -\gamma I & * \\ \mathcal{B}_w^T & \mathcal{D}_{zw}^T & -\gamma I \end{bmatrix} < 0. \tag{80}$$

This result, proposed in [22], gives a necessary and sufficient condition to compute the $\mathcal{H}_\infty$ norm of a stable fractional-order model whatever the commensurate order ν. As shown in [13], Theorem 5 can be simplified in the case $1 < \nu < 2$ as it is possible to set $Q = 0$ without adding conservatism, which leads to the following result.

Theorem 6. *Let the stable pseudo state space description* (64) *and a real positive number* $\gamma > \overline{\sigma}(\mathcal{D}_{zw})$. $\|T_{zw}^{\Sigma_f \star K_f}(s)\|_\infty < \gamma$ *iff the following LMI, where* $\theta = (1-\nu)\frac{\pi}{2}$, *are satisfied.*
Case 1 ($1 < \nu < 2$)*:* $\exists P = P^* \in \mathbb{C}^{n \times n}$ *such that*

$$\begin{bmatrix} \mathrm{Sym}\{\mathcal{A}\mathrm{e}^{\mathrm{j}\theta}P\} & * & * \\ \mathcal{C}_z\mathrm{e}^{\mathrm{j}\theta}P & -\gamma I & * \\ \mathcal{B}_w^T & \mathcal{D}_{zw}^T & -\gamma I \end{bmatrix} < 0. \tag{81}$$

Case 2 ($0 < \nu < 1$)*:* $\exists P = P^* \in \mathbb{C}^{n \times n}$, $Q = Q^* \in \mathbb{C}^{n \times n}$ *such that*

$$Q > 0, \quad \begin{bmatrix} \mathrm{Sym}\{\mathcal{A}(\mathrm{e}^{\mathrm{j}\theta}P + (1-\nu)Q)\} & * & * \\ \mathcal{C}_z(\mathrm{e}^{\mathrm{j}\theta}P + (1-\nu)Q) & -\gamma I & * \\ \mathcal{B}_w^T & \mathcal{D}_{zw}^T & -\gamma I \end{bmatrix} < 0. \tag{82}$$

In spite of the usefulness of Theorem 6 for analysis purposes, the necessity to verify the model's stability *a priori* limits its interest for synthesis purposes. In order to verify stability simultaneously, the modified version of Lemma 1 is used.

Lemma 2 ([22]). *If the set $\Lambda(\Phi, \Psi)$ of Lemma 1 is replaced by the set,*

$$\Upsilon(\Phi, \Psi) := \left\{ \lambda \in \mathbb{C} : \begin{bmatrix} \lambda \\ 1 \end{bmatrix}^* \Phi \begin{bmatrix} \lambda \\ 1 \end{bmatrix} \geq 0, \begin{bmatrix} \lambda \\ 1 \end{bmatrix}^* \Psi \begin{bmatrix} \lambda \\ 1 \end{bmatrix} \geq 0 \right\} \tag{83}$$

then condition (73) *is true* $\forall \lambda \in \Upsilon$ *if* $\exists P, Q > 0$ *satisfying LMI* (74).

Choosing matrices Φ and Ψ according to (76), the next theorem is obtained.

Theorem 7 ([22]). *Let the pseudo state space description* (64) *and a real positive number* $\gamma > \bar{\sigma}(\mathcal{D}_{zw})$. *The model is stable and* $\|T_{zw}^{\Sigma_f \star K_f}(s)\|_\infty < \gamma$ *if the following LMI, where* $\theta = (1 - \nu)\frac{\pi}{2}$, *are satisfied.*

Case 1 ($1 < \nu < 2$): $\exists P = P^* \in \mathbb{C}^{n \times n}$ *such that*

$$P > 0, \quad \begin{bmatrix} \mathrm{Sym}\{\mathcal{A} e^{j\theta} P\} & * & * \\ \mathcal{C}_z e^{j\theta} P & -\gamma I & * \\ \mathcal{B}_w^T & \mathcal{D}_{zw}^T & -\gamma I \end{bmatrix} < 0. \tag{84}$$

Case 2 ($0 < \nu < 1$): $\exists P = P^* \in \mathbb{C}^{n \times n}$, $Q = Q^* \in \mathbb{C}^{n \times n}$ *such that*

$$P > 0, \quad Q > 0, \quad \begin{bmatrix} \mathrm{Sym}\{\mathcal{A}(e^{j\theta} P + e^{-j\theta} Q)\} & * & * \\ \mathcal{C}_z (e^{j\theta} P + e^{-j\theta} Q) & -\gamma I & * \\ \mathcal{B}_w^T & \mathcal{D}_{zw}^T & -\gamma I \end{bmatrix} < 0. \tag{85}$$

Moreover, in case 1, *LMI condition* (84) *is also necessary.*

As shown in [13], the number of variables can be reduced without additional conservatism by choosing $X = P$ and $\overline{X} = Q$.

Theorem 8. *Let the pseudo state space description* (64) *and a real positive number* $\gamma > \bar{\sigma}(\mathcal{D}_{zw})$. *The model is stable and* $\|T_{zw}^{\Sigma_f \star K_f}(s)\|_\infty < \gamma$ *if the following LMI, where* $\theta = (1-\nu)\frac{\pi}{2}$, *are satisfied.*

Case 1 ($1 < \nu < 2$): $\exists P = P^* \in \mathbb{C}^{n \times n}$ *such that*

$$P > 0, \quad \begin{bmatrix} \mathrm{Sym}\{\mathcal{A} e^{j\theta} P\} & * & * \\ \mathcal{C}_z e^{j\theta} P & -\gamma I & * \\ \mathcal{B}_w^T & \mathcal{D}_{zw}^T & -\gamma I \end{bmatrix} < 0. \tag{86}$$

Case 2 ($0 < \nu < 1$): $\exists X = X^* \in \mathbb{C}^{n \times n}$ *such that*

$$X > 0, \quad \begin{bmatrix} \mathrm{Sym}\{\mathcal{A}(e^{j\theta} X + e^{-j\theta} \overline{X})\} & * & * \\ \mathcal{C}_z (e^{j\theta} X + e^{-j\theta} \overline{X}) & -\gamma I & * \\ \mathcal{B}_w^T & \mathcal{D}_{zw}^T & -\gamma I \end{bmatrix} < 0. \tag{87}$$

Moreover, in case 1, *LMI condition* (86) *is also necessary.*

In the case $0 < \nu < 1$, condition (87) exhibits the real term $(e^{j\theta}X + e^{-j\theta}\overline{X})$ which is more adapted to synthesis than the one of Theorem 7.

3.2 LMI formulations for the $\mathcal{H}_\infty$ control

LMI analysis conditions of Section 3.1 are now used to address the design of $\mathcal{H}_\infty$ control laws. The idea is to apply these conditions to the closed-loop model $\Sigma_f \star K_f$. The parameters of the control law K_f are now variables and obtained matrix inequalities are thus nonlinear. Linearizing change of variables must be determined in order to obtain synthesis LMI conditions.

Moreover, the control law must guarantee both the performances and the stability of the closed-loop system. Hence, only LMI conditions that both compute the $\mathcal{H}_\infty$ norm and verify the stability can be used.

3.2.1 Pseudo state feedback control law

In this section, Theorem 5 (based on the pseudo-Hamiltonian matrix) and Theorem 6 (based on the KYP lemma) are used for the design of a pseudo state feedback control law K_f of the form:

$$K_f: \quad u(t) = Kx(t), \tag{88}$$

where $K \in \mathbb{R}^{m \times n}$ is a constant matrix gain. The closed-loop model $\Sigma_f \star K_f$, with Σ_f given by (4) and K_f given by (88), writes as (64) with

$$\mathcal{A} = A + BK, \quad \mathcal{B}_w = B_w, \quad \mathcal{C}_z = C_z + D_{zu}K, \quad \mathcal{D}_{zw} = D_{zw}. \tag{89}$$

Consider Theorem 5 and multiply inequality (71) on the left and on the right by matrix $\operatorname{diag}\{X^{-1}, I, I\}$. Posing $P = X^{-1}$, the following LMI is obtained:

$$\begin{bmatrix} \operatorname{Sym}\{r\mathcal{A}P\} & \star & \star \\ \mathcal{B}_w^T & -\gamma^2 I & \star \\ r\,\mathcal{C}_z P & \mathcal{D}_{zw} & -I \end{bmatrix} < 0. \tag{90}$$

Replacing $\mathcal{A}$, $\mathcal{B}_w$, $\mathcal{C}_z$, and $\mathcal{D}_{zw}$ by their expressions given by (89) and using the linearizing change of variables $Q = KP$ allows to obtain the following theorem.

Theorem 9 ([16]). *Fractional order model Σ_f given by* (4) *is stabilizable by the pseudo state feedback control law* (88) *if $\exists\, P = P^* \in \mathbb{R}^{n \times n} > 0$ and $Q \in \mathbb{R}^{m \times n}$ s.t.*

$$\begin{bmatrix} \operatorname{Sym}\{rAP + rB_u Q\} & \star & \star \\ B_w^T & -\gamma^2 I & \star \\ rC_z P + rD_{zu} Q & D_{zw} & -I \end{bmatrix} < 0, \quad r = e^{j(1-\nu)\frac{\pi}{2}}. \tag{91}$$

The control law gain $K = QP^{-1}$ ensures that $\|T_{zw}^{\Sigma_f \star K_f}(s)\|_\infty < \gamma$.

Remark. Matrices P and Q of Theorem 9 are real although LMI (91) is deduced from complex LMI (71). This choice allows to design a real control law gain but introduces some conservatism. Hence, LMI (71) is a necessary and sufficient analysis condition for $1 < \nu < 2$ but LMI (91) is only a sufficient synthesis condition. Along the same lines, when $0 < \nu < 1$, the conservatism stemming from using real variables increases the one that was already in the analysis LMI.

Another design method is obtained by replacing the closed-loop pseudo state space matrices by their expressions given by (89) in LMI (86) and (87) of Theorem 8 and using the linearizing change of variables $Q = KP$ and $Y = K(\mathrm{e}^{\mathrm{j}\theta}X + \mathrm{e}^{-j\theta}\overline{X})$.

Theorem 10. *Fractional-order model Σ_f given by* (4) *is stabilizable by the pseudo state feedback control law* (88) *if the following LMI conditions are satisfied.*

Case 1 ($1 < \nu < 2$)*: if there exist matrices $P \in \mathbb{R}^{n\times n} > 0$ and $Q \in \mathbb{R}^{m\times n}$ s.t.*

$$\begin{bmatrix} \mathrm{Sym}\{A\mathrm{e}^{\mathrm{j}\theta}P + B_u\mathrm{e}^{\mathrm{j}\theta}Q\} & * & * \\ C_z\mathrm{e}^{\mathrm{j}\theta}P + D_{zu}\mathrm{e}^{\mathrm{j}\theta}Q & -\gamma I & * \\ B_w^T & D_{zw}^T & -\gamma \end{bmatrix} < 0, \tag{92}$$

then $K = QP^{-1}$ is a stabilizing controller gain s.t. $\|T_{zw}^{\Sigma_\mathrm{f}\star K_\mathrm{f}}(s)\|_\infty < \gamma$.

Case 2 ($0 < \nu < 1$)*: if there exist matrices $X \in \mathbb{C}^{n\times n} > 0$ and $Y \in \mathbb{R}^{m\times n}$ s.t.*

$$\begin{bmatrix} \mathrm{Sym}\{A(\mathrm{e}^{\mathrm{j}\theta}X + \mathrm{e}^{-j\theta}\overline{X}) + B_uY\} & * & * \\ C_z(\mathrm{e}^{\mathrm{j}\theta}X + \mathrm{e}^{-j\theta}\overline{X}) + D_{zu}Y & -\gamma I & * \\ B_w^T & D_{zw}^T & -\gamma I \end{bmatrix} < 0, \tag{93}$$

then $K = Y(\mathrm{e}^{\mathrm{j}\theta}X + \mathrm{e}^{-j\theta}\overline{X})^{-1}$ is a stabilizing controller gain s.t. $\|T_{zw}^{\Sigma_\mathrm{f}\star K_\mathrm{f}}(s)\|_\infty < \gamma$.

Results of Theorems 9 and 10 are applied to the vibration isolation of a bridge in [17]. They however require that all pseudo state variables are measured. Otherwise, a dynamic output feedback control law may be designed.

3.2.2 Dynamic output feedback control law

Let a fractional-order dynamic output feedback control law defined as

$$K_\mathrm{f} : \begin{cases} \mathfrak{D}_0^\nu x_K(t) = A_Kx_K(t) + B_Ky(t), \\ u(t) = C_Kx_K(t) + D_Ku(t), \end{cases} \quad \{x_K(t), y(t)\} = 0\ \forall\, t < 0. \tag{94}$$

The closed-loop model $\Sigma_\mathrm{f} \star K_\mathrm{f}$, with Σ_f given by (4) and K_f given by (94), writes as (64) with $\hat{x}(t) = \begin{bmatrix} x^T(t) & x_K^T(t)\end{bmatrix}^T$ and

$$\begin{gathered} \mathcal{A} = \begin{bmatrix} A + BD_KC & BC_K \\ B_KC & A_K \end{bmatrix}, \quad \mathcal{B}_w = \begin{bmatrix} B_w + BD_KD_{yw} \\ B_KD_{yw} \end{bmatrix}, \\ \mathcal{C}_z = \begin{bmatrix} C_z + D_{zu}D_KC & D_{zw}C_K \end{bmatrix}, \quad \mathcal{D}_{zw} = D_{zu} + D_{zu}D_KD_{yw}. \end{gathered} \tag{95}$$

The synthesis theorem is obtained by applying to closed-loop model (64) the linearizing change of variables proposed in [37] for integer order models.

Theorem 11 ([15]). *Fractional-order model Σ_f given by (4) is stabilizable by dynamic output feedback control law (94) if there exist matrices $Z = Z^T \in \mathbb{R}^{n\times n}$, $Y = Y^T \in \mathbb{R}^{n\times n}$, $\hat{A} \in \mathbb{R}^{n\times n}$, $\hat{B} \in \mathbb{R}^{n\times p}$, $\hat{C} \in \mathbb{R}^{m\times n}$, and $\hat{D} \in \mathbb{R}^{m\times p}$ s.t.*

$$\begin{aligned}&\begin{bmatrix} \mathrm{Sym}\{\bar{r}(AZ + B\hat{C})\} & \star & \star & \star \\ \bar{r}\hat{A} + r(A + B\hat{D}C)^T & \mathrm{Sym}\{\bar{r}(YA + \hat{B}C)\} & \star & \star \\ (B_w + B\hat{D}D_{yw})^T & (YB_w + \hat{B}D_{yw})^T & -\gamma^2 I & \star \\ r(C_z Z + D_{zu}\hat{C}) & r(C_z + D_{zu}\hat{D}C) & D_{zw} + D_{zu}\hat{D}D_{yw} & -I \end{bmatrix} < 0, \\ &\begin{bmatrix} Z & I \\ I & Y \end{bmatrix} > 0.\end{aligned} \tag{96}$$

Pseudo state space matrices of the control law are then given by:

$$\begin{aligned} D_K &= \hat{D}, \\ C_K &= (\hat{C} - D_K CX)M^{-T}, \\ B_K &= N^{-1}(\hat{B} - YBD_K), \\ A_K &= N^{-1}(\hat{A} - NB_K CX - YBC_K M^T - Y(A + BD_K C)X)M^{-T} \end{aligned} \tag{97}$$

where M and N are nonsingular matrices s.t. $MN^T = I - ZY$ that can be chosen as in Theorem 1. Moreover, control law (94) ensures that $\|T_{zw}^{\Sigma_f \star K_f}(s)\|_\infty < \gamma$.

When $0 < \nu < 1$, Theorem 11 suffers from the same conservatism as Theorem 5 on which it is based. Moreover, as for pseudo state feedback, matrix variables of LMI (96) are chosen to be real in order to obtain a real control law.

Theorem 8 proposes an analysis condition with low conservatism and involves real variables. Its extension to the dynamic output feedback synthesis remains however tedious as the linearizing change of variables used above is not adapted.

3.3 Remarks on the $\mathcal{H}_2$ control problem

In order to tackle problem (5) with $\mathcal{H}_2$ norm in the optimization criterion, one may consider to apply an approach similar to the one of Section 3.2. However, in spite of the analytical formulae in [25], no LMI formulation has been proposed yet for the computation of the $\mathcal{H}_2$ norm of commensurate fractional-order models.

In the case of integer order models, a way to obtain such a LMI formulation involves the grammians computation. This approach is reminded in Section 3.3.1 and its extension to fractional-order models is discussed in Section 3.3.2.

3.3.1 Reminder on the integer order case

Let a stable integer order model Σ_{i} described by its space description:

$$\Sigma_{\mathrm{i}}: \begin{cases} \dot{\tilde{x}}(t) = \tilde{A}\tilde{x}(t) + \tilde{B}_w w(t), \\ z(t) = \tilde{C}_z \tilde{x}(t), \end{cases} \tag{98}$$

whose impulse response $\tilde{g}(t)$ is given by

$$\tilde{h}(t) = \tilde{C}_z \mathrm{e}^{\tilde{A}t} \tilde{B}_w. \tag{99}$$

According to Definition (3), the $\mathcal{H}_2$ norm of Σ_{i} writes

$$\begin{aligned} \|T_{zw}^{\Sigma_{\mathrm{i}}}(s)\|_2^2 &= \int_{t=0}^{\infty} \mathrm{trace}(\tilde{h}^T(t)\tilde{h}(t))dt = \int_{t=0}^{\infty} \mathrm{trace}(\tilde{B}_w^T \mathrm{e}^{\tilde{A}^T t} \tilde{C}_z^T \tilde{C}_z \mathrm{e}^{\tilde{A}t} \tilde{B}_w)dt \\ &= \mathrm{trace}\left(\tilde{B}_w^T \int_{t=0}^{\infty} \mathrm{e}^{\tilde{A}^T t} \tilde{C}_z^T \tilde{C}_z \mathrm{e}^{\tilde{A}t} dt \tilde{B}_w \right) = \mathrm{trace}(\tilde{B}_w^T \tilde{W} \tilde{B}_w), \end{aligned} \tag{100}$$

where $\tilde{W}$ is the observability grammian:

$$\tilde{W} = \int_0^{\infty} \mathrm{e}^{\tilde{A}^T t} \tilde{C}_z^T \tilde{C}_z \mathrm{e}^{\tilde{A}t} dt. \tag{101}$$

Grammian $\tilde{W}$ can be computed as the single solution to Lyapunov equation:

$$\tilde{A}^T \tilde{W} + \tilde{W}\tilde{A} + \tilde{C}_z^T \tilde{C}_z = 0. \tag{102}$$

The fact that $\tilde{W}$ is solution of (102) can be shown by noticing that

$$\frac{d}{dt} \mathrm{e}^{\tilde{A}^T t} \tilde{C}_z^T \tilde{C}_z \mathrm{e}^{\tilde{A}t} = A^T \mathrm{e}^{\tilde{A}^T t} \tilde{C}_z^T \tilde{C}_z \mathrm{e}^{\tilde{A}t} + \mathrm{e}^{\tilde{A}^T t} \tilde{C}_z^T \tilde{C}_z \mathrm{e}^{\tilde{A}t} A. \tag{103}$$

Indeed, integrating w.r.t. time equation (103) between 0 and ∞ leads to

$$\int_0^{\infty} \frac{d}{dt} \mathrm{e}^{\tilde{A}^T t} \tilde{C}_z^T \tilde{C}_z \mathrm{e}^{\tilde{A}t} dt = \int_0^{\infty} (A^T \mathrm{e}^{\tilde{A}^T t} \tilde{C}_z^T \tilde{C}_z \mathrm{e}^{\tilde{A}t} + \mathrm{e}^{\tilde{A}^T t} \tilde{C}_z^T \tilde{C}_z \mathrm{e}^{\tilde{A}t} A)dt, \tag{104}$$

$$\mathrm{e}^{\tilde{A}^T t} \tilde{C}_z^T \tilde{C}_z \mathrm{e}^{\tilde{A}t}|_0^{\infty} = A^T \left(\int_0^{\infty} \mathrm{e}^{\tilde{A}^T t} \tilde{C}_z^T \tilde{C}_z \mathrm{e}^{\tilde{A}t} dt \right) + \left(\int_0^{\infty} \mathrm{e}^{\tilde{A}^T t} \tilde{C}_z^T \tilde{C}_z \mathrm{e}^{\tilde{A}t} dt \right) A. \tag{105}$$

Lyapunov equation (102) is finally obtained thanks to the convergence toward 0 of the term $\mathrm{e}^{\tilde{A}t}$ when $t \to \infty$, consequence of the model stability.

The $\mathcal{H}_2$ norm of Σ_{i} is thus given by (100) where $\tilde{W}$ is the solution of (102), which can be formulated as the LMI optimization problem:

$$\begin{aligned} \|T_{zw}^{\Sigma_{\mathrm{i}}}(s)\|_2^2 = \min_{\tilde{W}=\tilde{W}^T>0} \quad & \mathrm{trace}(\tilde{B}_w^T \tilde{W} \tilde{B}_w) \\ \text{s.t.} \quad & A^T \tilde{W} + \tilde{W}\tilde{A} + \tilde{C}_z^T \tilde{C}_z < 0. \end{aligned} \tag{106}$$

3.3.2 Extension to the fractional-order case

In order to extend to fractional-order models, the method presented in the previous paragraph, let us remind the impulse response of a fractional-order model Σ_f of pseudo state space description:

$$\Sigma_f : \begin{cases} \mathfrak{D}_0^\nu x(t) = Ax(t) + B_w w(t), \\ z(t) = C_z x(t) \end{cases} \tag{107}$$

that can be written, according to [39]:

$$h(t) = C_z t^{\nu-1} \mathrm{E}_{\nu,\nu}(At^\nu) B_w, \tag{108}$$

where E is the two-parameters Mittag-Leffler function defined as

$$\mathrm{E}_{\alpha,\beta}(z) = \sum_{k=0}^{\infty} \frac{z^k}{\Gamma(\alpha k + \beta)}. \tag{109}$$

According to Definition (3), $\mathcal{H}_2$ norm of Σ_f writes

$$\|T_{zw}^{\Sigma_f}(s)\|_2^2 = \int_{t=0}^{\infty} \mathrm{trace}(h^T(t)h(t))dt \tag{110}$$

$$= \mathrm{trace}\left(B_w^T \int_{t=0}^{\infty} \mathrm{E}_{\nu,\nu}(A^T t^\nu) t^{\nu-1} C_z^T C_z t^{\nu-1} \mathrm{E}_{\nu,\nu}(At^\nu) dt B_w \right) \tag{111}$$

$$= \mathrm{trace}(B_w^T W B_w) \tag{112}$$

where W may be defined as a pseudo observability grammian:

$$W = \int_{t=0}^{\infty} \mathrm{E}_{\nu,\nu}(A^T t^\nu) C_z^T t^{\nu-1} C_z t^{\nu-1} \mathrm{E}_{\nu,\nu}(At^\nu) dt. \tag{113}$$

Pseudo observability grammian denomination is chosen as W does not have the properties of an observability grammian in the integer order case (in particular concerning observability). The difficulty is thus to find a linear equation such as (102) that may be used to compute W. Indeed, the Mittag-Leffler function replaces the exponential function and does not have the same properties [31]. The authors of this chapter are currently working on this problem.

Bibliography

[1] O. Bachelier and B. Pradin, Bounds for uncertain matrix root-clustering in a union of subregions, *International Journal of Robust and Nonlinear Control*, **9**(6) (1999), 333–359.

[2] J.-L. Battaglia, O. Cois, L. Puigsegur, and A. Oustaloup, Solving an inverse heat conduction problem using a non-integer identified model, *International Journal of Heat and Mass Transfer*, **44**(14) (2001), 2671–2680.

[3] S. Boyd, B. Venkataramanan, and P. Kabamba, A bisection method for computing the h_∞ norm of a transfer matrix and related problems, *Mathematics of Control, Signals and Systems*, **2**(3) (1989), 207–219.

[4] M. Chevrié, C. Farges, J. Sabatier, F. Guillemard, and L. Pradere, H2-norm for mesh optimization with application to electro-thermal modeling of an electric wire in automotive context, *Communications in Nonlinear Science and Numerical Simulation*, **45** (2017), 176–189.

[5] M. Chevrié, J. Sabatier, C. Farges, and R. Malti, H2-norm computation of a class of implicit fractional transfer functions: application to approximation by integer order models, *International Journal of Dynamics and Control*, **5**(1) (2017), 95–101.

[6] M. Chilali, *Méthodes LMI Pour L'analyse Et la Synthèse Multi-critère*, PhD thesis, Paris IX—Dauphine, Paris, France, 1996.

[7] O. Cois, *Systèmes linéaires non entiers et identification par modèle non entier: application en thermique*, PhD thesis, University of Bordeaux, Bordeaux, France, 7 2002.

[8] K. S. Cole, Electrical conductance of biological systems, in *Proceedings Symposium on Quantitative Biology*, vol. 1, pp. 107–116, Cold Spring Harvor, New York, USA, 1933.

[9] A. Djouambi, A. Charef, and A. Besançon, Optimal approximation, simulation and analog realization of the fundamental fractional order transfer function, *International Journal of Applied Mathematics and Computer Science*, **17** (2007), 455–462.

[10] J. C. Doyle, Synthesis of robust controllers and filters, in *The 22nd IEEE Conference on Decision and Control*, Dec. 1983, pp. 109–114.

[11] Y. Ebihara, D. Peaucelle, and D. Arzelier, *S-Variable Approach to LMI-Based Robust Control*, 1st edition, Communications and Control Engineering, Springer-Verlag, London, 2015.

[12] O. Enacheanu, D. Riu, and N. Retière, Fractional order and fractal modelling of electrical networks, in *IFAC Workshop on Fractional Differentiation and Its Applications (FDA)*, Porto, Portugal, 7 2006, vol. 39, pp. 426–431.

[13] L. Fadiga, *Commande Hinfini à base de modèles non entiers*, PhD thesis, University of Bordeaux, 6 2014.

[14] L. Fadiga, C. Farges, J. Sabatier, and M. Moze, On computation of Hinfinity norm for commensurate fractional order systems, in *50th IEEE Conference on Decision and Control and European Control Conference (CDC-ECC)*, Orlando, Florida, USA, 12 2011, pp. 8231–8236.

[15] L. Fadiga, C. Farges, J. Sabatier, and K. Santugini, Hinfinity output feedback control of commensurate fractional order systems, in *12th European Control Conference (ECC)*, Zurich, Switzerland, 7 2013, pp. 4538–4543.

[16] L. Fadiga, J. Sabatier, and C. Farges, Hinfinity state feedback control of commensurate fractional order systems, in *6th IFAC Workshop on Fractional Differentiation and Its Applications (FDA)*, vol. 6, Grenoble, France, 2 2013, pp. 54–59.

[17] C. Farges, L. Fadiga, and J. Sabatier, Hinfinity analysis and control of commensurate fractional order systems, *Mechatronics*, **23**(7) (2013), 772–780.

[18] P. Gahinet and P. Apkarian, A linear matrix inequality approach to hinfinity control, *International Journal of Robust and Nonlinear Control*, **4**(4) (1994), 421–448.

[19] R. Hotzel and M. Fliess, Gain estimation for fractional linear systems, *IFAC Proceedings Volumes*, **31**(18) (1998), 221–226.

[20] T. Iwasaki and S. Hara, Generalized kyp lemma: unified frequency domain inequalities with design applications, *IEEE Transactions on Automatic Control*, **50**(1) (2005), 41–59.

[21] S. Liang, C. Peng, Z. Liao, and Y. Wang, State space approximation for general fractional order dynamic systems, *International Journal of Systems Science*, **45**(10) (2014), 2203–2212.

[22] S. Liang, Y.-H. Wei, J.-W. Pan, Q. Gao, and Y. Wang, Bounded real lemmas for fractional order systems, *International Journal of Automation and Computing*, **12**(2) (2015), 192–198.
[23] R. L. Magin, *Fractional Calculus in Bioengineering*, Begell House Publishers Inc., USA, 2006.
[24] F. Mainardi, *Fractional Calculus and Waves in Linear Viscoelasticity*, Imperial College Press, London, UK, 2010.
[25] R. Malti, M. Aoun, F. Levron, and A. Oustaloup, Analytical computation of the h2-norm of fractional commensurate transfer functions, *Automatica*, **47**(11) (2011), 2425–2432.
[26] R. Malti, M. Chevrié, C. Farges, and J. Sabatier, H2-norm of fractional transfer functions of implicit type, *Communications in Nonlinear Science and Numerical Simulation*, **26** (2015), 223–237.
[27] D. Matignon, B. D'Andréa-Novel, P. Depalle, and A. Oustaloup, Viscothermal losses in wind instruments: a noninteger model, in *International Symposium on the Mathematical Theory of Networks and Systems (MTNS)*, Regensburg, Germany, 8 1993.
[28] P. Melchior, B. Orsoni, O. Lavialle, A. Poty, and A. Oustaloup, Consideration of obstacle danger level in path planning using a* and fast-marching optimisation: comparative study, *Signal Processing*, **83**(11) (2003), 2387–2396.
[29] A. Oustaloup, *La dérivation non entière: théorie, synthèse et applications*, Traité des nouvelles technologies. Série Automatique, Hermès, France, Paris, 1995.
[30] A. Oustaloup, O. Cois, P. Lanusse, P. Melchior, X. Moreau, and J. Sabatier, The crone approach: theoretical developments and major applications, in *Proceedings of the 2nd IFAC Workshop on Fractional Differentiation and Its Applications*, vol. 39, pp. 324–354, 2006.
[31] I. Podlubny, *Fractional Differential Equations. Mathema*, Mathematics in Science and Engineering, vol. 198, Academic Press, 1999.
[32] S. Rodrigues, N. Munichandraiah, and A. K. Shukla, A review of state-of-charge indication of batteries by means of a.c. impedance measurements, *Journal of Power Sources*, **87**(1–2) (2000), 12–20.
[33] J. Sabatier, M. Aoun, A. Oustaloup, G. Grégoire, F. Ragot, and P. Roy, Fractional system identification for lead acid battery sate charge estimation, *Signal Processing*, **86**(10) (2006), 2645–2657.
[34] J. Sabatier, C. Farges, and L. Fadiga, Approximation of a fractional order model by an integer order model: a new approach taking into account approximation error as an uncertainty, *Journal of Vibration and Control*, **22**(8) (2016), 2069–2082.
[35] J. Sabatier, P. Lanusse, P. Melchior, and A. Oustaloup, *Fractional Order Differentiation and Robust Control Design*, Springer, 2015.
[36] E. Scalas, R. Gorenflo, and F. Mainardi, Fractional calculus and continuous-time finance, *Physica A: Statistical Mechanics and its Applications*, **284**(1) (2000), 376–384.
[37] C. Scherer, P. Gahinet, and M. Chilali, Multiobjective output-feedback control via lmi optimization, *IEEE Transactions on Automatic Control*, **42**(7) (1997), 896–911.
[38] S. Skogestad and I. Postlethwaite, *Multivariable Feedback Control: Analysis and Design*, John Wiley & Sons, 2005.
[39] B. M. Vinagre and Y. Chen, Lecture notes on fractional calculus applications in automatic control and robotics, in *41st IEEE Conference on Decision and Control (CDC)*, Las Vegas, USA, 12 2002, pp. 1–310.
[40] B. M. Vinagre, I. Podlubny, A. Hernandez, and V. Feliu, Some approximations of fractional order operators used in control theory and applications, *Fractional Calculus and Applied Analysis*, (2000), 231–248.
[41] D. Xue, C. Zhao, and Y. Chen, A modified approximation method of fractional order system, in *International Conference on Mechatronics and Automation*, pp. 1043–1048, 2006.
[42] Q. Yang, D. Chen, T. Zhao, and Y. Chen, Fractional calculus in image processing: a review, *Fractional Calculus and Applied Analysis*, **19**(5) (2016).

Hamed Taghavian and Mohammad Saleh Tavazoei

Stability analysis of discrete time distributed order LTI dynamic systems

Abstract: Discrete time LTI distributed order dynamic systems are introduced in this chapter. Necessary and sufficient conditions for stability of these systems with both certain and uncertain weight functions are discussed. At the end, by using the presented results, it is proved that asymptotical and BIBO stability criteria for continuous time LTI distributed order systems are equivalent.

Keywords: Stability, distributed order systems, discrete-time systems

MSC 2010: 39A30, 93D20, 93D05, 93C55, 93C05

1 Introduction

It was more than 300 years ago when the idea of differentiating with a non-integer order was realized and developed. However, it was less conceived as a contribution to the real world problems until recently. As more physical phenomena were described accurately using noninteger models [5, 6, 12, 15], fractional calculus drew more attention from the scientific world by showing its capability to model physical processes with long lasting memory. Later, fractional calculus was also proved beneficial in control systems field [4, 7, 10, 22]. Nowadays, fractional calculus is considered as a special case of distributed order calculus, in which a more general operator called the distributed order differential operator described by a weighted order integral of fractional derivatives is used [9]. Equations involving distributed order operators are emerging in different areas of science and technology [1, 11, 14]. For instance, one could refer to diffusion equations [17] as a very hot topic in this field. In control science, distributed order systems and controllers analysis is considered as a quite new topic as well. Contributions made to this field so far include [3, 8, 13, 23]. In spite of all the work that has been done in the distributed order systems field, still a lot is left be done. For example, a large part of the literature concerning distributed order LTI systems stability is devoted to BIBO stability analysis [9] (pp. 11–25). In contrast, there are less analytical studies regarding the behavior of these systems with initial conditions and their asymptotical stability analysis. In the present chapter, discrete time LTI distributed order systems are introduced, both in finite and infinite dimensional versions. Afterward, an asymptotical stability boundary curve is introduced on the complex plane, based on which necessary and sufficient conditions of asymptotical stability are de-

Hamed Taghavian, Mohammad Saleh Tavazoei, Sharif University of Technology, Azadi Avenue, Tehran, Iran, e-mails: taghavian@alum.sharif.edu, tavazoei@sharif.edu

https://doi.org/10.1515/9783110571745-005

rived for these systems. At the end, it is proved that a continuous time LTI distributed order system is asymptotically stable, if and only if it is BIBO stable, hoping to help to fill the gaps in asymptotical stability analysis of distributed order systems in the literature.

This chapter is organized as follows. In Section 2, we provide a short review on BIBO stability and some essential definitions. In Section 3, discrete time LTI distributed order systems are introduced. Section 4 comprises our main results on asymptotical stability and associated stability boundary curves. Section 5 is dedicated to numerical simulations and the chapter is finally concluded in Section 6.

2 Preliminaries

The fractional differentiation operator that is typically used in control systems is of Caputo type [16] (page 78). Nevertheless, there are also some other definitions proposed for fractional differentiation that are appropriate in different applications. For example, one may refer to the Riemann–Liouville [16] (p. 62) and Grunwald–Letnikov definitions of fractional derivative [16] (p. 43). In [2], Caputo came up with a more general definition of differentiation with a weighted integral over derivation order. This kind of operator in Caputo type is defined as follows:

$$ {}_0^C D_t^{w(\alpha)} x(t) = \int_0^1 w(\alpha) {}_0^C D_t^{\alpha} x(t) d\alpha \tag{1} $$

where $w(\alpha)$ ($\alpha \in [0,1]$) is called the weight function. In this chapter, we assume the weight function to be nonnegative $w(\alpha) \geq 0$ for all $\alpha \in [0,1]$ and $w(\alpha) \not\equiv 0$. Accordingly, a distributed order LTI system in pseudo state space is defined as follows:

$$ \begin{cases} {}_0^C D_t^{w(\alpha)} x(t) = Ax(t) + Bu(t), \\ y(t) = Cx(t) + Du(t), \end{cases} \quad x(0) = x_0. \tag{2} $$

In (2), $x(t) \in \mathrm{R}^n$ is called the pseudo state, $u(t) \in \mathrm{R}$ denotes the input and $y(t) \in \mathrm{R}$ denotes the output of the system. Also, constant matrices are defined as $A \in \mathrm{R}^{n\times n}$, $B \in \mathrm{R}^{n\times 1}$, $C \in \mathrm{R}^{1\times n}$, and $D \in \mathrm{R}$. We consider $x_e = 0$ as the equilibrium point of this system. Like any other dynamic system, stability analysis comes first in studying the behavior of these systems. Therefore, it is necessary to give a review on different stability definitions for such systems.

Definition 1 (BIBO stability). Distributed order system (2) is said to be BIBO stable if for any bounded input $u(t) \in L^\infty$ a bounded output $y(t) \in L^\infty$ is produced.

BIBO stability of system (2) is relatively well studied in the literature. The necessary and sufficient condition for this definition of stability may be stated in an appropriate way in the following lemma.

Lemma 1 ([20]). *System (2) is BIBO stable if and only if every eigenvalue of matrix A with a nonnegative imaginary part is located in left side of the parametric curve*

$$\phi = \{s = x_w(a) + jy_w(a) \mid -\infty < a < +\infty\} \tag{3}$$

where

$$\begin{cases} x_w(a) = \int_0^1 w(\alpha)\exp(a\alpha)\cos(\pi\alpha/2)d\alpha, \\ y_w(a) = \int_0^1 w(\alpha)\exp(a\alpha)\sin(\pi\alpha/2)d\alpha. \end{cases} \tag{4}$$

Besides Definition 1, there are several other definitions available for stability of a dynamic system. Among them, the following two definitions are presented for distributed order systems at the end of this chapter.

Definition 2 (Lyapunov stability). The equilibrium point $x_e = 0$ in distributed order dynamic system ${}_0^C D_t^{w(\alpha)} x(t) = Ax(t)$ is said to be Lyapunov stable, if for every $\varepsilon > 0$ there exists $\delta > 0$ such that if $\|x(0)\| < \delta$ then we have $\|x(t)\| < \varepsilon$ for every $t \geq 0$.

Definition 3 (Asymptotical stability). The equilibrium point $x_e = 0$ in distributed order dynamic system ${}_0^C D_t^{w(\alpha)} x(t) = Ax(t)$ is said to be asymptotically stable, if it is Lyapunov stable and there exists some $\delta > 0$ such that if $\|x(0)\| < \delta$ then $\lim_{t\to+\infty} x(t) = 0$. If the relation holds true for $\forall\delta > 0$, it is called globally asymptotically stable.

3 Discrete time distributed order LTI systems

In this section, we introduce discrete time LTI distributed order systems, described in both finite and infinite dimensional models. These systems are much like their corresponding fractional order counterparts in various aspects. As still analytic solutions of distributed order differential equations in the general sense remain unexplored, discretization methods are appealing for numerical applications. In addition to that, they also come useful in studying the behavior of the corresponding continuous time systems and digital control systems. In this chapter, we focus on the discrete time pseudo state space properties and its applications while avoiding a theoretical study of the convergence of the discrete numerical scheme and error analysis. We will consider continuous time LTI distributed order systems as defined in (2) and their discrete time versions.

GL derivatives are normally used for deriving discrete time fractional-order system models. Likewise in this chapter, we will use this definition to introduce discrete time versions of continuous time distributed order systems. Specifically, the following discrete approximation of distributed order derivatives will be used [21]:

$$ {}_0^C D_t^{w(\alpha)} z(kT) \approx \sum_{i=0}^{k+1} [(-1)^i z((k+1-i)T) P_w(i)] \tag{5}$$

in which

$$ P_w(i) = \int_0^1 w(\alpha) \frac{1}{T^\alpha} \binom{\alpha}{i} d\alpha \tag{6}$$

and $T > 0$ denotes the sampling time. Coefficients $(-1)^i P_w(i)/P_w(0)$ can be considered as the distributed order versions of Grunwald coefficients. This is due to the fact that they share analogous properties with the fractional Grunwald coefficients, which can be directly deduced from the following lemma.

Lemma 2 ([21]). *Let $P_w(i)$ be defined as* (6). *Then we have*

1. $P_w(0) > 0$,
2. $\forall i \geq 1 \rightarrow -(-1)^i P_w(i) > 0$,
3. $\sum_{i=1}^{+\infty} -(-1)^i P_w(i)/P_w(0) = 1$.

Using the variable change $z(t) = x(t) - x(0)$ makes system equations (2) take the form

$$ \begin{cases} {}_0^C D_t^{w(\alpha)} z(t) = Az(t) + Bu(t) + Ax_0, \\ y(t) = Cz(t) + Cx_0 + Du(t), \end{cases} \quad z(0) = 0. \tag{7}$$

Denote the discrete time pseudo states by $\tilde{z}_k$. Discretizing (7) by using (5) eventually yields

$$ \tilde{z}_{k+1} = \left(\frac{A}{P_w(0)} + \frac{P_w(1)}{P_w(0)} I \right) \tilde{z}_k - \sum_{i=2}^{k+1} (-1)^i \tilde{z}_{k+1-i} \frac{P_w(i)}{P_w(0)} + \frac{B}{P_w(0)} u_k + \frac{A}{P_w(0)} \tilde{x}_0 \tag{8}$$

where $z(0) = 0$, $k \geq 0$. Define $\tilde{x}_k = \tilde{z}_k + x_0$ and $Q_k = \sum_{i=0}^{k+1} (-1)^i \frac{P_w(i)}{P_w(0)}$. Writing (8) again in terms of $\tilde{x}_k$ gives

$$ \tilde{x}_{k+1} = \left(\frac{A}{P_w(0)} + \frac{P_w(1)}{P_w(0)} I \right) \tilde{x}_k - \sum_{i=2}^{k+1} (-1)^i \tilde{x}_{k+1-i} \frac{P_w(i)}{P_w(0)} + \frac{B}{P_w(0)} u_k + Q_k \tilde{x}_0 \tag{9}$$

where $k \geq 0$. We refer to (9) as the discrete time version of distributed order LTI system (2). This system could be used in computer simulations to assess the behavior of distributed order dynamics. Like discrete time fractional-order systems, system (9) can also be represented by the following time-varying infinite dimensional discrete state space:

$$\begin{pmatrix}\tilde{x}_{k+1} & \tilde{x}_k & \tilde{x}_{k-1} & \cdots & \tilde{x}_2 & \tilde{x}_1\end{pmatrix}^T = \begin{pmatrix} \frac{A+P_w(1)}{P_w(0)}I & -(-1)^2\frac{P_w(2)}{P_w(0)}I & -(-1)^3\frac{P_w(3)}{P_w(0)}I & \cdots & \cdots & -(-1)^{k+1}\frac{P_w(k+1)}{P_w(0)}I + Q_k I \\ I & 0 & 0 & \cdots & \cdots & 0 \\ 0 & I & 0 & \cdots & \cdots & 0 \\ \vdots & \vdots & \vdots & \ddots & \vdots & \vdots \\ 0 & 0 & 0 & \cdots & I & 0 \end{pmatrix} \times \begin{pmatrix}\tilde{x}_k & \tilde{x}_{k-1} & \tilde{x}_{k-2} & \cdots & \tilde{x}_1 & \tilde{x}_0\end{pmatrix}^T + \begin{pmatrix}\frac{B}{P_w(0)} & 0 & 0 & \cdots & 0 & 0\end{pmatrix}^T u_k. \tag{10}$$

As it can be seen, system (10) changes dimension on each time step and tends to an infinite dimensional system as $k \to +\infty$, which is a quite similar behavior to the fractional-order case. This is actually caused by the nonlocal nature of fractional and distributed order operators. In practice, however, truncated models are considered instead, which possess a finite amount of memory [18]. In order to find the truncated system with finite memory corresponding to (9), we need to suppress the number of the terms involved in the summation $\sum_{i=2}^{k+1}(-1)^i\tilde{x}_{k+1-i}\frac{P_w(i)}{P_w(0)}$ to $\bar{J}-1$, for some $\bar{J} > 0$. To this aim, define

$$b_{i,k} = \begin{cases} (-1)^{k+1}\frac{P_w(k+1)}{P_w(0)} + Q_k, & i = k+1, \\ (-1)^i\frac{P_w(i)}{P_w(0)}, & 2 \le i < k+1. \end{cases} \tag{11}$$

It is worth mentioning that according to Lemma 2 the term Q_k vanishes as $k \to \infty$. We could write the finite memory model from (9) as

$$x_{k+1} = \left(\frac{A}{P_w(0)} + \frac{P_w(1)}{P_w(0)}\right)x_k - \sum_{i=2}^{J} b_{i,k}x_{k+1-i} + \frac{B}{P_w(0)}u_k, \tag{12}$$

$$J = \min(k+1, \bar{J}). \tag{13}$$

It is clear that system (12) is equivalent to system (9) as $\bar{J} \to +\infty$. We can also represent system (12) in the following equivalent form:

$$\begin{pmatrix}x_{k+1} & x_k & \cdots & x_{k+3-J} & x_{k+2-J}\end{pmatrix}^T = \underbrace{\begin{pmatrix} \hat{A} & -b_{2,k}I & -b_{3,k}I & \cdots & -b_{J-1,k}I & -b_{J,k}I \\ I & 0 & 0 & \cdots & 0 & 0 \\ 0 & I & 0 & \cdots & 0 & 0 \\ \vdots & \vdots & \ddots & \cdots & \vdots & \vdots \\ 0 & 0 & 0 & \ddots & 0 & 0 \\ 0 & 0 & 0 & \cdots & I & 0 \end{pmatrix}}_{\tilde{A}_k} \begin{pmatrix} x_k \\ x_{k-1} \\ \vdots \\ x_{k+2-J} \\ x_{k+1-J} \end{pmatrix} + \begin{pmatrix} \frac{B}{P_w(0)} \\ 0 \\ 0 \\ \vdots \\ 0 \\ 0 \end{pmatrix} u_k,$$

$$\hat{A} = \left(\frac{A}{P_w(0)} + \frac{P_w(1)}{P_w(0)}I\right). \tag{14}$$

4 Asymptotical stability

In this section, asymptotical stability of distributed order systems is studied. Necessary and sufficient conditions for stability of infinite dimensional discrete time systems are not as straightforward to find as the classical systems. All the same, this is done for fractional-order systems in [19], where the characteristic equation of the finite memory version of fractional-order system is first found which is subsequently used to obtain the characteristic equation of its original system with infinite memory. On the same basis, a similar method was used for the scalar case of distributed order systems in [21]. A detailed proof for this method in the general matrix case is proposed in the following lemma.

Lemma 3. *Discrete time distributed order system (9) is asymptotically stable if and only if all the roots of the following equation is located inside the unit circle:*

$$z\int_0^1 \frac{w(\alpha)}{T^\alpha}(1-z^{-1})^\alpha d\alpha = \lambda_l. \tag{15}$$

For $l = 1, 2, \ldots, n$, in which λ_l denote the eigenvalues of matrix A.

Proof. At first consider the finite dimensional version of system (9) with representation (14). Note that in order to evaluate asymptotical stability of this system, we have to choose $i < k + 1$ in (11). By this assumption, the coefficients (11) become invariant with time. Thus in this case, it is revealed from system representation (14) that we are actually dealing with a classic linear time invariant system with the following dynamic matrix:

$$\tilde{A} = \begin{pmatrix} \hat{A} & -b_2 I & -b_3 I & \cdots & -b_{\bar{J}-1} I & -b_{\bar{J}} I \\ I & 0 & 0 & \cdots & 0 & 0 \\ 0 & I & 0 & \cdots & 0 & 0 \\ \vdots & \vdots & \ddots & \cdots & \vdots & \vdots \\ 0 & 0 & 0 & \ddots & 0 & 0 \\ 0 & 0 & 0 & \cdots & I & 0 \end{pmatrix}, \tag{16}$$
$$\hat{A} = \left(\frac{A}{P_w(0)} + \frac{P_w(1)}{P_w(0)} I\right)$$

for $\forall \bar{J} > 0$ where $b_i = (-1)^i \frac{P_w(i)}{P_w(0)}$. Therefore, system (12) is asymptotically stable if and only if all the roots of its characteristic equation which is obtained via equation $\det(zI - \tilde{A}) = 0$ are located inside the unit circle. On the other hand, this system tends to system (9) as $\bar{J} \to \infty$. Therefore, it is deduced that system (9) is asymptotically stable, if and only if all the roots of the characteristic equation $\det(zI - \tilde{A}) = 0$ as $\bar{J} \to \infty$ are located inside the unit circle. In order to calculate $\det(zI - \tilde{A})$, consider the matrix

$$zI - \tilde{A} = \begin{pmatrix} zI - \hat{A} & b_2 I & b_3 I & \cdots & b_{\bar{J}-1} I & b_{\bar{J}} I \\ -I & zI & 0 & \cdots & 0 & 0 \\ 0 & -I & zI & \cdots & 0 & 0 \\ \vdots & \vdots & \vdots & \ddots & \vdots & \vdots \\ 0 & 0 & 0 & \cdots & -I & zI \end{pmatrix}. \tag{17}$$

In order to derive a more coherent form of the characteristic equation, we will perform elementary operations. Starting from the last column, multiplying each column by z^{-1} and adding it to the previous column turns (17) into an upper triangular matrix. Therefore, by writing its determinant, the characteristic equation takes the form

$$\begin{aligned} \det(zI - \tilde{A}) &= \det\left(Iz^{\bar{J}} - \hat{A}z^{\bar{J}-1} + I \sum_{j=2}^{\bar{J}} b_j z^{\bar{J}-j} \right) \\ &= \det\left(z^{\bar{J}-1} \left(Iz - \hat{A} + \frac{I}{P_w(0)} \int_0^1 \frac{w(\alpha)}{T^\alpha} \sum_{j=2}^{\bar{J}} \binom{\alpha}{j} (-1)^j z^{1-j} d\alpha \right)\right). \end{aligned} \tag{18}$$

In order to obtain the characteristic equation of system (9), however, we need to consider the limit of (18) as $\bar{J} \to \infty$. Therefore, after some manipulation and using the extended Newton binomial expansion, we arrive at

$$\begin{aligned} &\lim_{\bar{J}\to+\infty} \det(zI - \tilde{A}) \\ &= \lim_{\bar{J}\to+\infty} \det\left(z^{\bar{J}-1} \left(Iz - \hat{A} + \frac{I}{P_w(0)} \int_0^1 \frac{w(\alpha)}{T^\alpha} (z(1 - z^{-1})^\alpha - z + \alpha) d\alpha \right)\right) \\ &= \lim_{\bar{J}\to+\infty} \det\left(z^{\bar{J}-1} \left(Iz - \hat{A} + \frac{P_w(1)}{P_w(0)} I - zI + \frac{I}{P_w(0)} \int_0^1 \frac{w(\alpha)}{T^\alpha} (z(1 - z^{-1})^\alpha) d\alpha \right)\right) \end{aligned} \tag{19}$$

and thereby

$$\begin{aligned} &\lim_{\bar{J}\to\infty} \det(zI - \tilde{A}) \\ &= \lim_{\bar{J}\to\infty} \det\left(z^{\bar{J}-1} \left(\frac{I}{P_w(0)} \int_0^1 \frac{w(\alpha)}{T^\alpha} (z(1 - z^{-1})^\alpha) d\alpha - \frac{A}{P_w(0)} \right)\right). \end{aligned} \tag{20}$$

Finally, the characteristic equation becomes

$$\det\left(\left(\int_0^1 \frac{w(\alpha)}{T^\alpha} (z(1 - z^{-1})^\alpha) d\alpha \right) I - A \right) = 0. \tag{21}$$

System (9) is asymptotically stable if and only if all the roots of (21) are located inside the unit circle. Equation (21) could be considered as the eigenvalue problem (15) which concludes the proof. □

Lemma 3 reduces the necessary and sufficient condition of asymptotical stability for the time-varying infinite dimensional system (9) to an eigenvalue problem based on the primary dynamic matrix A of the corresponding continuous time system with finite dimensions. Nonetheless, still utilizing the characteristic equation derived in Lemma 3 to determine the stability status in each case can be challenging in practice. One approach toward this problem is to obtain a stability boundary curve on the complex plane, based on which stability conditions could be expressed in a new graphical form. This was performed in [21] and is stated in the following theorem.

Theorem 1 ([21]). *Define*

$$\begin{cases} \theta(\Omega) = \frac{\pi-\Omega}{2}, \\ \rho(\Omega) = 2\sin\frac{\Omega}{2}. \end{cases} \tag{22}$$

Discrete time distributed order system (9) is asymptotically stable if and only if all the eigenvalues of matrix A are inside the boundary curve expressed by

$$\phi_D = \{x + jy \mid x = \cos(\Omega)i_c(\Omega) - \sin(\Omega)i_s(\Omega), y = \sin(\Omega)i_c(\Omega) + \cos(\Omega)i_s(\Omega)\} \tag{23}$$

in which $0 \le \Omega \le 2\pi$ *and*

$$\begin{cases} i_c(\Omega) = \int_0^1 w(\alpha)(\rho(\Omega)/T)^\alpha \cos(\theta(\Omega)\alpha)d\alpha, \\ i_s(\Omega) = \int_0^1 w(\alpha)(\rho(\Omega)/T)^\alpha \sin(\theta(\Omega)\alpha)d\alpha. \end{cases} \tag{24}$$

The argument (Ω) of the functions $\theta(\Omega)$, $\rho(\Omega)$, $i_c(\Omega)$, and $i_s(\Omega)$ are sometimes omitted in the rest of this chapter for convenience. It is also possible to express the asymptotical stability boundary curve obtained above (23) in the following form:

$$\phi_D = \{(i_c(\Omega) + ji_s(\Omega))\exp(j\Omega) \mid 0 \le \Omega \le 2\pi\}. \tag{25}$$

It is worth mentioning that it is sufficient to consider the frequency range $0 \le \Omega \le \pi$ in the boundary curve ϕ_D, as it can be shown that the rest of the boundary curve is produced by just a reflection with respect to real axis. Therefore, by defining the associated radius and phase functions as

$$\begin{aligned} R(\Omega) &= \sqrt{i_c^2(\Omega) + i_s^2(\Omega)}, \quad 0 \le \Omega \le \pi, \\ \phi(\Omega) &= \begin{cases} \Omega + \tan^{-1}(i_s(\Omega)/i_c(\Omega)), & 0 < \Omega \le \pi, \\ 0, & \Omega = 0 \end{cases} \end{aligned} \tag{26}$$

we come up with another representation of the boundary curve ϕ_D as

$$\{x + jy \mid x = \Re(R(\Omega)\exp(j\varphi(\Omega))), y = \pm\Im(R(\Omega)\exp(j\varphi(\Omega)))\} \tag{27}$$

where $0 \le \Omega \le \pi$. With aim to get a general outlook of the stability boundary curve ϕ_D on the complex plane, we need a more clear understanding of the behavior of the radius and phase functions. This is the main concern of the next lemma.

Lemma 4. *Let $R(\Omega)$ and $\phi(\Omega)$ be defined as (26). Then the following properties hold true for $\forall T > 0$:*
1. $R(0) = 0$;
2. $\phi(\pi) = \pi$, $0 \le \lim_{\Omega\to0^+} \phi(\Omega) \le \pi/2$;
3. *Functions $\phi(\Omega)$ and $R(\Omega)$ are increasing with respect to their argument Ω.*

Proof. 1. This property is obvious regarding the fact that $\rho(0) = 0$.
2. Relation $\phi(\pi) = \pi$ is immediately obtained from setting $\Omega = \pi$ in (26). Also the inequality $0 \le \lim_{\Omega\to0^+} \phi(\Omega) \le \pi/2$ is obvious from (26), since we have

$$i_s(\Omega), i_c(\Omega) \ge 0 \quad \text{for } 0 \le \Omega \le \pi. \tag{28}$$

3. Before starting the proof of the third part, derivatives of integrals (24) were found with respect to Ω as

$$\begin{cases} \frac{di_c}{d\Omega} = i_{s\alpha}(\Omega)/2 + \cot(\Omega/2)i_{c\alpha}(\Omega)/2, \\ \frac{di_s}{d\Omega} = -i_{c\alpha}(\Omega)/2 + \cot(\Omega/2)i_{s\alpha}(\Omega)/2 \end{cases} \tag{29}$$

in which

$$\begin{cases} i_{c\alpha}(\Omega) = \int_0^1 \alpha w(\alpha)(\rho(\Omega)/T)^\alpha \cos(\theta(\Omega)\alpha)d\alpha, \\ i_{s\alpha}(\Omega) = \int_0^1 \alpha w(\alpha)(\rho(\Omega)/T)^\alpha \sin(\theta(\Omega)\alpha)d\alpha. \end{cases} \tag{30}$$

Using (29), derivative of the radius function $R(\Omega)$ is then calculated as

$$\frac{dR}{d\Omega} = \frac{1}{2}(i_c i_{s\alpha} - i_s i_{c\alpha} + \cot(\Omega/2)i_c i_{c\alpha} + \cot(\Omega/2)i_s i_{s\alpha})/R(\Omega). \tag{31}$$

In order to prove the claim of this part about $R(\Omega)$, it is sufficient to prove $i_c i_{s\alpha} - i_s i_{c\alpha} \ge 0$, as all the other terms in (31) are nonnegative. To this aim, we can write this term in the following double integral form:

$$i_c i_{s\alpha} - i_s i_{c\alpha} = \int_0^1\int_0^1 \alpha q(\alpha)q(\beta)(\sin(\theta\alpha)\cos(\theta\beta) - \cos(\theta\alpha)\sin(\theta\beta))d\alpha d\beta \tag{32}$$

where $q(\alpha) \triangleq w(\alpha)(\rho/T)^\alpha$. Using the sine sum identity in (32) we obtain

$$i_c i_{s\alpha} - i_s i_{c\alpha} = \int_0^1\int_0^1 \alpha q(\alpha)q(\beta)\sin(\theta(\alpha - \beta))d\alpha d\beta. \tag{33}$$

Splitting the inner integral as

$$i_c i_{s\alpha} - i_s i_{c\alpha} = \int_0^1 \int_\beta^1 \alpha q(\alpha) q(\beta) \sin(\theta(\alpha - \beta)) d\alpha d\beta + \int_0^1 \int_0^\beta \alpha q(\alpha) q(\beta) \sin(\theta(\alpha - \beta)) d\alpha d\beta \tag{34}$$

and considering the equalities

$$\begin{aligned} i_c i_{s\alpha} - i_s i_{c\alpha} &= \int_0^1 \int_\beta^1 \alpha q(\alpha) q(\beta) \sin(\theta(\alpha - \beta)) d\alpha d\beta + \int_0^1 \int_\alpha^1 \alpha q(\alpha) q(\beta) \sin(\theta(\alpha - \beta)) d\beta d\alpha \\ &= \int_0^1 \int_\beta^1 \alpha q(\alpha) q(\beta) \sin(\theta(\alpha - \beta)) d\alpha d\beta + \int_0^1 \int_\beta^1 \beta q(\beta) q(\alpha) \sin(\theta(\beta - \alpha)) d\alpha d\beta \\ &= \int_0^1 \int_\beta^1 q(\alpha) q(\beta)(\alpha \sin(\theta(\alpha - \beta)) + \beta \sin(\theta(\beta - \alpha))) d\alpha d\beta \\ &= \int_0^1 \int_\beta^1 q(\alpha) q(\beta) \sin(\theta(\alpha - \beta))(\alpha - \beta) d\alpha d\beta \geq 0 \end{aligned} \tag{35}$$

proves that the radius function $R(\Omega)$ is increasing with respect to Ω, since the last line in (35) is clearly nonnegative due to the inequality $0 \leq \beta \leq \alpha \leq 1$ that holds inside the integral. In order to investigate the monotonicity of $\phi(\Omega)$, we calculate its derivative as $\frac{d\phi}{d\Omega} = 1 + (\frac{di_s}{d\Omega} i_c - \frac{di_c}{d\Omega} i_s)/(i_c^2 + i_s^2)$. By using (29), we get

$$\begin{aligned} \frac{d\phi}{d\Omega} &= \frac{i_c^2 + i_s^2 + (-i_{c\alpha} + \cot(\Omega/2) i_{s\alpha}) i_c/2 - (i_{s\alpha} + \cot(\Omega/2) i_{c\alpha}) i_s/2}{i_c^2 + i_s^2} \\ &= \frac{i_c(i_c - i_{c\alpha}/2) + i_s(i_s - i_{s\alpha}/2) + \cot(\Omega/2)(i_c i_{s\alpha} - i_s i_{c\alpha})/2}{i_c^2 + i_s^2}. \end{aligned} \tag{36}$$

Note that the terms $(i_c - i_{c\alpha}/2)$ and $(i_s - i_{s\alpha}/2)$ are clearly nonnegative according to (24) and (30) and the fact that $0 \leq \alpha \leq 1$ holds in their defining integral. On the other hand, the term $(i_c i_{s\alpha} - i_s i_{c\alpha})$ was proved to be non-negative in (35). Therefore, from (36) it is deduced that the phase function $\phi(\Omega)$ is increasing with respect to Ω as well. □

Distributed order systems can also suffer from uncertainty just like any other systems used for modeling physical processes. This includes distributed order system models with uncertain weight functions lying within a known range. For more information in this regard, see [20]. Therefore, in order to provide some robust stability conditions for dealing with uncertainties present in models describing real physical phenomena, a brief investigation is provided in the following.

Lemma 5. *Consider a bounded weight function in a way that* $0 < \underline{w} \le w(\alpha) \le \bar{w}$ *is satisfied for* $\forall \alpha \in [0,1]$. *There holds* $\lim_{\Omega\to 0} \phi(\Omega) = 0$.

Proof. In order to calculate $\lim_{\Omega\to 0} \phi(\Omega)$, note that

$$\begin{aligned}
&\lim_{\Omega\to 0} |i_s(\Omega)/i_c(\Omega)| \\
&\quad = \lim_{\Omega\to 0} \left| \int_0^1 w(\alpha)(\rho(\Omega)/T)^\alpha \sin(\theta\alpha) d\alpha \Big/ \int_0^1 w(\alpha)(\rho(\Omega)/T)^\alpha \cos(\theta\alpha) d\alpha \right| \\
&\quad \le \lim_{\Omega\to 0} \frac{\bar{w}}{\underline{w}} \int_0^1 (\rho(\Omega)/T)^\alpha \sin(\theta\alpha) d\alpha \Big/ \int_0^1 (\rho(\Omega)/T)^\alpha \cos(\theta\alpha) d\alpha \\
&\quad = \lim_{\Omega\to 0} -\frac{\bar{w}(\rho(\Omega)(\theta \cos\theta - \ln(\rho(\Omega)/T) \sin\theta)/T - \theta)}{\underline{w}(\rho(\Omega)(\ln(\rho(\Omega)/T) \cos\theta + \theta \sin\theta)/T - \ln(\rho(\Omega)/T))} \\
&\quad = \lim_{\Omega\to 0} \frac{\bar{w}\pi/2}{-\underline{w} \ln(\rho(\Omega)/T)} = 0.
\end{aligned} \tag{37}$$

Therefore, we have $\lim_{\Omega\to 0}(\Omega + \tan^{-1}(i_s(\Omega)/i_c(\Omega))) = 0$, from which it is deduced that $\lim_{\Omega\to 0} \phi(\Omega) = 0$. □

In fact, Lemma 4 gives a general outlook of the boundary curve (25) indicating that it is always a closed curve on the complex plane, which is also tangent to the real axis at the origin in case of a bounded weight function (due to Lemma 5). Therefore, the stability condition that this curve has to offer could be expressed in the following theorem.

Theorem 2. *Discrete time distributed order system* (9) *is asymptotically stable, if and only if for every eigenvalue* $\lambda_l = x_l + jy_l$ *of matrix* A *with nonnegative imaginary parts either of the conditions*

$$R(\Omega) = |\lambda_l| \to \phi(\Omega) < \measuredangle\, \lambda_l \tag{38}$$

or

$$\phi(\Omega) = \measuredangle\, \lambda_l \to R(\Omega) > |\lambda_l| \tag{39}$$

is satisfied where $0 \le \measuredangle\, \lambda_l \le \pi$.

It is worth mentioning that asymptotical stability of system (9) is dependent on the value of sampling time $T > 0$ as well as the weight function $w(\alpha)$ and dynamic matrix A. According to the third part of Lemma 4, functions $\phi(\Omega)$ and $R(\Omega)$ are increasing with respect to Ω which implies that a conservative stability boundary curve providing a sufficient condition of asymptotical stability may be obtained from (38) or (39) by replacing $\phi(\Omega)$ ($R(\Omega)$) with its upper (lower) bound. This is also directly deduced from asymptotical stability criteria (38) and (39). This concept is used in case of an uncertain bounded weight function in the following theorem to give a robust stability condition.

Theorem 3. *Consider an unknown bounded weight function in a way that* $0 < \underline{w} \leq w(\alpha) \leq \bar{w}$ *is satisfied for* $\alpha \in [0,1]$ *where* $\underline{w}$ *and* $\bar{w}$ *are known. Discrete time distributed order system* (9) *is asymptotically stable if for each eigenvalue* λ_l *of matrix A with a non-negative imaginary part (i. e.,* $0 \leq \measuredangle\ \lambda_l \leq \pi$*), either of the conditions*

$$g_c(\Omega)^2 + g_s(\Omega)^2 = |\lambda_l|^2/\underline{w}^2 \to \Omega + \tan^{-1}\left(\frac{\bar{w}g_s(\Omega)}{\underline{w}g_c(\Omega)}\right) < \measuredangle\ n\lambda_l \tag{40}$$

or

$$\frac{\underline{w}g_c(\Omega)\tan\Omega + \bar{w}g_s(\Omega)}{\underline{w}g_c(\Omega) - \bar{w}g_s(\Omega)\tan\Omega} = \tan(\angle\lambda_l) \to g_c(\Omega)^2 + g_s(\Omega)^2 > |\lambda_l|^2/\underline{w}^2 \tag{41}$$

is met for some $T > 0$, *in which*

$$\begin{cases} g_s(\Omega) = \frac{\ln(\rho/T)\sin(\theta)\rho/T - \theta(\cos(\theta)\rho/T - 1)}{\theta^2 + (\ln(\rho/T))^2}, \\ g_c(\Omega) = \frac{\theta\sin(\theta)\rho/T + \ln(\rho/T)(\cos(\theta)\rho/T - 1)}{\theta^2 + (\ln(\rho/T))^2} \end{cases} \tag{42}$$

and θ, ρ *are defined in* (22).

Proof. Integrals $i_c(\Omega)$ and $i_s(\Omega)$ are bounded by

$$\begin{aligned} \underline{w}\int_0^1 \exp(\alpha\ln(\rho/T))\cos(\theta\alpha)d\alpha \leq i_c(\Omega) \leq \bar{w}\int_0^1 \exp(\alpha\ln(\rho/T))\cos(\theta\alpha)d\alpha, \\ \underline{w}\int_0^1 \exp(\alpha\ln(\rho/T))\sin(\theta\alpha)d\alpha \leq i_s(\Omega) \leq \bar{w}\int_0^1 \exp(\alpha\ln(\rho/T))\sin(\theta\alpha)d\alpha. \end{aligned} \tag{43}$$

The integrals appeared above have the following explicit expressions:

$$\begin{cases} g_c(\Omega) = \int_0^1 \exp(\alpha\ln(\rho/T))\cos(\theta\alpha)d\alpha, \\ g_s(\Omega) = \int_0^1 \exp(\alpha\ln(\rho/T))\sin(\theta\alpha)d\alpha. \end{cases} \tag{44}$$

In which $g_c(\Omega)$ and $g_s(\Omega)$ are given by (42). Therefore, it is possible to write (43) in the following compact form:

$$\begin{aligned} \underline{w}g_c(\Omega) \leq i_c(\Omega) \leq \bar{w}g_c(\Omega), \\ \underline{w}g_s(\Omega) \leq i_s(\Omega) \leq \bar{w}g_s(\Omega). \end{aligned} \tag{45}$$

Accordingly, for the magnitude function $R(\Omega)$ and the phase function $\phi(\Omega)$ the following bounds are found:

$$\begin{aligned} &\sqrt{g_c^2(\Omega) + g_s^2(\Omega)}\underline{w} \leq R(\Omega), \\ &\phi(\Omega) \leq \Omega + \tan^{-1}(\bar{w}g_s(\Omega)/\underline{w}g_c(\Omega)). \end{aligned} \tag{46}$$

The new bounds appeared in (46) can be used on behalf of the original radius and phase functions in Theorem 2 to derive a robust condition of asymptotical stability as addressed in this theorem. □

At the end by using the results obtained in this section, we consider the continuous time distributed order system (2) again and prove that asymptotical and BIBO stability criteria for this system are the same in the following theorem. There has been either an inconsistent proof for this concept or the problem has been assumed to be true without a proof in the literature. A formal proof for this problem is therefore provided in the next theorem by comparing the BIBO and asymptotical stability boundary curves on the complex plane which turn out to be identical.

Theorem 4. *Continuous time distributed order system* ${}_0^C D_t^{w(\alpha)} x(t) = Ax(t)$ *is asymptotically stable if and only if system* (2) *is BIBO stable.*

Proof. It is obvious that asymptotical stability of system (9) is equivalent to asymptotical stability of system (2) as $T \to 0$. In the previous section, we found a curve on the complex plane as the asymptotical stability boundary curve of system (9). Here, we will find the curve to which this boundary curve approaches as $T \to 0$, which would be the asymptotical stability boundary curve of system (2). Finally, it will be shown that the resultant curve matches the BIBO boundary curve of system (2) given by (3) which proves this theorem. Let us define $a(\Omega) = \ln(\rho(\Omega)/T)$ in which $\rho(\Omega)$ is defined as (22), and rewrite the integrals (24) as

$$\begin{cases} i_c(\Omega) = \int_0^1 w(\alpha)\exp(a(\Omega)\alpha)\cos(\pi\alpha/2 - \Omega\alpha/2)d\alpha, \\ i_s(\Omega) = \int_0^1 w(\alpha)\exp(a(\Omega)\alpha)\sin(\pi\alpha/2 - \Omega\alpha/2)d\alpha. \end{cases} \tag{47}$$

Function $a(\Omega)$ is clearly continuous on $(0, \pi]$. Assume $0 < T < 1$ and define $\varepsilon = 2\sin^{-1}(T/2)$ and $\sigma = 2\sin^{-1}(\sqrt{T}/2)$. It can be shown that $a(\Omega)$ as a function of Ω, satisfy the following bounds in each frequency interval:

$$\begin{cases} -\infty < a(\Omega) \le 0, & 0 \le \Omega \le \varepsilon, \\ 0 < a(\Omega) \le \ln(1/\sqrt{T}), & \varepsilon < \Omega \le \sigma, \\ \ln(1/\sqrt{T}) < a(\Omega) \le \ln(2/T), & \sigma < \Omega \le \pi. \end{cases} \tag{48}$$

Noting the limit $\lim_{T\to 0} \varepsilon, \sigma \to 0$ it is observed that for small frequencies (i. e., $0 \le \Omega \le \sigma$) the variable $a(\Omega)$ sweeps the whole real axis (i. e., $-\infty < a < +\infty$), and for the rest of frequencies (i. e., $\sigma \le \Omega \le \pi$) the variable $a(\Omega)$ remains at infinity ($a = +\infty$). Therefore, it is revealed that the produced curve for $0 \le \Omega \le \sigma$ is equal to the BIBO boundary curve ϕ expressed in Lemma 1. After that in the range $\sigma \le \Omega \le \pi$, the radius of the curve is infinity $|(i_c + ji_s)\exp(j\Omega)| = +\infty$. On the other hand, according to Lemma 4 the phase angle satisfy $0 < \arg((i_c + ji_s)\exp(j\Omega)) \le \pi$. In other words, as Ω approaches π from σ,

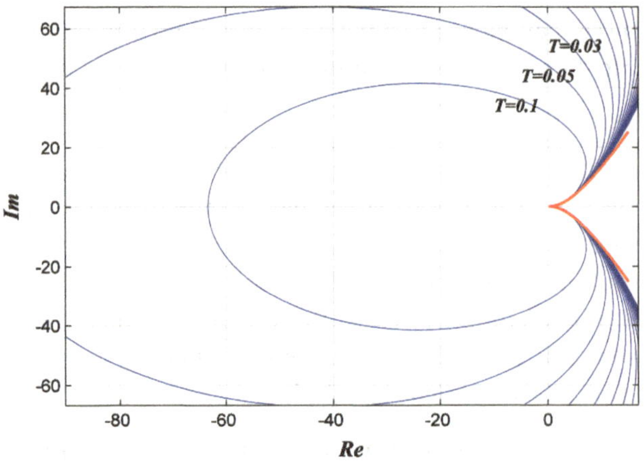

Figure 1: Asymptotical stability boundary curves of discretized versions of system (2) for several sample times are plotted in normal lines (blue) and its BIBO stability boundary curve is plotted in the bold line (red), all for the weight function $w(\alpha) = 10$.

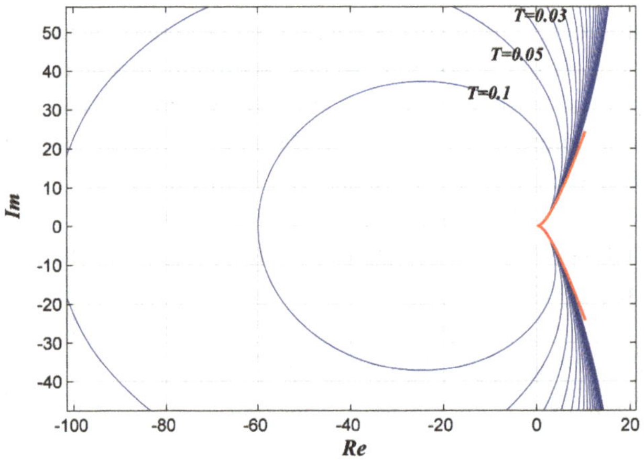

Figure 2: Asymptotical stability boundary curves of discretized versions of system (2) for several sample times are plotted in normal lines (blue) and its BIBO stability boundary curve is plotted in the bold line (red), all for the weight function $w(\alpha) = 5\alpha^2 + 10$.

the boundary curve travels on an arch with an infinite radius from the end of the curve ϕ to the negative end of the real axis. This indicates that the BIBO stability region is exactly identical to the asymptotical stability region. □

5 Numerical example

A numerical example is presented for evaluation of the validity of Theorem 4. In the following, system (2) is considered with three different weight functions. For each case, BIBO stability boundary curve of the continuous time system (ϕ) which is introduced in (3) and the asymptotical stability boundary curves of the corresponding discrete time versions of the same system (ϕ_D) given by (25), are plotted for several values of sample time T. As it is seen in Figures 1, 2, and 3, the asymptotical stabil-

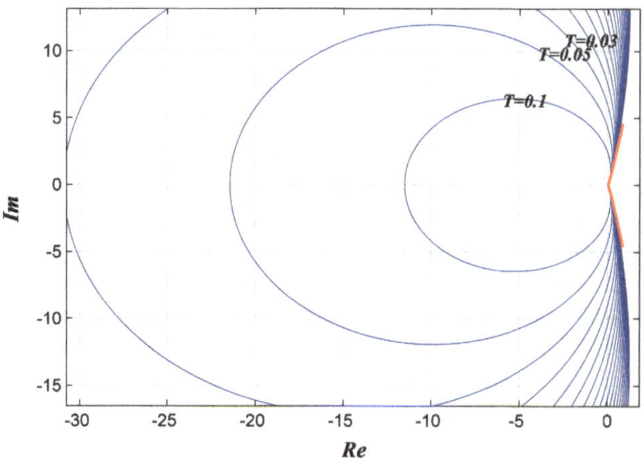

Figure 3: Asymptotical stability boundary curves of discretized versions of system (2) for several sample times are plotted in normal lines (blue) and its BIBO stability boundary curve is plotted in the bold line (red), all for the weight function $w(\alpha) = 5\alpha^5$.

ity boundary curve of the discrete time version of each system approaches the BIBO stability boundary curve of the corresponding continuous time version as $T \to 0$. The weight functions considered in these simulations are $w(\alpha) = 10$, $w(\alpha) = 5\alpha^2 + 10$ and $w(\alpha) = 5\alpha^5$ whose results are shown in Figures 1, 2, and 3, respectively.

6 Conclusion

In this chapter, we introduced discrete time distributed order LTI systems which can be used in computer simulations and digital control systems. Asymptotical stability of these systems were also analyzed with or without the presence of uncertainty. Based on the results, a formal proof was then provided for the equivalence of BIBO and asymptotical stability criteria for continuous time distributed order LTI systems, which was confirmed in numerical simulations.

Bibliography

[1] T. M. Atanackovic, On a distributed derivative model of a viscoelastic body, *Comptes Rendus. Mécanique*, **331**(10) (2003), 687–692.

[2] M. Caputo, *Elasticita e dissipazione (Elasticity and Anelastic Dissipation)*, Zanichelli, Bologna, 1969.

[3] P. L. T. Duong, E. Kwok, and M. Lee, Optimal design of stochastic distributed order linear siso systems using hybrid spectral method, *Mathematical Problems in Engineering*, **2015** (2015), Article ID 989542, 14 pages.

[4] K. Erenturk, Fractional-order pid and active disturbance rejection control of nonlinear two-mass drive system, *IEEE Transactions on Industrial Electronics*, **60**(9) (2013), 3806–3813.

[5] V. S. Erturk, G. Zaman, and S. Momani, A numeric–analytic method for approximating a giving up smoking model containing fractional derivatives, *Computers & Mathematics with Applications*, **64**(10) (2012), 3065–3074.
[6] M. Hasan and C. Drapaca, A fractional order model for local electric fields in tissues, in *Mechanics of Biological Systems and Materials*, vol. 7, pp. 75–79, Springer, 2015.
[7] C. Izaguirre-Espinosa, A. J. Munoz-Vazquez, A. Sanchez-Orta, V. Parra-Vega, and P. Castillo, Attitude control of quadrotors based on fractional sliding modes: theory and experiments, *IET Control Theory & Applications*, **10**(7) (2016), 825–832.
[8] B. B. Jakovljevic, M. R. Rapaic, Z. D. Jelicic, and T. B. Sekara, Optimization of distributed order fractional pid controller under constraints on robustness and sensitivity to measurement noise, in *Fractional Differentiation and Its Applications (ICFDA), 2014 International Conference on*, pp. 1–6, IEEE, 2014.
[9] Z. Jiao, Y. Q. Chen, and I. Podlubny, *Distributed-Order Dynamic Systems: Stability, Simulation, Applications and Perspectives*, SpringerBriefs in Control, Automation and Robotics, 2012.
[10] S. W. Khubalkar, A. S. Chopade, A. S. Junghare, and M. V. Aware, Design and tuning of fractional order pid controller for speed control of permanent magnet brushless dc motor, in *2016 IEEE First International Conference on Control, Measurement and Instrumentation (CMI)*, pp. 326–330, 2016.
[11] G. Lazovic, Z. Vosika, M. Lazarevic, J. Simic-Krstic, and D. Koruga, Modeling of bioimpedance for human skin based on fractional distributed-order modified cole model, *FME Transactions*, **42**(1) (2014), 74–81.
[12] D. Li, J. Cao, S. Zhou, and Y. Chen, Fractional order model of broadband piezoelectric energy harvesters, in *ASME 2015 International Design Engineering Technical Conferences and Computers and Information in Engineering Conference*, V009T07A019, American Society of Mechanical Engineers, 2015.
[13] Y. Li, H. Sheng, and Y. Q. Chen, On distributed order lead-lag compensator, *Proceedings of FDA*, **10** (2010), 18–20.
[14] F. Mainardi, A. Mura, R. Gorenflo, and M. Stojanovic, The two forms of fractional relaxation of distributed order, *Journal of Vibration and Control*, **13**(9–10) (2007), 1249–1268.
[15] A. E. Matouk, A. A. Elsadany, E. Ahmed, and H. N. Agiza, Dynamical behavior of fractional-order Hastings–Powell food chain model and its discretization, *Communications in Nonlinear Science and Numerical Simulation*, **27**(1) (2015), 153–167.
[16] I. Podlubny, *Fractional Differential Equations: an Introduction to Fractional Derivatives, Fractional Differential Equations, to Methods of Their Solution and Some of Their Applications*, vol. 198, Academic Press, 1999.
[17] T. Sandev, A. V. Chechkin, N. Korabel, H. Kantz, I. M. Sokolov, and R. Metzler, Distributed-order diffusion equations and multifractality: models and solutions, *Physical Review E*, **92**(4) (2015), 042117.
[18] R. Stanislawski and K. J. Latawiec, Normalized finite fractional differences: computational and accuracy breakthroughs, *International Journal of Applied Mathematics and Computer Science*, **22**(4) (2012), 907–919.
[19] R. Stanislawski and K. J. Latawiec, Stability analysis for discrete-time fractional-order lti state-space systems. Part i: new necessary and sufficient conditions for the asymptotic stability, *Bulletin of the Polish Academy of Sciences. Technical Sciences*, **61**(2) (2013), 353–361.
[20] H. Taghavian and M. S. Tavazoei, Robust stability analysis of distributed-order linear time-invariant systems with uncertain order weight functions and uncertain dynamic matrices, *Journal of Dynamic Systems, Measurement, and Control*, **139**(12) (2017), 121010.
[21] H. Taghavian and M. S. Tavazoei, Stability analysis of distributed order nonlinear dynamic systems, *International Journal of Systems Science*, **49**(3) (2018), 523–536.

[22] J. Viola and L. Angel, Fractional control and robustness analysis of an inverted pendulum system, in *Automatic Control (CCAC), 2015 IEEE 2nd Colombian Conference on*, pp. 1–6, IEEE, 2015.
[23] F. Zhou, Y. Zhao, Y. Li, and Y. Chen, Design, implementation and application of distributed order pi control, *ISA Transactions*, **52**(3) (2013), 429–437.

Manuel D. Ortigueira, Duarte Valério, and António M. Lopes

Continuous-time fractional linear systems: transient responses

Abstract: This chapter presents an introduction to fractional continuous-time linear systems, based on suitable fractional derivatives. The usual tools, impulse response, transfer function, and frequency responses are defined, and the manner of their computation is given, considering both commensurate and noncommensurate cases. Properties of the systems, like causality, periodicity, and stability are also studied. The formulation in the commensurate case is done in a full agreement with classic results that are recovered when the derivative order becomes equal to 1. Multiple-input, multiple-output systems are also studied from the state-space point of view.

Keywords: Fractional calculus, continuous-time signals and systems, transient responses, transfer function, stability, state-space

MSC 2010: 26A33, 93A99, 93D25

1 Introduction

Linear time-invariant (LTI) systems defined by constant coefficient linear differential equations are ubiquitous. Their mathematical manageability make them useful tools, even if they are not the most precise for studying or modeling natural or artificial systems.

Usually, LTI systems are written in the general form

$$\sum_{k=0}^{N} a_k D^{\alpha_k} y(t) = \sum_{k=0}^{M} b_k D^{\beta_k} x(t), \quad t \in \mathbb{R}, \tag{1}$$

where the symbol D represents the derivative operator that will be defined in Section 2, and the parameters $\alpha_k, \beta_k \in \mathbb{R}$ are the derivative orders. In the so-called commensu-

Acknowledgement: This work was supported by FCT, through IDMEC, under LAETA, project UID/EMS/50022/2013.

Manuel D. Ortigueira, CTS-UNINOVA and DEE of NOVA School of Science and Technology of NOVA University of Lisbon, Campus da FCT da UNL, Quinta da Torre, 2829–516 Caparica, Portugal, e-mail: mdo@fct.unl.pt
Duarte Valério, IDMEC, Instituto Superior Técnico, Universidade de Lisboa, Av. Rovisco Pais 1, 1049–001 Lisboa, Portugal, e-mail: duarte.valerio@tecnico.ulisboa.pt
António M. Lopes, UISPA–LAETA/INEGI, Faculty of Engineering, University of Porto, Rua Dr. Roberto Frias, 4200–465 Porto, Portugal, e-mail: aml@fe.up.pt

https://doi.org/10.1515/9783110571745-006

rate case, we have $\alpha_k = \beta_k = k\alpha$. When writing the second member with this format, we need derivatives of the input and, for studying the transient behavior, the corresponding initial conditions are necessary. We assume that $M \le N$ for stability reasons. In engineering applications, the transient analysis is studied by computing the output to some well-known input functions. Frequently, we use the impulse, $\delta(t)$, the Heaviside unit step, $\varepsilon(t)$, and the ramp $r(t) = t\varepsilon(t)$. Stability and initial-conditions are also considered.

1.1 Remarks

We assume that

- We work on $\mathbb{R}$;
- We use the two-sided Laplace transform (LT)

$$F(s) = \int_{\mathbb{R}} f(t)e^{-st}\mathrm{d}t, \tag{2}$$

  where $f(t)$ is any function defined on $\mathbb{R}$ and $F(s)$ is its transform, provided it has a nonempty region of convergence. Usually, the one-sided LT

$$F(s) = \int_0^\infty f(t)e^{-st}\mathrm{d}t,$$

  is used. However, it has some inconveniences:
  - It introduces some initial conditions (IC) that are, in general, not suitable for the system under study;
  - It does not allow the study of systems when the input is nonnull for $t < 0$.
  - It is useless for studying the systems with stochastic input/output, since the autocorrelation is a two-sided signal.
- The Fourier transform (FT) is obtained from the LT through the substitution $s = i\omega$ with $\omega \in \mathbb{R}$;
- The functions and distributions have Laplace and/or Fourier transforms;
- Usual properties of the Dirac delta distribution and its derivatives will be used;
- We will work with the usual convolution:

$$f(t) * g(t) = \int_{\mathbb{R}} f(\tau)g(t-\tau)\mathrm{d}\tau. \tag{3}$$

- The order of the fractional derivative is assumed to be any real or complex number;
- The multivalued expressions s^α and $(-s)^\alpha$ will be used in the follow-up. To obtain functions from them, we will use, as branch-cut lines, the negative real half-axis for the first, and the positive real half-axis for the second, and, in both cases, the first Riemann surface is chosen.

In this line of thought, the chapter is organized as follows. Section 2 introduces the fractional derivative formulations suitable for signals and systems. Section 3 studies the time responses of fractional LTI systems. Finally, Section 4 investigates the multi-input/output case and state variable representations.

2 On the fractional derivatives suitable for signals and systems

2.1 Causal derivatives

Not all fractional derivative definitions are interesting for studying LTI systems and for obtaining generalizations of classic tools and results [13]. As in most applications, the independent variable is time, and we need to work with causal (left) derivatives.

Let us assume that we are dealing with "well-behaved enough" functions—for instance, functions with LT. We define:

- The Grünwald–Letnikov derivative, [6, 14]

$$D_f^{\alpha} f(t) = \lim_{h\to 0^+} \frac{\sum_{k=0}^{\infty}(-1)^k \binom{\alpha}{k} f(t-kh)}{h^{\alpha}}. \tag{4}$$

  We call (4) the forward Grünwald–Letnikov[1] derivative. It is a causal derivative, as is stated by its impulse and step responses that we will compute later—see (24) and (25).

  If N is the integer part of α, then we define

- The regularized Liouville derivative, [8]

$$D_f^{\alpha} f(t) = \frac{1}{\Gamma(-\alpha)} \int_0^{\infty} \tau^{-\alpha-1} \left[f(t-\tau) - \varepsilon(\alpha) \sum_0^N \frac{(-)^m f^{(m)}(t)}{m!} \tau^m \right] d\tau. \tag{5}$$

- The Liouville–Caputo derivative, [11]

$$D_f^{\alpha} f(t) = \frac{1}{\Gamma(-\alpha+N)} \int_0^{\infty} \tau^{N-\alpha-1} f^{(N)}(t-\tau) d\tau. \tag{6}$$

- The Liouville derivative, [11]

$$D_f^{\alpha} f(t) = D_f^N \left[\frac{1}{\Gamma(-\alpha+N)} \int_0^{\infty} \tau^{N-\alpha-1} f(t-\tau) d\tau \right], \tag{7}$$

  that constitutes a derivative of the Riemann–Liouville type.

1 The terms forward and backward are used here in agreement to the way the time flows, from past to future or the reverse.

Remark 1. Let $f(t)$ be a function having LT, $F(s)$. The above derivatives verify

$$\mathcal{L}[D_f^\alpha f(t)] = s^\alpha F(s), \quad \text{for } \mathbb{R}(s) > 0, \tag{8}$$

generalizing a very important property of the LT.

Remark 2 (Classic Riemann–Liouville and Caputo derivatives). The commutative property of the convolution allows us to write from (6) and (7):

$$D_f^\alpha f(t) = D_f^N \left[\frac{1}{\Gamma(-\alpha+N)} \int_{-\infty}^{t} (t-\tau)^{N-\alpha-1} f(\tau) d\tau \right] \tag{9}$$

and

$$D_f^\alpha f(t) = \frac{1}{\Gamma(-\alpha+N)} \int_{-\infty}^{t} (t-\tau)^{N-\alpha-1} f^{(N)}(\tau) d\tau. \tag{10}$$

These expressions are the general formulations of Riemann–Liouville and Liouville–Caputo derivatives. Their usual formulations are obtained from the above relations assuming that the function is defined in a given interval $[a, b]$ [5, 15, 16]. Therefore, for $t \in [a, b]$ these derivatives are given by

$$^{\mathrm{RL}}D_f^\alpha f(t) = D_f^N \left[\frac{1}{\Gamma(-\alpha+N)} \int_{a}^{t} (t-\tau)^{N-\alpha-1} f(\tau) d\tau \right] \tag{11}$$

and

$$^{C}D_f^\alpha f(t) = \frac{1}{\Gamma(-\alpha+N)} \int_{a}^{t} (t-\tau)^{N-\alpha-1} f^{(N)}(\tau) d\tau. \tag{12}$$

These derivatives do not enjoy most of the important properties of the classic derivative, namely the results referring the derivatives of sinusoids that, following (11) and (12), are no longer sinusoids.

2.2 Anti-causal derivatives

Given that $\mathcal{L}[D_f^\alpha f(t)] = s^\alpha F(s)$, $\mathbb{R}(s) > 0$, there are derivatives for which the following relation is valid:

$$\mathcal{L}[D_b^\alpha f(t)] = s^\alpha F(s), \quad \mathbb{R}(s) < 0. \tag{13}$$

We will call them “backward” derivatives. They are defined as follows:

- The Grünwald–Letnikov derivative

$$D_b^{\alpha} f(t) = e^{-j\alpha\pi} \lim_{h\to 0^+} \frac{\sum_{k=0}^{\infty} (-1)^k \binom{\alpha}{k} f(t+kh)}{h^{\alpha}}. \tag{14}$$

We call (14) the backward Grünwald–Letnikov derivative, since the flow of time is reversed.

If N is again the integer part of α, then we have:

- The regularized Liouville derivative,

$$D_b^{\alpha} f(t) = e^{-j\alpha\pi} \frac{1}{\Gamma(-\alpha)} \int_0^{\infty} \tau^{-\alpha-1} \left[f(t+\tau) - \varepsilon(\alpha) \sum_0^N \frac{f^{(m)}(t)}{m!} \tau^m \right] d\tau. \tag{15}$$

- The Liouville–Caputo derivative, [4]

$$D_b^{\alpha} f(t) = e^{-j\alpha\pi} \frac{1}{\Gamma(-\alpha+N)} \int_0^{\infty} \tau^{N-\alpha-1} f^{(N)}(t+\tau) d\tau. \tag{16}$$

- The Liouville derivative,

$$D_b^{\alpha} f(t) = e^{-j\alpha\pi} D_f^N \left[\frac{1}{\Gamma(-\alpha+N)} \int_0^{\infty} \tau^{N-\alpha-1} f(t+\tau) d\tau \right], \tag{17}$$

that constitutes again a derivative of the Riemann–Liouville type.

Remark 3. In models where the independent variable is not time, we can remove the factor $e^{-j\alpha\pi}$. In fact, if such variable is space, there is no difference in going from left to right and vice versa. In these cases, we call the forward derivatives “left” derivatives, and call the backward derivatives, without the factor, “right” derivatives.

Remark 4. When a given formula or property is independent of the character, forward or backward, we can remove the subscript, f or b. The same will be done when the context is clear, as it will be in the computation of the outputs of LTI systems where we will consider causal systems only.

2.3 The distributional fractional derivative

Here, we will generalize the concept of fractional derivative in order to guarantee that all the properties of the above defined derivative remain valid even with generalized functions [3, 9]. In particular, the group properties should be valid:

$$D^{\gamma}\left[D^{\alpha+\beta} f(t)\right] = D^{\gamma+\alpha+\beta} f(t) = D^{\alpha+\beta+\gamma} f(t) = D^{\alpha}\left[D^{\beta+\gamma} f(t)\right], \tag{18}$$

with neutral element

$$D^{\alpha}[D^{-\alpha}f(t)] = D^0 f(t) = f(t) \quad \text{and} \quad D^{-\alpha}[D^{\alpha}f(t)] = D^0 f(t) = f(t). \tag{19}$$

In some derivatives, these properties do not remain valid, since commutativity does not hold. Here, we will extend the validity of formula (18) by a suitable generalized function definition.

The above properties are valid provided that all the involved derivatives exist. This may not happen in a lot of situations; for example, the derivative of the power function defined on $\mathbb{R}$. As these functions are very important, we will consider them with detail. Meanwhile, let us see how we can enlarge the validity of the above formulas. Let us consider formula (18) and a function $f(t)$ such that $D^{\alpha}f(t)$ exists, but may be discontinuous. In principle, we cannot ensure that we can apply (18) to obtain $D^{\alpha+\beta}f(t)$. To solve the problem, we define distribution as an integer order derivative of a continuous function [2]: $f(t) = D^n g(t)$, where n is a positive integer and $g(t)$ is continuous and with continuous fractional derivative of order $\alpha + \beta$. In this case, we can write

$$D^{\alpha+\beta}f(t) = D^{\alpha+\beta}D^n g(t) = D^n D^{\alpha+\beta}g(t). \tag{20}$$

So, we obtained the desired derivative by integer order derivation of the fractional derivative. The other properties are consequence of this one. An example will clarify the situation.

The results obtained allow us to obtain the derivative of any order of the continuous function $p(t) = t^{\beta}\varepsilon(t)$, with $\beta > 0$, where $\varepsilon(t)$ is the Heaviside unit step. As a continuous function, $p(t)$ is indefinitely (integer order) derivable in distributional sense. To compute the fractional derivative of $p(t)$, the easiest way is to use the LT, which is

$$P(s) = \mathcal{L}[t^{\beta}\varepsilon(t)] = \int_0^{\infty} t^{\beta}e^{-st}dt = \frac{1}{s^{\beta+1}}\int_0^{\infty}(st)^{\beta}e^{-st}dst. \tag{21}$$

Then

$$P(s) = \frac{\Gamma(\beta+1)}{s^{\beta+1}}, \tag{22}$$

for $\mathbb{R}(s) > 0$, $\mathbb{R}(\beta) > 0$. The transform of the fractional derivative of order α is given by $s^{\alpha}\frac{\Gamma(\beta+1)}{s^{\beta+1}}$. So,

$$D_f^{\alpha}t^{\beta}\varepsilon(t) = \frac{\Gamma(\beta+1)}{\Gamma(\beta-\alpha+1)}t^{\beta-\alpha}\varepsilon(t), \tag{23}$$

for $\mathbb{R}(s) > 0$, $\mathbb{R}(\beta-\alpha) > 0$, generalizing the integer order formula [17]. We can go ahead and establish the validity of (23) for any $\alpha \in \mathbb{C}$ and β is not a negative integer. To obtain it, we use the rule of the derivative of the product. With $\beta > 0$, $p(t)$ is a continuous function. So, we can compute the Nth order derivative to obtain a distribution. The derivative of $\varepsilon(t)$ is $\delta(t)$ that appears multiplied by a power that is zero at $t = 0$:

$$D(t^\beta \varepsilon(t)) = \beta t^{\beta-1}\varepsilon(t) + t^\beta \delta(t) = \beta t^{\beta-1}\varepsilon(t).$$

However, repeating the procedure, the second term is no longer null, but we remove it to give us the finite part. With these considerations, we conclude that (23) remains valid provided that $\alpha \in \mathbb{C}$ and $\beta \in \mathbb{C} - \mathbb{Z}^-$. In particular, we have:

– Causal step response

$$D_f^\alpha \varepsilon(t) = \frac{1}{\Gamma(1-\alpha)} t^{-\alpha}\varepsilon(t), \tag{24}$$

with $\mathcal{L}[D^\alpha \varepsilon(t)] = s^{\alpha-1}$, for $\mathbb{R}(s) > 0$ and, from it,

– Causal impulse response

As $\delta(t) = D\varepsilon(t)$,

$$D_f^\alpha \delta(t) = \frac{t^{-\alpha-1}}{\Gamma(-\alpha)}\varepsilon(t), \tag{25}$$

with $\mathcal{L}[D^\alpha \delta(t)] = s^\alpha$, for $\mathbb{R}(s) > 0$.

– Anti-causal step response

$$D_b^\alpha \varepsilon(-t) = -\frac{1}{\Gamma(1-\alpha)} t^{-\alpha}\varepsilon(-t), \tag{26}$$

with $\mathcal{L}[D^\alpha \varepsilon(t)] = s^{\alpha-1}$, for $\mathbb{R}(s) < 0$ and, from it,

– Anti-causal impulse response

As $\delta(t) = -D\varepsilon(-t)$,

$$D_b^\alpha \delta(t) = -\frac{t^{-\alpha-1}}{\Gamma(-\alpha)}\varepsilon(-t), \tag{27}$$

with $\mathcal{L}[D^\alpha \delta(t)] - s^\alpha$, for $\mathbb{R}(s) < 0$.

Relation (23) can be written in a more useful form:

$$D_f^\alpha \frac{t^\beta}{\Gamma(\beta+1)}\varepsilon(t) = \frac{t^{\beta-\alpha}}{\Gamma(\beta-\alpha+1)}\varepsilon(t). \tag{28}$$

The generalization of the above result for $\beta \in \mathbb{Z}^-$ is not useful in studying LTI systems. We will not consider it [6].

Example 1 (The general power function). We start by computing the fractional derivative of the constant function. Let $f(t) = 1$ for every $t \in \mathbb{R}$ and $\alpha \in \mathbb{R} - \mathbb{Z}^-$. From (4), we have

$$D^\alpha 1 = \lim_{t\to 0} \frac{\sum_{k=0}^\infty \binom{\alpha}{k}(-1)^k}{h^\alpha} = \begin{cases} 0, & \text{if } \alpha > 0, \\ \infty, & \text{if } \alpha < 0. \end{cases} \tag{29}$$

To prove it, we are going to consider the partial sum of the series

$$\begin{aligned}\sum_{k=0}^{n}\binom{\alpha}{k}(-1)^k &= \binom{\alpha-1}{n}(-1)^n \\ &= \frac{1}{\Gamma(1-\alpha)}\frac{-\Gamma(\alpha+n+1)}{\Gamma(n+1)} \to \frac{1}{\Gamma(1-\alpha)}\frac{1}{n^\alpha}.\end{aligned} \tag{30}$$

As $n \to \infty$, we obtain the limits shown in (29) [16]. So, the α order fractional derivative of 1 is the null function. If $\alpha < 0$, the limit is infinite. So, there is no fractional "primitive" of a constant.

This means that when working in the context defined by (4), two functions with the same fractional derivative are equal.

The example illustrates an interesting result: "there are no fractional derivatives of the power function defined in $\mathbb{R}$." In fact, suppose that there is a fractional derivative of t^n, $t \in \mathbb{R}$, $n \in \mathbb{N}^+$. From properties (18) to (20), we must have

$$D^\alpha t^n = n! D^\alpha D^{-n} 1 = D^{-n} D^\alpha 1, \tag{31}$$

since $D^n \frac{t^n}{n!} = 1$, implying that $t^n = n! D^{-n} 1$. This means that we must be careful when trying to generalize the Taylor series. We conclude also that we cannot compute the fractional derivative of a function by using directly its Taylor expansion.

The same result could be obtained directly from (4). It is enough to remark that a power function tends to infinite when the argument tends to $-\infty$.

3 Time responses of fractional LTI systems

3.1 In terms of the Mittag-Leffler function

The transfer function of a given system can be used to compute the impulse response. So, we must perform a LT inverse computation. In general, we use the inversion integral, but it can be very difficult, since the poles may be unknown and we may have branch points.

Now we return to (1). The general case is hard to solve because it is difficult to find the poles of the system. We will consider the simpler case where one of these two conditions holds:

- the α_n are rational numbers that we will write in the form p_n/q_n. Let p and q be the least common multiples of p_n and q_n; then $\alpha_n = np/q$, where n and q are positive integers. So, $\alpha_n = n.\alpha$, with $\alpha = 1/q$ (a differential equation with $\alpha = 1/2$ is said semidifferential). The coefficients and orders do not coincide necessarily with the previous ones, since some of the coefficients can be zero. For example, the equation $[aD^{1/2} + bD^{1/3}]y(t) = x(t)$ transforms into $[aD^{3.\frac{1}{6}} + bD^{2.\frac{1}{6}} + 0.D^{\frac{1}{6}}]y(t) = x(t)$;

– the α_n are irrational numbers multiples of a given α.

When comparing with the integer order case, we performed a substitution $s \to s^\alpha$. This implies that the interval $[0, \pi)$ is transformed into the interval $[0, \alpha\pi)$.

Then (1) assumes the form

$$\sum_{k=0}^{N} a_k D^{\alpha k} y(t) = \sum_{k=0}^{M} b_k D^{\alpha k} x(t) \tag{32}$$

and the corresponding transfer function is

$$H(s) = \frac{B(s)}{A(s)} = \frac{\sum_{k=0}^{M} b_k s^{k\alpha}}{\sum_{k=0}^{N} a_k s^{k\alpha}}. \tag{33}$$

Without losing generality, we will assume in the following that $0 < \alpha \leq 1$. With (33) we can perform the inversion quite easily, by following the steps:

1. Transform $H(s)$ into $H(u)$, by substitution of s^α for u;
2. The denominator polynomial in $H(u)$ is the indicial polynomial or characteristic pseudo-polynomial. We will call its roots "pseudo-poles." Some or even all may be true poles. Perform the expansion of $H(z)$ in partial fractions;
3. Substitute back z for s^α, to obtain the partial fraction decomposition:

$$H(s) = \sum_{k=0}^{N_p} \frac{A_k}{(s^\alpha - p_k)^{n_k}}, \tag{34}$$

where N_p is the number of pseudo-poles, p_k, and the n_k are the corresponding multiplicities.

We are going to invert a partial fraction using the form

$$F(s) = \frac{1}{(s^\alpha - p)}. \tag{35}$$

Remark 5. It is simple to observe that having solved the order one problem, the others are very easy. In fact, we verify that

$$\frac{1}{(s^\alpha - p)^2} = D_p\left[\frac{1}{s^\alpha - p}\right], \tag{36}$$

where D_p means the derivative in order to p. The generalization is simple, but it is not very useful.

We are going then to invert the partial fraction $F(s) = \frac{1}{s^\alpha - p}$ where p is any pseudo-pole. The inversion is easily performed using the geometric series:

$$\frac{1}{s^\alpha - p} = s^{-\alpha} \sum_{0}^{\infty} p^n s^{-\alpha n} = \sum_{1}^{\infty} p^{n-1} s^{-\alpha n}, \tag{37}$$

valid for $|ps^{-\alpha}| < 1$. Choosing the region of convergence $\mathbb{R}(s) > |ps^{-\alpha}|$, we will arrive at the causal inverse of $F(s)$:

$$f(t) = \sum_1^\infty p^{n-1} \frac{t^{n\alpha-1}}{\Gamma(n\alpha)} \varepsilon(t). \tag{38}$$

This function is called alpha-exponential and is normally expressed in terms of the Mittag-Leffler function (MLF). In fact, letting the MLF be defined by

$$E_{\alpha,\beta}(z) = \sum_0^\infty \frac{z^n}{\Gamma(\alpha n + \beta)}, \tag{39}$$

we can write

$$f(t) = t^{\alpha-1} E_{\alpha,\alpha}(pt^\alpha) \cdot \varepsilon(t). \tag{40}$$

Remark 6. The anti-causal solution can be easily obtained from (37) using (27).

The function $f(t)$ is the impulse response corresponding to the partial fraction $\frac{1}{s^\alpha - p}$. The step response can be obtained from (37) that gives successively

$$\mathcal{L}[r_\varepsilon(t)] = \frac{1}{s^\alpha - p} \cdot \frac{1}{s} = \sum_1^\infty p^{n-1} s^{-\alpha n-1}, \tag{41}$$

and

$$r_\varepsilon(t) = t^\alpha E_{\alpha,\alpha+1}(pt^\alpha) \cdot \varepsilon(t) = \frac{1}{p}[E_{\alpha,1}(pt^\alpha) - 1] \cdot \varepsilon(t). \tag{42}$$

With expressions (40) and (42), we are able to compute the impulse and step responses of any LTI systems defined by the transfer function (33).

Remark 7. Defining the ramp function by $r(t) = t \cdot \varepsilon(t)$ and knowing that its LT is $R(s) = \frac{1}{s^2}$ for $\mathbb{R}(s) > 0$, it is a simple task to show that the corresponding response is

$$r_r(t) = t^{\alpha+1} E_{\alpha,\alpha+2}(pt^\alpha) \cdot \varepsilon(t). \tag{43}$$

3.2 By integer/fractional decomposition

The solution supplied by the Taylor series, by MLF, or any series of the same type, masks the underlying structure of the transfer function. This limitation is revealed when we try to compute its inversion by using the Bromwich integral, or the Mellin's inverse formula. In fact, to obtain it, we must fix a branch cut line. As the transform must be analytic on the right half-complex plane we choose the left half-real axis. On

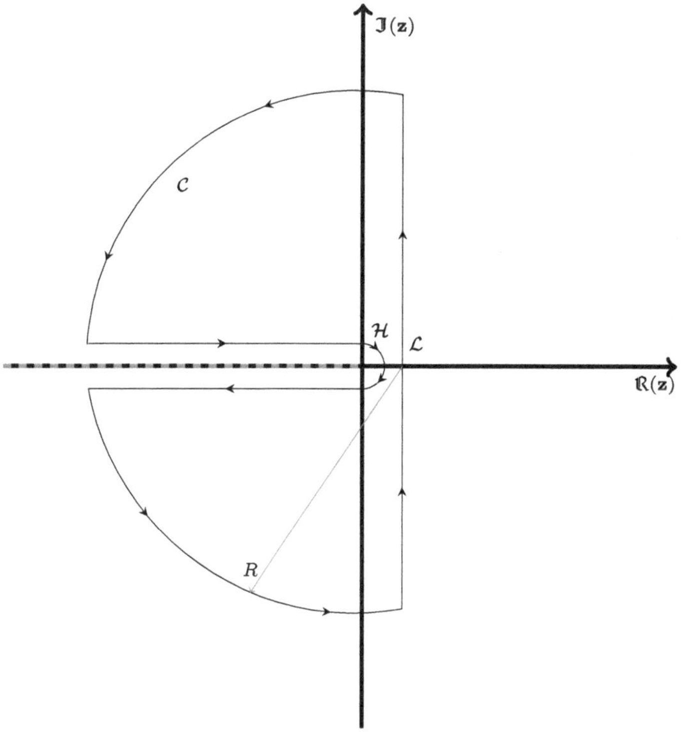

Figure 1: The path $\mathcal{C} \cup \mathcal{L} \cup \mathcal{H}$ used to invert the LT.

the other hand, function (35) is continuous from above on the branch cut line and verifies $\lim_{s\to\infty} H(s) = 0$, $|\arg(s)| < \pi$. We will assume that $\lim_{s\to 0} sH(s) = 0$. Let (γ_k, p_k), $k = 1, 2, \ldots, K$, be the pairs (order, root) such that $A(s) = \prod_1^K (s^{\gamma_k} - p_k)$. Let $K_0 \leq K$ be the number of poles. We remember that a given root p, corresponding to a given order γ, is a pole, if, when $s = |s|e^{i\theta}$ and $p = |p|e^{i\phi}$, we have $|s| = |p|^{1/\gamma}$ and $\theta = \phi/\gamma$. However, we have $|\phi| < \pi$ and, therefore, we only obtain a pole if $|\phi| < \pi/\gamma$. The $|\phi| = \pi$ case must be treated separately since it originates a point over the branch-cut line. The situations where $|\phi| > \pi$ do not originate poles. The corresponding responses are purely fractional.

In these conditions, we can use the integration path $\mathcal{C} \cup \mathcal{L} \cup \mathcal{H}$ in Figure 1 to compute the inverse LT. Let $u \in \mathbb{R}^+$ and consider $H(e^{i\pi}u)$ and $H(e^{-i\pi}u)$, the values of $H(s)$ immediately above and below the branch cut line. As the integral along the closed path is given by the residue theorem, we obtain

$$h(t) = \sum_{k=1}^{K_0} A_k e^{p^{1/\gamma_k} t} \varepsilon(t) + \frac{1}{2\pi i} \int_0^\infty [H(e^{-i\pi}u) - H(e^{i\pi}u)] e^{-\sigma t} \mathrm{d}u \cdot \varepsilon(t), \tag{44}$$

where $\varepsilon(t)$ is the unit step function and the constants A_k, $k = 1, 2, \ldots, K_0$, are the residues of (34) at p^{1/γ_k}.

Computing the LT of both sides in (44) we obtain

$$H(s) = H_i(s) + H_f(s), \tag{45}$$

where the integer order part is

$$H_i(s) = \sum_{k=1}^{K_0} \frac{A_k}{s - p^{1/\gamma_k}}, \quad \Re(s) > \max(\Re(p^{1/\gamma_k})), \tag{46}$$

and the fractional part is

$$H_f(s) = \frac{1}{2\pi i} \int_0^{\infty} [H(e^{-i\pi}u) - H(e^{i\pi}u)] \frac{1}{s+u} \mathrm{d}u, \tag{47}$$

valid for $\Re(s) > 0$.

3.3 Some reflections on stability

The above steps lead us to realize that:
- For $\gamma_k = 1$, $k = 1, 2, \ldots, K$, we have no fractional component. The transfer function is a sum of partial fractions and each one has a causal exponential for solution.
- For $\gamma_k < 1$, $k = 1, 2, \ldots, K$, we may have two components depending on the location of p_k in the complex plane:
  - If $|\arg(p_k)| > \pi \cdot \gamma_k$, $k = 1, 2, \ldots, K$, then we do not have the integer order component; it is a purely fractional system.
  - If $|\arg(p_k)| \leq \pi \cdot \gamma_k$, $k = 1, 2, \ldots, K$, for some k, then the system is a mixed character system in the sense that we have both components.
  - When $|\arg(p_k)| = \gamma_k \cdot \pi/2$, for some k, the integer order component is sinusoidal; however, the fractional component decreases to zero.
- The stability condition comes only from the integer order component. In fact, and as it is straightforward to verify, the integer order component is stable if $\gamma_k \cdot \pi/2 < |\arg(p_k)| < \pi \cdot \gamma_k$, $k = 1, 2, \ldots, K_0$, and unstable if $|\arg(p_k)| < \gamma_k \cdot \pi/2$, $k = 1, 2, \ldots, K_0$. The case $|\arg(p_k)| = \gamma_k \cdot \pi/2$ corresponds to a critically stable system (wide sense stable).

Concerning to the fractional part we can verify that $H(e^{-i\pi}\sigma) - H(e^{i\pi}\sigma)$ is a bounded function. Therefore, the integral in (44) is also bounded and decreases to zero as t goes to infinite.

In what follows, we will assume that we are dealing with stable systems. The above considerations allow us to conclude that we can have integer, fractional, and mixed behavior systems, namely:

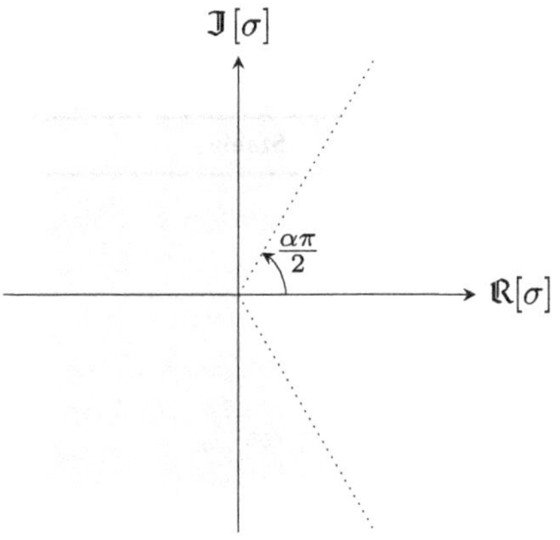

Figure 2: In grey: regions of the complex plane where the roots of $A(p)$ may lie in the case of a stable commensurate transfer function, $0 < \alpha < 1$.

- Classical integer order systems: they have impulse responses corresponding to linear combinations of exponentials. If they are strictly stable, the impulse responses go to zero very fast. They are short memory systems.
- New systems without poles: the exponential component vanishes. These are long memory systems. The arguments of the pseudo-polynomial roots have absolute values greater than π/γ, where γ is any derivative order smaller than 1.
- Mixed systems: they have both components. Some arguments of the pseudo-polynomial roots have absolute values larger than $\gamma \cdot \pi/2$.

Remark 8. These rules constitute what is known by Matignon's theorem. However, the approach followed here is more interesting, because it shows that the origin of instability is in the integer order part and not in the fractional one.

The regions where p may lie in the case of a stable commensurable transfer function are shown in Figure 2.

Remark 9. For integer systems, we reach the usual criterion of stability: all poles must have a negative real part.

Remark 10. Orders $\alpha > 2$ make the system necessarily unstable. For this reason, from here on, more often than not transfer functions will verify $0 < \alpha < 2$.

Example 2. Table 1 shows four applications of the above rules.

3.4 Causal periodic impulse response

In causal LTI system applications, the oscillatory behavior is very important, because often we are interested in removing, or attenuating, it. The transient behavior of such

Table 1: Stability of four systems verified by the above rules. The grey zone is where stable poles are found.

Transfer function	p	α	p in the complex plane	Stable
$\frac{1}{s-2s^{1/2}+4}$	$1 \pm \sqrt{3}i$	$\frac{1}{2}$	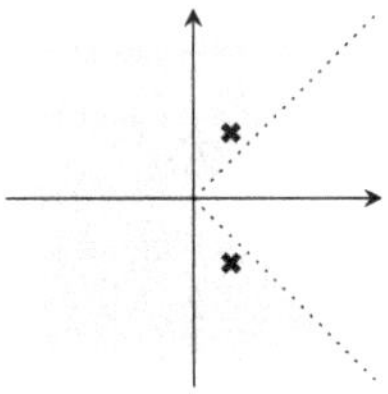	Yes
$\frac{1}{s^{4/3}-2s^{1/3}+4}$	$1 \pm \sqrt{3}i$	$\frac{1}{3}$	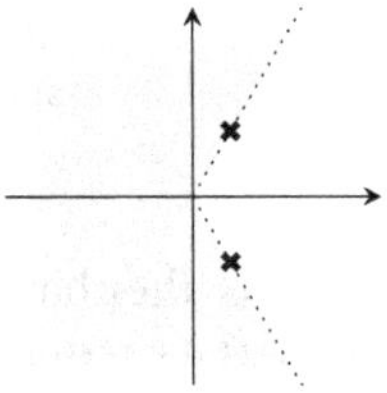	No (Yes, in wider sense)
$\frac{1}{s^{8/5}-2s^{4/5}+4}$	$1 \pm \sqrt{3}i$	$\frac{4}{5}$	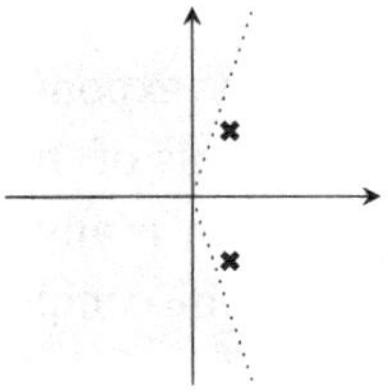	No
$\frac{1}{s^{12/5}+2s^{6/5}+4}$	$-1 \pm \sqrt{3}i$	$\frac{6}{5}$	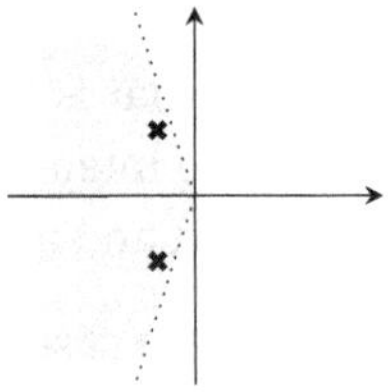	Yes

systems is embedded in the impulse or step responses. In the present case, we are interested in knowing under which conditions the impulse response is oscillatory and, in particular, when it is periodic. To study such problem, we consider the case where the transfer function can be decomposed in the form (34) [12]. A direct observation allows us to conclude, from the previous sub-section, that it is necessary to have

- $\arg(p_k) = \gamma_k \pi/2$ (pure oscillation condition),
- $n_k = 1$ (for stability reasons).

These conditions are not sufficient to ensure a causal periodic output. Let us proceed by considering one generic simple fraction. The impulse response, $h(t)$, corresponding to the simple system with transfer function $H(s) = \frac{1}{s^\gamma - p}$, where $|\arg p| = \gamma\pi/2$, is given

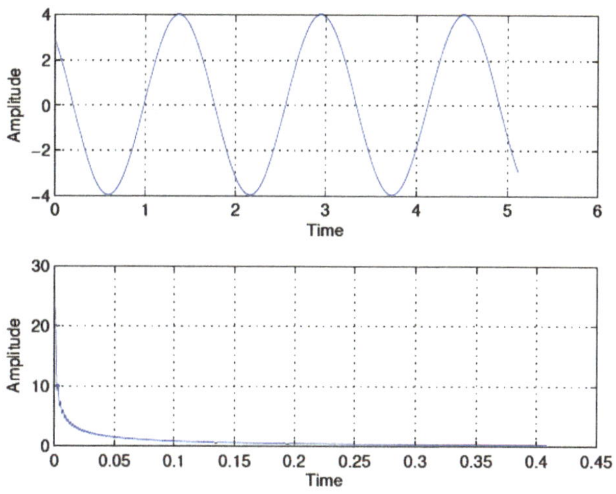

Figure 3: Integer (top) and fractional (bottom) parts of the impulse response for a system with $\alpha = 1/2$ and a pole $p = 2e^{i\pi/4}$.

by

$$h(t) = \frac{p^{1/\gamma-1}}{\gamma} e^{p^{1/\gamma}t}\varepsilon(t) + \frac{1}{2\pi i}\int_0^\infty \left[\frac{1}{e^{-i\pi\gamma}u - p} - \frac{1}{e^{i\pi\gamma}u - p}\right] e^{-ut}\mathrm{d}u \cdot \varepsilon(t). \tag{48}$$

This relation shows that:

1. The first term in (48) is sinusoidal for any value of γ;
2. The second term is not periodic for any value of γ;
3. The second therm is null iff $\gamma \in \mathbb{Z}$.

Figure 3 shows, for example, the integer and fractional parts of the impulse response for a system with $\alpha = 1/2$ and a pole $p = 2e^{i\pi/4}$.

3.5 The initial conditions

The free term of the output depends merely on the initial conditions (IC) that are the values assumed at the reference instant $t = t_0$ by the system variables associated with energy storage.

It is the structure of the system that should impose the IC and not the method of calculating the derivatives. The IC influence the output when we excite the system by means of a fresh input at instant $t = t_0$.

The IC problem in fractional LTI systems is not well solved with the one-sided LT. A suitable solution will be described next and is based on the following assumptions:

1. Equation (1) is defined for any $t \in \mathbb{R}$,
2. Our observation window is the unit step $\varepsilon(t - t_0)$,

3. The IC depend on the structure of the system and are independent of the tools that we adopt for the analysis,
4. The IC are the values assumed by the variables at the instant of opening the observation window.

These reasons point towards a simple procedure to rework the differential equation (1) to include the IC [7]:

$$\begin{aligned}\sum_{k=0}^{N} a_k[y(t)\cdot\varepsilon(t)]^{(\alpha_k)} &= \sum_{k=0}^{M} b_k[x(t)\cdot\varepsilon(t)]^{(\beta_k)}\\ &+\sum_{k=1}^{N} a_k\sum_{m=0}^{k-1} y^{(\alpha_m)}(t_0)\delta^{(\alpha_k-\alpha_m)}(t-t_0)\\ &-\sum_{k=1}^{M} b_k\sum_{m=0}^{k-1} x^{(\beta_m)}(t_0)\delta^{(\beta_k-\beta_m)}(t-t_0).\end{aligned} \tag{49}$$

Considering $\alpha_k = \beta_k = k$ and applying the two-sided LT to (49), we obtain an equation similar to the one obtained classically with the one-sided LT, provided that $t = 0$.

With this general formulation, we obtained a fractional version of the current LTI systems. In particular, we obtain for the commensurate case

$$\begin{aligned}\sum_{k=0}^{N} a_k[y(t)\cdot\varepsilon(t)]^{(\alpha k)} &= \sum_{k=0}^{M} b_k[x(t)\cdot\varepsilon(t)]^{(\alpha k)}\\ &+\sum_{k=1}^{N} a_k\sum_{m=0}^{k-1} y^{(\alpha m)}(t_0)\delta^{(\alpha k-\alpha m)}(t-t_0)\\ &-\sum_{k=1}^{M} b_k\sum_{m=0}^{k-1} x^{(\alpha m)}(t_0)\delta^{(\alpha k-\alpha m)}(t-t_0).\end{aligned} \tag{50}$$

We note that the initial instant does not need to be $t = 0$. We conclude that the above formulation is fully compatible with classic results. Obviously, we can apply the LT.

4 Impulse response of the general noncommensurate case

The solution of the general has been considered as a rather complex task and only particular cases have been addressed. We will present a simple algorithm for decomposing a given transfer function, not necessarily a rational function of s^α, but possibly assuming very general aspects, as it is the case of the function arctan(1/s). The approach is based on a negative power series expansion of the transfer function [10]. The term by term LT inversion of such series leads to a power series that generalizes well-known power series: MacLaurin, Taylor, and Puiseaux. The series is really a Dirichelet series because the power may be uncommensurate.

4.1 The general initial value theorem

The Abelian initial value theorem [18] is a very important result in dealing with the LT. This theorem relates the asymptotic behavior of a causal signal, $\phi(t)$, as $t \to 0^+$ to the asymptotic behavior of its LT, $\Phi(\sigma) = \mathcal{L}[\phi(t)]$, as $\sigma = \mathbb{R}(s) \to \infty$.

Theorem 1 (The initial-value theorem). *Assume that $\phi(t)$ is a causal signal such that, in some neighborhood of the origin, it is a regular distribution corresponding to an integrable function, and its LT is $\Phi(s)$, with region of convergence defined by $\mathbb{R}(s) > 0$. Also assume that there is a real number $\beta > -1$ such that $\lim_{t\to 0^+} \phi(t)t^\beta$ exists and is a finite complex value. Then*

$$\lim_{t\to 0^+} \frac{\phi(t)}{t^\beta} = \lim_{\sigma\to\infty} \frac{\sigma^{\beta+1}\Phi(\sigma)}{\Gamma(\beta+1)}, \tag{51}$$

where $\sigma \in \mathbb{R}^+$.

For proof, see [18] (Section 8.6, pp. 243–248).

As referred above, $\mathcal{L}[f^{(\alpha)}] = s^\alpha F(s)$ for $\mathbb{R}(s) > 0$. On the other hand attending to the usual initial value theorem

$$f(0+) = \lim_{\sigma\to\infty} \sigma F(\sigma),$$

we can write

$$\lim_{t\to 0^+} \frac{\phi(t)}{t^\beta} = \lim_{t\to 0^+} \phi^{(\beta)}(t) = \lim_{\sigma\to\infty} \sigma^{\beta+1}\Phi(\sigma). \tag{52}$$

Corollary 1.1. *Let $-1 < \alpha < \beta$. Then*

$$\lim_{t\to 0^+} \frac{\phi(t)}{t^\alpha} = 0.$$

If $\alpha > \beta$,

$$\lim_{t\to 0^+} \frac{\phi(t)}{t^\alpha} = \infty.$$

To prove the first part, we only have to write

$$\lim_{t\to 0^+} \frac{\phi(t)}{t^\alpha} = \lim_{t\to 0^+} \frac{\phi(t)}{t^\beta}\frac{t^\beta}{t^\alpha} = 0$$

and realize that the first factor has a finite limit given in (52) *and the second has zero as limit. The second part is immediately evident.*

This means that the function $\phi(t)$ has the behavior $\phi^{(\beta)}(0+)t^{\beta}$ near the origin.

4.2 The transfer function series representation

Let $H(s)$ be a transfer function and its associated region of convergence $\Re(s) > 0$. We are going to use the initial value theorem to obtain a decomposition of $H(s)$ into a sum of negative power functions plus an error term:

1. Define $R_0(s) = H(s)$ with inverse $r_0(t)$.
2. Let γ_0 be the real value such that

$$\lim_{\sigma\to\infty} \sigma^{\gamma_0} H(\sigma) = A_0,$$

where A_0 is finite and nonnull. Then, if

$$R_1(s) = H(s) - A_0 s^{-\gamma_0},$$

it is clear that $\lim_{\sigma\to\infty} \sigma^{\gamma_0} R_1(\sigma) = 0$ and $A_0 = \lim_{t\to 0^+} r_0(t) = h^{(\gamma_0-1)}(0^+)$.

3. Now, repeat the process. Let γ_1 be the real value such that

$$\lim_{\sigma\to\infty} \sigma^{\gamma_1} R_1(\sigma) = A_1,$$

where A_1 is finite and nonnull. Again introduce

$$R_2(s) = R_1(s) - A_1 s^{-\gamma_1} = H(s) - A_0 s^{-\gamma_0} - A_1 s^{-\gamma_1},$$

with $\lim_{\sigma\to\infty} \sigma^{\gamma_1} R_2(\sigma) = 0$ and $A_1 = \lim_{t\to 0^+} r_1^{(\gamma_1-1)}(t)$, with $r_1(t)$ as the inverse of $R_1(s)$.

4. In general, let γ_n be the real value such that

$$\lim_{\sigma\to\infty} \sigma^{\gamma_n} R_n(\sigma) = A_n,$$

where A_n is finite and nonnull. We arrive at the function:

$$R_n(s) = R_n(s) - A_{n-1} s^{-\gamma_{n-1}} = H(s) - \sum_0^{n-1} A_k s^{-\gamma_k} \tag{53}$$

and $\lim_{\sigma\to\infty} \sigma^{\gamma_n} R_n(\sigma) = 0$. As above $A_n = \lim_{t\to 0^+} r_{n-1}^{(\gamma_n-1)}(t) = r_{n-1}^{(\alpha_n)}(0^+)$ to be coherent with the initial value theorem: $\gamma_n = \alpha_n + 1$, for $n \in \mathbb{Z}_0^+$.

We can write

$$H(s) = \sum_{0}^{n-1} A_k s^{-\gamma_k} + R_n(s), \tag{54}$$

leading us to conclude that $H(s)$ can be expanded in a Laurent-like power series.

Theorem 2 (The generalized Laurent series). *It is easy to conclude that*[2] $|R_n(s)| = o\left(|\frac{1}{s^\gamma}|\right)$ *with* $\gamma > \gamma_n$, *allowing us to write what we can consider a generalized Laurent series:*

$$H(s) = \sum_{0}^{\infty} r_k^{(\alpha_k)}(0^+) s^{-\gamma_k}. \tag{55}$$

In the general case, it is not easy to state the convergence of this series. If $\alpha_k = k\alpha$, we can apply the theory of $\mathcal{Z}$ transform. In this case, if the sequence $A_k = r_k^{(k\alpha)}(0^+)$ is also of exponential order, and the series in (55) converges in the region that is the exterior of a circle with center at $s = 0$.

According to the development above, the successive functions $r_k^{(\alpha_k)}(t)$ come from $h(t)$ by removing the inverses of $r_m^{(\alpha_m)}(0^+)s^{-\gamma_m}$ for $m < k$. As it is easy to verify, the derivatives of order α_k of such terms are singular at the origin. So, the terms $r_k^{(\alpha_k)}(t)$ are equal to the analytic part of $h^{(\alpha_k)}(t)$. When the derivative orders are positive integers, this does not happen because the derivatives of the removed terms are zero or derivatives of the impulse $\delta(t)$, null for $t = 0^+$. In the following, we will represent them by $h_a^{(\alpha_k)}(t)$.

Theorem 3 (The generalized MacLaurin series). *Let* $h(t)$ *be the impulse response corresponding to* $H(s)$. *Then*

$$h(t) = \sum_{0}^{\infty} h_a^{(\gamma_k-1)}(0^+) \frac{t^{\gamma_k-1}}{\Gamma(\gamma_k)} \varepsilon(t), \tag{56}$$

where $\varepsilon(t)$ *is the Heaviside unit step function.*

That can be written as

$$h(t) = \sum_{0}^{\infty} h_a^{(\alpha_k)}(0^+) \frac{t^{\alpha_k}}{\Gamma(\alpha_k + 1)} \varepsilon(t). \tag{57}$$

In the commensurate case, $\alpha_k = \alpha k$, for $k \in \mathbb{Z}_0^+$, we obtain easily

$$h(t) = \sum_{0}^{\infty} h_a^{(\alpha k)}(0^+) \frac{t^{\alpha k}}{\Gamma(\alpha k + 1)} \varepsilon(t). \tag{58}$$

With $\alpha = 1$, we obtain the traditional MacLaurin series.

---

2 $o\left(|\frac{1}{s^\gamma}|\right)$ means, in the time domain, that the error in time is less than $\frac{t^{\alpha_n}}{\Gamma(\alpha_n+1)}$.

4.2.1 Examples

1. $H(s) = \frac{s^{\alpha-\beta}}{s^\alpha+1}$
   It is the LT of the generalized MLF. We are going to proceed as pointed above. We have

$$A_0 = 1 \quad \text{and} \quad \gamma_0 = \beta,$$
$$A_1 = -1 \quad \text{and} \quad \gamma_1 = \alpha + \beta.$$

   Repeating the process, we obtain

$$R_n(s) = (-1)^n \frac{1}{s^{\beta+n.\alpha}(s^\alpha + 1)},$$

   with

$$A_n = (-1)^n \quad \text{and} \quad \gamma_n = n.\alpha + \beta.$$

   This leads to

$$H(s) = s^{-\beta} \sum_0^\infty (-1)^n s^{-n\alpha}.$$

   Its inverse is easily obtained:

$$h(t) = \sum_0^\infty (-1)^n \frac{t^{n\alpha+\beta-1}}{\Gamma(n\alpha+\beta)}.\varepsilon(t),$$

   in agreement with well-known results [5].
2. $H(s) = \arctan(1/s)$
   We are going to proceed as above, but as the function is more involved, we are going to do all the steps with more detail. We have

$$\lim_{\sigma\to\infty} \sigma.\arctan(1/\sigma) = \lim_{\sigma\to\infty} \frac{\arctan(1/\sigma)}{1/\sigma} = \lim_{v\to 0} \frac{\arctan(v)}{v} = \lim_{v\to 0} \frac{\frac{1}{1+v^2}}{1} = 1.$$

   Then $A_0 = 1$ and $\gamma_0 = 1$ and $R_0(s) = \arctan(1/s) - 1/s$. In the second step, we obtain

$$\lim_{\sigma\to\infty} \sigma^2[\sigma.\arctan(1/\sigma) - 1/\sigma] = 0.$$

   So,

$$\lim_{\sigma\to\infty} \sigma^3[\arctan(1/\sigma) - 1/\sigma] = \lim_{v\to 0} \frac{\arctan(v) - v}{v^3} = \lim_{v\to 0} \frac{D^3[\arctan(v)]}{3!}$$
$$= \lim_{v\to 0} \frac{D^2[\frac{1}{1+v^2}]}{3!} = -\frac{2!}{3!} = -\frac{1}{3}.$$

We obtain $A_1 = -1/3$ and $\gamma_1 = 3$ and $R_1(s) = \arctan(1/s) - 1/s + 1/(3s^3)$. In the third step, we obtain as above:

$$\lim_{\sigma\to\infty} \sigma^4[\arctan(1/\sigma) - 1/\sigma + 1/(3\sigma^3)] = 0,$$

leading us to try the next power

$$\begin{aligned}
&\lim_{\sigma\to\infty} \sigma^5[\arctan(1/\sigma) - 1/\sigma + 1/(3\sigma^3)] \\
&= \lim_{v\to 0} \frac{\arctan(v) - v + (v^3)/3}{v^5} \\
&= \lim_{v\to 0} \frac{D^5[\arctan(v)]}{5!} = \lim_{v\to 0} \frac{D^4[\frac{1}{1+v^2}]}{5!} = -\frac{4!}{5!} = -\frac{1}{5},
\end{aligned}$$

leading to $A_2 = -1/5$ and $\gamma_2 = 5$ and $R_2(s) = \arctan(1/s) - 1/s + 1/(3s^3) - 1/(5s^5)$. Generalizing and assuming, we proceed indefinitely to obtain

$$H(s) = \sum_0^\infty (-1)^n \frac{1}{2n+1} s^{-2n-1}.$$

The inverse LT gives

$$h(t) = \sum_0^\infty (-1)^n \frac{1}{2n+1} \frac{t^{2n}}{(2n)!} .\varepsilon(t) = \frac{\sum_0^\infty (-1)^n \frac{t^{2n+1}}{(2n+1)!}}{t} .\varepsilon(t) = \frac{\sin(t)}{t} .\varepsilon(t),$$

that is the result shown in [1].

3. The following example illustrates a very difficult situation, although without practical application. Let $H(s) = \frac{1}{s^{\sqrt{3}} + s^{\sqrt{2}} + 1}$.
The coefficients of the generalized Laurent series are

$$A_n = (-1)^n,$$

and the powers

$$\gamma_n = (n + 1)\sqrt{3} - n\sqrt{2},$$

for $n = 0, 1, \ldots$
Although this is an academic example, it is not difficult to devise similar situations with rational powers. For that purpose, it is enough to use rational approximations for the square roots.

4.3 The rational format commensurate case

In the case of transfer functions of with constant coefficients, we have

$$G(s) = \frac{\sum_{k=0}^M b_k s^{\alpha k}}{\sum_{k=0}^N a_k s^{\alpha k}}, \tag{59}$$

where α is a positive real. We will assume that $M \le N$ and normalize a_N to 1.

In the following, we particularize the above algorithm for this case. We start by introducing a modified sequence of the numerator coefficients

$$b^0_{N-i} = \begin{cases} b_{M-i} & i \leq M, \\ 0 & i > M, \end{cases} \quad i = 0, \ldots, N \tag{60}$$

and represent the denominator in (59) by $D(s)$.

It is not difficult to conclude that

$$A_0 = b_M \quad \text{and} \quad \gamma_0 = (N - M)\alpha. \tag{61}$$

On the other hand, the repetition of the algorithm above leads to

$$R_n(s) = \frac{\sum_{k=0}^{N-1} b^n_k s^{\alpha k}}{s^{(N+n-M)\alpha} D(s)}, \tag{62}$$

where[3]

$$b^n_{N-k} = b^{n-1}_{N-1-k} - A_{n-1}.a_N - 1 - k, \quad k = 0, \ldots, N - 1 \tag{63}$$

and

$$A_n = b^n_N \quad \text{and} \quad \gamma_n = (N + n - M + 1)\alpha. \tag{64}$$

The computation can be done using a table similar to the Routh table used to test the stability of LTI systems. The table starts by creating the first two lines. In the first, we put the symmetric of the denominator coefficients in the second we put coefficients of numerator inserting zeros, if necessary, to have polynomials of equal order.

$-a_N = -1$	$-a_{N-1}$	...	...	...	$-a_1$	$-a_0$	
b^0_N	b^0_{N-1}	...	...	...	b^0_1	b^0_0	$A_0 = b^0_N$

To continue, assume that we want to compute the nth line:

1. Multiply the first line by $A_{n-1} = b^{n-1}_N$
2. Add to the nth line
3. Discard the first coefficient.

If one (or several) zeros appear in the beginning of the line, we slide the values to the left. This means that a power of s^α equal to the number of slides plus 1.

Let us see an example.

3 The upper letters are superscripts, not powers.

Example 3. $H(s) = \frac{1}{s^{2\alpha}+3s^{\alpha}+2}$

This is a transfer function studied in [7]. The corresponding impulse response was computed there and is given by

$$h(t) = \sum_{n=2}^{\infty} (-1)^n [2^{n-1} - 1] \frac{t^{n\alpha-1}}{\Gamma(n\alpha)} \varepsilon(t).$$

For this computation, the current procedure was followed:
1. Computation of the partial fraction decomposition
2. Inversion of the two fractions
3. Addition of the results.

We are going to obtain the same result using the new procedure we just describe.

−1	−3	−2		
0	0	1	$A_0 = 0$	we must slide the line 2 columns
1	0	0	$A_0 = 1$	$\gamma_0 = 2\alpha$
−3	−2	0	$A_1 = -3$	$\gamma_1 = 3\alpha$
7	6	0	$A_2 = 7$	$\gamma_2 = 4\alpha$
−15	−14	0	$A_3 = -15$	$\gamma_3 = 5\alpha$
31	30	0	$A_4 = 31$	$\gamma_4 = 6\alpha$

Example 4. $H(s) = \frac{1}{s^3+s^{\frac{1}{3}}+1}$

This example is more difficult than the previous one as the fraction decomposition involves a larger number of terms, of complex nature:

$$\begin{aligned} H(s) &= \frac{1}{s^3 + s^{\frac{1}{3}} + 1} \\ &= \left[(s^{2\alpha} - 2.02s^{\alpha} + 1.17) \cdot (s^{2\alpha} - 0.94s^{\alpha} + 1.13) \cdot (s^{2\alpha} + 0.56s^{\alpha} + 1.04)\right. \\ &\quad \left.\cdot (s^{2\alpha} + 1.58s^{\alpha} + 0.87) \cdot (s^{\alpha} + 0.82)\right]^{-1}\Big|_{\alpha=\frac{1}{3}}. \end{aligned}$$

The algorithm implementation as a computer program is straightforward, allowing us to compute the expansion (23) with as many terms as desired. The first ten coefficients corresponding to this example are presented in the following table ($\alpha = \frac{1}{3}$). The full computation of these coefficients is presented below.

n	0	1	2	3	4	5	6	7	8	9
$\gamma_n =$	9α	17α	18α	25α	26α	27α	33α	34α	35α	36α
$A_n =$	1	−1	−1	1	2	1	−1	−3	−3	−1

Here, we deal with the decomposition of the transfer function $H(s) = \frac{1}{s^3+s^{1/3}+1} = \frac{1}{s^{9\alpha}+s^{\alpha}+1}|_{\alpha=\frac{1}{3}}$ as series of the form $H(s) = \sum_0^{n-1} A_k s^{-\gamma_k} + R_n(s)$. The next table presents the computation of the series coefficients of $H(s)$ for $n = 10$:

−1	0	0	0	0	0	0	0	−1	−1		
1	0	0	0	0	0	0	0	0	0	$\gamma_0 = 9\alpha$	$A_0 = 1$
−1	−1	0	0	0	0	0	0	0	0	$\gamma_1 = 17\alpha$	$A_1 = -1$
−1	0	0	0	0	0	0	1	1	0	$\gamma_2 = 18\alpha$	$A_2 = -1$
1	2	1	0	0	0	0	0	0	0	$\gamma_3 = 25\alpha$	$A_3 = 1$
2	1	0	0	0	0	0	−1	−1	0	$\gamma_4 = 26\alpha$	$A_4 = 2$
1	0	0	0	0	0	−1	−3	−2	0	$\gamma_5 = 27\alpha$	$A_5 = 1$
−1	−3	−3	−1	0	0	0	0	0	0	$\gamma_6 = 33\alpha$	$A_6 = -1$
−3	−3	−1	0	0	0	0	1	1	0	$\gamma_7 = 34\alpha$	$A_7 = -3$
−3	−1	0	0	0	0	1	4	3	0	$\gamma_8 = 35\alpha$	$A_8 = -3$
−1	0	0	0	0	1	4	6	3	0	$\gamma_9 = 36\alpha$	$A_9 = -1$

5 Multiple input, multiple output systems and state variable representation

5.1 Fractional MIMO transfer function matrix

In previous sections, we considered single-input/single-output (SISO) systems. Most of the theory then presented can be generalized of the multiple-input/multiple-output (MIMO) cases, but can be somehow more involved. However, the state-space description of a system is similar for both SISO or MIMO systems.

Consider now a MIMO system with m inputs and p outputs, related by linear, time-invariant differential equations. The LT of these equations allows us to find $p \times m$ SISO transfer functions, each of them relating one input with one output when all other inputs are zero. These transfer functions can be collected into a matrix, that will, since the system is linear, relate all inputs to all outputs.

Definition 1 (Fractional MIMO transfer function matrix). A fractional MIMO transfer function matrix for a plant with m inputs and p outputs is a $p \times m$ matrix $\mathbf{G}$, the elements of which are SISO fractional transfer functions.

Definition 2 (State variables). Consider a system with a known dynamical behavior. Suppose that its inputs are known from some arbitrary time instant t on. The state variables of the system are those in a set, with as few elements as possible, such that, knowing them at instant t, it is possible to calculate the future behavior of the system.

Remark 11. As is well known, system state variables do not form a unique set. If $\mathbf{x}(t)$ is a $n \times 1$ vector with the system state variables, and $\mathbf{P}$ is a $n \times n$ invertible matrix, then the variables in vector $\mathbf{w}(t) = \mathbf{P}\mathbf{x}(t)$ also are state variables.

5.2 State-space representations of MIMO systems

Definition 3 (State-space representation of a LTI system). Given state-space representations LTI systems a MIMO system, wherein the outputs and the inputs are related by linear, time-invariant fractional differential equations, its state-space representation is

$$\begin{cases} D^{\boldsymbol{\alpha}}\mathbf{x}(t) = \mathbf{A}\mathbf{x}(t) + \mathbf{B}\mathbf{u}(t), \\ \mathbf{y}(t) = \mathbf{C}\mathbf{x}(t) + \mathbf{D}\mathbf{u}(t), \end{cases} \tag{65}$$

where
- the first equality is the state equation; it is the equation of dynamics;
- the second equality is the output equation; it is the observation equation;
- $\mathbf{x}(t) = [x_1(t)\, x_2(t) \, \cdots \, x_n(t)]^T$ is a vector with n state variables;
- $\boldsymbol{\alpha} = [\alpha_1 \, \alpha_2 \, \cdots \alpha_n]^T$ is a vector with n differentiation orders, all of them positive: $\alpha_k > 0$, $k = 1, 2, \ldots, n$;
- the derivative of vectorial order $\boldsymbol{\alpha}$ is defined as D

$$D^{\boldsymbol{\alpha}}\mathbf{x}(t) = [D^{\alpha_1}x_1(t)\, D^{\alpha_2}x_2(t) \, \cdots \, D^{\alpha_n}x_n(t)]^T; \tag{66}$$

- $\mathbf{A}$ is the $n \times n$ state matrix;
- $\mathbf{u}(t) = [u_1(t)\, u_2(t) \, \cdots \, u_m(t)]^T$ is a vector with m inputs;
- $\mathbf{B}$ is the $n \times m$ input matrix;
- $\mathbf{y}(t) = [y_1(t)\, y_2(t) \, \cdots \, y_p(t)]^T$ is a vector with p outputs;
- $\mathbf{C}$ is the $p \times n$ output matrix;
- $\mathbf{D}$ is the $p \times m$ direct transmission matrix.

In the state equation, $t = 0$ is assumed to be the initial instant. Anyway, we can use another one.

Define the diagonal matrix $\operatorname{diag}(s^{\boldsymbol{\alpha}})$ with diagonal elements s^{α_k}, $k = 1, 2, \ldots, n$. The Laplace transformation of (69) is

$$\begin{cases} \operatorname{diag}(s^{\boldsymbol{\alpha}})\mathbf{X}(s) = \mathbf{A}\mathbf{X}(s) + \mathbf{B}\mathbf{U}(s), \\ \mathbf{Y}(s) = \mathbf{C}\mathbf{X}(s) + \mathbf{D}\mathbf{U}(s) \end{cases} \tag{67}$$

$$\Rightarrow \begin{cases} \mathbf{X}(s) = (\operatorname{diag}(s^{\boldsymbol{\alpha}}) - \mathbf{A})^{-1}\mathbf{B}\mathbf{U}(s), \\ \mathbf{Y}(s) = \mathbf{C}\mathbf{X}(s) + \mathbf{D}\mathbf{U}(s) \end{cases}$$

and thus

$$\frac{\mathbf{Y}(s)}{\mathbf{U}(s)} = \mathbf{C}\big[\mathrm{diag}(s^{\alpha}) - \mathbf{A}\big]^{-1}\mathbf{B} + \mathbf{D}, \tag{68}$$

which is a MIMO fractional transfer function matrix.

Definition 4 (Commensurable state-space representation). The state space representation is said to be commensurable when all the orders α_k, $k = 1, 2, \ldots, n$ are equal, in which case (65) can be written thus:

$$\mathrm{diag}(s^{\alpha}) = s^{\alpha}\mathbf{I}. \tag{69}$$

5.3 State transition operator

Return back to the dynamic equation

$$D^{\alpha}\mathbf{x}(t) = \mathbf{A}\mathbf{x}(t) + \mathbf{B}\mathbf{u}(t)$$

and assume that the input is null for $t > 0$. Using the results introduced in Section 3.5, we can write

$$\mathrm{diag}(s^{\alpha})\mathbf{X}(s) - \mathrm{diag}(s^{\alpha-1})\mathbf{x}(0) = \mathbf{A}\mathbf{x}(t).$$

Therefore, the transition operator verifies

$$\mathbf{x}(t) = \Phi(0,t)\cdot\mathbf{x}(0) \tag{70}$$

and is given by the inverse LT,

$$\Phi(0,t) = \mathcal{L}^{-1}\{\mathrm{diag}(s^{\alpha-1})\cdot\big[\mathrm{diag}(s^{\alpha}) - \mathbf{A}\big]^{-1}\}, \tag{71}$$

that is nothing else than the multidimensional MLF

$$\Phi(0,t) = \sum_{0}^{\infty}\mathbf{A}^{n}\frac{t^{n\alpha}}{\Gamma(n\alpha+1)}\varepsilon(t) \tag{72}$$

and leads easily to the general state transition operator.

Definition 5 (State transition operator). The state at two instants t and τ are related by the "State transition operator," $\Phi(\tau,t)$,

$$\mathbf{x}(t) = \Phi(\tau,t)\cdot\mathbf{x}(\tau), \tag{73}$$

that is defined by

$$\Phi(\tau,t) = \Phi(0,t)\cdot\Phi^{-1}(0,\tau). \tag{74}$$

with $\Phi(0,0) = \mathbf{I}$.

It can be shown that this operator verifies the usual properties, namely the semi-group property. To finish this study, we only need to point out how we can compute $\Phi^{-1}(0,\tau)$. If we write $\Phi^{-1}(0,\tau) = \sum_0^\infty \mathbf{B}_n \frac{\tau^{n\alpha}}{\Gamma(n\alpha+1)}\varepsilon(\tau)$, we can show that $\mathbf{B}_0 = \mathbf{I}$ and

$$\mathbf{B}_n = \sum_{k=0}^{n} \binom{n\alpha}{k\alpha} \mathbf{A}^k \mathbf{B}_{n-k}, \quad n = 1,2,\dots \tag{75}$$

allowing us to compute $\Phi^{-1}(0,\tau)$ in a recursive way.

5.4 State-space representations of SISO systems

Consider the SISO commensurable transfer function given by (33), with $n = \max\{N, M\}$ and $a_n = 1$. Just as for the integer case $\alpha = 1$, there are infinite state-space representations, but some are particularly useful and have special names.

Definition 6 (Controllable canonical form). The controllable canonical form of (67) is

$$\begin{cases} s^\alpha \begin{bmatrix} X_1 \\ X_2 \\ \vdots \\ X_{n-1} \\ X_n \end{bmatrix} = \overbrace{\begin{bmatrix} 0 & 1 & 0 & \cdots & 0 \\ 0 & 0 & 1 & \cdots & 0 \\ \vdots & \vdots & \vdots & \ddots & \vdots \\ 0 & 0 & 0 & \cdots & 1 \\ -a_0 & -a_1 & -a_2 & \cdots & -a_{n-1} \end{bmatrix}}^{\mathbf{A}} \begin{bmatrix} X_1 \\ X_2 \\ \vdots \\ X_{n-1} \\ X_n \end{bmatrix} + \overbrace{\begin{bmatrix} 0 \\ 0 \\ \vdots \\ 0 \\ 1 \end{bmatrix}}^{\mathbf{B}} U \\ Y = \underbrace{\begin{bmatrix} b_0 - a_0 b_n & b_1 - a_1 b_n & \cdots & b_{n-1} - a_{n-1} b_n \end{bmatrix}}_{\mathbf{C}} \begin{bmatrix} X_1 \\ X_2 \\ \vdots \\ X_n \end{bmatrix} + \underbrace{b_n}_{\mathbf{D}} U \end{cases} \tag{76}$$

Definition 7 (Observable canonical form). The observable canonical form of (67) is

$$\begin{cases} s^\alpha \begin{bmatrix} X_1 \\ X_2 \\ \vdots \\ X_{n-1} \\ X_n \end{bmatrix} = \overbrace{\begin{bmatrix} 0 & 0 & \cdots & 0 & -a_0 \\ 1 & 0 & \cdots & 0 & -a_1 \\ \vdots & \vdots & \ddots & \vdots & \vdots \\ 0 & 0 & \cdots & 1 & -a_{n-1} \end{bmatrix}}^{\mathbf{A}} \begin{bmatrix} X_1 \\ X_2 \\ \vdots \\ X_n \end{bmatrix} + \overbrace{\begin{bmatrix} b_0 - a_0 b_n \\ b_1 - a_1 b_n \\ \vdots \\ b_{n-1} - a_{n-1} b_n \end{bmatrix}}^{\mathbf{B}} U \\ Y = \underbrace{\begin{bmatrix} 0 & 0 & \cdots & 0 & 1 \end{bmatrix}}_{\mathbf{C}} \begin{bmatrix} X_1 \\ X_2 \\ \vdots \\ X_{n-1} \\ X_n \end{bmatrix} + \underbrace{b_n}_{\mathbf{D}} U \end{cases} \tag{77}$$

Remark 12. Notice that both in (76) and in (76),
- **A** is a $n \times n$ matrix,
- **B** is a $n \times 1$ vector,
- **C** is a $1 \times n$ vector, and
- **D** is a scalar.

Remark 13. There are other interesting representations, namely the “diagonal canonical form” and the “Jordan canonical form.”

Table 2: Time responses of important transfer functions studied in this chapter, valid for $t > 0$.

Function	Unit impulse response	Unit step response	Unit slope ramp response	See above
s^{α}	$\frac{t^{-\alpha-1}}{\Gamma(-\alpha)}$	$\frac{t^{-\alpha}}{\Gamma(1-\alpha)}$	$\frac{t^{-\alpha+1}}{\Gamma(2-\alpha)}$	Subsection 2.3, equations (24)–(25)
$\frac{1}{s^{\alpha}\pm p}$	$t^{\alpha-1}E_{\alpha,\alpha}(\mp pt^{\alpha})$	$t^{\alpha}E_{\alpha,\alpha+1}(\mp pt^{\alpha})$	$t^{\alpha+1}E_{\alpha,\alpha+2}(\mp pt^{\alpha})$	Subsection 3.1, equations (40), (42)–(43)
$\frac{s^{\alpha-\beta}}{s^{\alpha}\pm p}$	$t^{\beta-1}E_{\alpha,\beta}(\mp pt^{\alpha})$	$t^{\beta}E_{\alpha,\beta+1}(\mp pt^{\alpha})$	$t^{\beta+1}E_{\alpha,\beta+2}(\mp pt^{\alpha})$	Subsection 4.2.1, example 1, cf. equation (39)

All the state-space representations of a same system have state matrices $\mathbf{A}$ with the same eigenvalues.

Theorem 4. *The roots of the polynomial $A(p) = p^n + \sum_{k=0}^{n-1} a_k p^k$, built with the denominator coefficients of transfer function $G(s)$, given by* (67), *are the eigenvalues of matrix* $\mathbf{A}$ *of any of the state-space representations of $G(s)$.*

Remark 14. Table 2 sums up the most important time responses studied in this chapter.

Bibliography

[1] R. V. Churchill, *Operational Mathematics*, 3rd edition, McGraw-Hill, 1972.

[2] J. C. Ferreira, R. F. Hoskins, and J. Sousa-Pinto, *Introduction to the Theory of Distributions*, CRC Press, 1997.

[3] I. M. Gel'fand and G. E. Shilov, *Generalized Functions. Volume I: Properties and Operations*, Academic Press, New York and London, 1964.

[4] R. Herrmann, *Fractional Calculus: An Introduction for Physicists*, World Scientific Publishing Co., Singapore, 2011.

[5] A. A. Kilbas, H. M. Srivastava, and J. J. Trujillo, *Theory and Applications of Fractional Differential Equations*, North-Holland Mathematics Studies, vol. 204, Elsevier, Amsterdam, 2006.

[6] M. D. Ortigueira, *Fractional Calculus for Scientists and Engineers*, Lecture Notes in Electrical Engineering, 2nd edition, Springer, Berlin, Heidelberg, 2011.

[7] M. D. Ortigueira and F. J. Coito, System initial conditions vs derivative initial conditions, *Computers & Mathematics with Applications*, **59**(5) (2010), 1782–1789.

[8] M. D. Ortigueira, R. L. Magin, J. J. Trujillo, and M. Pilar Velasco, A real regularised fractional derivative, *Signal, Image and Video Processing*, **6**(3) (2012), 351–358.

[9] M. D. Ortigueira, M. Rivero, and J. J. Trujillo, The incremental ratio based causal fractional calculus, *International Journal of Bifurcation and Chaos*, **22**(04) (2012), 1250078.

[10] M. D. Ortigueira, J. J. Trujillo, V. I. Martynyuk, and F. J. Coito, A generalized power series and its application in the inversion of transfer functions, *Signal Processing*, **107** (2015), 238–245.

[11] M. D. Ortigueira and J. Tenreiro Machado, Which derivative?, *Fractal and Fractional*, **1**(1) (2017), 1–13.
[12] M. D. Ortigueira, J. Tenreiro Machado, and J. J. Trujillo, Fractional derivatives and periodic functions, *International Journal of Dynamics and Control*, **5**(1) (2017), 72–78.
[13] M. D. Ortigueira and J. A. Tenreiro Machado, What is a fractional derivative?, *Journal of Computational Physics*, **293** (2015), 4–13.
[14] M. D. Ortigueira and J. J. Trujillo, A unified approach to fractional derivatives, *Communications in Nonlinear Science and Numerical Simulation*, **17**(12) (2012), 5151–5157.
[15] I. Podlubny, *Fractional Differential Equations: an Introduction to Fractional Derivatives, Fractional Differential Equations, to Methods of Their Solution and Some of Their Applications*, Elsevier, 1999.
[16] S. G. Samko, A. A. Kilbas, and O. I. Marichev, *Fractional Integrals and Derivatives: Theory and Applications*, Gordon and Breach Science Publishers, Amsterdam, 1993.
[17] D. Valério, J. J. Trujillo, M. Rivero, J. Tenreiro Machado, and D. Baleanu, Fractional calculus: a survey of useful formulas, *The European Physical Journal Special Topics*, **222**(8) (2013), 1827–1846.
[18] A. H. Zemanian, *Distribution Theory and Transform Analysis: an Introduction to Generalized Functions, with Applications*, Courier Corporation, 1965.

Duarte Valério, Manuel D. Ortigueira, J. A. Tenreiro Machado, and António M. Lopes

Continuous-time fractional linear systems: steady-state responses

Abstract: This chapter presents the frequency responses of several single-input, single-output fractional-order systems—commensurate, noncommensurate, and implicit. Both analytical expressions and the Bode, Nyquist, and Nichols diagrams are shown for several typical transfer functions. The chapter also addresses the Levy identification method for fractional orders, integer-order approximations of fractional-order systems (Crone, Carlson, Matsuda), and the Nyquist stability criterion for closed-loop commensurate systems.

Keywords: Fractional calculus, continuous-time signals and systems, steady-state responses, transfer function, stability, state-space

MSC 2010: 26A33, 93C80, 93D25, 93B30, 93B50

1 Introduction

This chapter presents the frequency response of fractional-order transfer functions (TF), that is to say, TF including noninteger powers of the Laplace variable s. We will only address the single-input, single-output case. Multiple-inputs and multiple-outputs lead to fractional-order matrices of TF, which are a straightforward generalization of the material here presented, or, as an alternative, a state-space representation can be used (see Chapter "Continuous-time fractional linear systems: transient responses" of Volume 6).

In the remainder of this Introduction, fractional TF are defined and typified. Section 2 gives the frequency responses of nine typical TF often found in applications.

Acknowledgement: This work was supported by FCT, through IDMEC, under LAETA, project UID/EMS/50022/2013.

Duarte Valério, IDMEC, Instituto Superior Técnico, Universidade de Lisboa, Av. Rovisco Pais 1, 1049–001 Lisboa, Portugal, e-mail: duarte.valerio@tecnico.ulisboa.pt
Manuel D. Ortigueira, CTS-UNINOVA and DEE of NOVA School of Science and Technology of NOVA University of Lisbon, Campus da FCT da UNL, Quinta da Torre, 2829–516 Caparica, Portugal, e-mail: mdo@fct.unl.pt
J. A. Tenreiro Machado, Institute of Engineering, Polytechnic of Porto, Dept. of Electrical Engineering, Rua Dr. António Bernardino de Almeida, 431, 4249–015 Porto, Portugal, e-mail: jtm@isep.ipp.pt
António M. Lopes, UISPA–LAETA/INEGI, Faculty of Engineering, University of Porto, Rua Dr. Roberto Frias, 4200–465, Porto, Portugal, e-mail: aml@fe.up.pt

https://doi.org/10.1515/9783110571745-007

Frequency responses are used in Section 3 for model identification, and in Section 4 to build approximations of fractional TF. Stability is addressed in Section 5. Finally, Section 6 sums up results for a stochastic input.

1.1 From the Laplace transform to explicit fractional-order transfer functions

Consider a linear, time-invariant differential equation with fractional derivatives, with the general form

$$\sum_{k=0}^{N} a_k D^{\alpha_k} y(t) = \sum_{k=0}^{M} b_k D^{\beta_k} x(t), \quad t \in \mathbb{R}, \tag{1}$$

where a_k and b_k are real constants, and α_k and β_k are the derivative orders. We adopt the definitions of operator D given in Chapter "Continuous-time fractional linear systems: transient responses" of Volume 6, chosen so that the derivative of an exponential is also an exponential. In particular, for the forward derivative we have [12]

$$D_f^{\alpha} e^{st} = s^{\alpha} e^{st} \quad \text{if } \mathbb{R}(s) > 0, \tag{2}$$

while for the backward derivative

$$D_b^{\alpha} e^{st} = s^{\alpha} e^{st} \quad \text{if } \mathbb{R}(s) < 0. \tag{3}$$

This property is very important. In fact, while s can be any complex when the order is integer, this does not happen in the fractional case: fractional derivatives impose a region of convergence tied to causality—the forward derivative is causal, while the backward is anti-causal. In what follows, we will use D_f^{α} only, omitting the subscript f or b.

Introducing the convolution between two functions $x(t)$ and $y(t)$, defined by

$$x(t) * y(t) = \int_{-\infty}^{\infty} x(\tau) y(t-\tau) \mathrm{d}\tau, \tag{4}$$

the impulse response of (1), $h(t)$, is the solution of

$$\sum_{k=0}^{N} a_k D^{\alpha_k} h(t) = \sum_{k=0}^{M} b_k D^{\beta_k} \delta(t), \tag{5}$$

where $\delta(t)$ is the Dirac distribution [2, 3, 13].

It can be easily verified that

$$x(t) = \delta(t) * x(t), \tag{6}$$

and

$$[D^\gamma h(t)] * x(t) = D^\gamma [h(t) * x(t)], \quad \gamma \in \mathbb{C}. \tag{7}$$

The solution of (1) is the convolution of $x(t)$ with the impulse response

$$y(t) = h(t) * x(t). \tag{8}$$

In particular, the solution of the differential equation (1) when $x(t) = e^{st}$, $s \in \mathbb{C}$, is given by

$$y(t) = H(s)e^{st}, \tag{9}$$

provided that $H(s)$ exists. As we are dealing with the forward fractional derivative, this solution exists on the right half-plane, $\mathbb{R}(s) > 0$. The eigenvalue $H(s)$ is the TF of the system defined by the differential equation (1), and is the Laplace transform of the impulse response:

$$H(s) = \int_{-\infty}^{\infty} h(\tau)e^{-s\tau} d\tau, \quad \mathrm{Re}(s) > 0. \tag{10}$$

Inserting (9) into (1), we conclude that

$$H(s) = \frac{B(s)}{A(s)} = \frac{\sum_{k=0}^{M} b_k s^{\beta_k}}{\sum_{k=0}^{N} a_k s^{\alpha_k}}. \tag{11}$$

Expressions such as (11) are called *explicit* (or rational) fractional-order TF.

For now, we will consider that the *characteristic pseudo-polynomial* in the denominator, $A(s)$, does not have a root at the particular value of s at hand. Later we will consider the cases where it does.

Example 1

Consider the differential equation

$$y''' + y'' - 4y' + 2y = x(t), \tag{12}$$

and let $x(t) = e^{-2t}$. The solution is given by

$$y(t) = \frac{1}{-2^3 + 2^2 + 8 + 2} e^{-2t} = \frac{1}{6} e^{-2t}. \tag{13}$$

Example 2

Consider the fractional differential equation

$$y^{(3\alpha)} + y^{(2\alpha)} - 4y^{(\alpha)} + 2y = x(t), \tag{14}$$

and let $\alpha = \frac{1}{2}$ and $x(t) = e^{2t}$. The solution is given by

$$y(t) = \frac{1}{2\sqrt{2} + 2 - 4\sqrt{2} + 2} e^{2t} = \frac{2 + \sqrt{2}}{4} e^{2t}. \tag{15}$$

If $x(t) = e^{-2t}$, then there is no solution, unless we use the backward derivative.

The above results are valid for $\Re(s) < 0$. We can extend the domain to include the case of a sinusoidal input ($x(t) = e^{st}$ with $\Re(s) = 0$) and, as the forward derivative of a constant is zero, also the $s = 0$ case.

1.2 Types of transfer functions

Explicit fractional-order TF are often divided in two types [16]:

- Fractional-order commensurate TF are given by the ratio of two polynomials in s^{α}, where α is the order of commensurability:

$$G(s) = \frac{\sum_{k=0}^{m} b_k s^{k\alpha}}{\sum_{k=0}^{n} a_k s^{k\alpha}}. \tag{16}$$

  The order of the polynomials in the numerator and the denominator are m and n, respectively.
- Fractional-order noncommensurate TF are given by

$$G(s) = \frac{\sum_{k=0}^{m} b_k s^{\beta_k}}{\sum_{k=0}^{n} a_k s^{\alpha_k}}, \tag{17}$$

  so that exponents β_k and α_k do not have a least common multiple α; if they do, then (17) reduces to (16).

Integer order TF are given by (16) with $\alpha = 1$, or by (17) with $\alpha_k, \beta_k \in \mathbb{N}$.

Explicit fractional-order TF have this name in opposition to implicit (or irrational) ones:

- Fractional-order implicit (or irrational) TF are input–output relations that cannot be written in the form of (17), and involve noninteger powers of s. An example would be $G(s) = \frac{1}{\sqrt{s+1}}$; see more examples in Sections 2.3, 2.6, 2.7, and 2.8 below.

Implicit TF correspond to differential equations with infinite dimension, rather than to those such as (1).

The root locus of commensurate TF can be obtained by means of the method presented in [8]. For explicit and implicit TF, we can adopt the algorithm proposed in [4, 15].

1.3 Pseudo-poles and pseudo-zeros of commensurate transfer functions

Commensurate TF, given by (16), can be given another form:

$$H(s) = \frac{B(s)}{A(s)} = b_M \frac{\prod_{k=1}^{M_z} (s^\alpha - z_k)^{m_k}}{\prod_{n=1}^{N_p} (s^\alpha - p_n)^{m_n}}, \tag{18}$$

where M_z and N_z are the number of distinct numerator and denominator roots, respectively, and the integers m_k and m_n denote the multiplicities of the roots.

The roots of the numerator will be called "pseudo-zeroes," while those of the denominator will be called "pseudo-poles." The use of the term "pseudo" is related to the fact that a zero/pole of $B(s)/A(s)$ may, or may not, be a zero/pole of the system. To understand this, let a be a root of $B(s)$ of multiplicity 1. Then we can write $B(s) = (s^\alpha - a) \cdot B_1(s)$. The zero exists if $s^\alpha - a = 0$. This only happens if $|\arg(a)| < \alpha\pi$. For the roots of $A(s)$, the situation is similar and allows us to formulate stability rules and decompose the impulse response in two components: one integer, and another fractional (see Chapter "Continuous-time fractional linear systems: transient responses" of Volume 6).

For our objectives, it is preferable to give another form to the TF:

$$H(s) = K_0 \frac{\prod_{k=1}^{M_z} \left(\left(\frac{s}{\zeta_k}\right)^\alpha + 1\right)^{m_k}}{\prod_{n=1}^{N_p} \left(\left(\frac{s}{\theta_n}\right)^\alpha + 1\right)^{m_n}}, \tag{19}$$

where K_0 is called *static gain*. As it is easy to verify, $\theta_k = (-p_k)^{1/\alpha}$ and $\zeta_k = (-z_k)^{1/\alpha}$. With these changes, (18) assumes a more classical form.

Assuming that all the coefficients a_k and b_k in (16) are real, then all values of z_k and p_k are either real or, being complex, appear in conjugate pairs. In this case, its is usual to join the corresponding terms:

$$\left(\left(\frac{s}{\theta}\right)^\alpha + 1\right)\left(\left(\frac{s}{\theta^*}\right)^\alpha + 1\right) = \left(\frac{s^\alpha}{|\theta|}\right)^2 + \left(\frac{s}{|\theta|}\right)^\alpha \frac{2\mathbb{R}(\theta^\alpha)}{|\theta|^\alpha} + 1. \tag{20}$$

In the following, we will assume multiplicities equal to 1.

1.4 The frequency response

The property (2) of the derivatives we are using can be extended to the imaginary axis yielding the following important result:

$$D^{\alpha} e^{i\omega_0 t} = (i\omega_0)^{\alpha} e^{i\omega_0 t} \quad \text{if } t, \omega_0 \in \mathbb{R}, \tag{21}$$

where $i = \sqrt{-1}$.

As $x(t) = \cos(\omega_o t) = \frac{1}{2} e^{i\omega_0 t} + \frac{1}{2} e^{-i\omega_0 t}$, we obtain

$$D^{\alpha} \cos(\omega_0 t) = (\omega_0)^{\alpha} \cos\left(\omega_0 t + \alpha \frac{\pi}{2}\right). \tag{22}$$

For the sine function, the result is similar: in general, and for a system defined by (1), if $x(t) = e^{i\omega_0 t}$, then we obtain immediately from (9)

$$y(t) = H(i\omega_0) e^{i\omega_0 t}. \tag{23}$$

When the input is $x(t) = \cos(\omega_o t)$, we have

$$x(t) = \cos(\omega_o t) = \frac{1}{2} e^{i\omega_0 t} + \frac{1}{2} e^{-i\omega_0 t}, \tag{24}$$

that leads to

$$y(t) = H(i\omega_0) \frac{1}{2} e^{i\omega_0 t} + H(-i\omega_0) \frac{1}{2} e^{-i\omega_0 t}. \tag{25}$$

Therefore, we have the following theorem.

Theorem 1 (Sinusoidal case). *The output of the system defined by* (1) *when* $x(t) = \cos(\omega_o t)$ *is given by*

$$y(t) = \left|H(\omega_0)\right| \cos(\omega_o t + \varphi(\omega_0)). \tag{26}$$

This is in fact true for implicit fractional TF as well.

Remark: When $H(i\omega_0) = 0$, $y(t)$ is identically null. This is the reason why we call *filters* the systems described by linear differential equations.

In engineering applications, function $H(i\omega) = |H(\omega)| e^{\varphi(\omega)}$ is called *frequency response* [14]. If the coefficients in (1) are real, then:

1. $|H(\omega)|$ is the *amplitude spectrum*, or *gain*, and is an even function,
2. $\varphi(\omega)$ is the *phase spectrum*, or simply *phase*, and is an odd function.

The gain is often given in decibels (dB), $A(\omega) = 20 \log_{10} |H(\omega)|$, and it can be seen that, for explicit fractional TF,

$$\begin{aligned} A(\omega) = 20 \log_{10}(K_0) + \sum_{k=1}^{M_z} m_k 20 \log_{10}\left|\left(\frac{i\omega}{\zeta_k}\right)^{\alpha} + 1\right| \\ - \sum_{k=1}^{N_p} n_k 20 \log_{10}\left|\left(\frac{i\omega}{\theta_k}\right)^{\alpha} + 1\right|, \end{aligned} \tag{27}$$

$$\varphi(\omega) = \arg(K_0) + \sum_{k=1}^{M_z} m_k \arg\left[\left(\frac{i\omega}{\zeta_k}\right)^{\alpha} + 1\right] - \sum_{k=1}^{N_p} n_k \arg\left[\left(\frac{i\omega}{\theta_k}\right)^{\alpha} + 1\right]. \tag{28}$$

The gain and phase are usually represented for positive frequencies, since, for real, physically realizable systems, the gain is an even function, while the phase is odd. Three graphical representations of frequency responses are given in what follows for each TF:

1. Bode diagrams show $A(\omega)$ and $\varphi(\omega)$ in two separate plots (often reduced to asymptotes), with frequencies in logarithmic scale in the abscissas, and $A(\omega)$ in dB;
2. polar plots show $\Im[H(\omega)]$ as a function of $\Re[H(\omega)]$;
3. Nichols diagrams show $A(\omega)$, usually in dB, as a function of $\varphi(\omega)$, in degrees.

The Nyquist diagram is similar to the polar plot, but shows the response for $\omega \in \mathbb{R}$ and not only for $\omega > 0$. Since $H(-i\omega)$ is the complex conjugate of $H(i\omega)$, the Nyquist diagram has the same curve found in the polar plot, and then another curve, symmetrical in relation to the real axis. The diagrams shown below are polar plots, for simplicity and clarity, but they allow finding easily the Nyquist diagram, necessary for the Nyquist criterion addressed in Section 5.

Example 3

Consider again the equation of Example 1, but change the second member:

$$y''' + y'' - 4y' + 2y = x'' - 4x \tag{29}$$

and assume that $x(t) = e^{i\pi t}$. Then

$$y(t) = \frac{\pi^2 + 4}{i\pi^3 + \pi^2 + 4i\pi - 2} e^{i\pi t}. \tag{30}$$

From this result, it would be immediate to compute the solution for $x(t) = \cos(\pi t)$ or $x(t) = \sin(\pi t)$. Remark that if $x(t) = e^{\pm i2t}$, then $y(t) = 0$.

2 Frequency responses of typical systems

In what follows, and for the reasons explained in Section 1.4, the frequency responses are given only for $\omega > 0$.

2.1 Frequency response of s^α

The frequency response of $G_1(s) = s^\alpha$, $\alpha \in \mathbb{R}$, is

$$G_1(j\omega) = (i\omega)^\alpha, \tag{31}$$

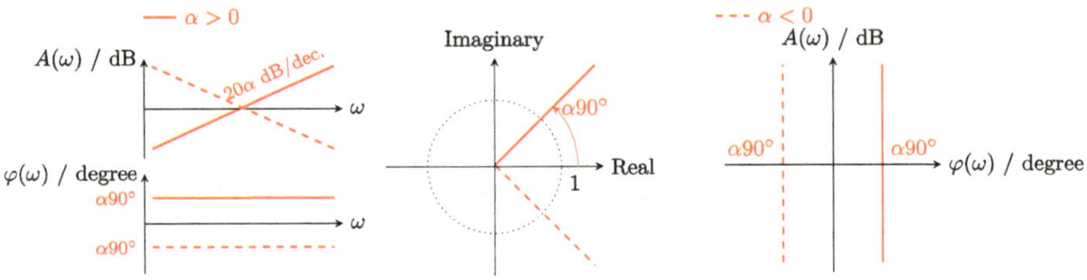

Figure 1: Frequency response of $G_1(s) = s^\alpha$, $\alpha \in \mathbb{R}$; left: Bode diagram; center: polar plot; right: Nichols diagram.

$$A_1(\omega) = 20\log_{10}\omega^\alpha = 20\alpha\log_{10}\omega, \tag{32}$$

$$\varphi_1(\omega) = \frac{\alpha\pi}{2}. \tag{33}$$

This corresponds to a straight line in all diagrams of Figure 1.

2.2 Frequency response of $((s/a)^\alpha + 1)^{\pm 1}$

The frequency response of $G_2(s) = ((\frac{s}{a})^\alpha + 1)^{\pm 1}$, $\alpha, a \in \mathbb{R}^+$, is

$$G_2(i\omega) = \left(\left(\frac{i\omega}{a}\right)^\alpha + 1\right)^{\pm 1},$$

$$\begin{aligned} A_2(\omega) &= 20\log_{10}\left[\sqrt{\left(\frac{\omega}{a}\right)^{2\alpha} + 2\left(\frac{\omega}{a}\right)^\alpha \cos\frac{\alpha\pi}{2} + 1}\right]^{\pm 1} \\ &= \pm 10\log_{10}\left[\left(\frac{\omega}{a}\right)^{2\alpha} + 2\left(\frac{\omega}{a}\right)^\alpha \cos\frac{\alpha\pi}{2} + 1\right], \end{aligned} \tag{34}$$

$$\varphi_2(\omega) = \pm\arctan\frac{(\frac{\omega}{a})^\alpha \sin\frac{\alpha\pi}{2}}{(\frac{\omega}{a})^\alpha \cos\frac{\alpha\pi}{2} + 1}. \tag{35}$$

The asymptotic expressions are useful for representing the diagrams. For low frequencies $\omega \ll a$, we have $G_2(i\omega) \approx 1$, while for high frequencies $\omega \gg a$, we have $G_2(i\omega) \approx \pm\left(\frac{i\omega}{a}\right)^\alpha$. At the corner frequency, $\omega = a\,\text{rad}\cdot\text{s}^{-1}$, $G_2(i\omega) = \left(1 + e^{i\frac{\alpha\pi}{2}}\right)^{\pm 1}$. Therefore, the asymptotic frequency response of $G_2(s)$ is given by

$$A_2(\omega) = \begin{cases} 0 & \omega \ll a, \\ 10\log_{10}(2 + 2\cos\frac{\alpha\pi}{2})^{\pm 1} & \omega = a, \\ \pm 20\alpha\log_{10}\omega \mp 20\alpha\log_{10}a & \omega \gg a, \end{cases} \tag{36}$$

$$\varphi_2(\omega) = \begin{cases} 0 & \omega \ll a, \\ \pm\frac{\alpha\pi}{4} & \omega = a, \\ \pm\frac{\alpha\pi}{2} & \omega \gg a. \end{cases} \tag{37}$$

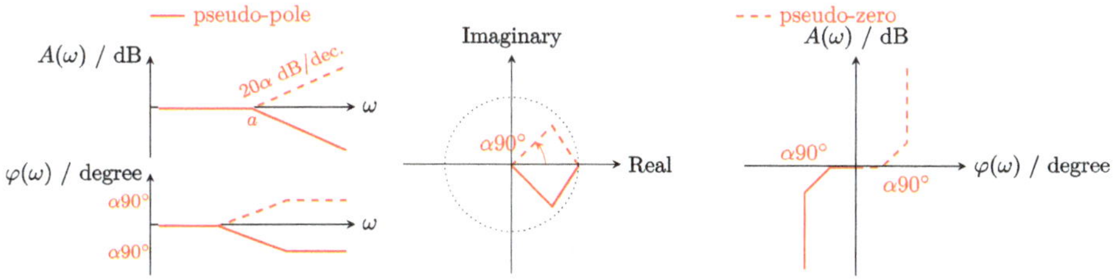

Figure 2: Asymptotic frequency response of $G_2(s) = \left(\left(\frac{s}{a}\right)^{\alpha} + 1\right)^{\pm 1}$, $a \in \mathbb{R}^+$, $\alpha \in]0, 2[$ and of $G_3(s) = \left(\frac{s}{a} + 1\right)^{\pm\alpha}$, $\alpha, a \in \mathbb{R}^+$; left: Bode diagram; center: polar plot; right: Nichols diagram.

Figure 2 shows the asymptotic frequency response, considering only the expressions for low and high frequencies.

Remark: Strictly speaking there is no need to study the case $|\alpha| > 1$. Calculating the derivative of $A_2(\omega)$, it can be seen that, if $1 < \alpha < 3 \vee 5 < \alpha < 7 \vee 9 < \alpha < 11\ldots$, the gain will have a peak at

$$\omega_{\text{peak}} = a\left(-\cos\frac{\alpha\pi}{2}\right)^{\frac{1}{\alpha}}, \tag{38}$$

$$A_2(\omega_{\text{peak}}) = \pm 20\log_{10}\left|\sin\frac{\alpha\pi}{2}\right|. \tag{39}$$

Otherwise, the gain increases or decreases monotonously. For the phase, $\arg[G_1(j\omega)]$, by calculating its derivative we verify that there are four possible cases. For the a pseudo-pole, they are as follows:

- If $0 < \alpha < 2 \vee 4 < \alpha < 6 \vee 8 < \alpha < 10 \vee \ldots$, the phase will decrease from 0 to $\lfloor\frac{\alpha}{4}\rfloor 2\pi - \alpha\frac{\pi}{2}$.
- If $2 < \alpha < 4 \vee 6 < \alpha < 8 \vee 10 < \alpha < 12 \vee \ldots$, the phase will increase from 0 to $(1 + \lfloor\frac{\alpha}{4}\rfloor)2\pi - \alpha\frac{\pi}{2}$.
- If $\alpha \in \{4, 8, 12, 16\ldots\}$, the phase is constant and equal to 0.
- If $\alpha \in \{2, 6, 10, 14\ldots\}$ the phase will have a discontinuity at $\omega = a\,\text{rad}\cdot\text{s}^{-1}$ and jumps from 0 to $\pm\pi$.

In the case of a pseudo-zero, the behavior is symmetrical.

See the two examples in Figure 3.

2.3 Frequency response of $(s/a + 1)^{\pm\alpha}$

The frequency response of the implicit fractional-order TF $G_3(s) = \left(\frac{s}{a} + 1\right)^{\pm\alpha}$, with $\alpha, a \in \mathbb{R}^+$, is

$$G_3(i\omega) = \left(\frac{i\omega + 1}{a}\right)^{\pm\alpha}, \tag{40}$$

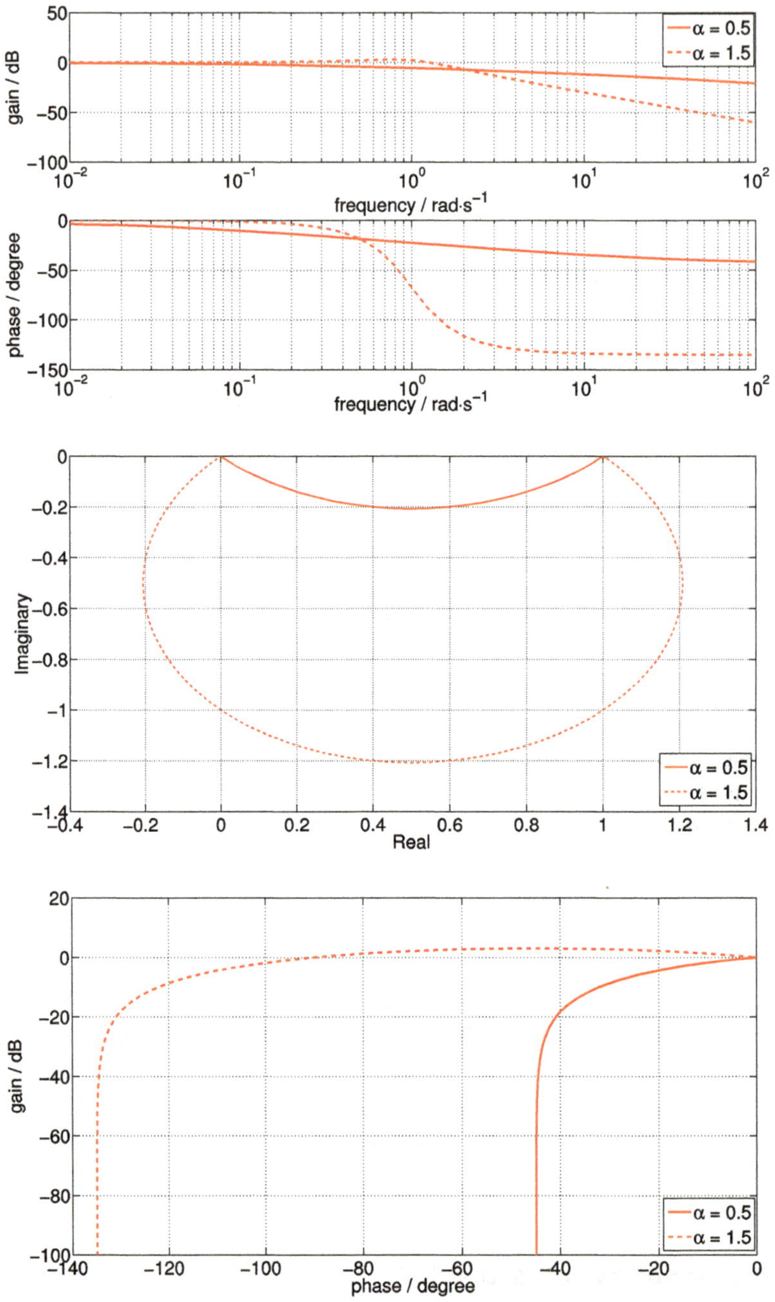

Figure 3: Frequency response of $G_2(s) = \frac{1}{s^\alpha+1}$, $\alpha = 0.5,\ 1.5$; top: Bode diagram; middle: polar plot; bottom: Nichols diagram.

$$A_3(\omega) = 20\log_{10}\left(1+\frac{\omega^2}{a^2}\right)^{\pm\frac{\alpha}{2}} = \pm 10\alpha\log_{10}\left(1+\frac{\omega^2}{a^2}\right), \tag{41}$$

$$\varphi_3(\omega) = \pm\alpha\arctan\frac{\omega}{a}. \tag{42}$$

There are no peaks in the gain, and the phase always decreases monotonously from 0 to $-\alpha\frac{\pi}{2}$. For low and high frequencies, the asymptotic approximations for $G_2(i\omega)$ are valid for $G_3(i\omega)$ as well. Therefore, the asymptotic frequency response of $G_3(s)$ is given by

$$A_2(\omega) = \begin{cases} 0 & \omega \ll a, \\ \pm 10\alpha\log_{10} 2 & \omega = a, \\ \pm 20\alpha\log_{10}\omega \mp 20\alpha\log_{10} a & \omega \gg a, \end{cases} \tag{43}$$

$$\varphi_2(\omega) = \begin{cases} 0 & \omega \ll a, \\ \pm\frac{\alpha\pi}{4} & \omega = a, \\ \pm\frac{\alpha\pi}{2} & \omega \gg a. \end{cases} \tag{44}$$

The asymptotic diagrams of Figure 2, considering only the expressions for low and high frequencies, apply to $G_3(s)$ as well. Figure 4 has the frequency responses of TF similar to those of Figure 3, so that the differences between them are more clear.

2.4 Frequency response of $\frac{1}{(s/a)^{2\alpha}+2\zeta(s/a)^{\alpha}+1}$

The frequency response of $G_4(s) = \frac{1}{(\frac{s}{a})^{2\alpha}+2\zeta(\frac{s}{a})^{\alpha}+1}$, with $\alpha, a, \zeta \in \mathbb{R}^+$, does not lend itself to simple analytical expressions of the gain and of the phase. Still it can be shown analytically [5] that $G_3(s)$ is stable only if

$$\zeta > -\cos\frac{\alpha\pi}{2} \wedge \zeta < 2. \tag{45}$$

Moreover, the gain may have two resonance frequencies, one resonance frequency, or no resonance frequency at all, depending on the values of α and ζ, as depicted in Figure 5 obtained numerically. This locus considers $0 < \alpha \leq 2$ only, since for $\alpha > 2$ the TF $G_4(s)$ is always unstable, as seen below in Section 5.1.

When $G_4(s)$ is stable, its phase decreases monotonously from 0 to $\alpha\pi$. Figure 6 shows the frequency responses of three stable TF illustrating the possible number of resonance frequencies.

2.5 Frequency response of $P + I/s^{\lambda} + Ds^{\mu}$ (fractional PID)

The frequency response of $G_5(s) = P + \frac{I}{s^{\lambda}} + Ds^{\mu}$, with $P, I, D, \lambda, \mu > 0$, is better studied making $P = K_p$, $I = \frac{K_p}{T_i^{\lambda}}$ and $D = K_d T_d^{\mu}$, for easier comparison with the irrational

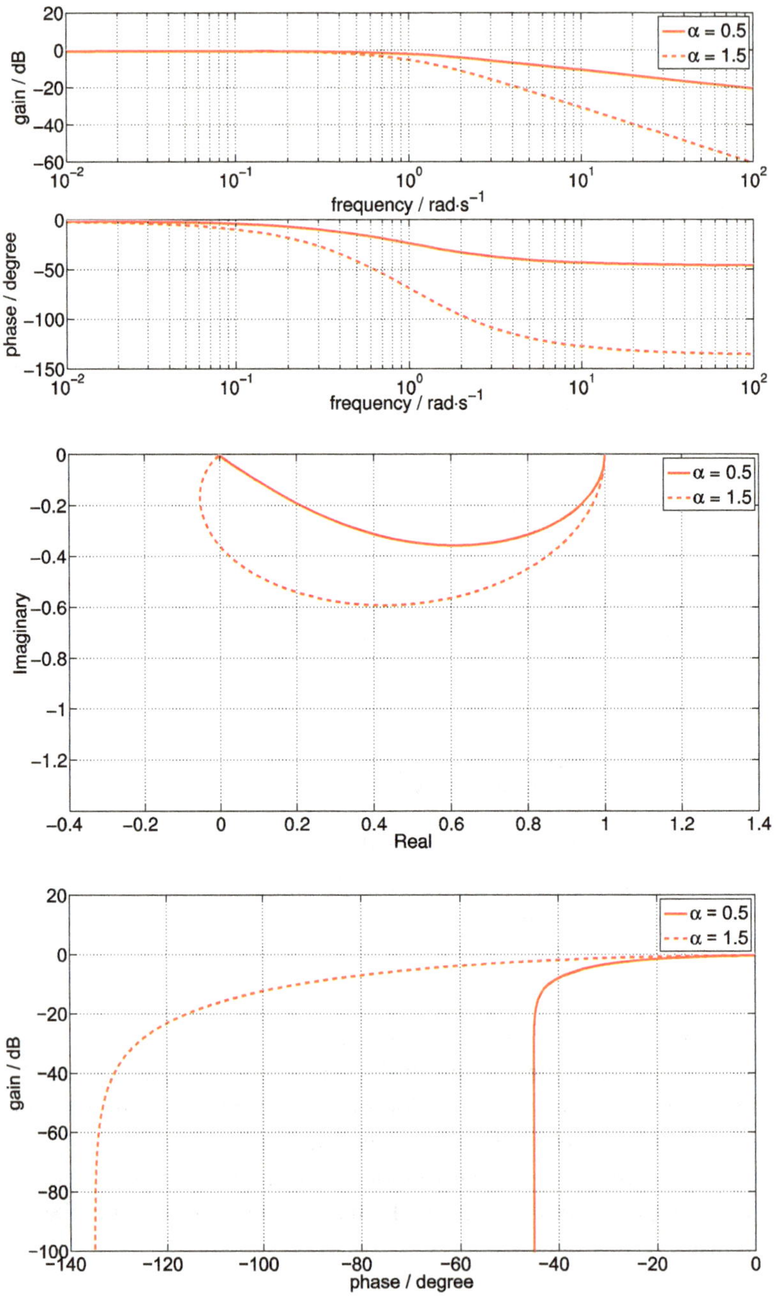

Figure 4: Frequency response of $G_3(s) = \frac{1}{(s+1)^{\alpha}}$, $\alpha = \{0.5, 1.5\}$; top: Bode diagram; middle: polar plot; bottom: Nichols diagram.

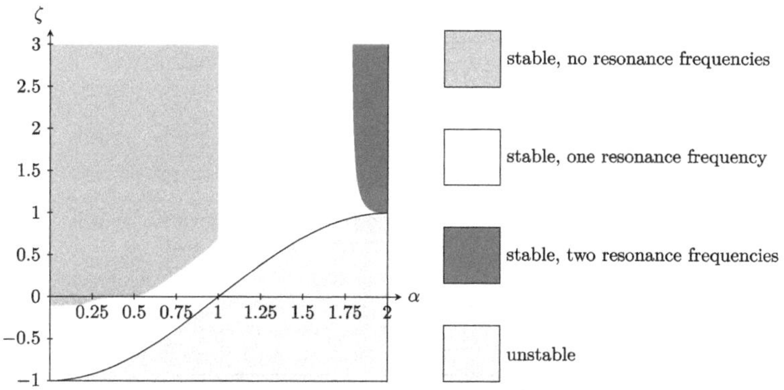

Figure 5: Different frequency responses of $G_4(s) = \frac{1}{(\frac{s}{a})^{2\alpha}+2\zeta(\frac{s}{a})^{\alpha}+1}$, $a, \zeta \in \mathbb{R}^+$, $0 < \alpha \leq 2$.

fractional PID. Thus we write

$$G_5(s) = K_p\left(1 + \frac{1}{(T_i s)^\lambda} + (T_d s)^\mu\right), \quad K_p, T_i, T_d, \lambda, \mu > 0, \tag{46}$$

$$G_5(i\omega) = K_p\left(1 + \frac{1}{(T_i i\omega)^\lambda} + (T_d i\omega)^\mu\right). \tag{47}$$

At low frequencies, the asymptotes of this frequency response are

$$G_5(i\omega) \approx \frac{K_p}{i^\lambda T_i \omega^\lambda}, \tag{48}$$

$$A_5(\omega) \approx 20 \log_{10} \frac{K_p}{T_i} - 20\lambda \log_{10} \omega, \tag{49}$$

$$\varphi_5(\omega) \approx \arg[i^{-\lambda}]. \tag{50}$$

At high frequencies, the asymptotes of this frequency response are

$$G_5(i\omega) \approx i^\mu K_p T_d \omega^\mu, \tag{51}$$

$$A_5(\omega) \approx 20 \log_{10}(K_p T_d) + 20\mu \log_{10} \omega, \tag{52}$$

$$\varphi_5(\omega) \approx \arg[i^\mu]. \tag{53}$$

In summary, the gain begins with a slope of -20λ dB per decade, and ends with a slope of 20μ dB per decade. The phase goes from $-\lambda\frac{\pi}{2}$ to either $\mu\frac{\pi}{2}$, or an angle with the same locus in the complex plane. This asymptotic behavior is shown if Figure 7. Moreover, Figure 8 shows the actual frequency responses for two particular cases, one with the zeros close to each other, and another with the zeros clearly separated.

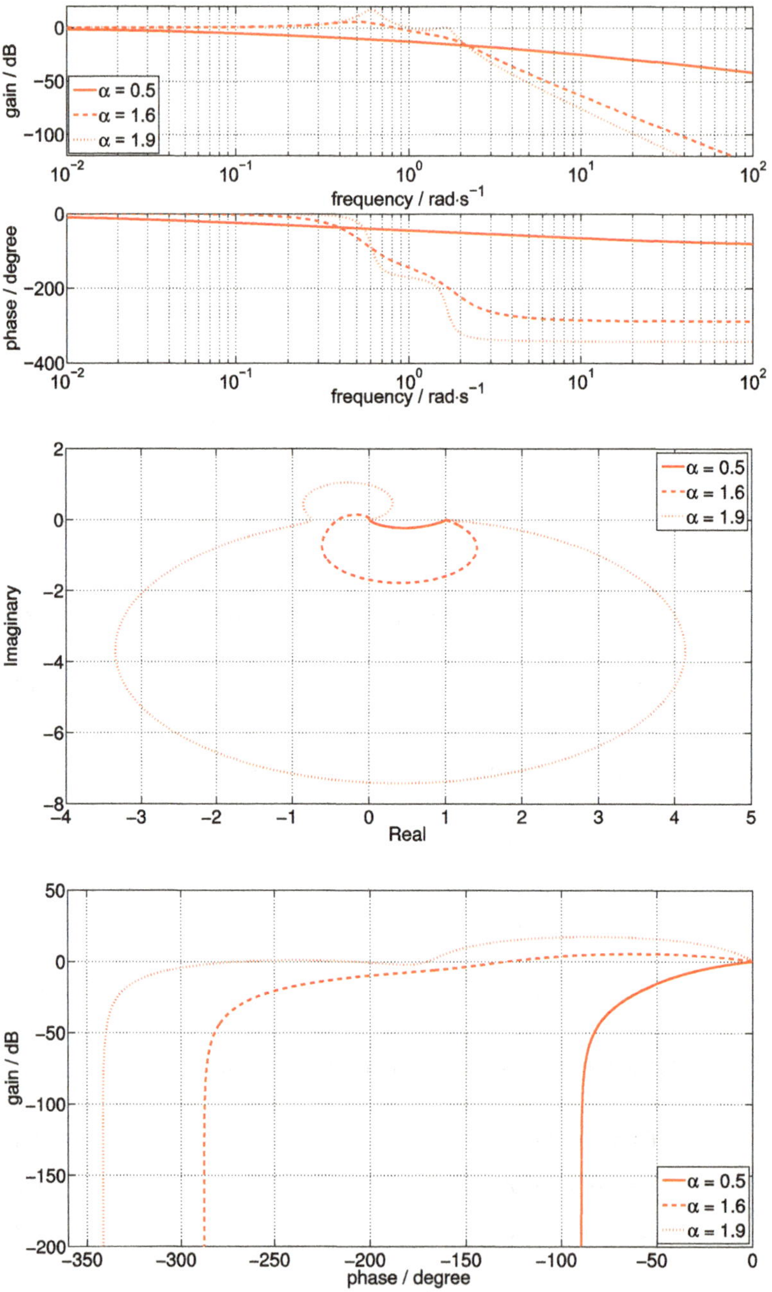

Figure 6: Frequency response of $G_4(s) = \frac{1}{s^{2\alpha}+3s^{\alpha}+1}$, $\alpha = \{0.5, 1.6, 1.9\}$; top: Bode diagram; middle: polar plot; bottom: Nichols diagram.

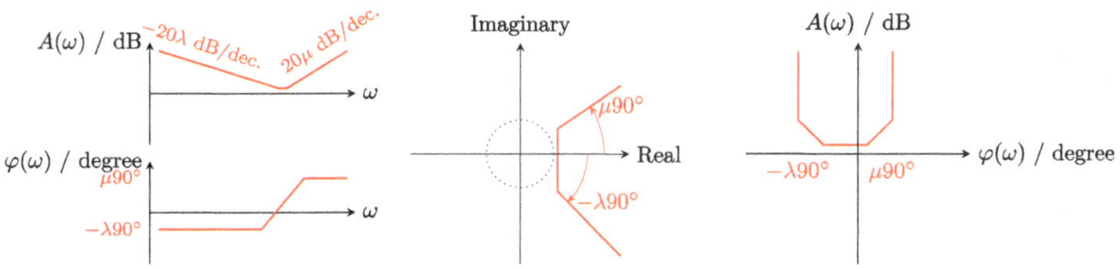

Figure 7: Asymptotic frequency response of $G_5(s) = P + \frac{I}{s^\lambda} + Ds^\mu = K_p\left(1 + \frac{1}{(T_is)^\lambda} + (T_ds)^\mu\right)$, $K_p, T_i, T_d, \lambda, \mu > 0$ and of $G_6(s) = K_p\left(1 + \frac{1}{T_is}\right)^\lambda (1 + T_ds)^\mu$, $K_p, T_i, T_d, \lambda, \mu > 0$; left: Bode diagram; center: polar plot; right: Nichols diagram.

2.6 Frequency response of $K_p(1 + \frac{1}{T_is})^\lambda(1 + T_ds)^\mu$ (implicit fractional PID)

The frequency response of the implicit fractional-order TF $G_6(s) = K_p\left(1 + \frac{1}{T_is}\right)^\lambda (1 + T_ds)^\mu$, $K_p, T_i, T_d, \lambda, \mu > 0$, is given by

$$G_6(i\omega) = K_p\left(1 + \frac{1}{T_i i\omega}\right)^\lambda (1 + T_d i\omega)^\mu. \tag{54}$$

At low and high frequencies, the asymptotic expansions (48)–(53), represented in Figure 7, are valid also for $G_6(s)$. The gain begins with a slope of -20λ dB per decade, and ends with a slope of 20μ dB per decade. In this case, the phase always goes up from $-\lambda\frac{\pi}{2}$ to $\mu\frac{\pi}{2}$ (see Figure 9).

2.7 Frequency response of $(\frac{\tau s+a}{s+a})^\alpha$ (fractional lead compensator)

The frequency response of the implicit fractional-order TF $G_7(s) = (\frac{\tau s+a}{s+a})^\alpha$, with $\alpha, a \in \mathbb{R}^+$, $\tau > 1$, is

$$G_7(i\omega) = \left(\frac{\tau i\omega + a}{j\omega + a}\right)^\alpha, \tag{55}$$

$$\begin{aligned} A_7(\omega) &= 20\log_{10}\left(\frac{a^2 + \tau^2\omega^2}{a^2 + \omega^2}\right)^{\frac{\alpha}{2}} \\ &= 10\alpha\log_{10}\left(\frac{a^2 + \tau^2\omega^2}{a^2 + \omega^2}\right), \end{aligned} \tag{56}$$

$$\varphi_7(\omega) = \alpha\left(\arctan\frac{\tau\omega}{a} - \arctan\frac{\omega}{a}\right). \tag{57}$$

For low frequencies $\omega \ll a$, we have $G_7(j\omega) \approx 1$, while for high frequencies $\omega \gg a$, we have $G_7(i\omega) \approx \tau^\alpha$. Joining these results with the values for the corner frequency

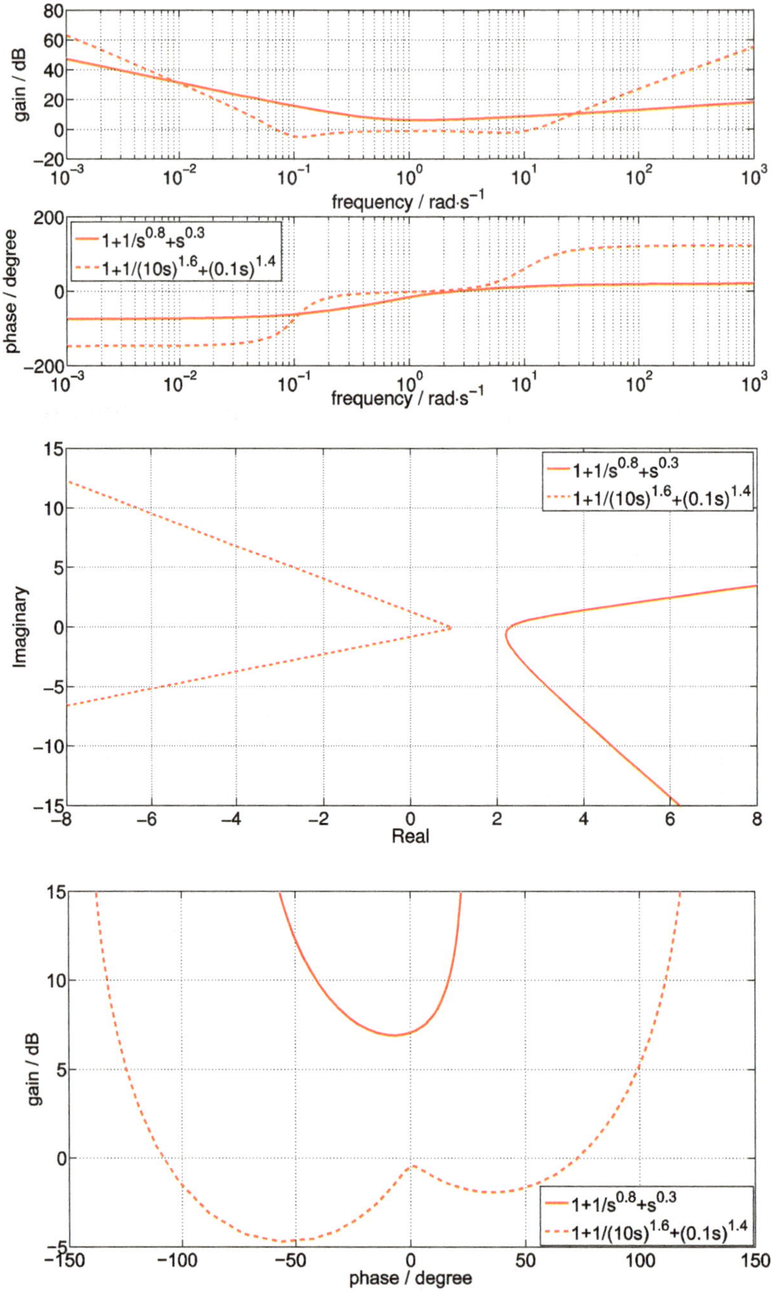

Figure 8: Frequency responses of $G_5(s) = 1 + \frac{1}{s^{0.8}} + s^{0.3}$ and $G_5(s) = 1 + \frac{1}{(10s)^{1.6}} + (0.1s)^{1.4}$; top: Bode diagram; middle: polar plot; bottom: Nichols diagram.

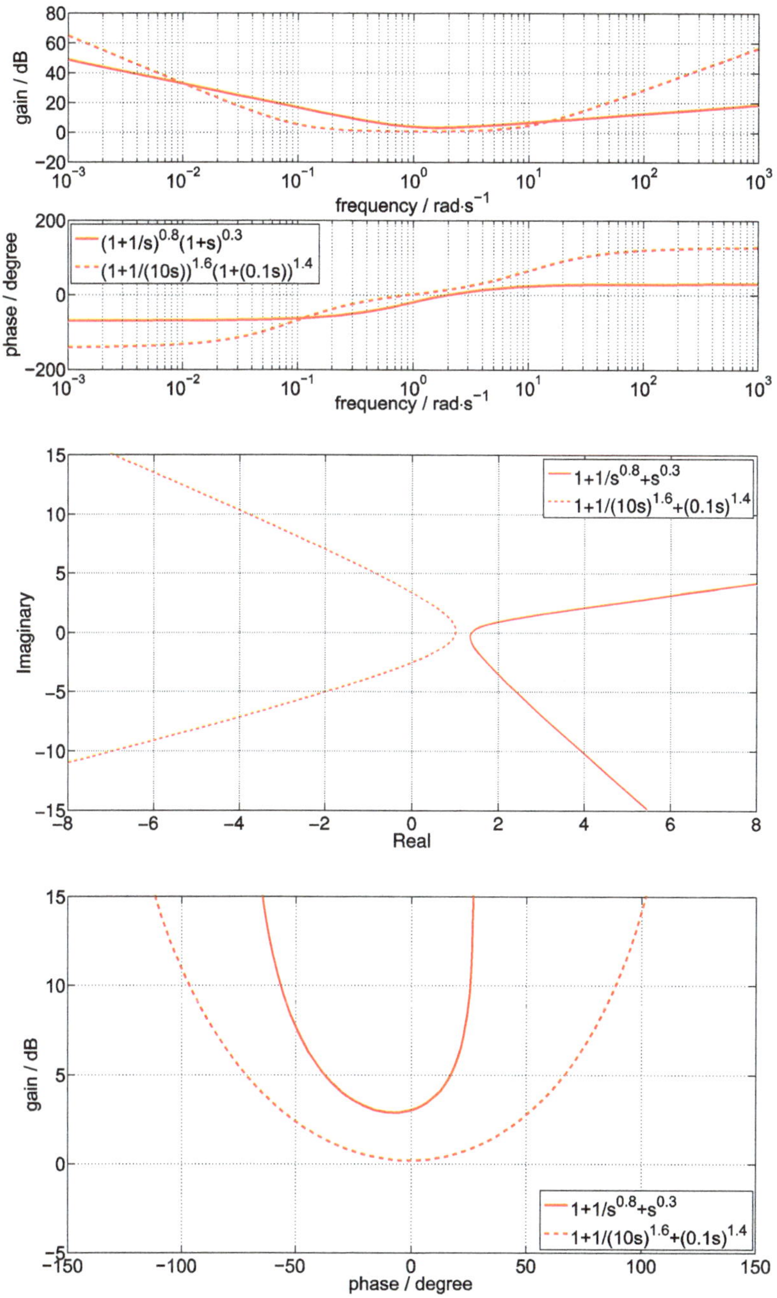

Figure 9: Frequency responses of $G_6(s) = (1 + \frac{1}{s})^{0.8}(1 + s)^{0.3}$ and $G_6(s) = (1 + \frac{1}{10s})^{1.6}(1 + 0.1s)^{1.4}$; top: Bode diagram; middle: polar plot; bottom: Nichols diagram.

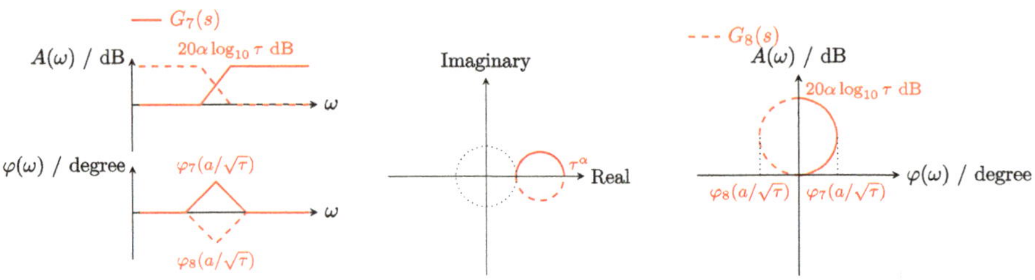

Figure 10: Asymptotic frequency response of $G_7(s) = (\frac{\tau s+a}{s+a})^\alpha$, $a \in \mathbb{R}^+, \tau > 1$ and $G_8(s) = \left(\frac{s+a}{s+\frac{a}{\tau}}\right)^\alpha$, $\alpha, a \in \mathbb{R}^+, \tau > 1$; left: Bode diagram; center: Nyquist diagram; right: Nichols diagram.

$\omega = \frac{a}{\sqrt{\tau}}$ rad $\cdot$ s^{-1}, the following asymptotic frequency response is obtained:

$$A_7(\omega) = \begin{cases} 0 & \omega \ll \frac{a}{\sqrt{\tau}}, \\ 10\,\alpha \log_{10} \tau & \omega = \frac{a}{\sqrt{\tau}}, \\ 20\,\alpha \log_{10} \tau & \omega \gg \frac{a}{\sqrt{\tau}}, \end{cases} \tag{58}$$

$$\varphi_7(\omega) = \begin{cases} 0 & \omega \ll \frac{a}{\sqrt{\tau}}, \\ \alpha \arctan \sqrt{\tau} - \alpha \arctan \frac{1}{\sqrt{\tau}} & \omega = \frac{a}{\sqrt{\tau}}, \\ 0 & \omega \gg \frac{a}{\sqrt{\tau}}. \end{cases} \tag{59}$$

See Figure 10 for the asymptotic frequency response.

2.8 Frequency response of $(\frac{s+a}{s+a/\tau})^\alpha$ (fractional lag compensator)

The frequency response of the implicit fractional-order TF $G_8(s) = \left(\frac{s+a}{s+\frac{a}{\tau}}\right)^\alpha$, with $\alpha, a \in \mathbb{R}^+, \tau > 1$, is

$$G_8(i\omega) = \left(\frac{i\omega + a}{i\omega + \frac{a}{\tau}}\right)^\alpha, \tag{60}$$

$$\begin{aligned} A_8(\omega) &= 20 \log_{10} \left(\frac{a^2 + \omega^2}{\frac{a^2}{\tau^2} + \omega^2}\right)^{\frac{\alpha}{2}} \\ &= 10\alpha \log_{10} \left(\frac{a^2 + \omega^2}{\frac{a^2}{\tau^2} + \omega^2}\right), \end{aligned} \tag{61}$$

$$\varphi_8(\omega) = \alpha \arctan \frac{\omega}{a} - \alpha \arctan \frac{\tau\omega}{a}. \tag{62}$$

For low frequencies $\omega \ll a$, we have $G_8(i\omega) \approx \tau^\alpha$, while for high frequencies $\omega \gg a$, we have $G_8(i\omega) \approx 1$. Joining these results with the values for the corner frequency

$\omega = \frac{a}{\sqrt{\tau}}$ rad · s^{-1}, the following asymptotic frequency response is obtained:

$$A_8(\omega) = \begin{cases} 20\,\alpha \log_{10} \tau & \omega \ll \frac{a}{\sqrt{\tau}}, \\ 10\,\alpha \log_{10} \tau & \omega = \frac{a}{\sqrt{\tau}}, \\ 0 & \omega \gg \frac{a}{\sqrt{\tau}}, \end{cases} \tag{63}$$

$$\varphi_8(\omega) = \begin{cases} 0 & \omega \ll \frac{a}{\sqrt{\tau}}, \\ \alpha \arctan \frac{1}{\sqrt{\tau}} - \alpha \arctan \sqrt{\tau} & \omega = \frac{a}{\sqrt{\tau}}, \\ 0 & \omega \gg \frac{a}{\sqrt{\tau}}. \end{cases} \tag{64}$$

See again Figure 10 for the asymptotic frequency response.

3 System identification from a frequency response

The problem of identifying a TF from the frequency response can be solved by means of Levy's identification method that represents a slight variation of the least-squares fit. The form that this method takes for commensurate fractional-order TF with a known fixed α is described in the follow-up [16, 17].

Let us consider a model such as (16), with known values of n, m, and α, so that the unknowns are the coefficients a_k and b_k. We will set $a_0 = 1$ (to make each TF unique) and let

$$N(i\omega_p) = \sum_{k=0}^{m} b_k (i\omega_p)^{k\alpha}, \tag{65}$$

$$D(i\omega_p) = 1 + \sum_{k=1}^{n} a_k (i\omega_p)^{k\alpha}, \tag{66}$$

so that the frequency response of the desired model is $G(i\omega) = \frac{N(i\omega)}{D(i\omega)}$. If the frequency response to be fitted is $G(i\omega_p)$, $p = 1, 2, \ldots, f$, (where f is the number of frequencies where the frequency response is known), then instead of minimizing the error

$$\varepsilon(i\omega_p) = G(i\omega_p) - \frac{N(i\omega_p)}{D(i\omega_p)} \tag{67}$$

(which is a nonlinear problem), the Levy's method minimizes the square of the norm of

$$E(i\omega_p) = \varepsilon(i\omega_p) D(i\omega_p) = G(i\omega_p) D(i\omega_p) - N(i\omega_p). \tag{68}$$

It is possible to give higher weight to the measurements of some frequencies (e. g., those with higher reliability) with a frequency-dependent weight $w(\omega_p)$ (if this is not desired, it is sufficient to set $w(\omega_p) = 1, \forall p = 1, 2, \ldots, f$). Therefore, we minimize $\sum_{p=1}^{f} w(\omega_p)|E(i\omega_p)|^2$

$$\begin{cases} \sum_{p=1}^{f} w(\omega_p)\{|G(i\omega_p)| \sum_{l=0}^{n}[a_l \cos((k-l)\alpha\frac{\pi}{2})\omega_p^{l\alpha}] \\ \quad - \sum_{l=0}^{m}[b_l \cos(\arg[G(i\omega_p)] + (k-l)\alpha\frac{\pi}{2})\omega_p^{l\alpha}]\} = 0, \quad k = 1 \ldots n, \\ \sum_{p=1}^{f} w(\omega_p)\{|G(i\omega_p)| \sum_{l=0}^{n}[a_l \cos(\arg[G(i\omega_p)] + (l-k)\alpha\frac{\pi}{2})\omega_p^{l\alpha}] \\ \quad - \sum_{l=0}^{m}[b_l \cos((k-l)\alpha\frac{\pi}{2})\omega_p^{l\alpha}]\} = 0, \quad k = 0, \ldots, m. \end{cases} \tag{69}$$

Solving this system of $n + m + 1$ equations give the desired coefficients.

As mentioned before, this method presumes (as it is well known) that the orders of the numerator and the denominator, n and m, are fixed in advance. Furthermore, when the order is fractional we also presume that α is fixed as well. Finding α along with the coefficients is not a linear problem. Therefore, the best solution is to proceed as for n and m: trying several values and keeping the best compromise between goodness of fit and complexity. The Nelder–Mead's simplex search method and heuristic optimization methods have been used for this.

4 Integer order approximations of systems based upon a frequency response

For practical purposes, fractional-order TF are often approximated by integer-order TF with frequency response that is "similar" in some sense. The three best performing methods are presented in the follow-up.

4.1 The CRONE or Oustaloup approximation

The CRONE approximation (French acronym for Noninteger Order Robust Control) of s^α, due to Alain Oustaloup, has N stable real poles and N stable real zeros, recursively placed within a frequency range $[\omega_l, \omega_h]$ [10, 11]:

$$s^\alpha \approx C \prod_{m=1}^{N} \frac{1 + \frac{s}{\omega_{z,m}}}{1 + \frac{s}{\omega_{p,m}}}, \tag{70}$$

$$\omega_{z,m} = \omega_l \left(\frac{\omega_h}{\omega_l}\right)^{\frac{2m-1-\alpha}{2N}}, \tag{71}$$

$$\omega_{p,m} = \omega_l \left(\frac{\omega_h}{\omega_l}\right)^{\frac{2m-1+\alpha}{2N}}. \tag{72}$$

The value of gain C is set to obtain the correct gain at $\omega = 1\,\mathrm{rad}\cdot s^{-1}$, which is $|(i\omega)^\alpha| = 1$, $\forall \alpha \in \mathbb{R}$ (i. e., 0 dB). If the frequency $\omega = 1\,\mathrm{rad}\cdot s^{-1}$ is outside the bandwidth $[\omega_l, \omega_h]$, then the correct gain at any other suitable frequency will have to be adjusted instead.

The recursion relation between poles and zeros is

$$\frac{\omega_{z,m+1}}{\omega_{z,m}} = \frac{\omega_{p,m+1}}{\omega_{p,m}} = \left(\frac{\omega_h}{\omega_l}\right)^{\frac{1}{N}}, \tag{73}$$

and causes them to be equidistant in a logarithmic scale of frequencies. For $\alpha \in\,]-1, 1[$, this approximation is good in the frequency range $[10\,\omega_l, \frac{\omega_h}{10}]$, but with some ripple both in the gain and phase, that decreases as N increases. It is a good practice to set N to at least the number of decades in $[\omega_l, \omega_h]$ for achieving good results. The approximation is poorer for $|\alpha| > 1$, but this can be avoided noting that $s^\alpha = s^{\lceil \alpha \rceil} s^{\alpha - \lceil \alpha \rceil}$ or $s^\alpha = s^{\lfloor \alpha \rfloor} s^{\alpha - \lfloor \alpha \rfloor}$, and that only the last term needs to be approximated. In fact, the second form may be used as well when $\alpha \in\,]-1, 0[$ to ensure the effect of an integer integrator.

Expressions (16) or (17) are approximated as linear combinations of fractional powers of s.

4.2 The Carlson approximation

The Carlson approximation of s^α, $\alpha = \frac{1}{n}$, $n \in \mathbb{N}$, is based on the Newton–Raphson iterative numerical method to find a root of an equation. After $k+1$ iterations, the approximation $G_{k+1}(s)$ of s^α is given by [1]

$$G_{k+1}(s) = G_k(s) \frac{\left(\frac{1}{\alpha} - 1\right) G_k^{\frac{1}{\alpha}}(s) + \left(\frac{1}{\alpha} + 1\right) s}{\left(\frac{1}{\alpha} + 1\right) G_k^{\frac{1}{\alpha}}(s) + \left(\frac{1}{\alpha} - 1\right) s}. \tag{74}$$

The first approximation $G_1(s)$ can be set to $G_1(s) = 1$ (better initial estimates are not needed). The order of the TF, as well as the range of frequencies where it is a good approximation, increase with k and depend on the order α being approximated. This range is centered on $\omega = 1\,\mathrm{rad}\cdot s^{-1}$, but multiplying all zeros and poles by a constant ω_c results in an approximation around the frequency $\omega = \omega_c\,\mathrm{rad}\cdot s^{-1}$ (the gain must be corrected accordingly). Poles and zeros are always stable, but they are not necessarily real.

Fixing the number of zeros and poles and the frequency range of interest is harder with this approximation than with the CRONE approximation. For a similar performance, the Carlson approximation often has to be of higher order than a CRONE approximation. Equations (16) or (17) are approximated as for the CRONE approximation, but all orders must be the inverse of an integer number. Consequently, this approximation is seldom a better choice than CRONE, though it may outperform it in some cases.

4.3 The Matsuda approximation

The Matsuda approximation of a given frequency behavior $G(i\omega)$, known at frequencies $\omega_0, \omega_1, \dots, \omega_N$ (which do not need to be ordered) can be used to approximate any fractional-order TF, commensurate, noncommensurate, or implicit [7]:

$$G(s) \approx d_0(\omega_0) + \frac{s-\omega_0}{d_1(\omega_1)+} \frac{s-\omega_1}{d_2(\omega_2)+} \frac{s-\omega_2}{d_3(\omega_3)+} \dots = \left[d_0(\omega_0); \frac{s-\omega_{k_1}}{d_k(\omega_k)} \right]_{k=1}^{N}, \tag{75}$$

$$d_0(\omega) = |G(i\omega)|, \tag{76}$$

$$d_k(\omega) = \frac{\omega - \omega_{k-1}}{d_{k-1}(\omega) - d_{k-1}(\omega_{k-1})}, \quad k = 1, 2, \dots, N. \tag{77}$$

This is a continued fraction interpolation with $\lceil \frac{N}{2} \rceil$ zeros and $\lfloor \frac{N}{2} \rfloor$ poles; therefore, $G(s)$ is causal only if N is even, and thus the number of frequencies $N+1$ is odd. The range of frequencies where the approximation is valid depends on how they are distributed. A logarithmic distribution of the frequencies is the simplest case, but better results may be obtained if a higher density of points ω_k is assigned where the desired frequency response varies more abruptly.

5 The Matignon and Nyquist criteria of stability for commensurate transfer functions

5.1 The Matignon theorem

It can be proved (see Chapter "Continuous-time fractional linear systems: transient responses" of Volume 6) that a commensurate TF (16) is stable if and only if the roots λ_k of polynomial $\sum_{k=0}^{n} a_k \lambda^k$, built with the coefficients of the denominator, verify

$$|\arg(\lambda_k)| > \frac{\alpha\pi}{2}, \quad \forall k, \tag{78}$$

where the phase $\arg(\lambda_k)$ is restricted to $[-\pi, +\pi]$ rad.

This result is known as Matignon theorem [6]. Notice that this means that no commensurate TF can be stable if $\alpha > 2$.

5.2 The Nyquist criterion

The Matignon theorem allows extending the Nyquist criterion of stability to commensurate TF.

Theorem 2. *Suppose that the denominator of the commensurate TF $G(s)$, given by* (16), *corresponds to a polynomial which according to condition* (78) *has P unstable poles.*

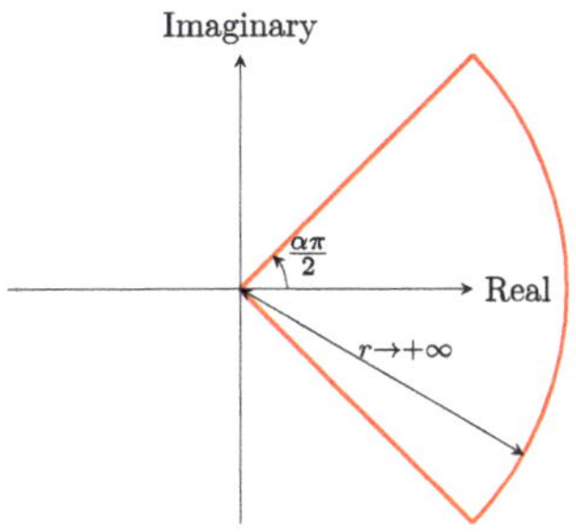

Figure 11: Complex contour used to build the Nyquist diagram and prove the Nyquist stability criterion. It is presumed that $0 < \alpha < 1$.

Let N be the number of turns that the Nyquist diagram of $G(s)$ encircles point $s = -\frac{1}{K}$: clockwise encirclements count as positive, counterclockwise encirclements as negative. Then, when $K \cdot G(s)$ is considered in closed loop with unit feedback, there will be $N + P$ unstable poles. Consequently, the closed-loop TF $\frac{K \cdot G(s)}{1+K \cdot G(s)}$ will be stable if and only if $N = -P$.

The proof is similar to that for integer order TF, considering only that the zone of the plane corresponding to unstable poles is different, and given by the Matignon theorem. Consequently, the complex contour which is being mapped is given by

$$\lim_{r\to+\infty} \{s \in \mathbb{C} : s = \varsigma e^{\pm\frac{i\alpha\pi}{2}}, \varsigma \in [0, r]\} \cup \left\{s \in \mathbb{C} : s = r e^{i\theta}, \theta \in \left[-\frac{\alpha\pi}{2}, +\frac{\alpha\pi}{2}\right]\right\}. \quad (79)$$

This contour is shown in Figure 11. If there are poles on this contour, circular indentations around them must be added to the contour, just as in the integer case.

Example 4

To study the stability of the plant,

$$G(s) = \frac{s + s^{\frac{1}{2}} - 6}{s^{\frac{1}{2}}(s+1)}, \quad (80)$$

we must consider the roots of polynomials $\lambda^2 + \lambda - 6$ (which are $\lambda = -3$ and $\lambda = 2$) and $\lambda^3 + \lambda$ (which are $\lambda = 0$ and $\lambda = \pm i$). Figure 12 shows on the top the corresponding map of poles and zeros, and a contour to build the Nyquist diagram. The Matignon theorem shows that the system is marginally stable in open-loop, since two of the roots of (78), $\lambda = \pm i$, fall outside the contour, and $\lambda = 0$ falls on the border. Even though this contour has arcs with finite radii, it suffices to plot the Nyquist diagram on the bottom of Figure 12, clearly showing that the diagram, that diverges to infinity, must be closed on the left. From here, the following table can be drawn, to see if a closed loop consisting of $G(s)$ and a gain K will be stable.

In fact, the closed loop will be stable only if $-1.4 < K < 0$, as can be easily verified computing its TF and using the Matignon theorem.

$-\frac{1}{K}$	$]-\infty, 0[$	$]0, 0.7[$	$]0.7, +\infty[$
K	$]0, +\infty[$	$]-\infty, -1.4[$	$]-1.4, 0[$
N	1	2	0
P	0	0	0
Stable in closed loop	No	No	Yes

6 Stochastic input

Consider a linear system having a TF without singularities on the imaginary axis (regular system). Assume that the input $x(t)$ to a given linear system is a stationary stochastic process with autocorrelation function $R_{xx}(t) = E[x(\tau + t)x(\tau)]$. The output is given by $y(t) = \int_{-\infty}^{t} h(\tau)x(t - \tau)d\tau$, and the corresponding autocorrelation reads

$$R_{yy}(t) = h(t) * h(-t) * R_{xx}(t). \tag{81}$$

Let $S(\cdot)$ represent the LT of the autocorrelation. We get

$$S_{yy}(s) = H(s)H(-s)S_{xx}(s),$$

that, in general, has an empty region of convergence, since $H(s)$ exists for $\mathbb{R}(s) > 0$. This leads us to define

$$S_{yy}(i\omega) = \lim_{s \to i\omega} H(s)H(-s) \cdot S_{xx}(i\omega) = \left|H(i\omega)\right|^2 S_{xx}(i\omega), \tag{82}$$

that relates the input with the corresponding output power spectral densities (or simply, spectra), $S_{xx}(i\omega)$ and $S_{yy}(i\omega)$, respectively. If the input is white noise, $n(t)$, then its autocorrelation function is $R_{nn}(t) = \sigma^2\delta(t)$, where σ^2 is its power average. Therefore, the output spectrum is

$$S_{yy}(i\omega) = \sigma^2\left|H(i\omega)\right|^2, \tag{83}$$

stating an important relation suitable for stochastic modeling and identification.

In particular, if $H(i\omega) = (i\omega)^\alpha$, then the system is a differintegrator and the output is called fractional noise. If the white noise input is Gaussian, then the output is *fractional Gaussian noise*. Due to (83), the fractional noise has as the spectrum

$$S_{yy}(i\omega) = \sigma^2\left|(i\omega)\right|^{2\alpha}. \tag{84}$$

If $\alpha > -\frac{1}{2}$, then the corresponding autocorrelation is

$$R_{yy}(t) = \sigma^2 \frac{|t|^{-\alpha-1}}{2\Gamma(-2\alpha)\cos(\alpha\pi)}. \tag{85}$$

The integer order cases can be treated using the reflection property of the gamma function [9].

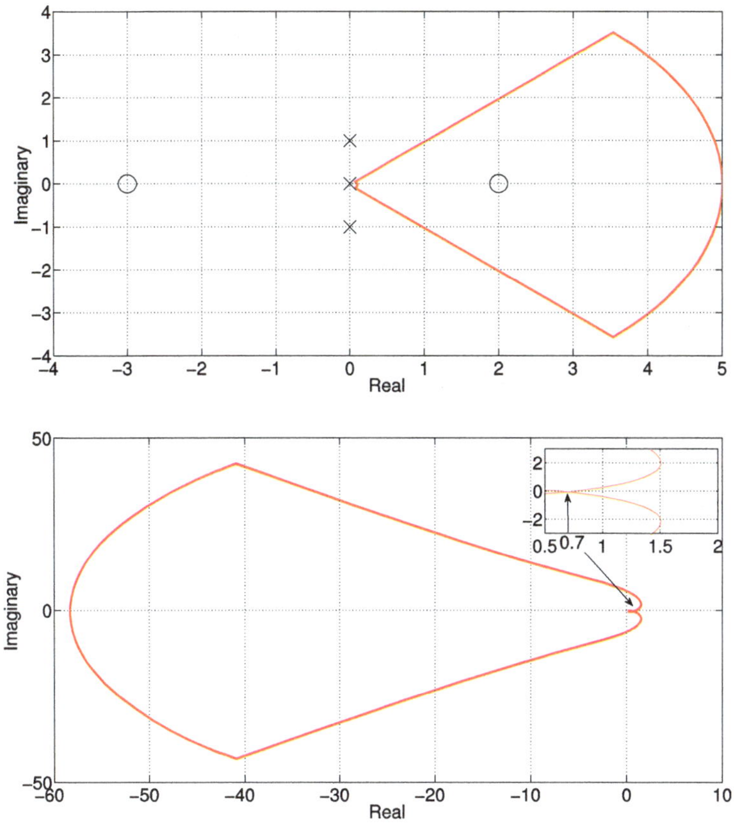

Figure 12: Top: pole-zero map of example 4, and complex contour for the Nyquist diagram; bottom: the Nyquist diagram obtained with the contour on the top.

Bibliography

[1] G. E. Carlson and C. A. Halijak, Approximation of fractional capacitors $(1/s)^{1/n}$ by a regular Newton process, *IEEE Transactions on Circuit Theory*, **7** (1964), 210–213.

[2] S. Haykin and B. Van Veen, *Signals and Systems*, John Wiley & Sons, New Delhi, 2007.

[3] T. Kailath, *Linear Systems*, Prentice-Hall, Englewood Cliffs, NJ, 1980.

[4] A. M. Lopes and J. A. Tenreiro Machado, Root locus practical sketching rules for fractional-order systems, *Abstract and Applied Analysis*, **2013** (2013), 102068.

[5] R. Malti, X. Moreau, and F. Khemane, Resonance of fractional transfer functions of the second kind, in *Fractional Differentiation and Its Applications*. IFAC, Ankara, 2008.

[6] D. Matignon, Stability properties for generalized fractional differential systems, *ESAIM. Proceedings*, **5** (1998), 145–158.

[7] K. Matsuda and H. Fujii, $\mathcal{H}_\infty$ optimized wave-absorbing control: analytical and experimental results, *Journal of Guidance, Control, and Dynamics*, **16**(6) (1993), 1146–1153.

[8] F. Merrikh-Bayat and M. Afshar, Extending the root-locus method to fractional-order systems, *Journal of Applied Mathematics*, **2008** (2008), 528934.

[9] M. D. Ortigueira, Fractional central differences and derivatives, *Journal of Vibration and Control*, **14**(9–10) (2008), 1255–1266.
[10] A. Oustaloup, *La commande CRONE: Commande Robuste d'Ordre Non Entier*, Hermès, Paris, 1991.
[11] A. Oustaloup, F. Levron, B. Matthieu, and F. M. Nanot, Frequency-band complex noninteger differentiator: characterization and synthesis, *IEEE Transactions on Circuits and Systems. I, Fundamental Theory and Applications*, **47**(1) (2000), 25–39.
[12] I. Podlubny, *Fractional Differential Equations: an Introduction to Fractional Derivatives, Fractional Differential Equations, to Methods of Their Solution and Some of Their Applications*, Academic Press, San Diego, 1999.
[13] M. J. Roberts, *Signals and Systems: Analysis Using Transform Methods and MATLAB*, McGraw-Hill Higher Education, New York, 2012.
[14] J. A. Tenreiro Machado, Fractional derivatives: probability interpretation and frequency response of rational approximations, *Communications in Nonlinear Science and Numerical Simulation*, **14**(9–10) (2009), 3492–3497.
[15] J. A. Tenreiro Machado, Root locus of fractional linear systems, *Communications in Nonlinear Science and Numerical Simulation*, **16**(10) (2011), 3855–3862.
[16] D. Valério and J. S. da Costa, *An Introduction to Fractional Control*, IET, Stevenage, 2013, ISBN 978-1-84919-545-4.
[17] D. Valério, M. D. Ortigueira, and J. S. da Costa, Identifying a transfer function from a frequency response, *ASME Journal of Computational and Nonlinear Dynamics*, **3**(2) (2008), 021207.

Inés Tejado, Blas M. Vinagre, and Dominik Sierociuk

State space methods for fractional controllers design

Abstract: This chapter looks at linear and optimal strategies of state space theory within the framework of fractional-order control. More precisely, it shows new alternatives of control when introducing fractional-order dynamics into integer-order control systems. Illustrative examples of application are given to demonstrate the benefits derived by such a fact in comparison with classical control strategies.

Keywords: State space, fractional, state feedback, output feedback, optimal control, Kalman filter, design

MSC 2010: 93B52, 26A33, 49N05, 93B51

1 Introduction

Despite important progress as regards offering new possibilities of control and improving the behavior of traditional integer-order controllers, the expected results of fractional-order dynamics in state space are still not sufficiently studied and exploited.

This chapter presents a comprehensive tutorial on the main issues involved in designing classical state space control strategies derived by introducing fractional-order dynamics into integer-order systems. More precisely, the performance of traditional controllers in state space, such as, state feedback, output feedback and optimal control, will be extended.

It is important to remark that, in the context of this chapter, in order to emphasize the benefits of fractional-order control in classical problems, systems to be controlled are assumed to be of integer order; fractional dynamics will be introduced into such systems by the concept of system augmentation.

2 State space fundamentals

This section is devoted to recall fundamentals of state space models for fractional-order systems, that is, the representation and analysis of both continuous and discrete

Inés Tejado, Blas M. Vinagre, Industrial Engineering School, University of Extremadura, Avenida de Elvas, s/n, 06006, Badajoz, Spain, e-mails: itejbal@unex.es, bvinagre@unex.es
Dominik Sierociuk, Institute of Control and Industrial Electronics, Warsaw University of Technology, Koszykowa 75, 00-662 Warsaw, Poland, e-mail: dominik.sierociuk@ee.pw.edu.pl

https://doi.org/10.1515/9783110571745-008

systems. It is important to remark that this chapter considers only systems that are linear and time invariant (LTI), and of commensurate order.

2.1 Continuous-time state space model

A state space representation of a multivariable system with l inputs, p outputs and n state variables can be given of the form:

$$\begin{aligned}\mathcal{D}^{\alpha}\boldsymbol{x}(t) &= \boldsymbol{A}\boldsymbol{x}(t) + \boldsymbol{B}\boldsymbol{u}(t),\\ \boldsymbol{y}(t) &= \boldsymbol{C}\boldsymbol{x}(t) + \boldsymbol{D}\boldsymbol{u}(t),\end{aligned} \tag{1}$$

where $\boldsymbol{\alpha} = [\alpha_0, \alpha_1, \alpha_2, \ldots, \alpha_n]$, $\boldsymbol{u}(t) \in \mathbb{R}^l$ is the input vector, $\boldsymbol{x}(t) \in \mathbb{R}^n$ is the state vector, $\boldsymbol{y}(t) \in \mathbb{R}^p$ is the output vector, $\boldsymbol{A} \in \mathbb{R}^{n\times n}$ is the state matrix, $\boldsymbol{B} \in \mathbb{R}^{n\times l}$ is the input matrix, $\boldsymbol{C} \in \mathbb{R}^{p\times n}$ is the output matrix, and $\boldsymbol{D} \in \mathbb{R}^{p\times l}$ is the direct transmission matrix. The above notation establishes that fractional operator $\mathcal{D}^{\alpha_i}$ is applied only to state variable x_i of $\boldsymbol{x}(t)$ in (1). When $\alpha_i \equiv \alpha$, $1 \le i \le n$, $0 < \alpha \le 1$, the state equation (1) becomes

$$\mathcal{D}^{\alpha}\boldsymbol{x}(t) = \boldsymbol{A}\boldsymbol{x}(t) + \boldsymbol{B}\boldsymbol{u}(t), \tag{2}$$

where now operator $\mathcal{D}^{\alpha}$ means that all the states are α-differentiated (the output equation of (1) remains).

Applying the Laplace transform to (2) with zero initial conditions and doing some manipulations, it is possible to obtain the system transfer function matrix $\boldsymbol{G}(s)$ (of dimension $p \times l$) as

$$\boldsymbol{G}(s) = \frac{\boldsymbol{Y}(s)}{\boldsymbol{U}(s)} = \boldsymbol{C}(s^{\alpha}\boldsymbol{I} - \boldsymbol{A})^{-1}\boldsymbol{B} + \boldsymbol{D}, \tag{3}$$

where $\boldsymbol{I}$ is the identity matrix of dimension $n \times n$. Notice that the elements of $\boldsymbol{G}(s)$ are transfer functions with the following common polynomial denominator:

$$\phi(s^{\alpha}) = \det(s^{\alpha}\boldsymbol{I} - \boldsymbol{A}). \tag{4}$$

Then, it can be easily derived from (4) that the poles of system (2) can be obtained as

$$p_i = \lambda_i^{1/\alpha}, \tag{5}$$

where λ_i, $1 \le i \le n$, are the eigenvalues of matrix $\boldsymbol{A}$.

The stability condition for commensurate-order systems in the bounded input-bounded output (BIBO) sense is given by

$$|\arg(\lambda_i)| > \alpha\frac{\pi}{2}, \quad 1 \le i \le n, \tag{6}$$

or, in the case of a rational commensurate-order system ($\alpha = 1/q$), by

$$|\arg(\lambda_i)| > \frac{\pi}{2q}, \quad 1 \le i \le n. \tag{7}$$

The solution of the state equation of commensurate-order systems is given by

$$\boldsymbol{x}(t) = \mathcal{E}_{\alpha,1}(\boldsymbol{A}t^{\alpha})\boldsymbol{x}(0) = \boldsymbol{\Phi}(t)\boldsymbol{x}(0), \tag{8}$$

where $\boldsymbol{\Phi}(t)$ is the state transition matrix, and $\mathcal{E}_{\alpha,1}$ is the one-parameter Mittag-Leffler function, which performs the same roll as the exponential function for integer-order systems. From this solution, two important remarks must be taken into account:

1. In classical state models, the state at instant t includes all the information needed to calculate the future behavior of the system provided whenever future inputs were known. This is not true for fractional-order models since the fractional-order derivative has "memory" from t up to the lower limit that defines the fractional operator. Then, in order to determine the future behavior of a fractional-order system not only the value of the state at instant t is required, but also of all the values of the state in the interval $[0, t]$. This condition is reduced to the knowledge of only $\boldsymbol{x}(t)$ in the case of integer-order derivatives because the fractional-order differentiation operator collapses into a local operator that only depends on the value of the state at the immediate surroundings of t. Then $\boldsymbol{x}(t)$ denotes a pseudo-state.
2. Matrix $\boldsymbol{\Phi}(t)$ in (8) is not a state transition matrix in the usual sense. Actually, it means $\boldsymbol{\Phi}(t, 0)$, in which the second argument denotes the lower limit of the integral that defines the fractional-order derivative used in the state model. Then, $\boldsymbol{\Phi}(t)$ is referred to as the state pseudo-transition matrix.

Likewise, the controllability and observability conditions for a commensurate-order system are the same as for that of integer order, only having its state description based on the operator $\mathcal{D}^{\alpha}$ as follows:

Definition 1. A system is controllable if it is possible to establish a non-restricted control vector which can lead the system from an initial state, $\boldsymbol{x}(t_0)$, to another final state, $\boldsymbol{x}(t_f)$, in a finite time $t_0 \le t \le t_f$.

Then, a commensurate-order system is controllable if and only if the rank of the controllability matrix

$$\mathcal{C} = [\boldsymbol{B}\ \boldsymbol{AB}\ \boldsymbol{A}^2\boldsymbol{B}\ \cdots\ \boldsymbol{A}^{n-1}\boldsymbol{B}] \tag{9}$$

is n.

Definition 2. A system is observable if any state, $\boldsymbol{x}(t_0)$, can be determined from the observation of $y(t)$ in a finite interval of time $t_0 \le t \le t_f$.

Thus, a commensurate-order system is observable if and only if the observability matrix

$$\mathcal{O} = \begin{bmatrix} \boldsymbol{C} \\ \boldsymbol{CA} \\ \vdots \\ \boldsymbol{CA}^{n-1} \end{bmatrix} \tag{10}$$

is of full rank n. The demonstration of the controllability and observability conditions for such systems can be found in Monje et al. [16].

2.2 Discrete-time LTI state space model

Now, state space model defined above for continuous-time systems is extended to those of discrete-time based on the Grünwald–Letnikov's definition of the fractional-order operator (Monje et al. [16], Podlubny [19], Oldham and Spanier [17]).

Let assume that the standard integer-order state space model

$$\dot{\boldsymbol{x}}(t) = \boldsymbol{Ax}(t) + \boldsymbol{Bu}(t), \tag{11}$$

is discretized with sampling period T by a numerical method. The simplest way of approximation of the first-order derivative is calculating its first-order forward difference[1] as follows:

$$\dot{\boldsymbol{x}}(kT) \approx \frac{\boldsymbol{x}((k+1)T) - \boldsymbol{x}(kT)}{T}. \tag{12}$$

Substituting derivative (12) in (11) results in

$$\frac{\boldsymbol{x}((k+1)T) - \boldsymbol{x}(kT)}{T} \approx \boldsymbol{Ax}(t) + \boldsymbol{Bu}(t), \tag{13}$$

which yields the following approximate equivalent discrete state space model:

$$\boldsymbol{x}((k+1)T) = (\boldsymbol{A}T + \boldsymbol{I})\boldsymbol{x}(kT) + \boldsymbol{B}T\boldsymbol{u}(kT), \tag{14}$$

$$\boldsymbol{y}(kT) = \boldsymbol{Cx}(kT) + \boldsymbol{Du}(kT). \tag{15}$$

Let generalize the previous integer-order discrete state space model to fractional order using the Grünwald–Letnikov's definition. Let us lead the fractional-order operator one sample in order to get a model where $\boldsymbol{x}((k+1)T)$ appears, similar to the

1 It is worth mentioning that the first-order backward difference is not adequate to obtain discrete-time model that relates $\boldsymbol{x}((k+1)T)$ with previous states and inputs (see, e. g., Monje et al. [16]).

forward difference used in the integer case. Then, the fractional-order derivative can be obtained as

$$\begin{aligned}\mathcal{D}^{\alpha}\boldsymbol{x}(kT) &\approx \frac{1}{T^{\alpha}}\sum_{i=0}^{k+1}(-1)^{i}\binom{\alpha}{i}x((k+1-i)T)\\ &= \frac{1}{T^{\alpha}}\left(x((k+1)T)-\binom{\alpha}{i}x(kT)+\sum_{i=2}^{k+1}(-1)^{i}\binom{\alpha}{i}x((k+1-i)T)\right),\end{aligned}\tag{16}$$

which, substituted in (1), yields for $k = 0$

$$\boldsymbol{x}((k+1)T) = (\boldsymbol{A}T^{\alpha} + \alpha\boldsymbol{I})\boldsymbol{x}(kT) + \boldsymbol{B}T^{\alpha}\boldsymbol{u}(kT),\tag{17}$$

and for $k \geq 1$

$$\begin{aligned}\boldsymbol{x}((k+1)T) = (\boldsymbol{A}T^{\alpha} + \alpha\boldsymbol{I})\boldsymbol{x}(kT) - \sum_{i=2}^{k+1}(-1)^{i}\binom{\alpha}{i}x((k+1-i)T)\\ + \boldsymbol{B}T^{\alpha}\boldsymbol{u}(kT).\end{aligned}\tag{18}$$

Hence, the fractional discrete-time LTI state space model is given by equations (17) and (18) combined with (15). In this model, the state at instant $(k+1)T$ depends on all the previous states until $kT = 0$, as can be expected from the considerations about the "memory" of fractional-order operators.

An alternative way of expressing the above mentioned model is based on using an expanded state formed by the actual state and all the past states from the initial instant as follows (Dzieliński and Sierociuk [7]):

$$\begin{bmatrix}\boldsymbol{x}((k+1)T)\\ \boldsymbol{x}(kT)\\ \boldsymbol{x}((k-1)T)\\ \vdots\end{bmatrix} = \tilde{\boldsymbol{A}}\begin{bmatrix}\boldsymbol{x}(kT)\\ \boldsymbol{x}((k-1)T)\\ \boldsymbol{x}((k-2)T)\\ \vdots\end{bmatrix} + \tilde{\boldsymbol{B}}u(kT),\tag{19}$$

$$y(kT) = \tilde{\boldsymbol{C}}\begin{bmatrix}\boldsymbol{x}(kT)\\ \boldsymbol{x}((k-1)T)\\ \boldsymbol{x}((k-2)T)\\ \vdots\end{bmatrix} + \boldsymbol{D}u(kT),$$

where

$$\tilde{\boldsymbol{A}} = \begin{bmatrix}(\boldsymbol{A}T^{\alpha}+\alpha\boldsymbol{I}) & -\boldsymbol{I}(-1)^{2}\binom{\alpha}{2} & -\boldsymbol{I}(-1)^{3}\binom{\alpha}{3} & \cdots\\ \boldsymbol{I} & \boldsymbol{0} & \boldsymbol{0} & \cdots\\ \boldsymbol{0} & \boldsymbol{I} & \boldsymbol{0} & \cdots\\ \vdots & \vdots & \vdots & \vdots\end{bmatrix},$$

$$\tilde{\boldsymbol{B}} = \begin{bmatrix} \boldsymbol{B}T^{\alpha} \\ \mathbf{0} \\ \mathbf{0} \\ \vdots \end{bmatrix}, \quad \text{and} \quad \tilde{\boldsymbol{C}} = \begin{bmatrix} \boldsymbol{C} & \mathbf{0} & \mathbf{0} & \cdots \end{bmatrix},$$

being $\mathbf{0}$ the null matrix of dimension $n \times n$. As can be observed, model given by (19) is an infinite-dimensional discrete-time system.

Omitting the sampling time T, model (19) can be also written as follows (Sierociuk et al. [24]):

$$\begin{bmatrix} x_{k+1} \\ x_k \\ x_{k-1} \\ \vdots \end{bmatrix} = \mathbb{A} \begin{bmatrix} x_k \\ x_{k-1} \\ x_{k-2} \\ \vdots \end{bmatrix} + \mathbb{B} u_k, \tag{20}$$

$$y_k = \mathbb{C} \begin{bmatrix} x_k \\ x_{k-1} \\ x_{k-2} \\ \vdots \end{bmatrix} + \mathbb{D} u_k,$$

where

$$\mathbb{A} = \begin{bmatrix} (A + \boldsymbol{I}\alpha) & -\boldsymbol{I}(-1)^2\binom{\alpha}{2} & -\boldsymbol{I}(-1)^3\binom{\alpha}{3} & \cdots \\ \boldsymbol{I} & \mathbf{0} & \mathbf{0} & \cdots \\ \mathbf{0} & \boldsymbol{I} & \mathbf{0} & \cdots \\ \vdots & \vdots & \vdots & \vdots \end{bmatrix},$$

$$\mathbb{B} = \begin{bmatrix} \boldsymbol{B} \\ \mathbf{0} \\ \mathbf{0} \\ \vdots \end{bmatrix}, \quad \mathbb{C} = \tilde{\boldsymbol{C}}, \quad \text{and} \quad \mathbb{D} = \boldsymbol{D}.$$

Likewise, the m-finite form of a discrete-time fractional-order system (with the new mth elements state vector) can be defined as follows (Sierociuk et al. [24]):

$$\mathbb{X}_{k+1} = \mathbb{A}_m \mathbb{X}_k + \mathbb{B}_m u_k + \mathbb{I}\omega_k - \mathbb{I} \sum_{j=m+1}^{k+1} (-1)^j \Upsilon_j x_{k+1-j}, \tag{21}$$

$$y_k = \mathbb{C}_m \mathbb{X}_k, \tag{22}$$

where

$$\mathbb{X}_k = \begin{bmatrix} x_k \\ x_{k-1} \\ \vdots \\ x_{k-m+1} \end{bmatrix}, \quad \mathbb{I} = \begin{bmatrix} I \\ 0 \\ \vdots \\ 0 \end{bmatrix},$$

$$\mathbb{A}_m = \begin{bmatrix} (A+\Upsilon_1) & -(-1)^2\Upsilon_2 & \cdots & -(-1)^m\Upsilon_m \\ I & \cdots & 0 & 0 \\ \vdots & \vdots & \vdots & \vdots \\ 0 & \cdots & I & 0 \end{bmatrix},$$

$$\mathbb{B}_m = \begin{bmatrix} B \\ 0 \\ \vdots \\ 0 \end{bmatrix}, \quad \mathbb{C}_m = \begin{bmatrix} C & 0 & \cdots & 0 \end{bmatrix}.$$

It should be remarked that the formulations given by (20) and (21) will be used in Section 5.2.

Other discrete-time state space models for fractional-order systems can be found in the literature. The interested reader can find more discrete-time models for fractional-order systems in, for example, Monje et al. [16], Ostalczyk [18].

Solution of the state equation, stability, and controllability and observability conditions of discrete-time LTI state space system given by (19) have no difference with the integer case (refer to Monje et al. [16], Dzieliński and Sierociuk [7] for more details).

3 Introducing fractional dynamics into integer-order systems: system augmentation concept

According to Theorem 1 in Sierociuk and Vinagre [26], the integer-order state space system

$$\begin{aligned} \dot{\boldsymbol{x}}(t) &= \boldsymbol{A}\boldsymbol{x}(t) + \boldsymbol{B}u(t) + \boldsymbol{E}d(t), \\ y(t) &= \boldsymbol{C}\boldsymbol{x}(t) + \boldsymbol{D}u(t) \end{aligned} \tag{23}$$

with initial conditions $x(0) = x_0$ and matrices $\boldsymbol{A} \in \mathbb{R}^{n\times n}$, $\boldsymbol{B} \in \mathbb{R}^{n\times m}$, $\boldsymbol{E} \in \mathbb{R}^{n\times m}$, $\boldsymbol{C} \in \mathbb{R}^{r\times n}$, $\boldsymbol{D} \in \mathbb{R}^{r\times m}$, can be rewritten in the form of fractional (rational) order system as

$$\begin{aligned} \mathcal{D}^{\alpha}\mathbf{x}(t) &= \tilde{\boldsymbol{A}}\mathbf{x}(t) + \tilde{\boldsymbol{B}}u(t) + \tilde{\boldsymbol{E}}d(t), \\ y(t) &= \tilde{\boldsymbol{C}}\mathbf{x}(t) + \boldsymbol{D}u(t) \end{aligned} \tag{24}$$

where $\alpha = 1/q$ is the fractional order ($q \in \mathbb{Z}^+$), and

$$\tilde{\boldsymbol{A}} = \left[\begin{array}{c|c} \mathbf{0}_{p\times n} & \boldsymbol{I}_{p\times p} \\ \hline \boldsymbol{A} & \mathbf{0}_{n\times p} \end{array}\right], \quad \tilde{\boldsymbol{B}} = \left[\begin{array}{c} \mathbf{0}_{p\times m} \\ \hline \boldsymbol{B} \end{array}\right], \quad \tilde{\boldsymbol{E}} = \left[\begin{array}{c} \mathbf{0}_{p\times m} \\ \hline \boldsymbol{E} \end{array}\right],$$

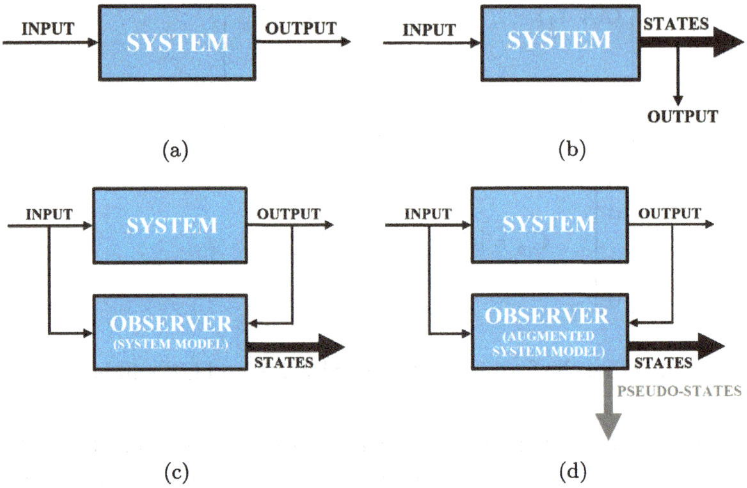

Figure 1: Block diagrams of: (a) Input-output system (b) Input-output system with states (c) Conventional observer (d) Observer based on the augmented system.

$$\tilde{C} = \left[\begin{array}{c:c} C & 0_{p\times r}\end{array}\right], \quad \mathbf{x}(t) = \begin{bmatrix} \boldsymbol{x}(t) \\ x_{a,1}(t) \\ \vdots \\ x_{a,q-2}(t) \\ x_{a,q-1}(t) \end{bmatrix},$$

being $x_{a,1}(t) = \mathcal{D}^{\alpha}x(t)$, $x_{a,i}(t) = \mathcal{D}^{\alpha}x_{a,i-1}(t)$ for $i = 2, \ldots, q-1$, the components of the augmented state vector (i. e., the pseudo-states) with initial conditions $x(0) = \left[x_0^T \quad x_{a,1}^T(0) \quad \cdots \quad x_{a,q-1}^T(0)\right]^T$ ($(-)^T$ denotes the transpose operator).

The idea behind this method is illustrated in Figure 1. Basically, the advantage of state space based methods is due to the use of more information about the system to be controlled compared to input-output based methods. System augmentation, or in general fractional derivatives, can be seen as having additional information available for control.

As an example, let consider the velocity of a servo system in state space form given by

$$\begin{aligned} \dot{x}(t) &= ax(t) + bu(t) + d(t), \\ y(t) &= x(t) \end{aligned} \tag{25}$$

with $a = -1/\tau$, and $b = k/\tau$, where τ and k are the time constant and the gain of the system, respectively.

Considering $q = 2$, the state vector for velocity in the form of (24) is

$$\mathbf{x}(t) = \left[x(t) \quad x_{a,1}(t)\right]^T = \left[x_1(t) \quad x_2(t)\right]^T,$$

with $x_2(t) = \mathcal{D}^{0.5}x_1(t)$. Thus, the augmented system is given by the following matrices:

$$\tilde{\boldsymbol{A}} = \begin{bmatrix} 0 & 1 \\ a & 0 \end{bmatrix}, \quad \tilde{\boldsymbol{B}} = \begin{bmatrix} 0 \\ b \end{bmatrix}, \quad \tilde{\boldsymbol{C}} = \begin{bmatrix} 1 & 0 \end{bmatrix}, \quad \tilde{\boldsymbol{E}} = \begin{bmatrix} 0 \\ 1 \end{bmatrix}.$$

For position, the state vector is

$$\mathbf{x}(t) = \begin{bmatrix} \boldsymbol{x}^T(t) & x_{a,1}^T(t) \end{bmatrix}^T = \begin{bmatrix} x_1(t) & x_2(t) & x_3(t) & x_4(t) \end{bmatrix}^T,$$

with $x_3(t) = \mathcal{D}^{0.5}x_1(t)$ and $x_4(t) = \mathcal{D}^{0.5}x_2(t)$. The matrices are the following:

$$\tilde{\boldsymbol{A}} = \begin{bmatrix} 0 & 0 & 1 & 0 \\ 0 & 0 & 0 & 1 \\ 0 & 1 & 0 & 0 \\ 0 & a & 0 & 0 \end{bmatrix}, \quad \tilde{\boldsymbol{B}} = \begin{bmatrix} 0 \\ 0 \\ 0 \\ b \end{bmatrix}, \quad \tilde{\boldsymbol{C}} = \begin{bmatrix} 1 & 0 & 0 & 0 \end{bmatrix}, \quad \tilde{\boldsymbol{E}} = \begin{bmatrix} 0 \\ 0 \\ 0 \\ 1 \end{bmatrix}.$$

4 Linear control

This section presents the algorithms for the extension of linear control design in state space for fractional-order augmented systems, that is, state and output feedback.

4.1 State feedback

For the case of commensurate-order systems, state feedback control follows the same rules used for integer-order systems, but it must be highlighted the different dynamic implications of the pole locations in the modified complex plane. For the case of integer-order systems (the most common), augmented state representation of the system can be used. Both cases imply that the control law uses not only the state vector of the integer-order system but also fractional-order derivatives of its components (i. e., the pseudo-states), as shown in Figure 2. And in both cases, those pseudo-states are obtained by the necessary observer which includes the fractional model (commensurate or augmented) of the system. In what follows, the more common case of augmented system will be considered.

The control law is given by

$$u(t) = -\boldsymbol{K}\mathbf{x}(t) = -\begin{bmatrix} \boldsymbol{K}_i & K_{a,1} & \cdots & K_{a,q-1} \end{bmatrix} \begin{bmatrix} \boldsymbol{x}(t) \\ x_{a,1}(t) \\ \vdots \\ x_{a,q-1}(t) \end{bmatrix}, \tag{26}$$

where $\boldsymbol{K}_i$ are the controller gains corresponding to the state vector of the integer-order system, whereas $K_{a,i}$, for $i = 1, \dots, q-1$, are those connected with the pseudo-states.

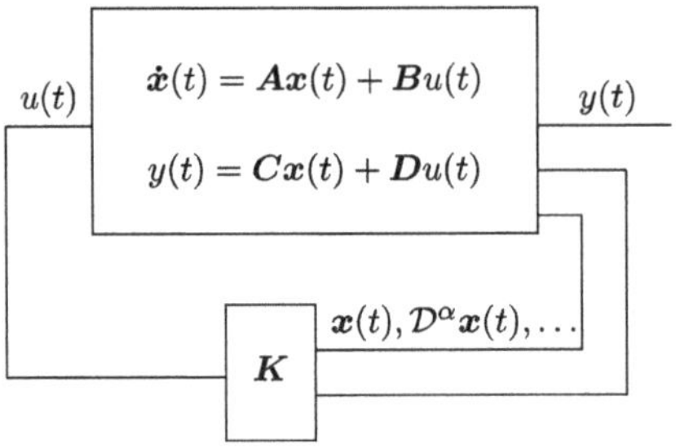

Figure 2: Scheme of state feedback control with system augmentation.

4.1.1 Pole placement

By using the control law given by (26), the closed-loop system matrix is given as follows:

$$\tilde{\boldsymbol{A}} - \tilde{\boldsymbol{B}}\boldsymbol{K} = \begin{bmatrix} \boldsymbol{0} & \boldsymbol{I} & \cdots & \boldsymbol{0} \\ \vdots & \vdots & \vdots & \vdots \\ \boldsymbol{0} & \boldsymbol{0} & \cdots & \boldsymbol{I} \\ \boldsymbol{A} - \boldsymbol{B}\boldsymbol{K}_i & -\boldsymbol{B}\boldsymbol{K}_{a,1} & \cdots & -\boldsymbol{B}\boldsymbol{K}_{a,q-1} \end{bmatrix}, \tag{27}$$

being the closed-loop characteristic polynomial

$$\begin{aligned} \lambda(w)_c &= \det(\boldsymbol{I}w - \tilde{\boldsymbol{A}} + \tilde{\boldsymbol{B}}\boldsymbol{K}) \\ &= w^{nq} + \lambda_{nq-1}w^{nq-1} + \cdots + \lambda_1 w + \lambda_0, \end{aligned} \tag{28}$$

where $w = s^{\frac{1}{q}}$. It is possible to obtain regulator $\boldsymbol{K}$ by choosing the closed-loop characteristic polynomial.

4.1.2 Ackerman's formula

To obtain the controller matrix $\boldsymbol{K}$, the Ackerman's formula can be used. It is easy to check that, when $\boldsymbol{z} = \boldsymbol{P}\boldsymbol{x}$, the controllability matrix is also transformed by the same matrix $\boldsymbol{P}$ according to relation $\mathbb{S}_z = \boldsymbol{P}\mathbb{S}_x$, where $\mathbb{S}_z$ is a controllability matrix of the system after state vector transformation, $\mathbb{S}_x$ is a controllability matrix of the original system, and $\boldsymbol{P} \in \mathbb{R}^{np \times np}$ is a nonsingular matrix, so $\det(\boldsymbol{P}) \neq 0$. When the system in state variable $\boldsymbol{z}$ is assumed to be in controller canonical form and using Cayley–Hamilton theorem, it is easy to show, analogically to the case of integer-order systems, that the matrix $\boldsymbol{K}$ can be obtained by using the redefined Ackerman's formula.

Lemma 1. *The Ackerman's formula for augmented system* (24) *is given by the following relation:*

$$\boldsymbol{K} = [0 \quad \cdots \quad 0 \quad 1]\, \mathbb{S}^{-1} \lambda_c(\boldsymbol{A}) \tag{29}$$

where $\lambda_c(\boldsymbol{A})$ is a matrix polynomial with the coefficients of the closed-loop characteristic polynomial, and $\mathbb{S}$ is a controllability matrix.

In the practical implementation of the fractional-order augmented state feedback control, two main problems can be found:

1. The control law uses the augmented states variables, which are not able to be measured directly from the plant. These state variables have to be estimated. As a solution, a fractional-order augmented observer can be used (see Section 4.2).
2. The above algorithms require knowledge of the closed-loop characteristic polynomial (closed-loop dynamics). This is a quite strong requirement because the influence of the poles location on the system dynamics is more complicated than for the integer-order case. In order to do easier and more intuitive the choosing of the closed-loop dynamics, the linear quadratic regulator (LQR) algorithm for the fractional-order augmented system, presented in Section 5.1, can be used.

4.2 Output feedback

The pseudo-states $x_{a,i}(t)$, for $i = 1, \ldots, q-1$, are not available directly from the system. Even, as usual, not all the states are available by measurement. Consequently, a observer is needed including the augmented model of the system.

The fractional-order observer for augmented fractional-order system (24) is given by the following linear fractional-order dynamic system:

$$\mathcal{D}_t^{\alpha}\hat{\mathbf{x}}(t) = \boldsymbol{F}\hat{\mathbf{x}}(t) + \boldsymbol{G}u(t) + \boldsymbol{H}y(t), \tag{30}$$

where $\boldsymbol{G} = \tilde{\boldsymbol{B}} - \boldsymbol{H}\boldsymbol{D}$ and $\boldsymbol{F} = \tilde{\boldsymbol{A}} - \boldsymbol{H}\tilde{\boldsymbol{C}}$, being $\boldsymbol{H}$ the unknown observer gain.

The estimation error equation is given as follows:

$$D_t^{\alpha}\tilde{\mathbf{x}}(t) = (\tilde{\boldsymbol{A}} - \boldsymbol{H}\tilde{\boldsymbol{C}})\tilde{\mathbf{x}}(t), \tag{31}$$

where $\tilde{\mathbf{x}}(t) = \mathbf{x}(t) - \hat{\mathbf{x}}(t)$ is the estimation error.

The design procedure is the same as for an integer-order observer: the matrix $\boldsymbol{H}$ should be chosen in order to guarantee asymptotic stability of the error equation and obtain the dynamics of the observer faster than that of the system.

Using the vector with the pseudo-states $\boldsymbol{x_a}(t)$, it is possible to obtain the estimated fractional (rational) order derivative of the output from the following relation:

$$\mathcal{D}_t^{\alpha}\hat{y}(t) = \boldsymbol{C}D_t^{\alpha}\hat{\mathbf{x}}(t) = C\hat{\boldsymbol{x}}_{\boldsymbol{a}}(t). \tag{32}$$

5 Optimal control

Recent advances in fractional-order control have enable many works in optimal control. In this section, we will consider the extension of some special types of optimal state space controllers for fractional-order augmented systems. More specifically, design methods for the extensions of LQR, Kalman filter (KF), and linear quadratic Gaussian (LQG) regulator will be described next.

5.1 Linear quadratic regulator

The LQR algorithm considers the following cost function:

$$J = \int_0^{T_f} \frac{1}{2}[x^T(t)Qx(t) + u^T(t)Ru(t)]dt, \tag{33}$$

where Q and R are positive semidefinite and positive definite matrices, respectively.

The extension of conventional LQR for fractional-order systems has the form of the following theorem.

Theorem 1. *The time invariant LQR type controller for a fractional-order system in state space and cost function given by* (33) *with assumption that the infinite time horizon is taken into consideration and the system is in the half of control process, is given as*

$$u(t) = R^{-1}B^T Px(t), \tag{34}$$

where P is a solution of the following algebraic Riccati equation:

$$A^T P + PA + PBR^{-1}B^T P + Q = 0. \tag{35}$$

It should be noted that matrices A and B refer to those of the commensurate system or, what is more usual, of the fractional augmented system. Further details of this strategy can be found in Sierociuk and Vinagre [25].

Other contributions in the literature have been already devoted to the extension of the classical LQR to fractional order. Refer to, for example, Arabi and Merrikh-Bayat [1], Liang et al. [14], Li and Chen [13], Shafieezadeh et al. [20].

5.2 Kalman filter

Even for variable-order systems, a KF can be obtained with infinite-dimensional matrices (or limited to some predefined, usually high value). However, it could be problematic in practical implementation and numerically complicated.

Theorem 2 (Sierociuk and Dzielinski [22]). *For discrete-time fractional-order system* (20), *the simplified KF, also known as fractional KF* (FKF), *is given by the following set* *of equations:*

$$\Delta^{\Upsilon}\tilde{x}_{k+1} = A\hat{x}_k + Bu_k, \tag{36}$$

$$\tilde{P}_k = (A + \Upsilon_1)P_{k-1}(A + \Upsilon_1)^T + Q_{k-1} + \sum_{j=2}^{k} \Upsilon_j P_{k-j} \Upsilon_j^T, \tag{37}$$

$$\hat{x}_k = \tilde{x}_k + K_k(y_k - C\tilde{x}_k), \tag{38}$$

$$P_k = (I - K_k C)\tilde{P}_k, \tag{39}$$

where

$$\tilde{x}_{k+1} = \Delta^{\Upsilon}\tilde{x}_{k+1} - \sum_{j=1}^{k+1}(-1)^j \Upsilon_j \hat{x}_{k+1-j}, \tag{40}$$

$$K_k = \tilde{P}_k C^T (C\tilde{P}_k C^T + R_k)^{-1}, \tag{41}$$

with initial conditions

$$x_0 \in \mathbb{R}^N, \quad P_0 = \mathrm{E}[(\tilde{x}_0 - x_0)(\tilde{x}_0 - x_0)^T],$$

and v_k and ω_k are assumed to be independent and with zero mean value.

The previous formulation of the FKF can be improved, and its computational load reduced, by assuming that only a fixed number m of past samples of the state vector is taken for estimation.

Theorem 3 (Sierociuk et al. [24]). *For discrete-time fractional-order system in m-finite form given by* (21)*, the Kalman filter algorithm* (*referred to as ExFKF*) *is given by the following set of equations:*

$$\tilde{\mathbb{X}}_{k+1} = \mathbb{A}_m \hat{\mathbb{X}}_k + \mathbb{B}_m u_k - \mathbb{I} \sum_{j=m+1}^{k+1} (-1)^j \Upsilon_j \hat{x}_{k+1-j}, \tag{42}$$

$$\tilde{\mathbb{P}}_k = \mathbb{A}_m \mathbb{P}_{k-1} \mathbb{A}_m^T + \mathbb{Q}_{k-1} + \sum_{j=m+1}^{k} \mathbb{I}\Upsilon_j P_{k-j} \Upsilon_j^T \mathbb{I}^T, \tag{43}$$

$$\hat{\mathbb{X}}_k = \tilde{\mathbb{X}}_k + \mathbb{K}_k(y_k - \mathbb{C}_m \tilde{\mathbb{X}}_k), \tag{44}$$

$$\mathbb{P}_k = (I - \mathbb{K}_k \mathbb{C}_m)\tilde{\mathbb{P}}_k \tag{45}$$

where

$$\mathbb{K}_k = \tilde{\mathbb{P}}_k \mathbb{C}_m^T (\mathbb{C}_m \tilde{\mathbb{P}}_k \mathbb{C}_m^T + R_k)^{-1}, \tag{46}$$

$$\mathbb{I} = \begin{bmatrix} I \\ 0 \\ \vdots \\ 0 \end{bmatrix}, \quad \mathbb{Q}_k = \mathbb{I} Q_k \mathbb{I}^T, \tag{47}$$

$$\mathbb{P}_k = \begin{bmatrix} P_{k,k} & P_{k,k-1} & \cdots & P_{k,k-m} \\ P_{k-1,k} & P_{k-1,k-1} & \cdots & P_{k-1,k-m} \\ \vdots & \vdots & \vdots & \vdots \\ P_{k-m,k} & P_{k-m,k-1} & \cdots & P_{k-m,k-m} \end{bmatrix},$$

with initial conditions

$$\hat{\mathbb{X}}_0 \in \mathbb{R}^{mN}, \quad \mathbb{P}_0 = \mathrm{E}[(\hat{\mathbb{X}}_0 - \mathbb{X}_0)(\hat{\mathbb{X}}_0 - \mathbb{X}_0)^T]$$

and noises v_k and ω_k are assumed to be independent and with zero expected value, and matrices $\mathbb{P}_k$ and R_k are positive defined.

Other formulations for the FKF can be found in the literature. For example, an improved version of the FKF was proposed by Sierociuk and Dzielinski [22] to estimate over lossy networks, or its extension to variable-order systems in Sierociuk et al. [23], Sierociuk and Ziubinski [27].

As an example, let consider a discrete-time fractional-order system in state space given by the following matrices:

$$A = \begin{bmatrix} 0 & -0.1 \\ 1 & 0.2 \end{bmatrix}, \quad B = \begin{bmatrix} 0.2 \\ 0.3 \end{bmatrix}, \quad C = \begin{bmatrix} 0.4 & 0.3 \end{bmatrix}, \quad \mathcal{N} = \begin{bmatrix} 0.5 & 0.5 \end{bmatrix}^T, \tag{48}$$

with noises such that $E[v_k v_k^T] = 0.0021$ and $E[\omega_k \omega_k^T] = \left[\begin{smallmatrix} 0.0980 & -0.0044 \\ -0.0044 & 0.0949 \end{smallmatrix}\right]$. Let consider the FKF and the ExFKF with

$$P_0 = 100I, \quad x_0 = [0, 0]^T,$$

with matrices Q_k and R_k equal to the values of the noises variances, and $m = 20$ for the ExFKF.

A comparison of the estimation results obtained with the FKF and the smoothing results when applying the ExFKF for the case $\hat{x}_{k-8}|k$ is plotted in Figure 3(a) for state x_1, and in Figure 3(b) for x_2. Figure 4 shows the system output when applying the ExFKF. Notice that the input considered was a square signal of amplitude 1.

It is important to mention that the FSST Toolkit (Sierociuk [21]) was used in the MATLAB/Simulink environment for a part of the simulations.

5.3 Linear quadratic Gaussian with loop transfer recovery

The methodolgy of LQG with loop transfer recovery (LTR), referred to as LQG/LTR, was originally developed by Kwakernaak [12], Doyle and Stein [4], Athans [2], Stein

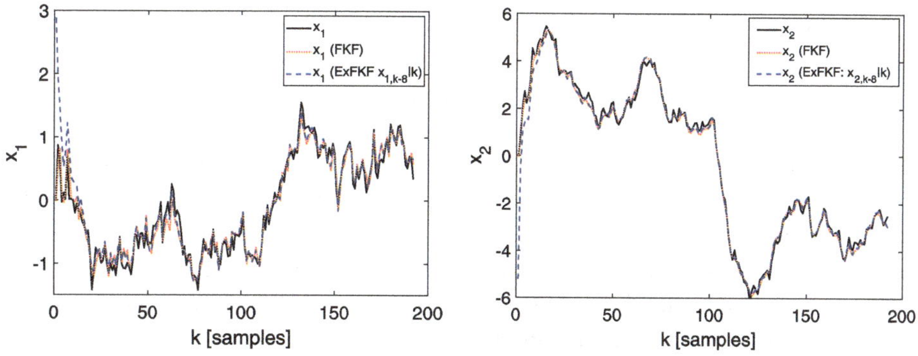

(a) (b)

Figure 3: Estimation results of discrete-time fractional-order system given by matrices (48) when applying the FKF and the ExFKF for: (a) State x_1 (b) State x_2.

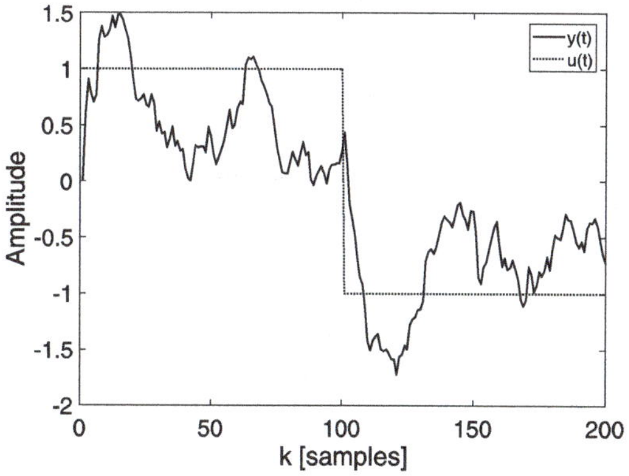

Figure 4: Output of discrete-time fractional-order system given by matrices (48) when applying the ExFKF with $m = 20$.

and Athans [28]. Its advantages lie in its systematic design procedure to solve design issues such as robustness, stability, and compensation between system performance and allowable control power. Despite that this technique does not guarantee a suitable performance for nonminimum phase plants (Athans [2]), it has been widely adopted in industrial and academic environments since provides control results comparable to more sophisticated methods as model predictive control, and better than classical PI control. Moreover, the LQG/LTR controller can provide good performance and guarantee stability in some practical circumstances where the dynamics of the controlled plant may not be exactly modeled, and there may be system disturbances and measurement noises (Maciejowski [15]). This methodology, because it involves the former concepts and developments, will be used now for an application case.

5.3.1 Generalities

The LTR method is a heuristic tool that was created in order to introduce robustness into the LQG, whereby confers valuable properties on the control system: nominal stability, stability-robustness to modeling errors, and good performance (Athans [2]).

In practice, LTR is the last step in a three-stage design procedure for obtaining the dynamic controller (Kim et al. [11]): (1) specification of feedback controlled system performances; (2) design of the corresponding target loop—will be called target filter loop (TFL)—using state feedback; and (3) recover the target loop by an admissible output based compensator. It is worth mentioning that KF and LQR are dual problems, so any of them can be used for TFL or LTR stages (Jabbar [9], Hespanha [8]). Furthermore, it should be noted that LQG/LTR is a control design method that uses LQG control structure, but it is not an optimal control design method, and not even a stochastic design method (it uses a fictitious KF with no reference to noise).

The problem to be solved is the design of a feedback dynamic controller $C(s)$ for the process $P(s)$ in order to obtain some specified performances for the closed-loop (see Figure 5(a)). The expression of the controller to be obtained will be

$$C(s) = K(sI - A + HC + BK)^{-1}H, \tag{49}$$

where A, B, and C are the matrices of the system $P(s)$ in state space, I is the identity matrix, and H and K are the unknown filter and control gain matrices, respectively, which will be designed next. The structure of the controller is illustrated in Figure 5(b) (de Souza et al. [3]).

In the design of target loop, both the KF and the LQR possess good robustness and frequency domain properties, and any of them can be used as a candidate for loop shaping in the TFL stage. The case of using KF as TFL is shown in Figure 6(a). The open-loop has the form

$$G_F(s) = C\Phi(s)H, \tag{50}$$

where $\Phi(s) = (sI - A)^{-1}$, being A the one obtained by incorporating the variable scaling and/or augmentation dynamics to the process $P(s)$ (such as integrators for no steady-state error) needed to meet performance specifications (Athans [2]). This augmented

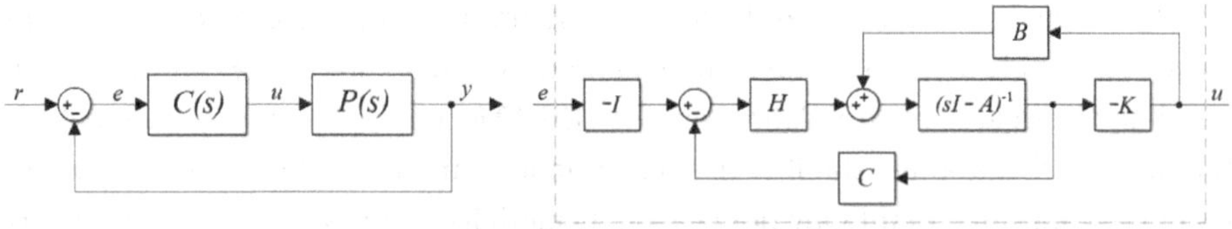

(a) (b)

Figure 5: Block diagrams of the control problem: (a) Controlled system (b) Details of LQG/LTR controller, $C(s)$.

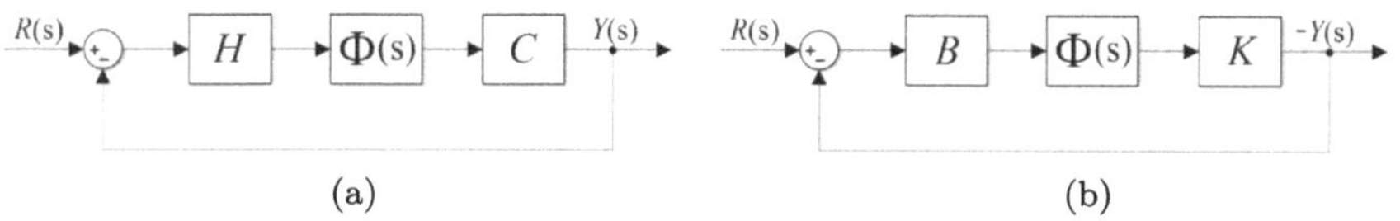

Figure 6: Block diagrams of target filter loop (dual problem): (a) using KF (b) using LQR.

dynamics should be moved from the design plant to the input side of the compensator $C(s)$ once the design process is terminated. (The dual problem would use the LQR loop shown in Figure 6(b) as TFL.)

The goal of this second step is to obtain the filter gain matrix H to meet the design objectives for TFL, which can be calculated solving the algebraic Riccati equation:

$$AP + PA^T + LL^T - \frac{1}{\sigma}PC^T CP = 0, \tag{51}$$

$$H = \frac{1}{\sigma}PC^T, \tag{52}$$

being L the matrix of the process noise and $\sigma > 0$ a scalar parameter, which can be viewed as design parameters. The advantage of using this methodology for calculating H is to guarantee that the TFL will never amplify disturbances at any frequency and never go unstable to multiplicative model errors (Athans [2]).

After solving the KF gain matrix that leads to the desired filter loop-shapes, the goal of this step is to obtain the control gain matrix K by means of tuning a common LQR whose cost function is given as

$$J = \int_0^\infty [x^T Qx + u^T Ru]dt, \tag{53}$$

where $Q > 0$ and $R > 0$ are the state weight and the control weight matrix, respectively. As usual, in order to obtain K, previously it is necessary to calculate G by solving the algebraic Riccati equation:

$$GA + A^T G + C^T C - \frac{1}{\rho}GBB^T G = 0. \tag{54}$$

Once G is calculated, we can obtain K by the following equation:

$$K = \rho^{-1}B^T G. \tag{55}$$

Letting $Q = C^T C$ and $R = \rho I$, it can be proved that

$$\lim_{\rho \to 0}(C(s)P(s)) = C\Phi(s)H. \tag{56}$$

The process for calculating K is iteratively repeated for different values of the design parameter ρ until obtaining the desired recovering of the loop.

Other contributions on the extension of the classical LQG to fractional order have been already reported in the literature. Refer to, for example, Duda [5], Duncan and Pasik-Duncan [6], Kiani-B et al. [10].

Let us show the LQG/LTR design procedure with an example taken from Hespanha [8]: the aircraft roll-dynamics. Denoting θ as the roll-angle, τ as the applied torque and $\omega = \dot{\theta}$ as the roll-rate, the dynamics of the aircraft is given as

$$\dot{\boldsymbol{x}}(t) = \begin{bmatrix} 0 & 1 & 0 \\ 0 & -0.875 & -20 \\ 0 & 0 & -50 \end{bmatrix} \boldsymbol{x}(t) + \begin{bmatrix} 0 \\ 0 \\ 50 \end{bmatrix} u(t), \qquad (57)$$

$$y(t) = \begin{bmatrix} 1 & 0 & 0 \end{bmatrix} \boldsymbol{x}(t)$$

being the state vector $x = \begin{bmatrix} \theta & \omega & \tau \end{bmatrix}^T$.

In this case, LQR loop is used as TFL. Figure 7(a) shows the Bode plots of the open-loop gain $\hat{H}(s) = K(sI - A)^{-1}B$ for several LQR controllers obtained for this system with $R = \rho I$ and

$$Q = \begin{bmatrix} 1 & 0 & 0 \\ 0 & 1 \times 10^{-4} & 0 \\ 0 & 0 & 0 \end{bmatrix}$$

and several values of the parameter ρ. It can be seen that ρ allows one to move the whole magnitude Bode plot up and down, obtaining higher crossover frequencies (faster systems) as ρ decreases. It is also interesting to observe that the loop behaves like an integrator cascaded with a lag compensator, with a high frequency slope of −20 dB/dec. The corresponding step responses for the closed-loop system (in accordance with Figure 6(b)) are plotted in Figure 7(b). As can be observed, the smaller the values of ρ, the faster the responses.

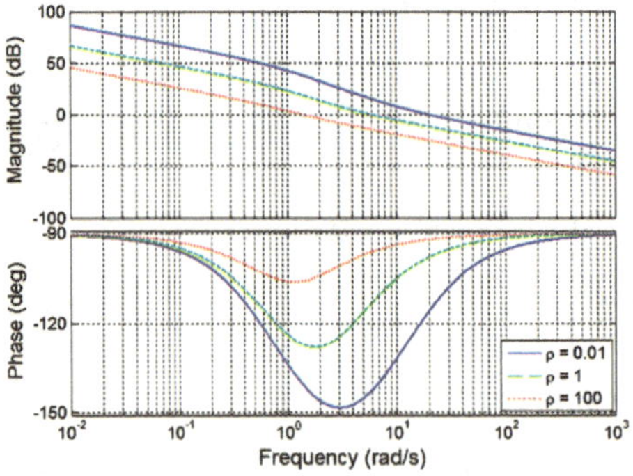

(a)

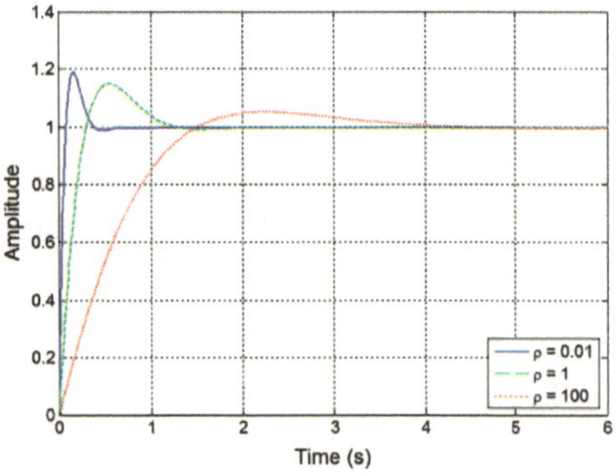

(b)

Figure 7: Simulations of LQR loop for system (57) for different values of ρ: (a) Bode plots (b) Closed-loop step responses.

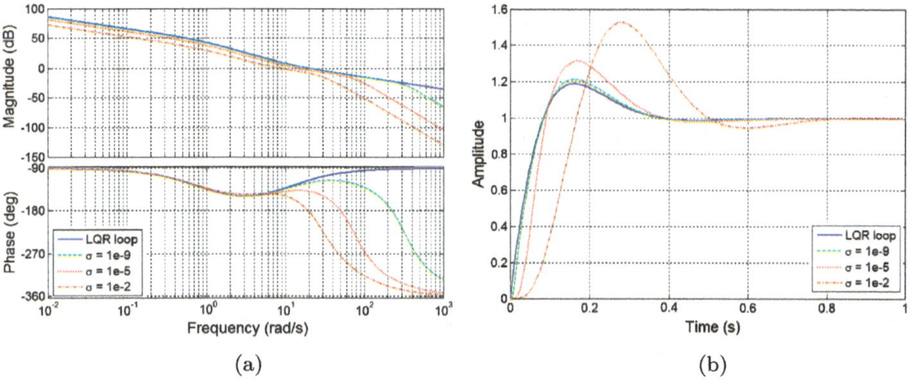

Figure 8: Simulations of LQG/LTR controller for system (57) for different values of σ: (a) Bode plots (b) Closed-loop step responses.

Then, KF gains have to be designed. Figure 8(a) compares the Bode plots of the open-loop gain for the state feedback LQR controller (target loop) with the open-loop gain for several output feedback LQG/LTR controllers for different values of the parameter σ and with $\rho = 0.01$. It is observed that the range of frequencies over which the open-loop gain of the LQG/LTR controller matches that of the target loop increases as σ decreases. It can be also seen that the designed LQG/LTR controllers behave better than the target loop since they exhibit higher slope of the magnitude curve at high frequencies. The corresponding step responses of the system with these LQG/LTR controllers are plotted in Figure 8(b). As can be observed, the smaller the values of σ, the closer the response to the LQR target loop.

5.3.2 Fractional LTR method

This section presents the effects of introducing fractional order dynamics into the LTR design method by means of the concept of system augmentation proposed in Sierociuk and Vinagre [26]. It is shown that this fact naturally leads to: (1) the use of an approximated fractional integrator as target loop, and (2) the use of a fractional KF for state estimation.

As in the case of integer-order systems, for obtaining perfect steady-state response to command signals, we can use a feedforward gain K_r as the inverse of the DC gain of the closed-loop system. But, to obtain good performance in the presence of uncertainty and load perturbations, it is needed to add an integral action into the feedback controller, which can be done introducing a new state defined as $\mathcal{D}^{\alpha}z = y - r$. It is known that a fractional-order integrator with order $\alpha = 1/q$, $0 < \alpha < 1$, can eliminate the same kind of steady-state errors than an integrator of order 1.

The use of this fractional model of the system obtained by applying system augmentation, naturally leads to: (1) TFL with fractional dynamics, and (2) the use of fractional-order observers (fractional KFs) for state (or pseudo-state) estimation.

Consider the servo model in Sierociuk and Vinagre [26] given in state space form by

$$\begin{aligned} \dot{x}(t) &= ax(t) + bu(t), \\ y(t) &= x(t) \end{aligned} \tag{58}$$

where $y(t)$ is the rotation speed of the servo (in rpm), $u(t)$ is the control signal (in V) and with $a = -2.381$ and $b = 26.204$. Thus, the augmented system for $q = 2$ is

$$\begin{aligned} {}_0^C\mathcal{D}_t^{0.5}\boldsymbol{x}(t) &= \begin{bmatrix} 0 & 1 \\ -2.381 & 0 \end{bmatrix} \boldsymbol{x}(t) + \begin{bmatrix} 0 \\ 26.204 \end{bmatrix} u(t), \\ y(t) &= \begin{bmatrix} 1 & 0 \end{bmatrix} \boldsymbol{x}(t). \end{aligned} \tag{59}$$

In what follows, we are going to proceed in a similar way than in the previous example: firstly, the LQR controller (target loop) will be obtained—the behavior in the frequency domain of that controller for different values of the parameter ρ—and then, once selected the best performance for the servomotor, the loop will be recovered with the LTR controller—again, the effect on the capacity of recovering the loop with the design parameter σ will be studied. For the design, it was considered that $R = \rho I$ and $Q = \boldsymbol{C}^T\boldsymbol{C}$.

In the case of using the feedforward gain K_r for perfect tracking, the results are presented in Figures 9 and 10. Figures 11 and 12 correspond to the case of integral control.

Figures 9(a) and 11(a) show the Bode plots of the open-loop gain $\hat{H}(s) = K(s\boldsymbol{I} - \tilde{\boldsymbol{A}})^{-1}\tilde{\boldsymbol{B}}$ for several LQR controllers obtained for this system for several values of the

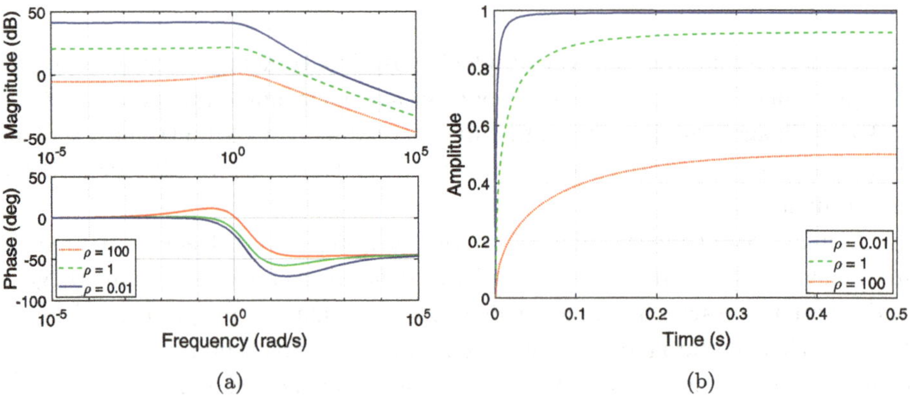

Figure 9: Simulations of LQR loop for system (58) with feedforward gain K_r for different values of ρ: (a) Bode plots (phase curves) (b) Closed-loop step responses.

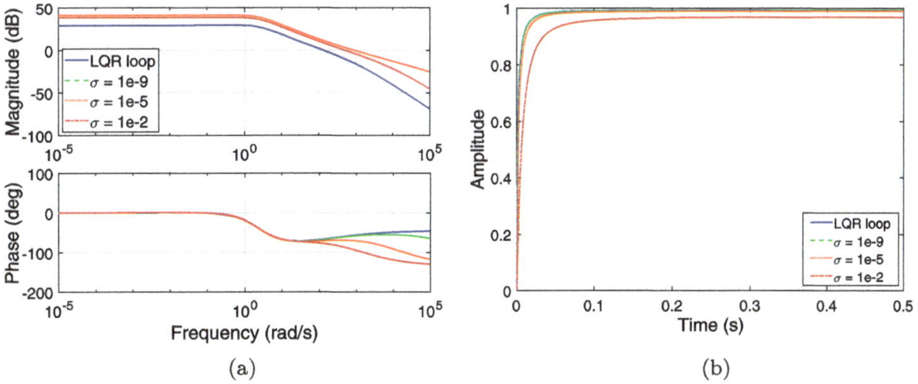

Figure 10: Simulations of fractional LQG/LTR controller for system (58) with feedforward gain K_r for different values of σ: (a) Bode plots (phase curves) (b) Closed-loop step responses.

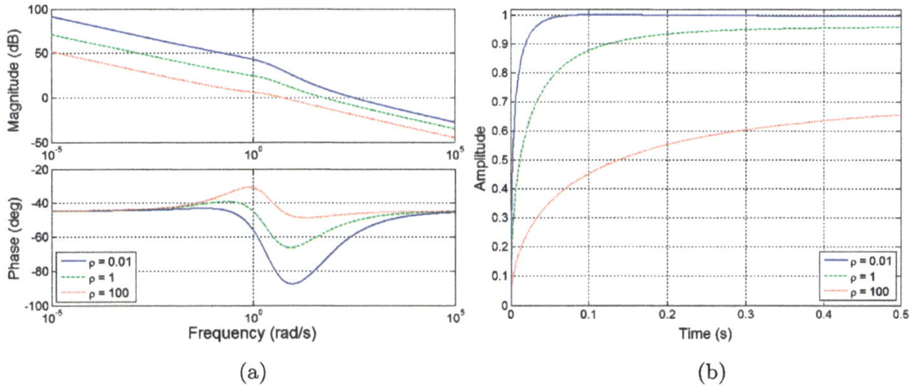

Figure 11: Simulations of LQR loop for system (58) with integral control for different values of ρ: (a) Bode plots (b) Closed-loop step responses.

parameter ρ. Notice that matrices $\tilde{\boldsymbol{A}}$ and $\tilde{\boldsymbol{B}}$ refer to the matrices of the fractional augmented system. As expected, again the parameter ρ allows us to move the whole magnitude Bode plot up and down, allowing to specify a desired crossover frequency (see the magnitude curve in Figure 9(a)). The gain magnitude falls at $-20/q = -10$ dB/dec at high frequencies, behaving, in the case of integral action (Figure 11(a)), like a fractional integrator of order $1/q = 0.5$ cascaded with a led-lag compensator. The main differences with the integer-order case are that the high frequency slope is lower, and the phase margin is larger. Figures 9(b) and 11(b) show the corresponding step responses for the state feedback fractional LQR controllers. Again, the smaller the values of ρ, the faster the responses.

Then, KF gains are designed to obtain the controller $C(s)$. Figures 10(a) and 12(a) show the Bode plots of the open-loop gain for the state feedback LQR controller (tar-

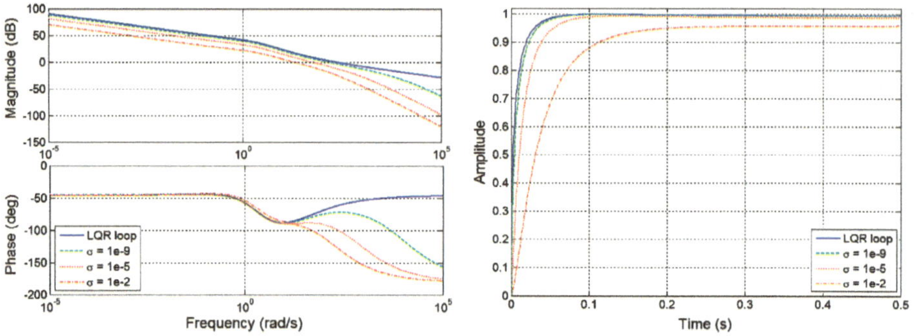

(a) (b)

Figure 12: Simulations of fractional LQG/LTR controller for system (58) with integral control for different values of σ: (a) Bode plots (b) Closed-loop step responses.

get loop) and the open-loop gain for several output feedback LQG/LTR controllers for different values of the parameter σ, and with $\rho = 0.01$. As in the integer case, one can observe that the range of frequencies over which the open-loop gain of the LQG/LTR controller matches that of the target loop increases as σ decreases. Furthermore, the designed LQG/LTR controllers behave better than the target loop since they exhibit higher slope of the magnitude curve at high frequencies. The step responses of the servomotor corresponding to these LQG/LTR controllers are represented in Figures 10(b) and 12(b). Similar to the integer case, the smaller the values of σ, the closer the response to the LQR target loop.

It is worth mentioning that a modification of the function `bodeFr` from *Ninteger* toolbox for MATLAB (Valério [29]) was used to obtain the system frequency responses, as well as step responses were obtained with solving numerically the fractional differential equation corresponding to the closed-loop system with the function `fode_sol` in Monje et al. [16].

Similar results can be obtained in Simulink by using the block diagrams in Figures 13(a) and 13(b). Figure 14 shows the internal structure of the observer including augmented fractional model for the system. Step and disturbance responses are presented in Figure 15. As can be observed, the control strategy with the integral control is more robust against disturbances that the one with feedforward gain.

It should be remarked that, in order to preserve the integral action, fractional integrals are evaluated by cascading an integrator with a fractional differentiator of order 0.5 approximated by CRONE method in the simulations.

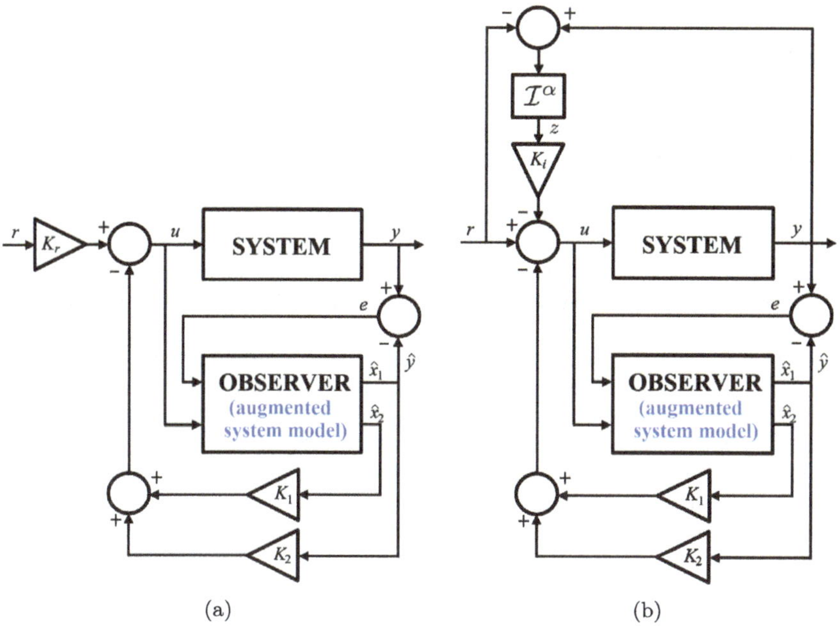

Figure 13: Control block diagrams of system (58) with: (a) Feedforward gain K_r (b) Integral control.

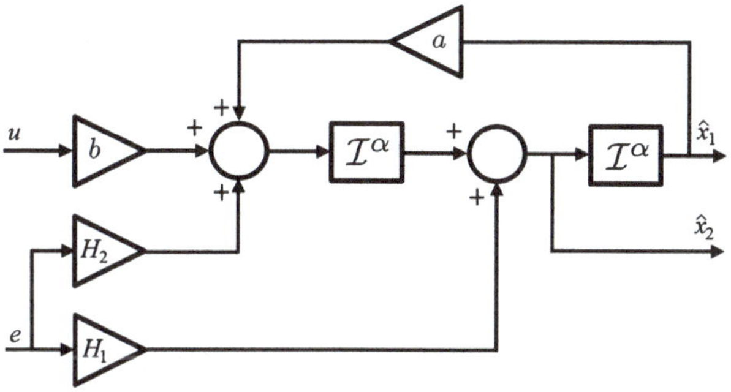

Figure 14: Control block diagram of system (58): observer structure.

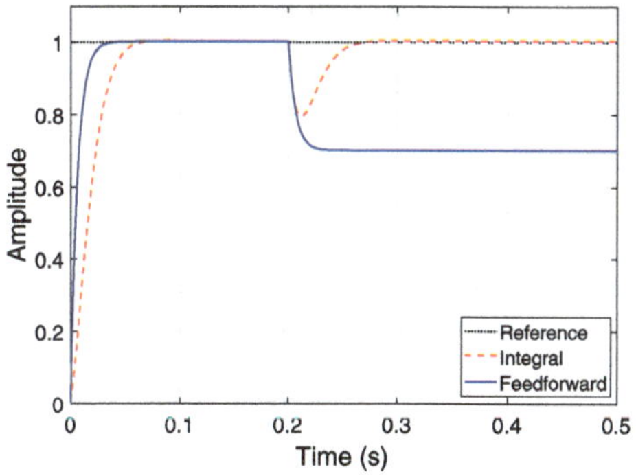

Figure 15: Responses to a step reference $r = 1$ with step load perturbation of amplitude $d = 3r$ at time $t_d = 0.2$ s.

Bibliography

[1] S. H. Arabi and F. Merrikh-Bayat, A practical method for designing linear quadratic regulator for commensurate fractional-order systems, *Journal of Optimization Theory and Applications*, **174**(2) (2017), 550–566.

[2] M. Athans, A tutorial on the LQG/LTR method, in *Proceedings of the American Control Conference*, pp. 1289–1296, 1986.

[3] E. C. de Souza, J. J. da Cruz, and N. Maruyama, The LQG/LTR methodology for position control of unmanned underwater vehicles, in *Proceedings of the XV Brazilian Congress of Automatica*, 2004.

[4] J. C. Doyle and G. Stein, Multivariable feedback design: concepts for a classical/modern synthesis, *IEEE Transactions on Automatic Control*, **26**(1) (1981), 1–14.

[5] Z. Duda, State estimation in a decentralized discrete time LQG control for a multisensor system, *The Journal of Polish Academy of Sciences*, **27**(1) (2017), 29–39.

[6] T. E. Duncan and B. Pasik-Duncan, Linear-quadratic fractional Gaussian control, *SIAM Journal on Control and Optimization*, **51**(6) (2013), 4504–4519.

[7] A. Dzieliński and D. Sierociuk, Stability of discrete fractional order state-space systems, *Journal of Vibration and Control*, **14**(9–10) (2008), 1543–1556.

[8] J. P. Hespanha, *Lecture Notes on LQR/LQG Controller Design*, 2005.

[9] F. Jabbar, Loop shaping and recovery LQR, Technical report, University of California, Irvine, 2014.

[10] A. Kiani-B, A. Karimpour, N. Pariz, and I. Veisi, Robust stability of the fractional order LQG controllers, in *Proceedings of the 17th Iranian Conference on Electrical Engineering (ICEE 2009)*, pp. 237–242, 2009.

[11] B. H. Kim, Y. M. Park, M. S. Choi, and J. W. Lee, LQG/LTR robust controller of cogeneration plant for disturbance rejection in electric frequency and steam pressure, *Electrical Power & Energy Systems*, **18**(4) (1996), 239–250.

[12] H. Kwakernaak, Optimal low-sensitivity linear feedback systems, *Automatica*, **5** (1969), 279–285.

[13] Y. Li and Y. Q. Chen, Fractional order linear quadratic regulator, in *Proceedings of the 2008 IEEE/ASME International Conference on Mechatronic and Embedded Systems and Applications (MESA 2008)*, pp. 363–368, 2008.

[14] S. Liang, S.-G. Wang, and Y. Wang, Representation and LQR of exact fractional order systems, in *Proceedings of the IEEE 53rd Annual Conference on Decision and Control (CDC'14)*, pp. 6908–6913, 2014.

[15] J. M. Maciejowski, *Multivariable Feedback Design*, Addison-Wesley, 1989.

[16] C. A. Monje, Y. Q. Chen, B. M. Vinagre, D. Xue, and V. Feliu, *Fractional-Order Systems and Controls. Fundamentals and Applications*, Springer, 2010.

[17] K. B. Oldham and J. Spanier, *The Fractional Calculus*, Academic Press, 1974.

[18] P. Ostalczyk, *Discrete Fractional Calculus*, vol. 4, World Scientific, 2016.

[19] I. Podlubny, *Fractional Differential Equations*, vol. 198, Academic Press, 1999.

[20] A. Shafieezadeh, K. Ryan, and Y. Q. Chen, Fractional order LQR for optimal robust control of a simple structure, in *Proceedings of the 2007 ASME International Design Engineering Technical Conferences and Computers and Information in Engineering Conference (DETC2007)*, 2007.

[21] D. Sierociuk, *Fractional Order Discrete State-Space System Simulink Toolkit User Guide*, Warsaw University of Technology, 2006.

[22] D. Sierociuk and A. Dzielinski, Fractional Kalman filter algorithm for the states, parameters and order of fractional system estimation, *International Journal of Applied Mathematics and Computer Science*, **16**(1) (2006), 129–140.

[23] D. Sierociuk, M. Macias, W. Malesza, and G. Sarwas, Dual estimation of fractional variable order based on the unscented fractional order Kalman filter for direct and networked measurements, *Circuits, Systems, and Signal Processing*, **35**(6) (2016), 2055–2082.
[24] D. Sierociuk, I. Tejado, and B. M. Vinagre, Improved fractional Kalman filter and its application to estimation over lossy networks, *Signal Proceesing*, **91** (2011), 542–552.
[25] D. Sierociuk and B. M. Vinagre, Infinite horizon state-feedback LQR controller for fractional systems, in *Proceedings of the 49th IEEE Conference on Decision and Control (CDC'10)*, pp. 6674–6679, 2010a.
[26] D. Sierociuk and B. M. Vinagre, State and output feedback fractional control by system augmentation, in *Proceedings of the 4th IFAC Workshop Fractional Differentiation and its Applications (FDA'10)*, 2010.
[27] D. Sierociuk and P. Ziubinski, Variable order fractional Kalman filters for estimation over lossy network, in *Advances in Modelling and Control of Non-Integer-Order Systems*, pp. 285–294, Springer, 2015.
[28] G. Stein and M. Athans, The LQG/LTR procedure for multivariable feedback control design, *IEEE Transactions on Automatic Control*, **32**(2) (1987), 105–114.
[29] D. Valério, *Ninteger Toolbox for MATLAB*, University of Lisbon, 2005.

Emmanuel A. Gonzalez

Posicast control of fractional-order systems

Abstract: In this chapter, Posicast control and its application to fractional-order underdamped systems is described. The half-cycle Posicast controller applied to the fractional-order system is demonstrated. Moreover, the three-step Posicast controller is also proposed.

Keywords: Posicast control, fractional-order system, Matlab

MSC 2010: 26A33, 37N35

1 Background and introduction

Posicast is an input-shaping method that was designed to "shape" the step response of an underdamped system by eliminating its overshoot. This method was introduced by O. M. Smith [17, 18] where a feedforward controller is implemented as shown in Figure 1. The feedforward controller serves as an input-shaping filter where the step input is reshaped in the form of a "staircase" with each step having a certain magnitude which is a fraction of the step input. The transition from one step to another depends on the time parameters of the step response.

The Posicast control method is probably best described as a gantry problem as shown in Figure 2. Assume a gantry with a swinging bob connected to the end of the pole attached in a trolley that has the capability to slide from Position X to Position Z. The objective is to move the bob from Positions X (Event #1) to Z (Event #4) by sliding the trolley but without causing any oscillations on the bob's position. Posicast control is done through the succeeding sequence of events: In Event #1, both the trolley and bob are located at Position X. Event #2 shows an instant push of the trolley from Position X to Position Y, allowing the bob to swing from Positions X to Z. Once the bob reaches Position Z (Event #3), the trolley is pushed again toward Position Z (Event #4), which prevents oscillations.

The use of the Posicast control technique is simple, straightforward, and easy to implement. This resulted in several research outputs applying such methodology to underdamped systems. For example, Cook in 1966 [2] explored the use of Posicast control in compensating a lightly-damped fourth-order system and related them to flexible structures [3]. Vibration control applications of Posicast control was also introduced since 1990 [4, 16]. Practical implementations were also presented such as the application of Posicast in a Z-source current-type inverter [14], buck converter [5],

Emmanuel A. Gonzalez, Schindler Elevator Corporation, 1530 Timberwolf Dr, Holland, OH, 42538-9161, USA, e-mail: emmgon@gmail.com

https://doi.org/10.1515/9783110571745-009

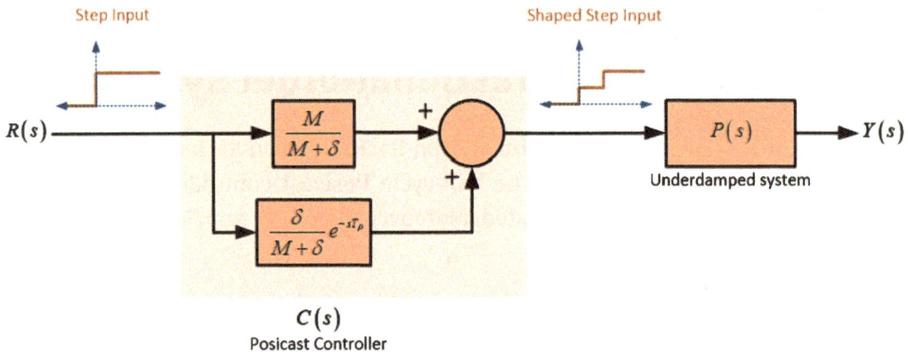

Figure 1: Feedforward configuration of a Posicast controller $C(s)$ on an underdamped system $P(s)$.

various power electronic applications [7], single ended primary inductor for renewable energy electronic applications [13], and resonance reduction for satellite control systems [6] as examples.

The Posicast control method used by Smith and Cook were known to have sensitivity issues. By applying the control methodology in feedback form [5, 10–12], such sensitivity issues could be addressed which were presented by Hung and colleagues. However, in most noncritical underdamped systems, the use of Posicast control, being a simple as it may sound, is just as effective as compared to other control methods.

The study of the application of Posicast control for fractional-order underdamped systems has just started in 2011 when Gonzalez, Hung, and Co [8] published a paper applying a two-step feedforward Posicast controller on a two-term fractional-order system. In 2013, another paper was published by Gonzalez et al. [9], applying a three-step feedforward Posicast controller on a two-term fractional-order system.

2 Fractional-order underdamped systems

In order to create an effective Posicast control mechanism for fractional-order underdamped systems, it is important first to understand how a fractional-order system can have output damped oscillations that are stable.

First, let us establish the step response of a fractional-order system of the form

$$P(s) = \frac{K}{Ts^{\alpha} + 1}, \tag{1}$$

where, $K, T, \alpha > 0$. To be able to use the results from other literature, let us consider first a fractional-order transfer function of the form

$$G(s) = \frac{1}{as^{\alpha} + b}, \tag{2}$$

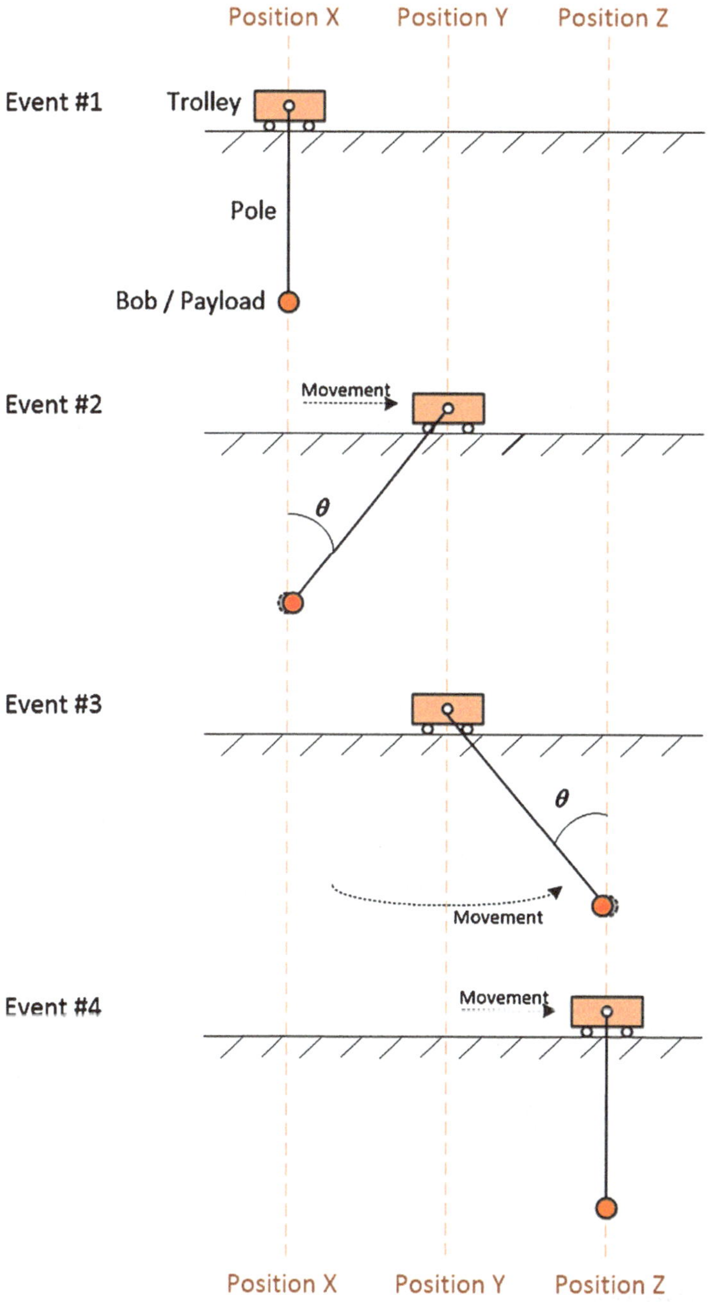

Figure 2: A gantry problem with a bob attached to end of the pole which is connected to trolley. The objective is to move the bob from Positions X to Z by moving the trolley without causing oscillations in the position of the bob upon reaching Position Z.

where, $a, b, \alpha > 0$. Podlubny in his book [15] presented the solution for (2) for a unit-step function in the form of

$$y(t) = \frac{1}{a}\mathcal{E}_0\left(t, -\frac{b}{a}; \alpha, \alpha + 1\right), \tag{3}$$

where, the Podlubny–Mittag-Leffler function is defined as

$$\mathcal{E}_k(t, y; \alpha, \beta) = t^{\alpha k + \beta - 1} E_{\alpha,\beta}^{(k)}(y t^{\alpha}), \tag{4}$$

for $k = 0, 1, 2, \ldots$. The term $E_{\alpha,\beta}^{(k)}(z)$ denotes the kth derivative of the Mittag-Leffler function which results in

$$E_{\alpha,\beta}^{(k)}(z) = \sum_{l=0}^{\infty} \frac{(l+k)! z^l}{l! \Gamma(\alpha l + \alpha k + \beta)}. \tag{5}$$

By multiplying $K > 0$ to (2), the relationship $a = T$ and $b = 1$ are then obtained when compared to (1). Substituting these values in (3) results in

$$y(t) = \frac{K}{T}\mathcal{E}_0\left(t, -\frac{1}{T}; \alpha, \alpha + 1\right), \tag{6}$$

with

$$y(t) = \frac{K}{T} t^{\alpha} \sum_{l=0}^{\infty} \frac{(-\frac{1}{T} t^{\alpha})^l}{\Gamma(\alpha l + \alpha + 1)}. \tag{7}$$

being its expanded form through (4) and (5).

For orders $1 < \alpha < 2$, Tavazoei [19, 20] has shown that the step response will always be stable and will have a nonmonotonic response in the form of an oscillation, which is basically the same characteristics of a second-order system having an underdamped response. The first peak time of (7) can be identified by equating the first derivative of (7) with zero, resulting in the equation

$$\begin{aligned} \frac{d}{dt} y(t) = 0 = \frac{K}{T}\Bigg\{ & t^{\alpha} \sum_{l=0}^{\infty} \frac{1}{\Gamma(\alpha l + \alpha + 1)} \left(-\frac{1}{T}\right)^l \alpha l t^{\alpha l - 1} \\ & + \alpha t^{\alpha - 1} \sum_{l=0}^{\infty} \frac{1}{\Gamma(\alpha l + \alpha + 1)} \left(-\frac{1}{T}\right)^l t^{\alpha l} \Bigg\} \end{aligned} \tag{8}$$

and solving for $T_P = t$.

3 Half-cycle Posicast

The half-cycle Posicast could be the most widely-known Posicast controller because of its simplistic approach in solving an underdamped control system problem. The Gantry problem in the first section is exactly how half-cycle Posicast works.

In half-cycle Posicast, the step input is divided into two steps. The first step in the "staircase" is a fraction of the step magnitude. Having this signal to the plant $P(s)$ allows the output response to reach the ideal setpoint of $M > 0$, where M is the magnitude of the step input. If the input signal is a unit-step input, then $M = 1.0$. Once the output signal reaches the setpoint M, then signal to the plant $P(s)$ is then reshaped to have a value of M which constitutes the second step. This method is also called as a two-step Posicast control method.

In the case of a second-order underdamped system of the form,

$$P(s) = \frac{K\omega_n^2}{s^2 + 2\zeta\omega_n s + \omega_n^2}, \tag{9}$$

where, $K, \omega_n > 0$ and $0 < \zeta < 1$, the duration of the first step should be equal to $T_P = \pi/(\omega_n\sqrt{1-\zeta^2})$ secs, while its magnitude should be at $M/(M+\delta)$, where $\delta > 0$ is the difference between the overshoot and the setpoint M, that is, $\delta = y(T_P) - M$. Figure 3 shows an example of the unit-step response of the underdamped second-order system

$$P(s) = \frac{1}{s^2 + s + 1} \tag{10}$$

with and without a Posicast controller. The peak time of (10) is determined as

$$T_P = \frac{\pi}{\omega_n\sqrt{1-\zeta^2}} = \frac{\pi}{\sqrt{1-0.5^2}} = 3.64 \text{ secs} \tag{11}$$

with an overshoot of

$$OS = 1 + \delta = \exp\left(-\frac{\pi\zeta}{\sqrt{1-\zeta^2}}\right) = \exp\left(-\frac{0.5}{\sqrt{1-0.5^2}}\right) = 1.1601 \text{ units.} \tag{12}$$

This then makes $\delta = 0.1601$.

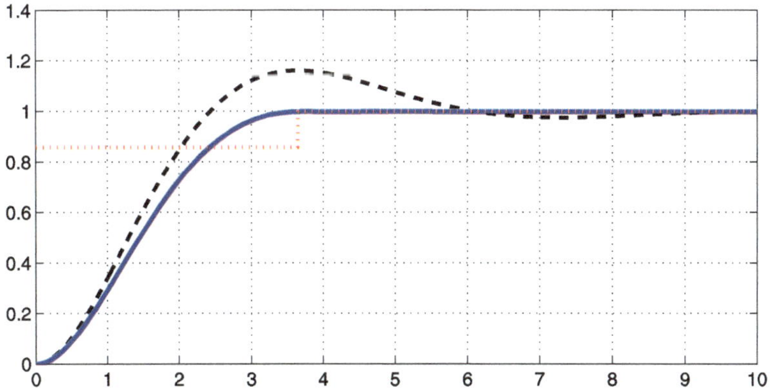

Figure 3: Step response of a second-order system $P(s) = 1/(s^2 + s + 1)$ with and without a Posicast controller. Red dotted line is the reshaped step input. Black dashed line is the unit step response of the plant (without controller). Blue solid line is the unit step response of the system with a half-cycle Posicast controller.

The half-cycle Posicast controller is defined as

$$C(s) = \frac{M}{M+\delta} + \frac{\delta}{M+\delta}e^{-sT_P}, \tag{13}$$

where $M > 0$ is the magnitude of the step input. By using the values obtained from the characteristics of the second-order system's step response, the transfer function of the Posicast controller is then obtained as

$$C(s) = \frac{1}{1+0.1601} + \frac{0.1601}{1+0.1601}e^{-3.64s}. \tag{14}$$

Let us now extend this approach for fractional-order systems of the form in (1). The peak overshoot $T_P > 0$ can be obtained by solving (8) numerically. Let us consider the plant

$$P(s) = \frac{1}{s^{1.75}+1}, \tag{15}$$

where the following relationships can be obtained: $K, T = 1$ and $\alpha = 1.75$. The resulting unit-step response is

$$y(t) = t^{1.75}\sum_{l=0}^{\infty}\frac{(-t^{1.75})^l}{\Gamma(1.75l+2.75)}, \tag{16}$$

which is shown in Figure 4.

Using any graphical or numerical technique, the overshoot is obtained at a value of $\delta = 0.5833$ while the peak time is obtained as $T_P = 2.98$ secs. By using the half-cycle Posicast structure in (13), the transfer function of the Posicast controller determined as

$$C(s) = \frac{1}{1+0.5833} + \frac{0.5833}{1+0.5833}e^{-2.98s}. \tag{17}$$

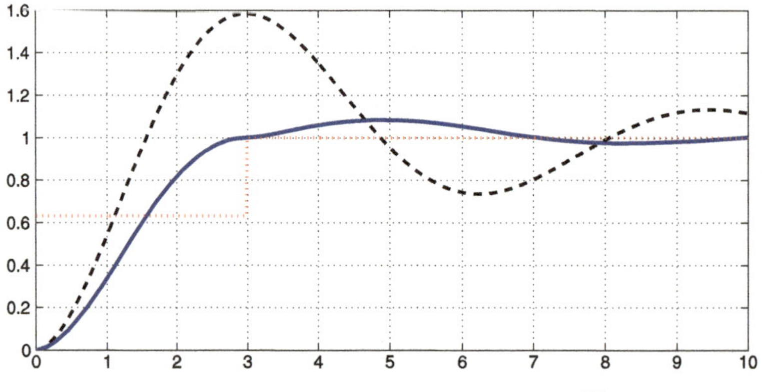

Figure 4: Step response of a fractional-order system $P(s) = 1/(s^{1.75}+1)$ with and without a Posicast controller. Red dotted line is the reshaped step input. Black dashed line is the unit step response of the plant (without controller). Blue solid line is the unit step response of the system with a half-cycle Posicast controller.

The step response of the system with a half-cycle Posicast controllers is shown also in Figure 4. It can be seen that the step response has improved drastically with the application of a half-cycle Posicast. There may be minimal overshoot at a certain extent at around 4.8 secs but such value may be practically acceptable for some industrial applications.

The following code can be used to generate the step response of a fractional-order system using the half-cycle Posicast controller. It is important to note that the function `fotf` used in this code shall only work through the creation of a class and additional m-files in MATLAB [1].

```
a=[1 1]; b=1;
nb=0; na=[1.75 0];
G=fotf(a,na,b,nb)
t=0:0.01:10;
yorig=step(G,t);
plot(t,yorig,'k--','LineWidth',2.5);
hold on
grid on
u1=(1/1.5833)*ones(1,length(0:0.01:2.98));
u2=ones(1,length(0:0.01:10)-length(u1));
u=[u1 u2];
yHCP=lsim(G,u,t);
plot(t,yHCP,'-','LineWidth',3);
plot(t,u,'r:','Linewidth',2.5);
```

4 Three-step Posicast

Similar to the half-cycle Posicast controller, the three-step Posicast controller [9] shapes in the input signal by dividing it into three steps rather than two. The way in determining the magnitude of the first step is the same with half-cycle Posicast where the input signal is attenuated at a value of $M/(M+\delta)$ until the first peak time, $T_P > 0$, is reached. The next step further increases the magnitude of the step input which would not be equal to the step input magnitude, $M > 0$. The duration of the second step can be assumed to be equal to the time it would take to reach the first undershoot from the peak time of the step response if there is no controller. Once that duration is reached, the full magnitude of the input signal is then input into the plant. The resulting three-step controller is then defined as

$$C(s) = \frac{M}{M+\delta} + \gamma\frac{\delta}{M+\delta}e^{-sT_P} + (1-\gamma)\frac{\delta}{M+\delta}e^{-sT_U}, \tag{18}$$

where, $0 < \gamma < 1$ is a parameter that defines the height of the second step, and $T_U > 0$ is the time of the first undershoot, following the condition that $T_U > T_P > 0$. If $\gamma = 1$, then the last term of (18) becomes zero and the controller results in a half-cycle Posicast controller. On the other hand, if $\gamma = 0$, then the second term of (18) becomes zero, and the controller results in a Posicast controller with two steps; however, having the second step initiated upon reaching T_U and not T_P.

The optimal way in determining the value of γ is still an open research.

Bibliography

[1] Y. Q. Chen, I. Petráš, and D. Xue, Fractional order control—a tutorial, in *American Control Conference*, pp. 1397–1411, 2009.

[2] G. Cook, An application of half-cycle Posicast, *IEEE Transactions on Automatic Control*, **AC-11** (1966), 556–559.

[3] G. Cook, Control of flexible structures via Posicast, in *Proceedings of the 1986 Southeastern Symposium on Systems Theory*, Knoxville, TN, pp. 31–35, 1986.

[4] G. Cook, Discussion on 'Pre-shaping command inputs to reduce system vibration', *Transactions of the ASME Journal of Dynamical Systems, Measurement, and Control*, **115**(2) (1993), 309–310.

[5] Q. Feng, R. M. Nelms, and J. Y. Hung, Posicast-based digital control of the buck converter, *IEEE Transactions on Industrial Electronics*, **53** (2006), 759–767.

[6] R. A. Fowell, Robust resonance reduction using staggered Posicast filters, U.S. Patent 5610848, Hughes Aircraft Company, March 11, 1999.

[7] E. A. Gonzalez, Posicast control in power electronics, in *AUN/SEED-Net Fieldwise Seminar on Power Engineering*, Hotel Intercontinental Manila, Philippines, March 21–22, 2007.

[8] E. A. Gonzalez, J. Y. Hung, and C. B. Co, Posicast control of two-term fractional-order systems, in *5th International Conference on Humanoid, Nanotechnology, Information Technology, Communication and Control, Environment, and Management (HNICEM)*, March 9–13, 2011.

[9] E. A. Gonzalez, J. Y. Hung, L. Dorčák, J. Terpák, and I. Petráš, Posicast control of a class of fractional-order processes, *Central European Journal of Physics*, **11**(6) (2013), 868–880.

[10] J. Y. Hung, Application of Posicast principles in feedback control, in *2002 IEEE International Symposium on Industrial Electronics*, L'Aquila, Italy, pp. 500–504, 2002.

[11] J. Y. Hung, Feedback control with Posicast, *IEEE Transactions on Industrial Electronics*, **50**(1) (2003), 94–99.

[12] J. Y. Hung, Posicast control past and present, *IEEE Multidisciplinary Engineering Education Magazine*, **2**(1) (2007), 7–11.

[13] M. Kalantar and G. S. M. Mousavi, Posicast control within feedback structure for a DC-DC single ended primary inductor converter in renewable energy applications, *Applied Energy*, **87** (2010), 3110–3114.

[14] P. C. Loh, C. J. Gajanayake, D. M. Vilathgamuwa, and F. Blaabjerg, Evaluation of resonant damping techniques for Z-source current-type inverter, *IEEE Transactions on Power Electronics*, **23**(4) (2008), 2035–2043.

[15] I. Podlubmy, *Fractional Differential Equations*, Mathematics in Science and Engineering, vol. 198, Academic Press, San Diego, 1999.

[16] N. C. Singer and W. P. Seering, Preshaping command inputs to reduce system vibration, *Transactions on the ASME Journal of Dynamical Systems, Measurement, and Control*, **112** (1990), 76–82.

[17] O. J. M. Smith, Posicast control of damped oscillatory systems, *Proceedings of the IRE*, **45** (1957), 1249–1255.

[18] O. J. M. Smith, *Feedback Control Systems*, pp. 331–345, McGraw-Hill, 1958.

[19] M. S. Tavazoei, On monotonic and nonmonotonic step responses in fractional order systems, *IEEE Transactions on Circuits and Systems—II: Express Briefs*, **58**(7) (2011), 447–451.

[20] M. S. Tavazoei, Overshoot in the step response of fractional-order control systems, *Journal of Process Control*, **22** (2012), 90–94.

Aleksei Tepljakov, Eduard Petlenkov, and Juri Belikov

FOMCON toolbox for modeling, design and implementation of fractional-order control systems

Abstract: Development of a coherent Computer Aided Control System Design (CACSD) toolset is an important problem, because the solution usually targets a broad spectrum of users, including research scientists, educators, and control engineers. Intrinsically, a CACSD tool must play a substantial role in enabling an efficient transfer of technology from academia to industry thereby ensuring sustainable technological advancement. This means that, on one hand, the use of the toolset should be made as intuitive as possible, and on the other—advanced features for the power user should also be made available. In the case of fractional-order calculus and related control system applications, one of the existing solutions is the so-called FOMCON toolbox for Mathworks MATLAB/Simulink environment. The toolbox provides the user with a full set of features for working with fractional-order transfer functions, including time and frequency domain identification of dynamic models, fractional-order PID controller design, and analog and digital implementation methods. A limited set of features is also implemented for MIMO models. Furthermore, all major features are accompanied with graphical user interfaces focusing on the model based control design workflow. This makes the use of the toolbox straightforward for engineers working on particular control design problems and also for educators who wish to illustrate the control design workflow. In the present contribution, the structure of the toolbox and its major features are described, and several relevant examples are provided.

Keywords: Modeling, identification, model based control design, fractional PID control

MSC 2010: 93B51, 37N30, 93B30

1 Introduction

Design of efficient industrial control applications is essential for reducing resource consumption and energy waste thereby ensuring ecological sustainability. However, even though advanced industrial control techniques are readily available, proportional-integral-derivative (PID) controllers massively enjoy unrivaled popu-

Aleksei Tepljakov, Eduard Petlenkov, Department of Computer Systems, Tallinn University of Technology, Tallinn, Estonia, e-mail: aleksei.tepljakov@taltech.ee
Juri Belikov, Department of Software Science, Tallinn University of Technology, Tallinn, Estonia

https://doi.org/10.1515/9783110571745-010

larity serving as an integral part of the majority of industrial process control loops. One of the reasons for the success of the common PID controller is its relative simplicity coupled with its applicability to a wide range of industrial control problems [1]. However, it is also a commonly acknowledged fact that only a fraction of the existing PI/PID controller based loops are tuned to achieve optimal performance [22]. Meanwhile, contemporary industrial applications are growing in complexity and demand a more advanced control scheme to be employed to ensure robustness and efficiency [37].

All of this leads to the necessity of employing a new generation of industrial controllers having all of the above mentioned desirable qualities. Toward that end, FOPID controllers [26] are seen as excellent candidates for the new industrial revolution because they have the capacity to considerably improve the performance of industrial control loops [15, 19, 20, 27, 39]. The basic FOPID controller has two additional "tuning knobs"—the orders of integration and differentiation—in addition to the usual P, I, and D term gains. This extended set of parameters allows, on one hand, to efficiently compensate for effects in control systems that exhibit memory-like or self-similarity phenomena [19, 37], and on the other—to ensure robust performance of the control loop [27]. Recent research results suggest that a significant advantage of FOPID controllers when applied to industrial problems is seen in the potential reduction of the control effort, which also results in reduction of wasted energy [8, 17]. At the same time, the practical effect of the two additional parameters can be relatively easily understood, so FOPID controllers can still be operated relatively easily. Furthermore, because they are generically compatible with conventional PID controllers, effective input shaping methods, such as the retuning method can be employed [31]. This approach allows to integrate FOPID controllers into existing industrial PID controller loops without breaking the latter thereby potentially reducing process control downtime, and thus also significantly lowering integration costs. Combining the retuning approach with an adaptation mechanism in certain cases is capable of substantially improving robust performance. For example, in [30] the accumulated control error based performance index for the FOPID control system that includes an adaptation mechanism is decreased by 81.7 % compared to the same system with adaptation disabled.

To facilitate the transition to the new improved controllers in industrial applications, several problems must be solved first [5]—including modeling, controller design, and reliable implementation—issues that are all within the scope of the FOMCON toolbox for MATLAB. Furthermore, the prospective applications of FO control in the industry need to be supported not only by a set of computing tools, but also by corresponding workflows. In FOMCON, one of the key design aspects has always been related to the *model based control design* workflow. In this sense, the toolset provides the means to conduct time- and frequency domain based identification of fractional-order models, model-based design of FOPID controllers, and real-life implementation of the latter in analog and digital form.

This workflow also serves as motivation for the present chapter which is organized as follows. In Section 2, a concise overview of FOMCON toolbox is provided. In Section 3, some fundamental fractional-order modeling issues are addressed. Then, in Section 4 dynamic fractional-order model identification from input/output data is discussed. Section 5 is concerned with FOPID controller design. In Section 6, real-life implementation of fractional-order controllers is addressed. Finally, in Section 7 conclusions are drawn.

2 Overview of the toolbox

Modern-day applications of fractional-order modeling tend to rely on numerical tools [19], a good summary of which is provided in [14]. FOMCON toolbox for MATLAB also falls into this category. Historically, the toolbox derived basic functionality from a kernel—a MATLAB object appropriately named `fotf`—containing the functions provided by the FOTF toolbox [4]. Additional functionality has been added as well, for example, symbolic operations with FO transfer functions [14, 28] and the root locus computation algorithm [16]. Furthermore, a set of core utilities has been added. These contain, for example, fractional operator approximation facilities. In addition, FOMCON also reuses some frequency-domain identification functions from the Ninteger toolbox [35] as well as has several convenience functions to transfer data to the CRONE toolbox [25]. This relation is shown in Figure 1.

The modules of the toolbox are depicted in Figure 2. They are as follows:

- Core module—FO system analysis (`fotf` kernel and utilities);
- Identification module (system identification based on experimental input/output data in both time and frequency domains);
- Control module (FOPID controller design, tuning and optimization tools, as well as some additional features);
- Implementation module (continuous and discrete time approximations, implementation of corresponding analog and digital filters).

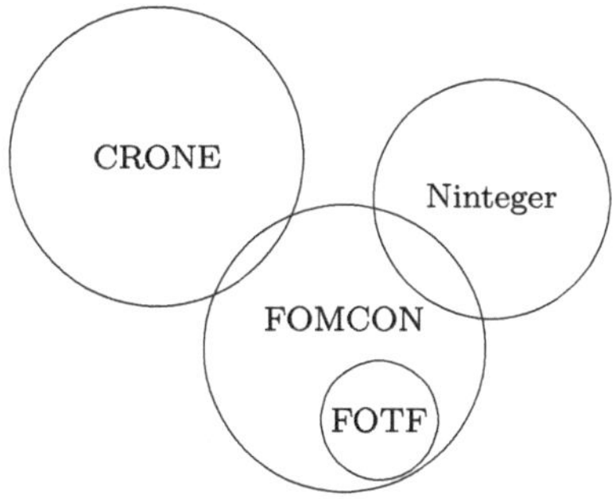

Figure 1: Relation of FOMCON toolbox for MATLAB/Simulink to similar packages.

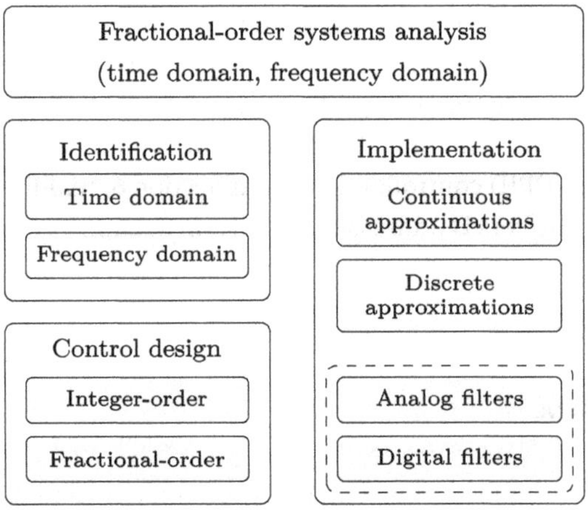

Figure 2: Modular structure of the FOMCON toolbox.

All the modules are interconnected. Most features are supported by graphical user interfaces. A Simulink blockset is also provided in the toolbox allowing more complex modeling tasks to be carried out. General approach to block construction was used where applicable. The following blocks—in several variants depending on the modeling or control design situation—are currently realized:

- General fractional-order operators: fractional integrator and differentiator with initial states;
- Continuous and discrete time fractional transfer function and state space with initial states;
- Continuous and discrete time FOPID controller;
- Discrete time retuning controller blocks [31].

The toolbox currently relies on the following MATLAB or third party products:

- Control System toolbox—required for most features;
- Optimization toolbox—required for time domain identification and conventional PID tuning, and also partially for fractional-order PID tuning;
- Nelder–Mead algorithm based function for nonlinear optimization subject to bounds and constraints [24];
- Ninteger toolbox frequency domain identification functions [35].

It is also possible to export fractional-order systems to the CRONE toolbox format [25]. This feature requires the object-oriented CRONE toolbox to be installed separately.

FOMCON toolbox is open source and is hosted on and developed through GitHub [29].

In the following section, we review some fundamental facilities implemented in the `fotf` kernel and associated utilities on which further discussion is based.

3 Fractional-order modeling and analysis

In FOMCON, the core object of study is a FDE of the form

$$\begin{aligned} & a_n D^{\alpha_n} y(t) + a_{n-1} D^{\alpha_{n-1}} y(t) + \cdots + a_0 D^{\alpha_0} y(t) \\ & = b_m D^{\beta_m} u(t) + b_{m-1} D^{\beta_{m-1}} u(t) + \cdots + b_0 D^{\beta_0} u(t) \end{aligned} \tag{1}$$

the Laplace transform of which under zero initial conditions yields the so-called fractional-order transfer function (FOTF) representation

$$G(s) = \frac{a_n s^{\alpha_n} + a_{n-1} s^{\alpha_{n-1}} + \cdots + a_0 s^{\alpha_0}}{b_m s^{\beta_m} + b_{m-1} s^{\beta_{m-1}} + \cdots + b_0 s^{\beta_0}}, \tag{2}$$

where $(a_i, b_j) \in \mathbb{R}^2$ and $(\alpha_i, \beta_j) \in \mathbb{R}_+^2$. The model in (2) can also be extended with the input–output delay term that allows to model the effect of delayed input $u(t) = u_d(t - L)$, $L \in \mathbb{R}_+$. Thus, the general form of the object is

$$G(s) = \frac{a_n s^{\alpha_n} + a_{n-1} s^{\alpha_{n-1}} + \cdots + a_0 s^{\alpha_0}}{b_m s^{\beta_m} + b_{m-1} s^{\beta_{m-1}} + \cdots + b_0 s^{\beta_0}} \mathrm{e}^{-Ls}, \tag{3}$$

where it is usual to take $\beta_0 = \alpha_0 = 0$ so that the static gain of the system is obtained as $K = b_0/a_0$.

Assuming the system in (2) is of commensurate order, that is, all the orders of derivation are integer multiples of a base order γ such that $\alpha_k, \beta_k = k\gamma$, $\gamma \in \mathbb{R}_+$, the system can be expressed as

$$\sum_{k=0}^{n} a_k D^{k\gamma} y(t) = \sum_{k=0}^{m} b_k D^{k\gamma} u(t). \tag{4}$$

For modeling dynamics systems with multiple inputs and outputs, one can use the notion of the matrix of transfer functions given by

$$\mathbf{G} = \begin{bmatrix} \frac{Z_{11}(s)}{P(s)} & \frac{Z_{12}(s)}{P(s)} & \cdots & \frac{Z_{1m}(s)}{P(s)} \\ \frac{Z_{21}(s)}{P(s)} & \frac{Z_{22}(s)}{P(s)} & \cdots & \frac{Z_{2m}(s)}{P(s)} \\ \vdots & & \ddots & \vdots \\ \frac{Z_{p1}(s)}{P(s)} & & \cdots & \frac{Z_{pm}(s)}{P(s)} \end{bmatrix}, \tag{5}$$

where $Z_{ij}(s)/P(s) = G_{ij}(s)$ denotes the transfer from the jth input to the ith output, $j = 1, 2, \ldots, m$ and $i = 1, 2, \ldots, p$. For systems of commensurate order modeled by (4) or (5), a state space model can be derived by converting a matrix of transfer functions in (5) to the conventional representation

$$\begin{aligned} D^{\gamma} \mathbf{x} &= \mathbf{A}\mathbf{x} + \mathbf{B}\mathbf{u} \\ \mathbf{y} &= \mathbf{C}\mathbf{x} + \mathbf{D}\mathbf{u}. \end{aligned} \tag{6}$$

Stability analysis of (6) is carried out using Matignon's stability theorem [4, 18], while controllability and observability properties can be assessed in the usual way [23]. Time-domain simulations of systems described by models in (3) and (5) are performed using the method introduced in [4, 19] and are based on the Grünwald–Letnikov definition

$$ {}_a\mathscr{D}_t^\alpha f(t) = \lim_{h\to 0} \frac{1}{h^\alpha} \sum_{k=0}^{[\frac{t-a}{h}]} (-1)^k \binom{\alpha}{k} f(t-kh), \tag{7}$$

where $[\cdot]$ means the integer part, h is the step size. Frequency domain analysis is performed such that s in (3) and (5) is substituted with $j\omega$, where ω denotes the frequency of interest in rad/s, and doing an N-point frequency sweep $\omega \in \{\omega_0, \omega_1, \dots, \omega_N\}$.

Fractional-order system approximations play a key role in FOMCON and Oustaloup's recursive filter approximation is used in most cases, although other methods are also available. Oustaloup's method uses conventional transfer functions to represent a band-limited approximation of a fractional-order operator. In order to approximate a fractional differentiator of order α or a fractional integrator of order $(-\alpha)$ by a conventional transfer function, one computes the zeros and poles of the latter using the following equations:

$$ s^\alpha \approx K \prod_{k=1}^{N} \frac{s+\omega_k'}{s+\omega_k}, \tag{8}$$

where

$$ \omega_k' = \omega_b \cdot \omega_u^{(2k-1-\alpha)/N}, \tag{9}$$

$$ \omega_k = \omega_b \cdot \omega_u^{(2k-1+\alpha)/N}, \tag{10}$$

$$ K = \omega_h^\alpha, \quad \omega_u = \sqrt{\omega_h/\omega_b}, \tag{11}$$

and N is the order of approximation in the valid frequency range $(\omega_b; \omega_h)$. For fractional orders $\alpha \geq 1$, it holds

$$ s^\alpha = s^n s^\gamma, \tag{12}$$

where $n = \alpha - \gamma$ denotes the integer part of α and s^γ is obtained by the Oustaloup approximation by using (8). Therefore, every operator in (2) may be approximated using (12) and substituted by the obtained approximation, yielding a conventional integer-order transfer function. For digital implementations, the obtained approximation may be converted to its discrete-time equivalent using a suitable method. In FOMCON, the modified Oustaloup filter is also available as an option in several algorithms [39].

From a purely practical perspective, the use of initial states with zero dynamics in modeling is often useful. Moreover, the solution should also work in real-time applications. For this reason, one may choose to use a state-space integer-order approximation of a system described by (2) and initialize the states such that the initial output matches the desired initial state and all dynamics at that point are assumed to be zero. Thus, for the following form of the state-space approximation,

$$\begin{bmatrix} \dot{\mathbf{x}} \\ \mathbf{y} \end{bmatrix} = \begin{bmatrix} \mathbf{A} & \mathbf{B} \\ \mathbf{C} & \mathbf{D} \end{bmatrix} \begin{bmatrix} \mathbf{x} \\ \mathbf{u} \end{bmatrix} \tag{13}$$

we take $\dot{\mathbf{x}} = \mathbf{0}$, $\mathbf{y} = \mathbf{x}_0$, and $\mathbf{D} = \mathbf{0}$. Then the steady-state initialization of the states $\tilde{\mathbf{x}}$ and input $\mathbf{u}$ is computed by means of

$$\begin{bmatrix} \tilde{\mathbf{x}} \\ \mathbf{u} \end{bmatrix} = \begin{bmatrix} \mathbf{A} & \mathbf{B} \\ \mathbf{C} & \mathbf{0} \end{bmatrix}^{-1} \begin{bmatrix} \mathbf{0} \\ \mathbf{x}_0 \end{bmatrix}. \tag{14}$$

This is a straightforward solution for initializing the states $\tilde{\mathbf{x}}$ of the approximation of (5) assuming a steady-state initial state with zero dynamics. Consider now an example.

Example 3.1. The task is to obtain a time domain response for a transfer function

$$G(s) = \frac{-2s^{0.63} + 4}{2s^{3.501} + 3.8s^{2.42} + 2.6s^{1.798} + 2.5s^{1.31} + 1.5} \tag{15}$$

with the initial condition $y_0 = 5$ and constant input $u(t) = 0$ in the time range $t \in [0, 100]$. In order to proceed with solving the task, one could consider taking the following steps. First, the approximation of (15) is obtained of the form (8). The approximation is converted to state space form and the method in (14) is applied. Then the approximate model is simulated using the MATLAB control system toolbox function `lsim` where the initial conditions $\tilde{\mathbf{x}}$ are specified. The resulting response is shown in Figure 3.

In the following chapters, we will focus on the fractional model based controller design workflow.

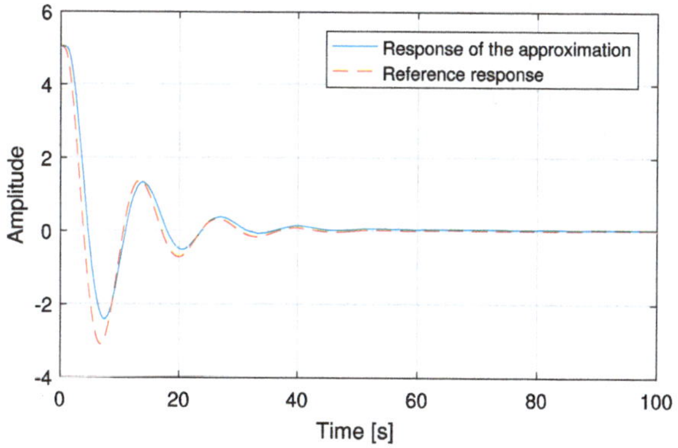

Figure 3: Response of the approximated implementation of the system in (15) compared to a reference response obtained using a Caputo definition based solver in [38].

4 Dynamic model identification

In FOMCON, the black box and gray box modeling approaches are considered in the context of dynamic system continuous-time model identification. Furthermore, current identification facilities are geared toward single input, single output system identification in the form of transfer functions in (3).

We begin with reviewing time domain identification methods implemented in FOMCON toolbox. The main features are:

- Commensurate and noncommensurate order system identification;
- Approximation of fractional systems by conventional process models;
- Grey box (parametric) identification which is applicable to closed-loop identification problems.

In what follows, we shall briefly review the main time domain model identification approach.

Suppose that experimental data is collected from a general single input, single output nonlinear system $\Psi : \mathscr{I} \to \mathscr{O}$, where $(\mathscr{I}, \mathscr{O}) \subset \mathbb{R}^2$ denote the measured input and output signals, respectively, such that

$$z(t) = \Psi(v(t)) + \mathfrak{N}, \tag{16}$$

where $z(t)$ denotes the system output, and $v(t)$ denotes the system input, and $\mathfrak{N}$ denotes measurement noise, and is represented by a data set holding the samples from the system input $u_k = v(kt_s)$ and output $y_k = z(kt_s) + \mathfrak{N}$ under a uniform sample rate $t_s = t_{k+1} - t_k \equiv \text{const}$:

$$Z_N = \{u_0, y_0, u_1, y_1, \dots, u_N, y_N, t_s\}, \tag{17}$$

where $k = 0, 1, \dots, N$. Since zero initial conditions are assumed, then if $z(0) = y_0 \neq 0$, then the offset is removed from each of the collected output samples by means of

$$y_k = y_k - y_0, \quad k = 0, 1, \dots, N. \tag{18}$$

In the case of fractional-order systems, the key issue is the evaluation of the response of the model [6, 12]. Two methods are considered for the purpose of model simulation: the Grünwald–Letnikov definition based solver and Oustaloup's recursive filter approximation technique.

The identification method is based on the minimization of model output error ε in the least-square sense. Mathematically, the problem is stated as an optimization problem

$$\min_{\theta} \varepsilon^2, \tag{19}$$

where $\varepsilon = y_k - \hat{y}_k$, and $\hat{y}_k = \hat{\Psi}(u_k, \Theta)$ denotes the response of the estimated system model $\hat{\Psi}$ under the input signal u_k, $k = 0, \ldots, N$, and Θ is the vector of estimated parameters of the model in (3). In order to better condition the optimization problem (19), the parameter vector Θ may contain various combinations of the model parameters, for example, model orders only, or polynomial term coefficients only. This choice also depends on the identification problem statement and prior analysis [28]. Also, by first identifying a coherent general fractional model for a data set comprising several captured input and output signals, the pole polynomial can be fixed and MIMO system identification can be carried out.

In the FOMCON implementation of the identification algorithms as a MATLAB function `fid`, the following two algorithms are used for nonlinear least-squares estimation of model parameters Θ:

1. Trust Region Reflective algorithm [7]—this method handles search variable bounds and is intended for solving large-scale problems [2]. Bounds are given for parameter sets as $\Theta_b = \{\Theta_{\min}, \Theta_{\max}\}$.
2. Levenberg–Marquardt algorithm [21]—the method is quite robust and is commonly used for black box model identification [9]. However, the conventional implementation does not handle bound constraints. Therefore, to correctly identify model orders $\alpha_p \geqslant 0$ and $\beta_z \geqslant 0$ and the delay $L \geqslant 0$ a coordinate transform is required [28]. In the new coordinates z_j, the corresponding parameters are given as

$$\Theta_j = z_j^2, \quad j = 0, 1, 2, 3 \ldots. \tag{20}$$

For gray box (parametric) identification, one could use the FOMCON function `pfid`. The main idea is to represent the model in (3) as

$$G(s) = \frac{Z(p,q)}{P(p,q)} D(q), \tag{21}$$

where $Z(p,q)$ is the zero polynomial, $P(p,q)$ is the pole polynomial, and $D(q)$ is the delay term, parameterized such that

$$\theta = \{p_1, p_2, \ldots, q_1, q_2, \ldots\}, \tag{22}$$

where some of the parameters are known *a priori* and $q_j \geqslant 0$. Parametric identification is very useful if the structure of the model is well known, for example, in case of a closed control loop.

To assess the quality of the obtained dynamic model, residual analysis can be carried out. Residuals are given by a vector containing the model output error

$$\varepsilon = y_r - y_m, \tag{23}$$

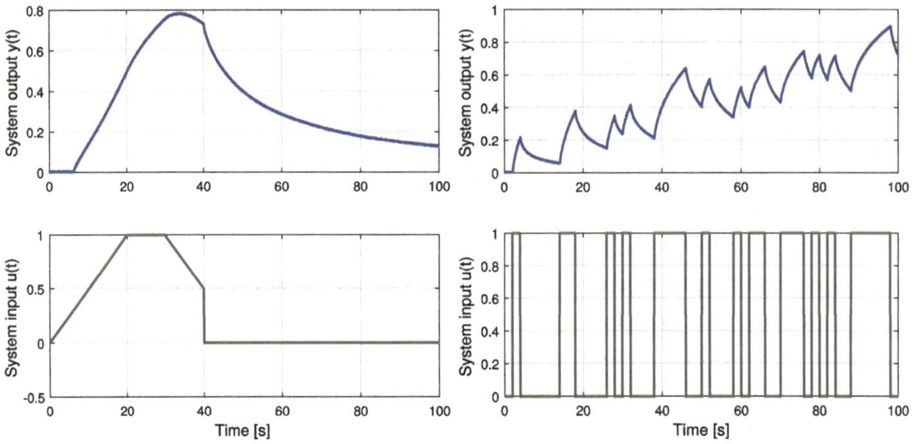

(a) Process dataset 1.

(b) Process dataset 2.

Figure 4: Experimental datasets for Example 4.1.

where y_r is the experimental plant output, and y_m is the identified model output. Various statistics can be derived from analyzing the residuals, such as percentage fit, maximum absolute error, mean squared error, autocorrelation (whiteness test). See [28] for more technical details. Consider now an example.

Example 4.1. The task is to identify a process model from the data shown in Figure 4(a) and validate the model using the data shown in Figure 4(b). In both cases, the datasets are represented by the FOMCON specific `fidata` object. The data originated from an academic example, where the studied system has the form

$$\xrightarrow{u(t)} \boxed{\text{Input nonlinearity}} \xrightarrow{f(u(t))} \boxed{G(s)} \xrightarrow{y(t)} \boxed{\text{Output nonlinearity}} \xrightarrow{h(y(t))}$$

and where the linear system is given by

$$G(s) = \frac{2.15}{15s^{0.75} + 2s^{0.25} + 0.75} e^{-0.1s}, \tag{24}$$

the input nonlinearity function $f(\cdot)$ introduces a dead zone into the control signal in the range $u_d \in [0, 0.3]$ and saturates the input to $u_s = [-1, 1]$, while the output nonlinearity limits the output of the system to the range $y_s \in [0, 2]$. The output signal is also corrupted with white noise of power 10^{-7}. In what follows, we assume that this information is not known and proceed with identification based on educated guesses. This way, we shall validate the identification algorithm using the provided academic example.

To solve the task, the graphical FOMCON time domain identification tool `fotfid` is used. First, by observing the first dataset, one can conclude that the studied system does not exhibit high order dynamics, so it is reasonable to start with a relatively low order model. Second, there seems to be either a delay or nonlinearity affecting the dynamics. The most likely nonlinearity is a deadzone. The system may also have a transport delay. Therefore, to identify a linear model from the response one could trim away the nonlinearity and model it separately later, if needed. The can be done in `fotfid` by first plotting the identification dataset and then choosing from the menu *Data→Trim*. One could keep the response in the interval $t \in [6, 40]$ s. Since we are concerned with identifying a process model, in the identification tool options we choose to estimate explicitly both the static gain K and the delay L, setting the latter to 0.1. The initial guess model could be chosen as $1/(s+1)$. Free identification is employed meaning that both orders of the model and coefficients of polynomial terms are estimated. Levenberg–Marquardt algorithm is used. As a result, the following fractional-order model is obtained:

$$G(s) = \frac{1.766}{12.967s^{0.67915} + 1} e^{-0.1s}. \tag{25}$$

The results of the whiteness test for the first dataset is shown in Figure 5(a), while model validation using the second dataset is shown in Figure 5(b). One can conclude that although the model can be improved, the current result is satisfactory for proceeding with model based control design. We shall extend this example in Section 5.

We now proceed with discussing frequency domain identification methods for fractional-order systems implemented in FOMCON toolbox. Frequency domain identification has several advantages. Most notably, this data is usually available for industrial modeling problems since conducting time domain experiments results in

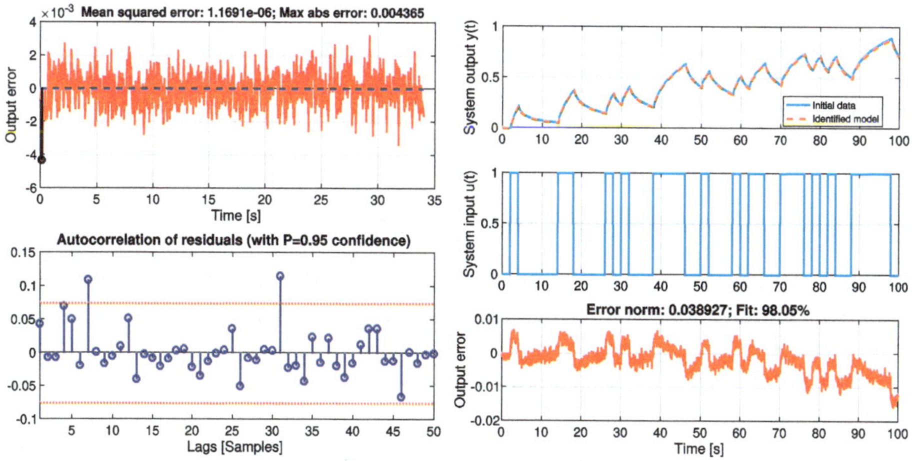

(a) Residual analysis of the model in (25) using the first experimental dataset.

(b) Validation of the model in (25) using the second experimental dataset.

Figure 5: Analysis of the identification results.

additional costs making this approach infeasible. In addition, since time domain simulation is not performed in this case, associated optimization problems can be solved more efficiently and in a much shorter time. In FOMCON, the following features related to frequency domain identification are currently available:

- Commensurate transfer function identification based on algorithms by Hartley, Levy, and Vinagre [11, 35, 36];
- Best fit algorithm for choosing an optimal commensurate order and pseudo-orders of the fractional transfer function [28];
- Grey box (parametric) identification as implemented in the `pfid` function of FOMCON toolbox.

We now discuss the use of the latter feature. The goal is to identify a transfer function model of the form (3). In case of frequency domain identification, the optimization problem uses an objective function based on the Euclidean norm

$$\|\rho\| = \sqrt{\sum_{k=1}^{N} |\rho(k)|^2}, \tag{26}$$

where $\rho(k)$ is the complex frequency response error computed from the experimental frequency response $\rho_\epsilon(k)$ and the response of the identified model $\rho_{\hat{G}}(k)$ at the kth frequency data point, $k = 1, 2, \ldots, N$:

$$\rho(k) = \rho_\epsilon(k) - \rho_{\hat{G}}(k). \tag{27}$$

Same optimization algorithms as in time domain identification are employed including the Nelder–Mead simplex method [13]. The structure of the identified model is determined by equations (21) and (22).

Example 4.2. In this example, frequency domain data (including noise) is generated using the model

$$G(s) = \frac{1}{0.8s^{2.2} + 0.5s^{0.9} + 1}. \tag{28}$$

The data represented by a FOMCON `ffidata` object is shown in Figure 6 (blue solid line). For identification, we consider a gray box model parameterized as

$$G_p(s) = \frac{p_1}{p_2 s^{q_1} + p_3 s^{q_2} + 1}. \tag{29}$$

Therefore, the parameters to be identified are $\Theta = \{p_1, p_2, p_3, q_1, q_2\}$. Furthermore, the q_j parameters are bounded such that

$$q_1 \in (0, 2], \quad q_2 \in [1, 3].$$

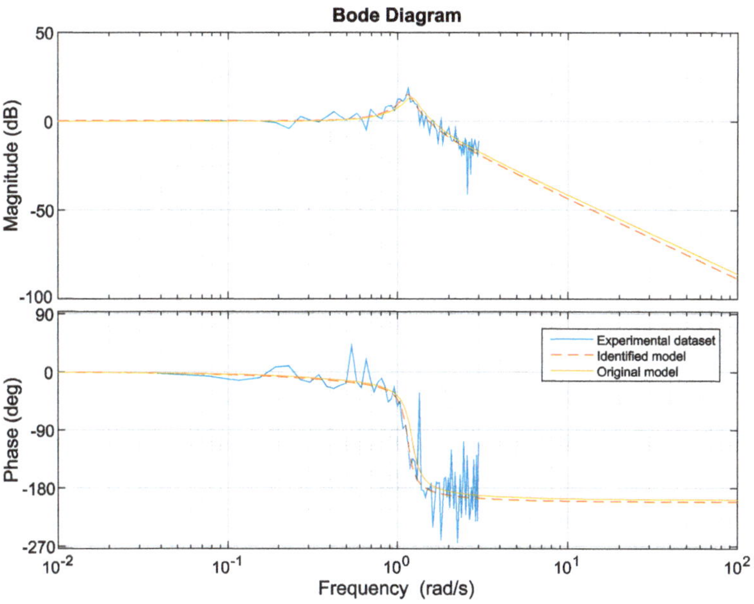

Figure 6: Results of frequency domain identification of a transfer function using the gray box approach.

After correctly supplying the data to the `pfid` function, the Trust Region Reflective algorithm is run. The resulting transfer function model is shown in Figure 6 (red dashed line) and is of the form

$$\hat{G}(s) = \frac{1.0439}{0.5974s^{0.8647} + 0.9570s^{2.2406} + 1}, \tag{30}$$

which is very close to the original model. This concludes the example. After identifying the process model, one can proceed with model based control design which is the subject of the next section.

Example 4.3. In this example, we will illustrate the use of FOMCON toolbox for modeling a real-life control object, namely, an ion polymer metal composite actuator (IPMC). The actuator was manufactured in the Intelligent and Smart Systems laboratory at the University of Tartu. It is controlled from a personal computer using an operational amplifier circuit. When a current is applied to the actuator, it begins to bend. The displacement of the IPMC sample is measured with a laser sensor. The experiments are conducted in a clear room. Complete details about the setup can be found in [28]. The task is to identify a model for the displacement of the IPMC based on the input signal. The task is carried out using `fotfid` graphical interface. In order to ascertain that the object under study indeed possesses FO dynamics, a conventional first-order transfer

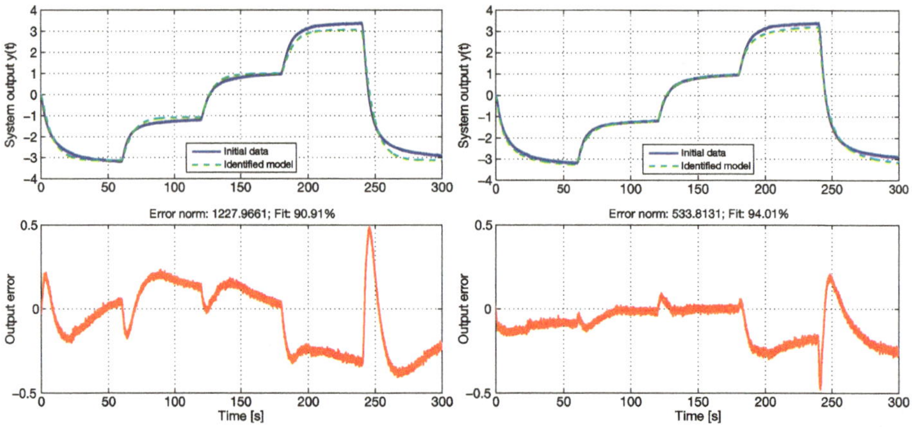

(a) Integer-order model of the IPMC actuator displacement.

(b) Fractional-order model of the IPMC actuator displacement.

Figure 7: IPMC actuator: identification results.

function is identified as well using the same tool. The following models are obtained:

$$G_1(s) = \frac{2.0688}{7.5959s + 1} \tag{31}$$

and

$$G_2(s) = \frac{2.2993}{4.8317s^{0.7797} + 1}. \tag{32}$$

The results of identification are presented graphically in Figure 7. It can be seen that the studied process does indeed possess FO dynamics. Therefore, a FO compensator would be required to control it. This example is continued in the next section.

5 Fractional-order PID control system design

Due to the importance of conventional PID controllers in industrial control applications [1], in the FOMCON toolbox we focus primarily on FOPID controllers. The notion of a fractional PID controller was introduced by Podlubny in [26]. It has been confirmed the FOPID controllers outperform their integer-order counterparts [3, 15, 39].

In the Laplace domain, the parallel form of the FOPID controller is given by

$$C_{\mathrm{FOPID}}(s) = K_p + K_i s^{-\lambda} + K_d s^{\mu}. \tag{33}$$

In FOMCON, we are mostly concerned with optimization-based controller design [32]. As a control object, we consider a nonlinear dynamic system of the form

$$\dot{x}(t) = f(x(t), u(t)), \quad y = h(x(t)), \tag{34}$$

where $x(t) \in \mathbb{R}^n$, $u(t) \in \mathbb{R}$, $f(x,u) \in \mathbb{R}^n \times \mathbb{R} \to \mathbb{R}^n$, and $h : \mathbb{R}^n \to \mathbb{R}$ and a FOPID controller of the form

$$u(t) = K_p e(t) + K_i D^{-\lambda} e(t) + K_d D^{\mu} e(t), \tag{35}$$

which corresponds to the parallel form of the FOPID controller in (33), and $e(t) = r(t) - y(t)$ is the error signal, and $r(t)$ is the desired control reference. The improvement of control system performance in the time domain is equivalent to the problem of minimizing $e(t)$, that is, solving an optimization problem

$$\min_{\Theta_c} J_c(\cdot), \tag{36}$$

where $\Theta_c = [K_p\ K_i\ K_d\ \lambda\ \mu]$ and $J_c(\cdot)$ is a cost function formulated based on the desired performance specifications. In FOMCON, if the Nelder–Mead algorithm is used for optimization of the control loop, the joint cost function consists of two parts

$$J_c = J_m + J_p, \tag{37}$$

where J_m is a performance metric in the time domain obtained by means of simulating the control system and computing the metric, and the J_p term contains nonlinear constraints applied as exponential penalty. The latter term usually contains deviations from desired frequency domain specifications. Details are provided in [28]. In the time domain, one of the predefined performance metrics is used. For example, one may use the so-called integral time-absolute error metric (ITAE) defined as

$$\mathrm{ITAE} = \int_0^t |e(t)|\,\mathrm{d}t. \tag{38}$$

In the frequency domain, based on the open-loop frequency response $F(j\omega) = C(j\omega)G_p(j\omega)$, where $C(j\omega)$ is the FOPID controller, some commonly used robustness specifications include

- Gain margin A_m, where

$$A_m = 1 - |F(j\omega_g)|, \quad \arg(F(j\omega_g)) = -\pi. \tag{39}$$

- Phase margin φ_m and gain crossover (critical) frequency ω_c, where

$$\arg(F(j\omega_c)) = -\pi + \varphi_m, \quad |F(j\omega_c)| = 1. \tag{40}$$

- Robustness to gain variations of the plant:

$$\left.\frac{\mathrm{d}\arg(F(j\omega))}{\mathrm{d}\omega}\right|_{\omega=\omega_c} = 0. \tag{41}$$

The FOMCON tool that performs FOPID controller optimization is called `fpid_optimize` and it has a graphical user interface called `fpid_optim`. The tool can be used to obtain suboptimal FOPID controller settings for FOPID control loops with nonlinear systems in (34) using Simulink as well as for control loops with fractional-order linear approximations. The latter are also used for deriving the frequency domain characteristics of the optimized control loop.

One more remarkable feature of FOMCON toolbox is the availability of a retuning controller Simulink blockset [31]. The retuning method can serve a very useful purpose in industrial control where conventional PID controllers are used but cannot deliver state-of-the-art performance. In this case, the retuning method can be used such that does not break the existing PID control loop, but rather integrates the FOPID controller dynamics using input shaping. This method can be thus very useful for industrial applications [30]. The main idea of the method is illustrated in Figure 8.

For example, a FOPID retuning controller for a conventional PID control loop has the form

$$C_R(s) = \frac{K_2 s^\beta + K_1 s^\alpha - K_d s^2 + (K_0 - K_p)s - K_i}{K_d s^2 + K_p s + K_i}, \tag{42}$$

where the orders are $-1 < \alpha < 1$ and $1 < \beta < 2$. The resulting $\mathrm{PI}^\lambda \mathrm{D}^\mu$ controller from a classical PID controller with parameters K_p, K_i, K_d has the following coefficients:

$$K_p^* = K_0, \quad K_i^* = K_1, \quad K_d^* = K_2. \tag{43}$$

The orders of fractional-order integration and differentiation are

$$\lambda^* = 1 - \alpha, \quad \mu^* = \beta - 1. \tag{44}$$

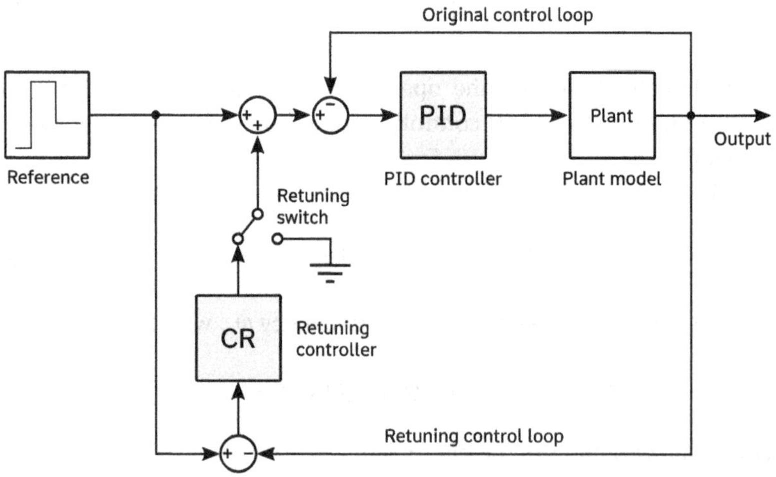

Figure 8: Retuning the control loop with the original PID control loop and the secondary loop utilizing input shaping to introduce FOPID controller dynamics into an existing control system.

Consider now two examples that illustrate the use of the optimization method and the application of the resulting controller to an existing conventional PID control loop. We then conclude the section with one more example of real-life application of the tool.

Example 5.1. In this example, we proceed with tuning a FOPID controller for the plant discussed in Example 4.1. For this purpose, we shall use the `fpid_optim` tool of FOMCON toolbox. Once the tool is launched, the name of the LTI fractional system model must be inserted into the corresponding field and all other configuration parameters properly set. In this case, the initial gains/orders of the FOPID controller are chosen as $K_p = K_i = K_d = 1$ and $\lambda = \mu = 0.5$. Oustaloup approximations are used throughout the simulation to represent the plant and controller. The simulation time is set to 100 s. Since the control signal is limited to $u_s \in [-1, 1]$, Simulink is used for simulations since this saturation is a nonlinear effect. ITAE performance metric in (38) is used, while the frequency domain specifications are gain margin $A_m = 10$ dB and phase margin $\varphi_m = 60°$. In addition, the robustness to gain variations specification must be fulfilled.

We proceed with the optimization based on the Nelder–Mead solver in two stages for the reason that if all specifications are set at once coupled with the fact that the frequency domain specifications are added to the cost function as exponential penalties, then the cost function would tend toward infinity and the optimization process will likely fail. So in the first stage, we only set the gain and phase margin specifications and proceed with optimization. In the second stage, the robustness to gain variations specification is added. The critical frequency ω_c is set based on that of the open loop control system obtained in the first stage. The results after the first stage in the time domain are shown in Figure 9(a). The evolution of the controller parameters during this stage can be easily viewed from the process user interface launched by `fpid_optim` automatically and that accompanies the optimization procedure. The corresponding graphs can be seen in Figure 9(b).

During the first stage, it was discovered that after 100 iterations the critical frequency ω_c was about equal to 2 rad/s. It was slightly decreased to $\omega_c = 1.5$ rad/s for the second stage. The optimizer was then run once more. After another 100 iterations, the result achieved in the frequency domain can be seen in Figure 10. It can be clearly observed that the robustness to gain variations specification has been fulfilled compared to the result obtained after stage one.

The FOPID controller obtained after the second stage has the following form:

$$C(s) = 8.2784 + 0.914s^{-0.49547} + 1.8864s^{0.70596}. \tag{45}$$

Once a coherent controller is found, the whole configuration of the `fpid_optim` tool can be saved as a MATLAB .mat file so that it can be recalled later or used in a special type of Simulink block that supports taking this .mat file directly and implements a FOPID controller for further simulations.

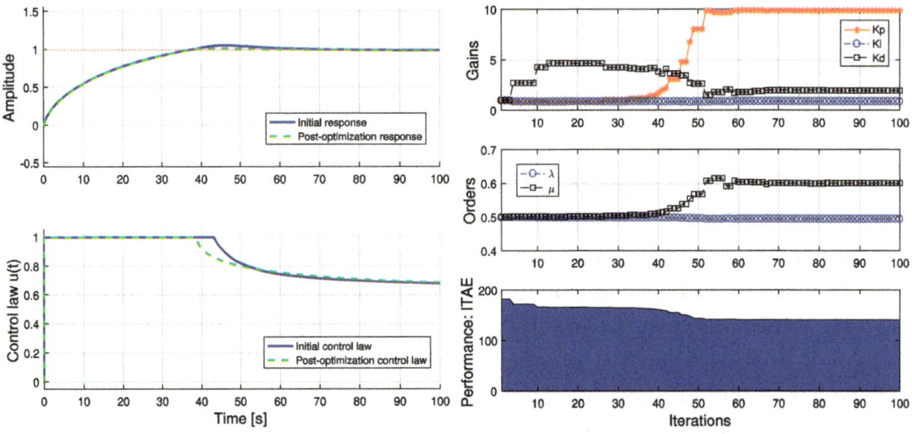

(a) Time domain response of the FOPID control system before and after first stage of optimization.

(b) Evolution of the FOPID controller parameters over the span of the iterative optimization process.

Figure 9: Results of the first stage of FOPID controller optimization based tuning.

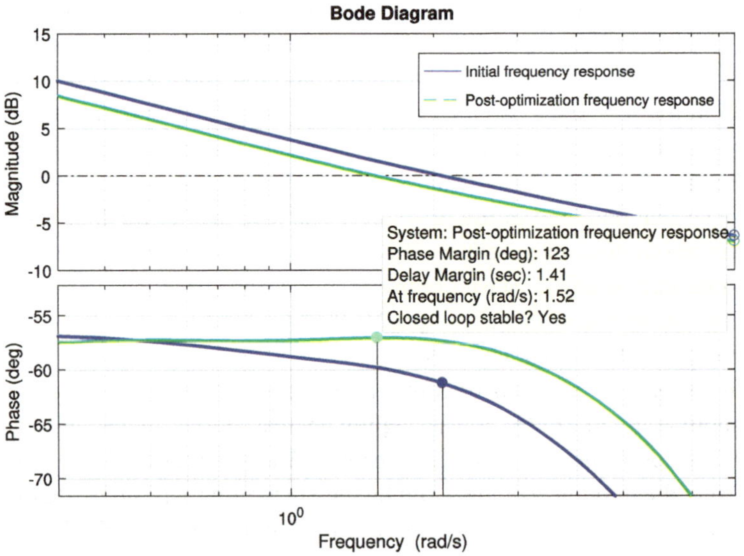

Figure 10: Results of the second stage of FOPID controller optimization based tuning: robustness to gain variations specification is fulfilled.

Example 5.2. In this example, a retuning controller of the form (42) is applied to an existing PID control loop. The control object is the plant in (24) that also has the input and output nonlinearities described in Example 4.1. The original PID controller has the parameters $K_p = K_i = 0.5$ and $K_d = 0.1$. The control schematic is shown in

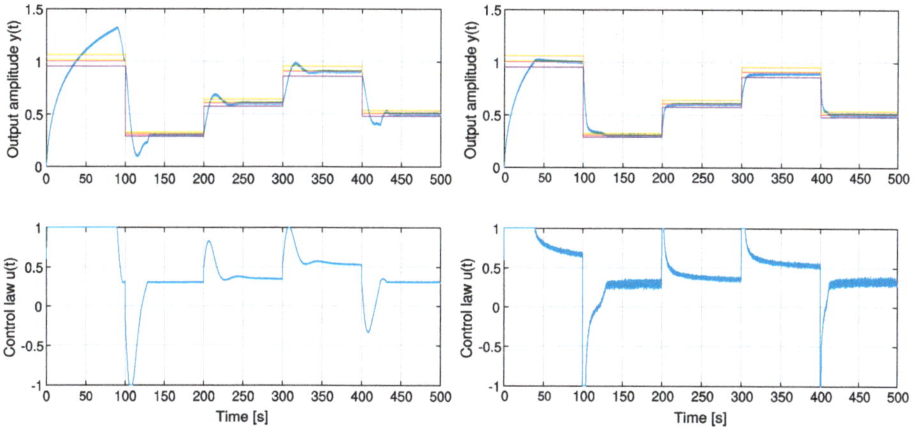

(a) Retuning controller disabled. (b) Retuning controller enabled.

Figure 11: Evaluation of the control system performance with and without input shaping based on FO retuning control.

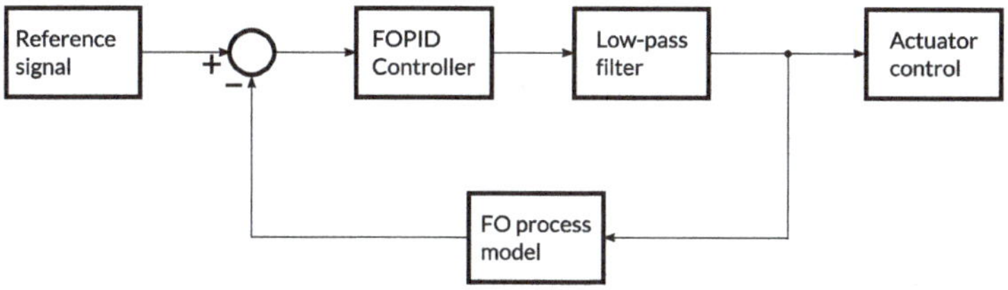

Figure 12: FOPID control of the IPMC actuator using a reference model.

Figure 8. With the loop disabled, the control loop performance can be observed in Figure 11(a). In order to run the input shaping via retuning control, both the gains of the original PID controller and the parameters of the FOPID controller must be set in the corresponding Simulink block. Once this is done, the block is enabled and simulation of the control system is repeated. The improved performance of the retuned control loop can be observed in Figure 11(b).

Example 5.3. In this example, we extend the results from Example 4.3 by designing a FOPI compensator for the IPMC actuator. The corresponding experimental control loop is shown in Figure 12. To obtain the FOPI controller, the design parameters are chosen as phase margin 100° and critical frequency $\omega_c = 1.25\,\text{rad/s}$. In addition, the robustness to gain variations specification must be fulfilled. Once again, the `fpid_optim` tool is used to obtain the parameters of the suboptimal controller. The resulting controller has the following form:

$$C(s) = 1.7887 + 1.3617s^{-0.51513}. \tag{46}$$

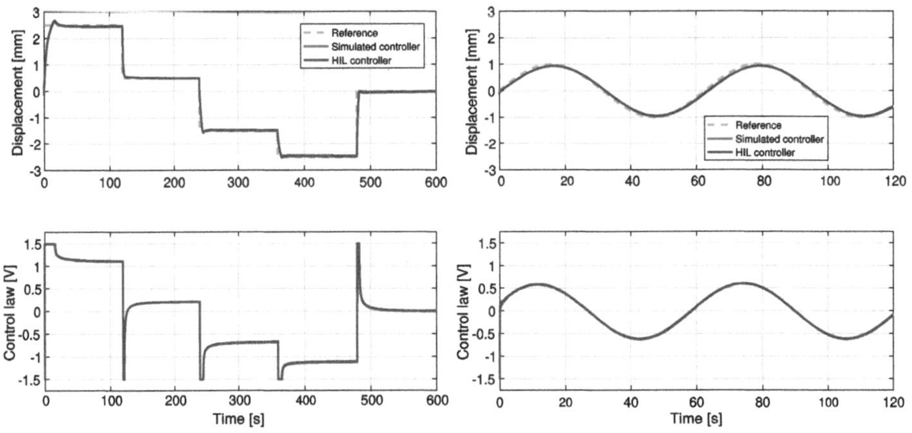

(a) Piecewise constant set point. (b) Sine wave tracking.

Figure 13: Evaluation of the IPMC control system performance.

The controller is then implemented according to the configuration shown in Figure 12. In addition, it is also implemented in a hardware prototype [28]. Experimental results for piecewise constant set point and sine wave tracking are presented in Figure 13. It can be seen that the designed control loop is performing well.

6 Implementation of fractional-order controllers

FOMCON toolbox offers a multitude of specialized tools for obtaining analog and digital implementations of fractional-order systems. In the following section, we shall review these tools, focusing on the digital implementation due to the relative importance of this approach for contemporary control systems [28].

In terms of analog implementation, FOMCON toolbox provides several approximation methods, for example, Oustaloup's approximation method discussed previously, and also a specific class `frac_rcl` that is used for generating, storing, and processing fractance circuits. An object of this class contains references to:

- Particular network structures, however complex, which return computed network transfer functions (impedance values);
- Corresponding implementations.

The idea is that a single structure can have several different implementations which depend on particular constraints of the desired application. This idea is illustrated in Figure 14.

Once a fractance object of this class is created, class methods can be used to extract relevant data about the circuit, change the approximation parameters based on

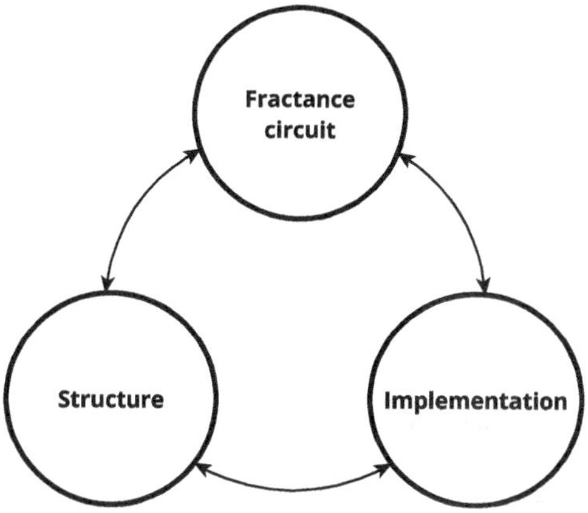

Figure 14: Fractance, network structure, and implementation relations.

particular electrical components comprising the resulting network, or to convert the object to, for example, an LTI zero-pole-gain model of class zpk. See [28, 33] for discussion and relevant examples.

For digital approximation, Oustaloup's recursive filter method in (8) is revisited. Suppose that we are given a sampling interval $T_s \in \mathbb{R}_+$. Then, given the approximation frequency range $[\omega_b, 2/T_s]$ rad/s, order of approximation $\nu \in \mathbb{Z}_+$ and fractional power $\alpha \in [-1, 1] \subset \mathbb{R}$, we proceed to compute $(2\nu+1)$ zeros and $(2\nu+1)$ poles. We consider the zero-pole matching equivalents method for obtaining a discrete-time equivalent of a continuous time transfer function [10]. For both zeros and poles, we have

$$\sigma_k' = \mathrm{e}^{-T_s\omega_k'}, \quad \sigma_k = \mathrm{e}^{-T_s\omega_k} \tag{47}$$

which is a direct mapping of continuous zeros and poles to their discrete-time equivalents. We notice that once the mapping is done, we need to compute the gain of the resulting discrete-time system $H_\alpha(z)$ at the central frequency $\omega_u = \sqrt{\omega_b\omega_h}$. Then the resulting system can be implemented as an IIR filter. The next step is to transform this representation into second-order section form to improve computational stability. Consider the set of discrete-time zeros (poles) that we have obtained

$$z = \{\sigma_{-\nu}, \sigma_{-\nu+1}, \ldots, \sigma_0, \ldots, \sigma_\nu, \sigma_{\nu-1}, \sigma_\nu\}. \tag{48}$$

Due to the specifics of the generation method, this is an ordered set. To arrive at the second-order section form for the zero (pole) polynomial, we proceed as follows. We have $2\nu + 1$ zeros (poles), so there are $\nu + 1$ second-order sections (including a single first-order section). Therefore, we have the polynomial

$$h(z) = (1 - \sigma_\nu z^{-1}) \cdot \prod_{k=0}^{\nu-1} \zeta(z) \tag{49}$$

in the variable z, where $\zeta(z) = 1 + (c_k + d_k)z^{-1} + (c_k \cdot d_k)z^{-2}$, $c_k = -\sigma_{-v+2k}$ and $d_k = -\sigma_{-v+2k+1}$. So finally we arrive at the final form of the approximation

$$\tilde{H}^{\alpha}(z) = \frac{(\omega_b\omega_h)^{\frac{\alpha}{2}}}{|H_a(e^{j\sqrt{\omega_b\omega_h}T_s})|}\prod_{k=1}^{v}\frac{1 + b_{0k}z^{-1} + b_{1k}z^{-2}}{1 + a_{0k}z^{-1} + a_{1k}z^{-2}}. \tag{50}$$

Digital implementation of the fractional-order PID controller may be obtained as

$$H_{PI^{\lambda}D^{\mu}}(z) = K_p + K_i H_I^{-\lambda}(z) + K_d H_D^{\mu}(z), \tag{51}$$

where K_p, K_i, and K_d are gains of the parallel form of the controller, $H_I^{\lambda}(z)$ corresponds to a discrete-time approximation of a fractional-order integrator of order λ and $H_D^{\mu}(z)$ corresponds to a discrete-time approximation of a fractional-order differentiator of order μ, and $\lambda, \mu \in [0, 1]$. Finally, to ensure adequate control loop performance near the set point, we implement the fractional-order integrator as

$$H_I(z) = H^{1-\lambda}(z) \cdot H_I(z), \tag{52}$$

where $H^{1-\lambda}(z)$ is computed using the method presented above, and

$$H_I(z) = \frac{T_s}{(1 - z^{-1})}. \tag{53}$$

Consider now an example.

Example 6.1. In this example, we obtain a digital approximation of the FOPID controller obtained previously in Example 5.1. For this purpose, the FOPID design tool `fpid` is first started. Then the saved controller design configuration file is used to restore the parameters of the FOPID controller. After this, the *Realize controller* option is selected. This opens up the so called `impid` tool that facilitates the implementation of FOPID controllers. Here, the Oustaloup filter is chosen for performing the continuous time approximation with parameters desired frequency band $\omega = [0.001, 1000]$ rad/s and approximation order $N = 5$. The matched pole-zero equivalents is then chosen for obtaining the discrete-time approximation with sample time set to $T_s = 0.01$ s. The resulting approximation is then saved to MATLAB workspace. Finally, running the `d2sos` function in FOMCON results in the IIR SOS coefficient arrays being displayed in MATLAB command line:

$$\begin{aligned} b = {} & \{+1.0000000000, -0.7251385040, +0.0000000000\}, \\ & \{+1.0000000000, -0.0312921101, +0.0000201642\}, \\ & \{+1.0000000000, -0.4812633669, +0.0443351495\}, \\ & \{+1.0000000000, -1.4452364660, +0.4931167901\}, \\ & \{+1.0000000000, -1.8018969442, +0.8085431615\}, \\ & \{+1.0000000000, -1.9541737227, +0.9541742114\}, \end{aligned} \tag{54}$$

$$\{+1.0000000000, -1.9999328005, +0.9999328016\},$$
$$\{+1.0000000000, -1.9997757994, +0.9997758119\},$$
$$\{+1.0000000000, -1.9992551301, +0.9992552688\},$$
$$\{+1.0000000000, -1.9719804027, +0.9721774793\},$$
$$\{+1.0000000000, -1.9915771183, +0.9915949036\},$$
$$\{+1.0000000000, -1.9975024171, +0.9975039794\},$$
$$a = \{-0.0002450747, +0.0000000000\}, \tag{55}$$
$$\{-0.0943408183, +0.0000616569\},$$
$$\{-0.6335988028, +0.0632370735\},$$
$$\{-1.3772361764, +0.4555252525\},$$
$$\{-1.7910734718, +0.7993613081\},$$
$$\{-1.9374843841, +0.9382118038\},$$
$$\{-1.9819389791, +0.9819993377\},$$
$$\{-1.9948350749, +0.9948400027\},$$
$$\{-1.9985272941, +0.9985276945\},$$
$$\{-1.9995804287, +0.9995804612\},$$
$$\{-1.9998804933, +0.9998804959\},$$
$$\{-1.9999742770, +0.9999742770\},$$

and

$$b_0 = 55.1154928235. \tag{56}$$

The corresponding frequency domain response of the implemented controller can be seen in Figure 15.

7 Conclusion

In this chapter, the open source FOMCON toolbox for MATLAB was presented. The toolbox comprises several modules that together implement the model based control design workflow. First, given process input/output data, an integer- or fractional-order model can be identified and validated using the tools included in the toolbox. Both time domain and frequency domain identification methods are supported. The model can be obtained by means of either black or gray box modeling approaches. Once a coherent model is found, it can be used to design a controller. In FOMCON toolbox, FOPID controllers have most support in terms of design and implementation tools

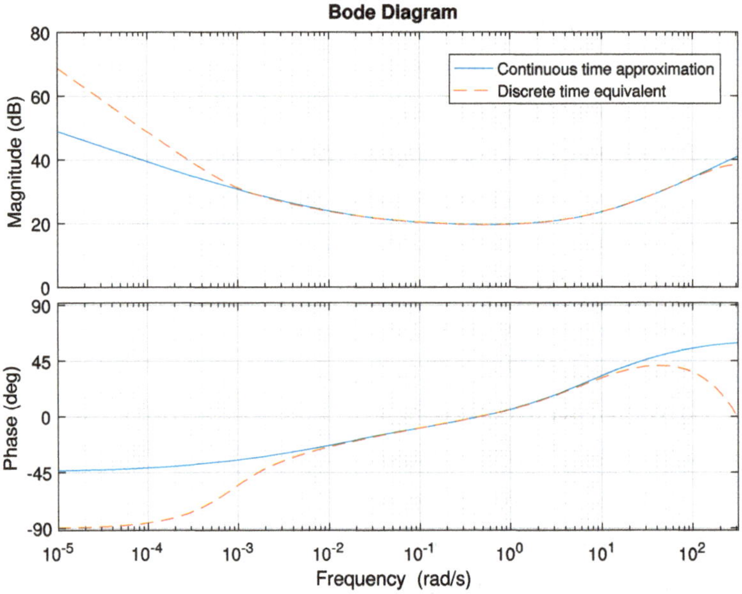

Figure 15: Frequency domain characteristics of the implemented FOPID controller.

due to their importance in contemporary industrial applications. When a suitable controller is obtained, it can be realized using either the analog or digital approximation methods. In this chapter, we are more concerned with the latter, since digital implementation allows for direct transfer of the results obtained in the MATLAB/Simulink environment to real life distributed control system solutions.

To illustrate the features of the toolbox, various examples were provided throughout the chapter. These highlighted several important advantages of the proposed control methods. The IPMC control example clearly showed the applicability of FOMCON toolbox to real life problems, while the retuning controller example demonstrated the benefits of introducing fractional-order control dynamics through input shaping.

In its present form, the toolbox is suitable for use by control engineers and practitioners as well as control system educators [34]. Further development of the toolbox shall be focused on developing efficient MIMO modeling and control design methods.

Bibliography

[1] K. Åström and T. Hägglund, *Advanced PID Control*, The Instrumentation, Systems, and Automation Society (ISA), 2006.

[2] M. A. Branch, T. F. Coleman, and Y. Li, A subspace, interior, and conjugate gradient method for large-scale bound-constrained minimization problems, *SIAM Journal on Scientific Computing*, **21**(1) (1999), 1–23.

[3] M. Čech and M. Schlegel, The fractional-order PID controller outperforms the classical one, in *Process control 2006*, pp. 1–6, Technical University, 2006.
[4] Y. Chen, I. Petráš, and D. Xue, Fractional order control – a tutorial, in *Proc. ACC '09. American Control Conference*, pp. 1397–1411, 2009.
[5] Y. Q. Chen, Ubiquitous fractional order controls?, *IFAC Proceedings Volumes*, **39**(11) (2006), 481–492.
[6] O. Cois, A. Oustaloup, T. Poinot, and J.-L. Battaglia, Fractional state variable filter for system identification by fractional model, in *Proc. of the 2001 European Control Conference*, pp. 417–433, 2001.
[7] T. F. Coleman and Y. Li, An interior trust region approach for nonlinear minimization subject to bounds, Tech. Rep., Department of Computer Science, Cornell University, Ithaca, NY, USA, 1993.
[8] F. A. de Castro, N. D. Bernardes, M. A. de S. L. Cuadros, and G. M. de Almeida, Comparison of fractional and integer PID controllers tuned by genetic algorithm, in *2016 12th IEEE International Conference on Industry Applications (INDUSCON)*, 2016.
[9] F. Declercq and R. De Keyser, Comparative study of neural predictors in model based predictive control, in *Neural Networks for Identification, Control, Robotics, and Signal/Image Processing, 1996. Proceedings, International Workshop on*, pp. 20–28, 1996.
[10] G. Franklin, J. Powell, and M. Workman, *Digital Control of Dynamic Systems*, Ellis-Kagle Press, 2006.
[11] T. T. Hartley and C. F. Lorenzo, Fractional-order system identification based on continuous order-distributions, *Signal Processing*, **83**(11) (2003), 2287–2300.
[12] A. Khadhraoui, K. Jelassi, J.-C. Trigeassou, and P. Melchior, Identification of fractional model by least-squares method and instrumental variable, *Journal of Computational and Nonlinear Dynamics*, **10**(5) (2015), 1–10.
[13] J. C. Lagarias, J. A. Reeds, M. H. Wright, and P. E. Wright, Convergence properties of the Nelder–Mead simplex method in low dimensions, *SIAM Journal on Optimization*, **9**(1) (1998), 112–147.
[14] Z. Li, L. Liu, S. Dehghan, Y. Chen, and D. Xue, A review and evaluation of numerical tools for fractional calculus and fractional order controls, *International Journal of Control*, **90**(6) (2016), 1165–1181.
[15] Y. Luo and Y. Chen, Stabilizing and robust fractional order PI controller synthesis for first order plus time delay systems, *Automatica*, **48**(9) (2012), 2159–2167.
[16] J. T. Machado, Root locus of fractional linear systems, *Communications in Nonlinear Science and Numerical Simulation*, **16**(10) (2011), 3855–3862.
[17] L. Marinangeli, F. Alijani, and S. H. HosseinNia, Fractional-order positive position feedback compensator for active vibration control of a smart composite plate, *Journal of Sound and Vibration*, **412**(Supplement C) (2018), 1–16.
[18] D. Matignon, Generalized fractional differential and difference equations: stability properties and modeling issues, in *Proc. of Math. Theory of Networks and Systems Symposium*, pp. 503–506, 1998.
[19] C. A. Monje, Y. Q. Chen, B. M. Vinagre, D. Xue, and V. Feliu, *Fractional-Order Systems and Controls: Fundamentals and Applications*, Advances in Industrial Control, Springer Verlag, 2010.
[20] C. A. Monje, B. M. Vinagre, V. Feliu, and Y. Q. Chen, Tuning and auto-tuning of fractional order controllers for industry applications, *Control Engineering Practice*, **16**(7) (2008), 798–812.
[21] J. J. Moré, The Levenberg–Marquardt algorithm: implementation and theory, in G. Watson (ed.) *Numerical Analysis*, Lecture Notes in Mathematics, vol. 630, pp. 105–116, Springer Berlin Heidelberg, 1978.

[22] A. O'Dwyer, *Handbook of PI and PID Controller Tuning Rules*, 3rd ed., Imperial College Press, 2009.
[23] K. Ogata, *Modern Control Engineering*, Prentice Hall, New Jersey, 2010.
[24] R. Oldenhuis, Optimize. MathWorks File Exchange. [Online], 2009, http://www.mathworks.com/matlabcentral/fileexchange/24298-optimize.
[25] A. Oustaloup, P. Melchior, P. Lanusse, O. Cois, and F. Dancla, The CRONE toolbox for Matlab, in *Proc. IEEE Int. Symp. Computer-Aided Control System Design CACSD 2000*, pp. 190–195, 2000.
[26] I. Podlubny, L. Dorčák, and I. Kostial, On fractional derivatives, fractional-order dynamic systems and $PI^{\lambda}D^{\mu}$-controllers, in *Proc. 36th IEEE Conf. Decision and Control*, vol. 5, pp. 4985–4990, 1997.
[27] J. Sabatier, P. Lanusse, P. Melchior, and A. Oustaloup, *Fractional Order Differentiation and Robust Control Design*, Springer Netherlands, 2015.
[28] A. Tepljakov, *Fractional-Order Modeling and Control of Dynamic Systems*, Springer-Verlag GmbH, 2017.
[29] A. Tepljakov, FOMCON toolbox for MATLAB. Retrieved on 10.04.2018. [Online], 2018, https://github.com/AlekseiTepljakov/fomcon-matlab.
[30] A. Tepljakov, B. B. Alagoz, E. Gonzalez, E. Petlenkov, and C. Yeroglu, Model reference adaptive control scheme for retuning method-based fractional-order PID control with disturbance rejection applied to closed-loop control of a magnetic levitation system, *Journal of Circuits, Systems, and Computers*, 27(11) (2018), 1850176.
[31] A. Tepljakov, E. A. Gonzalez, E. Petlenkov, J. Belikov, C. A. Monje, and I. Petráš, Incorporation of fractional-order dynamics into an existing PI/PID DC motor control loop, *ISA Transactions*, **60** (2016), 262–273.
[32] A. Tepljakov, E. Petlenkov, and J. Belikov, A flexible MATLAB tool for optimal fractional-order PID controller design subject to specifications, in *Proceedings of the 31st Chinese Control Conference*, pp. 4698–4703, 2012.
[33] A. Tepljakov, E. Petlenkov, and J. Belikov, Efficient analog implementations of fractional-order controllers, in *Proceedings of the 14th International Carpathian Control Conference (ICCC)*, 2013.
[34] A. Tepljakov, E. Petlenkov, E. A. Gonzalez, and I. Petras, Design of a MATLAB-based teaching tool in introductory fractional-order systems and controls, in *2017 IEEE Frontiers in Education Conference (FIE)*, 2017.
[35] D. Valério, Toolbox ninteger for MatLab, v.2.3. [Online], 2005, http://web.ist.utl.pt/duarte.valerio/ninteger/ninteger.htm.
[36] D. Valério, M. D. Ortigueira, and J. S. da Costa, Identifying a transfer function from a frequency response, *Journal of Computational and Nonlinear Dynamics*, **3**(2) (2008), 1–7.
[37] B. J. West, *Fractional Calculus View of Complexity*, Taylor & Francis Inc, 2015.
[38] D. Xue, *Fractional-Order Control Systems: Fundamentals and Numerical Implementations*, Fractional Calculus in Applied Sciences and Engineering, De Gruyter, 2017.
[39] D. Xue, C. Zhao, and Y. Chen, Fractional order PID control of a DC-motor with elastic shaft: a case study, in *2006 American Control Conference*, pp. 3182–3187, 2006.

Dingyü Xue

FOTF Toolbox for fractional-order control systems

Abstract: Handy and reliable computer tools are helpful for the researchers to handle their research problems.

In Section 1, a brief review to MATLAB toolboxes in fractional-order systems is given, and the main toolbox involved in this chapter, FOTF Toolbox, is briefly addressed. The use of the major functions in the FOTF Toolbox is described and demonstrated through examples.

In Section 2, numerical functions for computing Mittag-Leffler function, fractional-order derivatives and integrals are presented. The Grünwald–Letnikov and Caputo definitions are both considered. In particular, $o(h^p)$ high precision algorithms are proposed and implemented in the FOTF Toolbox, with much higher accuracy than any other existing algorithms and tools. MATLAB solvers for linear and fractional-order differential equations are provided. In Section 3, a Simulink based blockset, `fotflib`, is introduced and its applications in solving fractional-order differential equations are addressed. A unified modeling strategy for nonlinear Caputo equation with any complexity is introduced, and illustrated through examples. In Section 4, two classes for linear fractional-order control components are designed and demonstrated. With these classes, complicated linear system models can be constructed, and time and frequency domain analysis can be carried out easily, as if one is processing integer-order models. In Section 5, several controller design examples are proposed, where an optimal fractional-order PID controller design function and interface are proposed first, followed by two examples in multivariable fractional-order controller design: pseudodiagonalization controller design and parameter optimization controller design. Simulation methods of closed-loop multivariable control systems are also demonstrated.

Keywords: MATLAB Toolbox, high-precision algorithm, fractional-order system, fractional-order PID controller, multivariable system

MSC 2010: 26A33, 34A08, 93C83, 46N40, 34C60

1 Introduction

There are several widely used MATLAB Toolboxes available in fractional calculus and fractional-order control, among the researchers in the fractional calculus community. A survey on some of the numerical tools can be found in [5]. Most of the tools listed

Dingyü Xue, Northeastern University, Shenyang, China, e-mail: xuedingyu@mail.neu.edu.cn

https://doi.org/10.1515/9783110571745-011

in [5] are just single MATLAB functions, solving one or a few specific problems. Only four of them can be considered as toolboxes. They are summarized below in the order of the time of their first appearance.

(1) *CRONE Toolbox* [7], developed by the CRONE team in France led by Professor Alain Oustaloup, is a MATLAB and Simulink Toolbox dedicated to the applications of noninteger derivatives in engineering and science. The work started in 1990s and it is useful in solving problems in fractional-order identification, robust control analysis, and design. Unfortunately, the code are encrypted in MATLAB's pseudo-code format, and it is not possible to modify or extend the code in the toolbox. CRONE is the French abbreviation of *Commande Robuste d'Ordre Non Entier*, in English, "noninteger order robust control." Multivariable systems can be handled by the toolbox.
(2) *Ninteger Toolbox* [12], developed by Professor Duarte Valério of Portugal since 2005, is a noninteger control toolbox for MATLAB intended to help with developing fractional-order controllers and assessing their performance. The kernels of the toolbox are the facilities of fractional-order system identification, and approximations with integer-order objects. The toolbox applies only to single-variable control systems.
(3) *FOTF Toolbox* [1, 13] is a control toolbox for fractional-order systems developed by Professor Dingyü Xue of China. It was first released under the name of FOTF in 2006, although several functions and Simulink blocks were released by the author since 2004. Now, all of its components are upgraded such that multivariable systems are fully supported. Also, the low-level fractional calculus computation facilities are fully upgraded to high-precision algorithms, which makes the computational support much more efficient and reliable. The theoretical background and programming details of the toolbox are fully addressed in [13].
(4) *FOMCON Toolbox* [11], developed Aleksei Tepljakov of Estonia, started by collecting the existing functions and classes in FOTF and Ninteger Toolboxes, to build a toolbox for solving problems in the identification, analysis, and design of fractional-order control systems. Since the earlier versions of my FOTF class were used as its basis, it is restricted to single-variable systems.

Besides, the readers are also recommended to search for possible new toolboxes from MathWorks' "File Exchange" web site. However, care must be taken to select the reliable tools, since the standard of implementation may be significantly different, and some are of very poor quality and may lead to erroneous results.

In this chapter, the facilities of the FOTF Toolbox are presented, while the algorithms involved are not given. Interested readers may refer to [13] for details of the algorithms. The FOTF Toolbox can be downloaded from: https://cn.mathworks.com/matlabcentral/fileexchange/60874-fotf-toolbox?s_tid=srchtitle.

2 Numerical computation in fractional calculus

In this section, the numerical functions for solving fractional calculus problems are presented. The functions in the FOTF Toolbox, involving Mittag-Leffler functions, fractional derivatives and integrals, linear and nonlinear fractional-order differential equations are all presented, demonstrated by illustrative examples.

2.1 Mittag-Leffler functions

Mittag-Leffler functions are commonly used in fractional calculus area, and here, a generalized Mittag-Leffler function with four parameters is used

$$E_{\alpha,\beta}^{\gamma,q}(z) = \sum_{k=0}^{\infty} \frac{(\gamma)_{kq}}{\Gamma(\alpha k+\beta)} \frac{z^k}{k!}, \tag{1}$$

where, $\alpha, \beta, \gamma \in \mathbb{C}$, and for any $z \in \mathbb{C}$, the convergent conditions are $\mathbb{R}(\alpha) > 0$, $\mathbb{R}(\beta) > 0$, $\mathbb{R}(\gamma) > 0$, and $q \in \mathbb{N}$, where $(\gamma)_k$ is the Pochhammer symbol.

It is known that Mittag-Leffler functions with one, two, and three parameters are special cases of the four-parameter one.

In the FOTF Toolbox, function `ml_func()` is presented, which can be used in evaluating the integer-order derivatives of Mittag-Leffler functions, with the following syntax $\mathbf{f}$=`ml_func(`$\mathbf{v}$,$\mathbf{z}$,n,ϵ_0`)`, where, for Mittag-Leffler functions with different numbers of parameters, $\mathbf{v}$ can be set to α, $[\alpha,\beta]$, $[\alpha,\beta,\gamma]$, $[\alpha,\beta,\gamma,q]$. When $n = 0$, the values of Mittag-Leffler function are returned. In the function, the parameters are expressed in the first argument in a vector, ϵ_0 is the error tolerance, with the default value of 2.2204×10^{-16}.

The algorithm of accumulation with truncations is used in the function, and it may fail in certain cases. Igor Podlubny's `mlf()` function is embedded in the `ml_func()` function. If it fails to converge, `mlf()` is called automatically, and the speed in computation may be low.

2.2 Fractional-order derivatives and integrals

The most widely used definitions in fractional calculus are Grünwald–Letnikov, Riemann–Liouville and Caputo definitions. For a wide variety of signals, Grünwald–Letnikov and Riemann–Liouville definitions are equivalent. They are not distinguished in the chapter.

The commonly used MATLAB functions for fractional-order are often $o(h)$ ones, whose accuracy heavily depends upon the selection of step size h. Such a function is implemented in `glfdiff()` function of FOTF Toolbox. The syntax of such a function is $\mathbf{y}_1$=`glfdiff(`$\mathbf{y}$`,`$\mathbf{t}$`,`γ`)`, which $\mathbf{y}$ can either be a vector of the samples of a signal, or a function handle of the function $y(t)$; the argument $\mathbf{t}$ is the equally-spaced time vector, γ is the order, which can be positive, zero or negative. When $\gamma = 0$, the original function is returned to $\mathbf{y}_1$, while if $\gamma < 0$, the $-\gamma$th order integral is returned in vector $\mathbf{y}_1$.

Algorithms are proposed in [13], aiming at finding $o(h^p)$ high precision numerical solutions of fractional-order derivatives and integrals under Grünwald–Letnikov and Caputo definitions. The implemented functions are provided in FOTF Toolbox, summarized below:

(1) *Grünwald–Letnikov definition.* Function `glfdiff9()` can be used and the syntax is similar to `glfdiff()`, with $\mathbf{y}_1$=`glfdiff9(`$\mathbf{y}$`,`$\mathbf{t}$`,`γ`,`p`)`, where p is the expected accuracy.
(2) *Caputo definition.* Function `caputo9()` can be used to evaluate Caputo derivatives of order γ with $\mathbf{y}_1$=`caputo9(`$\mathbf{y}$`,`$\mathbf{t}$`,`γ`,`p`)`.

The evaluation of Grünwald–Letnikov derivatives are relatively easier, compared with Caputo one, where the initial values must be considered. Therefore, the following examples are made in the evaluations of Caputo derivatives only.

Example 1. Compute the 0.6th order Caputo derivative of signal e^{-t} with a step size $h=0.01$.

Solutions: The reason for the choice of e^{-t} is that the analytical results for Caputo derivative is known, such that the assessment of the accuracy is possible.

It is known that the ${}_0^{\mathrm{C}}D_t^{0.6}e^{-t} = -t^{0.4}E_{1,1.4}(-t)$. The following MATLAB commands can be used to evaluate the 0.6th order derivatives for different selections of h, and the maximum errors are shown in Table 1.

```
>> t0 = 0.5:0.5:5; q = 1; gam = q-0.6; t = 0:0.01:5;
   y0 = -t0.^0.4.*ml_func([1,1.4],-t0,0,eps);
   y = exp(-t); ii = [51:50:501]; T = [];
   for p=1:6, y1 = caputo9(y,t,0.6,p); T = [T [y1(ii)-y0']]; end
   max(abs(T))
```

The accuracy of the result for $p = 1$ is the same as the traditional $o(h)$ results. It can be seen that with properly selected p's, the accuracy can be significantly improved. If $p = 6$ is selected for this example, the results are 10^{10} times more accurate than the $o(h)$ results.

Table 1: Maximum errors for $h = 0.01$.

$p = 1$	$p = 2$	$p = 3$	$p = 4$	$p = 5$	$p = 6$
0.0018	1.19×10^{-5}	8.89×10^{-8}	7.07×10^{-10}	5.85×10^{-12}	5.3×10^{-13}

Table 2: Maximum errors for $h = 0.1$.

$p = 4$	$p = 5$	$p = 6$	$p = 7$	$p = 8$	$p = 9$
4.53×10^{-6}	1.98×10^{-7}	3.07×10^{-9}	8.17×10^{-11}	2.97×10^{-12}	2.48×10^{-10}

Example 2. Consider the previous example. What if $h = 0.1$?
Solutions: If a larger step-size $h = 0.1$ is used, the following statements can be issued, and the maximum errors are given in Table 2, with $p = 8$ a good choice:

```
>> t = 0:0.1:5; y = exp(-t); ii=[6:5:51]; T = [];
   for p=4:9, y1 = caputo9(y,t,0.6,p); T = [T [y1(ii)-y0']]; end
   max(abs(T))
```

It can be seen from the comparisons that, even the step-size is selected as a large value as $h = 0.1$, the accuracy of the computation results are still very high. Therefore, the algorithm and implementation here are effective and efficient.

Three more comments regarding the `glfdiff9()`, `caputo9()` solvers are:
(1) The argument **y** is the function can also be specified as anonymous functions, for instance, y=`@(t)exp`$(-t)$.
(2) The MATLAB functions also work for $y < 0$, where fractional-order integrals are taken automatically.
(3) For a signal whose Caputo derivative is not analytically known, how to select p and h? Generally speaking, h can be selected to a moderate one, and should not be chosen as very small ones. The selection p can be made as follows. If a certain value of p is selected, $p + 1$ should also be selected. If the errors between p and $p + 1$ are large, larger p's can be probed, until the errors are acceptable.

2.3 Higher order derivative evaluations

If the order of derivative expected are very high, for instance, $\alpha = 5.6$, most existing numerical algorithms may fail. A possible solution of the algorithm is introduced such that satisfactory approximations can be obtained.

Example 3. For the given function $f(t) = e^{-t}$, generate samples, from which, compute 5.6th order Caputo derivative, and validate the results.
Solutions: The analytical solution is ${}_0^{C}D_t^{5.6}e^{-t} = t^{0.4}E_{1,1.4}(-t)$. select $h = 0.1$, generate a set of samples, from which the computed results obtained. It can be seen from the generated plot that the result obtained diverges, which is wrong.

Other feasible methods must be explored.

```
>> t0 = 0.5:0.5:5; t = 0:0.1:5; y = exp(-t);
   for p=1:9, y1 = caputo9(y,t,5.6,p); plot(t,y1); hold on, end
```

Table 3: Maximum errors in finding 5.6th order Caputo derivatives.

$p = 1$	$p = 2$	$p = 3$	$p = 4$	$p = 5$	$p = 6$
0.036706	0.0022979	0.00015914	1.3936×10^{-5}	7.5657×10^{-6}	6.9727×10^{-6}

Theorem 1. *Caputo derivative can be obtained with*

$$ {}_{t_0}^{\mathrm{C}}\mathcal{D}_t^{\gamma} y(t) = {}_{t_0}^{\mathrm{RL}}\mathcal{D}_t^{-(\lceil\gamma\rceil-\gamma)}\left[y^{(\lceil\gamma\rceil)}(t)\right]. \tag{2} $$

The physical interpretation of the theorem is that, the γth order Caputo derivative of signal $y(t)$ is expected, the integer order derivative $y^{\lceil\gamma\rceil}(t)$ should be used, then the expected result can be obtained by taking $(\lceil\gamma\rceil-\gamma)$th order Riemann–Liouville integral.

If high precision solutions of numerical integer-order derivatives are available, the function `glfdiff()` can be used to evaluate high order fractional derivatives with ease. High precision MATLAB implementation of integer-order derivatives is available in [14]. Such a function can be tried, together with `glfdiff9()`, in the evaluation of high order derivatives.

Example 4. Solve the problem in Example 3 numerically.
Solutions: Sixth-order derivative can be evaluated under the same step size, and based on the derivative samples, 0.4th order Riemann–Liouville integral can be evaluated under different selections of p, such that the 5.6th order derivatives can be found. Compared with analytical solutions, the maximum computational errors from the new algorithm can be obtained, as shown in Table 3.

```
>> t0 = 0.5:0.5:6.5; h = 0.1; t = 0:h:6.5; y = exp(-t);
   [y01,t1] = num_diff(y,h,6,6); ii = [6:5:length(t1)];
   y0 = t1.^0.4.*ml_func([1,1.4],-t1,eps); T=[];
   for p = 1:6,
      y1=glfdiff9(y01,t1,-0.4,p); err=y0-y1; T=[T abs(err(ii).')];
   end, max(T)
```

2.4 Linear fractional-order differential equations

General form of linear fractional-order differential equation is [9]

$$ \begin{aligned} & a_n\mathcal{D}_t^{\beta_n}y(t) + a_{n-1}\mathcal{D}_t^{\beta_{n-1}}y(t) + \cdots + a_1\mathcal{D}_t^{\beta_1}y(t) + a_0\mathcal{D}_t^{\beta_0}y(t) \\ & \quad = b_1\mathcal{D}_t^{\gamma_1}u(t) + b_2\mathcal{D}_t^{\gamma_2}u(t) + \cdots + b_m\mathcal{D}_t^{\gamma_m}u(t) \end{aligned} \tag{3} $$

where, for equations with zero initial conditions, Riemann–Liouville and Caputo differential equations are the same, however, if the initial conditions are not zero, the two types of equations should be considered separately.

If the maximum order of the equation $\lceil\max(\alpha_i)\rceil = q$, there should be q initial conditions known for the Caputo differential equations $y(0), y'(0), \ldots, y^{(q-1)}(0)$, such that Caputo equations have unique solutions.

Closed-form solution of Riemann–Liouville and Caputo equations can be solved. Several solvers of linear fractional-order differential equations are provided in the FOTF Toolbox, with details of the algorithms and MATLAB implementations presented in Reference [13]:

(1) *Riemann–Liouville equations.* The function `fode_sol()` can be used to solve linear Riemann–Liouville fractional-order differential equations, with the syntax **y**=`fode_sol(`**a**`,`$\mathbf{n}_\mathrm{a}$`,`**b**`,`$\mathbf{n}_\mathrm{b}$`,`**u**`,`**t**`)`, with accuracy of $o(h)$. The $o(h^p)$ high precision version of the function is given in `fode_sol9()`, with the syntax **y**=`fode_sol9(`**a**`,` $\mathbf{n}_\mathrm{a}$`,`**b**`,`$\mathbf{n}_\mathrm{b}$`,`**u**`,`**t**`,`p`)`.

(2) *Caputo equations.* High order Caputo differential equations can be solved with **y**=`fode_caputo9(`**a**`,`$\mathbf{n}_\mathrm{a}$`,`**b**`,`$\mathbf{n}_\mathrm{b}$`,`$\mathbf{y}_0$`,`**u**`,`**t**`,`p`)` function, where $\mathbf{y}_0$ is the initial condition vector, **u** is a vector with input samples, p is the precision specification such that the accuracy of the algorithm of $o(h^p)$.

Example 5. Solve Caputo fractional-order differential equation

$$y'''(t) + \frac{1}{16}{}_0^{\mathrm{C}}D_t^{2.5}y(t) + \frac{4}{5}y''(t) + \frac{3}{2}y'(t) + \frac{1}{25}{}_0^{\mathrm{C}}D_t^{0.5}y(t) + \frac{6}{5}y(t) = \frac{172}{125}\cos\frac{4t}{5}$$

with initial conditions $y(0) = 1$, $y'(0) = 4/5$, $y''(0) = -16/25$, $0 \leqslant t \leqslant 30$. It is known the analytical solution is $y(t) = \sqrt{2}\sin(4t/5 + \pi/4)$. Find the numerical solutions.

Solutions: With the given equation and initial conditions, the following statements can be used to compute the Caputo equation solution with the following statements.

The maximum error is 3.11×10^{-6}, indicating the solution is reliable.

```
>> a = [1 1/16 4/5 3/2 1/25 6/5]; na = [3 2.5 2 1 0.5 0];
   b = 1; nb = 0; t = [0:0.1:30]; u = 172/125*cos(4*t/5);
   y0 = [1 4/5 -16/25]; y = sqrt(2)*sin(4*t/5+pi/4);
   y1 = fode_caputo9(a,na,b,nb,y0,u,t,5);
   max(abs(y-y1)), plot(t,y,t,y1)
```

2.5 Nonlinear differential equations

Explicit nonlinear Caputo fractional-order differential equation is expressed as

$${}_0^{\mathrm{C}}D_t^{\alpha}y(t) = f(t, y(t), {}_0^{\mathrm{C}}D_t^{\alpha_1}y(t), \ldots, {}_0^{\mathrm{C}}D_t^{\alpha_{n-1}}y(t)) \tag{4}$$

where, $q = \lceil\alpha_n\rceil$, the initial conditions are

$$y(0) = y_0, \quad y'(0) = y_1, \quad y''(0) = y_2, \quad \ldots, \quad y^{(q-1)}(0) = y_{q-1}. \tag{5}$$

A high precision prediction–corrector algorithm and implementation are provided in [13], and provided in the FOTF Toolbox. The syntaxes of the functions are

$$[\mathbf{y},\mathbf{t}]=\texttt{nlfep}(\texttt{fun},\alpha,\mathbf{y}_0,t_\mathrm{n},h,p,\epsilon),\quad \mathbf{y}=\texttt{nlfec}(\texttt{fun},\alpha,\mathbf{y}_0,\mathbf{y}_\mathrm{p},\mathbf{t},p,\epsilon)$$

where, fun is used to express the right-hand side of the differential equation, which can either be an anonymous function, or an M function. The vector α is composed of all the orders in the equation, including the highest order on the left-hand side. The initial condition vector is composed of $\mathbf{y}(0) = [y(0), \dots, y^{\lceil\alpha\rceil-1}]$. The argument t_n is h is the step size, ϵ is the error tolerance, p is the precision specification, such that the accuracy is $o(h^p)$, and $p \leqslant \lceil\alpha\rceil$.

Example 6. Solve the nonlinear Caputo equation given by [2]

$${}_0^\mathrm{C}\mathcal{D}_t^{1.455}y(t) = -t^{0.1}\frac{E_{1,1.545}(-t)}{E_{1,1.445}(-t)}\mathrm{e}^t y(t)\,{}_0^\mathrm{C}\mathcal{D}_t^{0.555}y(t) + \mathrm{e}^{-2t} - [y'(t)]^2$$

where, $y(0) = 1$, $y'(0) = -1$, with the known analytical solution $y(t) = \mathrm{e}^{-t}$.
Solutions: The original problem came from [2], however, the original one has typos, where e^{-t} is not the analytical solution. The Mittag-Leffler function should be replaced by the function with two parameters so as to satisfy the analytical solution e^{-t}. The traditional algorithms may be inaccurate and time consuming, and it may take hours of time.

With the new high precision prediction–corrector solver, one may select $\alpha = [1.455, 0.555, 1]$, and assign $\mathbf{y}_0 = [1, -1]$. Anonymous function can be use to express Caputo equation in vectorized format, and the equation can be solved with the following statements:

```
>> f = @(t,y,Dy)-t.^0.1.*ml_func([1,1.545],-t).*exp(t)./...
      ml_func([1,1.445],-t).*y.*Dy(:,1)+exp(-2*t)-Dy(:,2).^2;
   alpha = [1.455,0.555,1]; y0 = [1,-1]; tn = 1; h = 0.01;
   err = 1e-8; p = 1; [yp1,t] = nlfep(f,alpha,y0,tn,h,p,err);
   p = 2; tic, [y2,t] = nlfec(f,alpha,y0,yp1,t,p,err); toc
   max(abs(y2-exp(-t)))
```

The execution time for the above code is only 2.33 s, with maximum error of 3.9337×10^{-5}. Further, if $h = 0.0001$ is selected, the maximum error is as low as 6.8857×10^{-9}, and the execution time is 62.05 s.

3 FOTF blockset and applications

The numerical evaluations of fractional-order derivatives so far studied are based on the assumption that $f(t)$ is a given function. While in real applications, this is not always true, since the signal may come from other parts of the system. Other methods should be introduced to compute fractional-order derivatives. For instance, filters can be designed to simulate the behaviors of fractional-order operators.

Filters can be continuous or discrete. Continuous filters can be used to approximate Laplace operator s^γ. The output of the filter can be regarded as the Riemann–Liouville derivative of the input signal.

Oustaloup filter is a good example in the continuous filters. The mathematical model and MATLAB function of the filter is presented first in the section, and based on the function, an FOTF blockset is designed, composed of essential Simulink blocks for modeling fractional-order systems. A universal strategy for modeling Caputo fractional-order systems of any complexity is presented through examples.

3.1 Oustaloup filter

The mathematical form of an Oustaloup filter is presented in [8]

$$G_{\rm f}(s) = K\prod_{k=1}^{N}\frac{s+\omega_k'}{s+\omega_k} \tag{6}$$

with $(\omega_{\rm b}, \omega_{\rm h})$ is the interested frequency interval, and the poles, zeros and the gain in (7) can be computed from

$$\omega_k' = \omega_{\rm b}\omega_{\rm u}^{(2k-1-\gamma)/N}, \quad \omega_k = \omega_{\rm b}\omega_{\rm u}^{(2k-1+\gamma)/N}, \quad K = \omega_{\rm h}^\gamma, \tag{7}$$

with $\omega_{\rm u} = \sqrt{\omega_{\rm h}/\omega_{\rm b}}$.

Based on the algorithm, a MATLAB function `ousta_fod()` can be designed with the syntax G_1=`ousta_fod(`γ`,`N`,`$\omega_{\rm b}$`,`$\omega_{\rm h}$`)`, where γ is the order of derivatives, which can either be positive, zero or negative. Argument N is the order of the filter. If the parameters of the Oustaloup filter are properly chosen, the output of the filter may approximate the Riemann–Liouville derivative $D_t^\gamma f(t)$ of its input signal $f(t)$.

3.2 Other blocks

A FOTF blockset is designed and provided in the FOTF Toolbox. Type `fotflib` command, a Simulink group is shown in Figure 1.

The “fractional operator” block is in fact the Oustaloup filter, of course other filters can also be selected in the block. The “Caputo operator” block was used for a particular example in [13], not recommended for further applications. Better solutions of Caputo derivatives will be illustrated next.

In the FOTF blockset, “approximate FOTF model” is recommended to model fractional-order transfer function, or even transfer function matrix, while “FOTF ma-

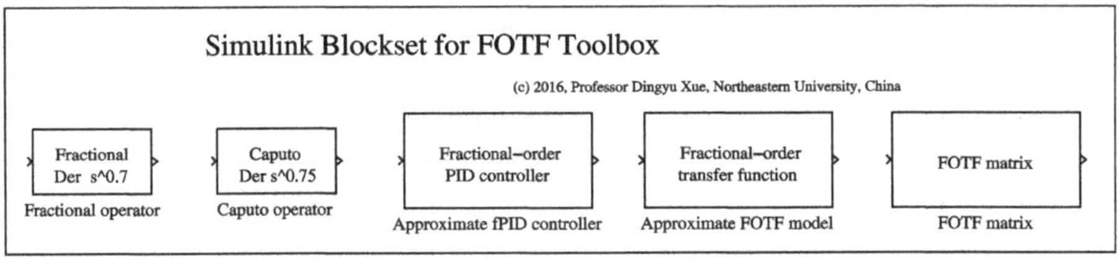

Figure 1: A block library for multivariable FOTF blocks.

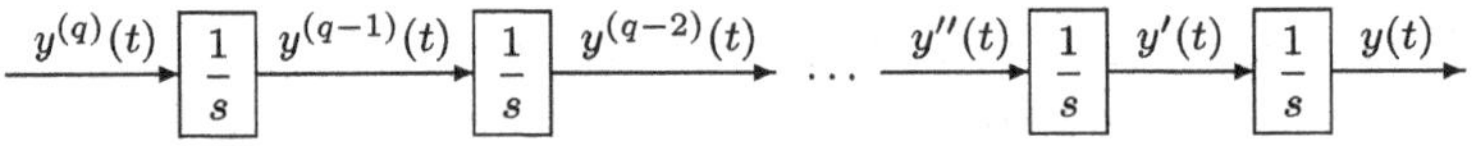

Figure 2: Integrator chain for integer-order derivatives.

trix" block is not recommended. The "approximate fPID controller" block can be used to model fractional-order PID controllers. The remaining blocks regarding control systems in the blockset will be used later.

3.3 Universal modeling scheme for fractional-order differential equations

In the previous sections, predictor–corrector solvers in MATLAB is presented. However, there are limitations in using the solves in control applications, since before using the solvers, an explicit form of the equation must be established. In control systems, this approach has apparent difficulties, since it is usually not possible to write out the explicit differential equation from complicated nonlinear control systems. Block diagram based solution is a possible and universal choice for dealing with the problems.

Block diagram modeling is an effective way in describing complicated systems. Here, two kinds of fractional-order operators are considered.

(1) *Solutions of nonlinear equations with zero initial conditions.* Since in the systems, only Riemann–Liouville derivatives are involved, and Oustaloup filters can be used directly to approximate Riemann–Liouville derivatives and integrals, if the parameters of the filters are well chosen.
(2) *Solutions of nonlinear Caputo equations with nonzero initial conditions.* In order to describe Caputo derivatives, the integer-order derivatives of the signal $y(t)$ can be established, as shown in Figure 2. All the initial conditions should be assigned to the corresponding integrators as initial conditions.

It has been indicated that Oustaloup filter cannot be used to deal with Caputo derivatives directly. Let us recall Theorem 1. It can be seen that high order Caputo derivatives can be constructed in an alternative way. For instance, if 3.4th order Caputo derivative is expected, 4th order signal can be used as the input signal of an Oustaloup filter, with order −0.6, the output of the filter can be regarded as the 3.4th order Caputo derivative.

There is another theorem which is equally important in Caputo differential equation modeling problems.

Theorem 2. *Taking* $(\lceil \gamma \rceil - \gamma)$*th order Riemann–Liouville derivatives to both sides, it can be shown that*

$$ {}^{\mathrm{RL}}_{t_0}\mathscr{D}_t^{\lceil \gamma \rceil - \gamma}\left[{}^{\mathrm{C}}_{t_0}\mathscr{D}_t^{\gamma} y(t)\right] = y^{(\lceil \gamma \rceil)}(t). \tag{8} $$

Having all the Caputo derivative signals, complicated fractional-order system models can easily be constructed with Simulink. In theory, nonlinear Caputo equations of any complexity can be modeled by the strategy.

In order to assess the efficiency of the nonlinear Caputo equation solvers, a set of benchmark problems are proposed in [15], and the accuracy of the solutions in all the equations are higher than the existing ones. Two examples are given here to show the modeling strategy with Simulink environment.

Example 7. Solve again the nonlinear Caputo differential equation in Example 6 with Simulink environment.

Solutions: For the simplicity of Simulink modeling, the nonlinear Caputo equation can be rewritten in the explicit form

$$ {}^{\mathrm{C}}_{0}D_t^{1.455} y(t) = -t^{0.1}\frac{E_{1,1.545}(-t)}{E_{1,1.445}(-t)}\mathrm{e}^t y(t)\,{}^{\mathrm{C}}_{0}D_t^{0.555} y(t) + \mathrm{e}^{-2t} - \left[y'(t)\right]^2 $$

where $y(0) = 1$, $y'(0) = -1$. Since the highest order of derivative is 1.455, then select $q = 2$. Two integer-order integrators should be connected in series to define the key signals $y(t)$, $y'(t)$ and $y''(t)$. The initial conditions should be accordingly set to the two integrators. The signal ${}^{\mathrm{C}}_{0}D_t^{0.555} y(t)$ can be established from (2), which is generated from $y'(t)$, and the signal is fed to a −0.445th order Oustaloup filter, whose output is ${}^{\mathrm{C}}_{0}D_t^{0.555} y(t)$. With the key signals, the right-hand side of the equation can be constructed with low-level blocks, such that ${}^{\mathrm{C}}_{0}D_t^{1.455} y(t)$ signal can be generated. A closed loop is needed to complete the modeling. If ${}^{\mathrm{C}}_{0}D_t^{1.455} y(t)$ signal is fed into a 0.445th order Oustaloup filter, the signal $y''(t)$ can be constructed. Since $y''(t)$ is the starting point in the integrator chain, the two signals can be joined up together to form the overall model of the whole equation, shown in Figure 3.

With the Oustaloup filter, assign the filter with the following parameters, the simulation results can be obtained. Compared with the analytical solution e^{-t}, the maximum error is 1.1636×10^{-4}, and time elapse is 0.21 s.

```
>> N = 18; ww = [1e-7 1e4];
   tic, [t,x,y] = sim('c10mexp2s'); toc, max(abs(y-exp(-t)))
```

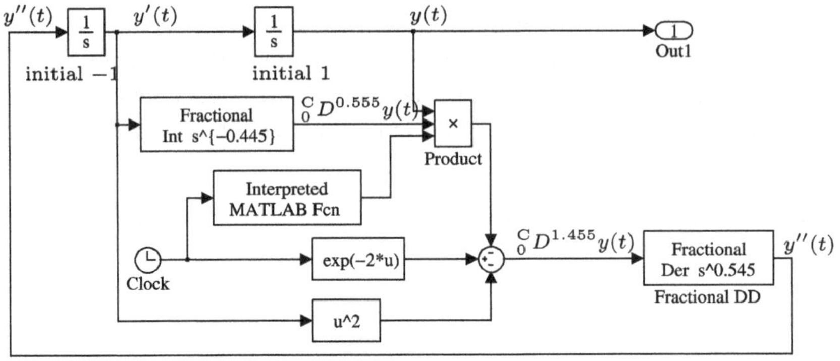

Figure 3: Simulink model.

Simulation results with better accuracy can be obtained by selecting larger frequency range and high order in the Oustaloup filter. For instance, selecting $(10^{-8}, 10^{7})$ rad/s, $N = 35$, the maximum error can be reduced to 1.353×10^{-7}, and the time elapse is only 4.7 s.

Example 8. Solve the following implicit Caputo differential equation:

$$
\begin{aligned}
&{}_0^{C}D_t^{0.2}y(t)\,{}_0^{C}D_t^{1.8}y(t) + {}_0^{C}D_t^{0.3}y(t)\,{}_0^{C}D_t^{1.7}y(t) \\
&= -\frac{t}{8}\left[E_{1,1.8}\left(-\frac{t}{2}\right)E_{1,1.2}\left(-\frac{t}{2}\right) + E_{1,1.7}\left(-\frac{t}{2}\right)E_{1,1.3}\left(-\frac{t}{2}\right)\right]
\end{aligned}
$$

where $y(0) = 1$, $y'(0) = -1/2$, with the analytical solution $y(t) = \mathrm{e}^{-t/2}$.
Solutions: The standard form of the implicit Caputo equation can be expressed as

$$
\begin{aligned}
&{}_0^{C}D_t^{0.2}y(t)\,{}_0^{C}D_t^{1.8}y(t) + {}_0^{C}D_t^{0.3}y(t)\,{}_0^{C}D_t^{1.7}y(t) \\
&+ \frac{t}{8}\left[E_{1,1.8}\left(-\frac{t}{2}\right)E_{1,1.2}\left(-\frac{t}{2}\right) + E_{1,1.7}\left(-\frac{t}{2}\right)E_{1,1.3}\left(-\frac{t}{2}\right)\right] = 0.
\end{aligned}
$$

The key signals $y(t)$, $y'(t)$ and $y''(t)$ can be defined through the integrator chain, based on them, the Caputo derivative signals $D^{0.2}y(t)$, $D^{0.3}y(t)$, $D^{1.7}y(t)$, and $D^{1.8}y(t)$ can be constructed. The left-hand side of the equation can be constructed, and feeds into the block "Algebraic Constraint," the output signal obtained is $D^{1.8}y(t)$. If the signal is connected to a 0.2th order Oustaloup filter, signal $y''(t)$ can be formed, and connected to $y''(t)$ signal. The constructed Simulink model of implicit Caputo equation is shown in Figure 4.

With the following Oustaloup filter parameters, the numerical solution of implicit Caputo equation can be obtained, with maximum error of 3.8182×10^{-5}, and time elapse of 334.8 s. The execution speed is relatively slow since there exists algebraic loop, and in each simulation step, algebraic equation is solved.

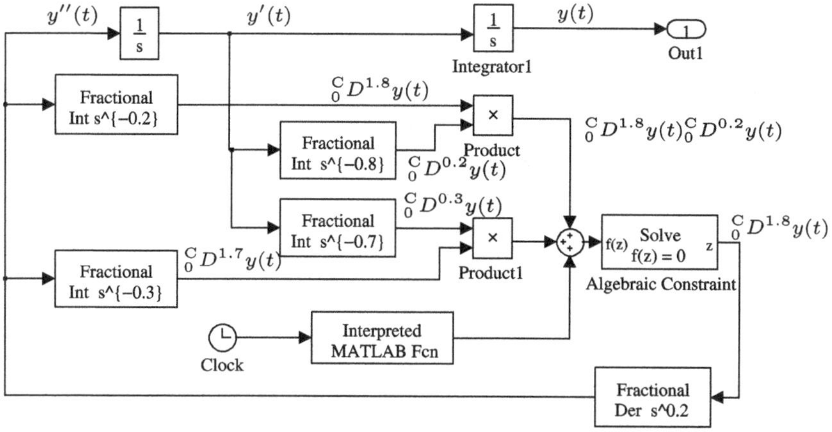

Figure 4: Simulink model of implicit system (c10mimps.slx).

```
>> ww = [1e-5 1e5]; n = 30;
   tic, [t,x,y] = sim('c10mimps'); toc, max(abs(y-exp(-t/2)))
```

4 Fractional-order linear models

In traditional linear control systems, the fundamental models are transfer functions and state space models. The models can be easily be expressed and analyzed by the Control System Toolbox. Similar to the successful and widely accepted tools, a MATLAB Toolbox, FOTF toolbox, is designed. The name for FOTF was chosen quite a few years ago, and its current version exhibits much more than the original name.

Two classes, `fotf` and `foss`, are designed to describe the two kinds of linear fractional-order linear systems, with object oriented programming technique. In this section, a brief description to the two categories of fractional-order models is given.

4.1 Fractional-order transfer functions

A typical single-input single-output fractional-order transfer function with constant delay can be expressed as

$$G(s) = \frac{b_1 s^{\gamma_1} + b_2 s^{\gamma_2} + \cdots + b_m s^{\gamma_m}}{a_1 s^{\eta_1} + a_2 s^{\eta_2} + \cdots + a_{n-1} s^{\eta_{n-1}} + a_n s^{\eta_n}} \mathrm{e}^{-Ts}. \tag{9}$$

It can be seen that the fractional-order transfer function can be described by the ratio of two pseudo-polynomials with a delay constant T.

Compared with the integer-order transfer functions, apart from the numerator and denominator coefficients, the orders should also be declared. Therefore, normally four vectors and a delay constant can be used to describe uniquely the fractional-order transfer function model in (9), simply denoted as $(\mathbf{a}, \boldsymbol{\eta}, \mathbf{b}, \boldsymbol{\gamma}, T)$.

For multivariable fractional-order systems, like in integer-order systems, an FOTF matrix can be considered as the standard model.

Multivariable linear fractional-order systems can be described by an FOTF matrix expressed as

$$\mathbf{G}(s) = \begin{bmatrix} g_{11}(s) & \cdots & g_{1p}(s) \\ \vdots & \ddots & \vdots \\ g_{q1}(s) & \cdots & g_{qp}(s) \end{bmatrix}, \tag{10}$$

where $g_{ij}(s)$ are fractional-order transfer functions defined in (9).

In the `fotf` class, the MATLAB files are all stored in the `@fotf` in the toolbox. The fields in the toolbox are: `den`, `num` for denominator and numerator coefficient vectors $(\mathbf{a},\mathbf{b})$, `nd`, `nn` for the respected orders $(\boldsymbol{\eta},\boldsymbol{\gamma})$, `ioDelay` is the time delay constant T.

The syntaxes to enter an FOTF object in MATLAB is:

(1) Direct input with G=`fotf`$(\mathbf{a},\boldsymbol{\eta},\mathbf{b},\boldsymbol{\gamma},T)$, where the four vectors can be entered into MATLAB first, then use such a command to create a single variable FOTF object G into MATLAB environment. The argument T is optional, and if it is omitted, the delay is set to zero.
(2) Convection, with the command G=`fotf`(a), where, a can be a double precision constant, an SS or a TF object in single-variable or multivariable format, an FOSS object. No matter what is the data type of a, the returned variable G is the corresponding FOTF object.
(3) Laplace operator, with G=`fotf('s')`.

If a single-variable FOTF objects $g_{11}, g_{12}, \dots, g_{nm}$ are specified in MATLAB, the following commands can be used, in a standard matrix input format, to construct a multivariable FOTF matrix $\mathbf{G}$:

$$\mathbf{G}\texttt{=[}g_{11}\texttt{,}\dots\texttt{,}g_{1m}\texttt{;}\ g_{21}\texttt{,}\dots\texttt{,}g_{2m}\texttt{;}\ \dots\texttt{;}\ g_{n1}\texttt{,}\dots\texttt{,}g_{nm}\texttt{]}$$

Example 9. Enter the FOTF model into MATLAB workspace

$$G(s) = \frac{0.8s^{1.2} + 2}{1.1s^{1.8} + 1.9s^{0.5} + 0.4}\,\mathrm{e}^{-0.5s}.$$

Solutions: The following MATLAB commands can be used:

```
>> G = fotf([1.1,1.9,0.4],[1.8,0.5,0],[0.8,2],[1.2,0],0.5)
```

The coefficients, orders, and delay constant of the fractional-order transfer function can be assigned to the relevant fields, such that the FOTF object can be established and displayed.

Example 10. Input the fractional-order transfer function into MATLAB

$$G(s) = \frac{(s^{0.3}+3)^2}{(s^{0.2}+2)(s^{0.4}+4)(s^{0.4}+3)}$$

Solutions: A Laplace operator is declared first, then the following commands can be used to input it into MATLAB environment:

```
>> s = fotf('s'); G = (s^0.3+3)^2/(s^0.2+2)/(s^0.4+4)/(s^0.4+3)
```

Example 11. Input the following multivariable FOTF matrix:

$$\mathbf{G}(s) = \begin{bmatrix} e^{-0.5s}/(1.5s^{1.2}+0.7) & 2e^{-0.2s}/(1.2s^{1.1}+1) \\ 3/(0.7s^{1.3}+1.5) & 2e^{-0.2s}/(1.3s^{1.1}+0.6) \end{bmatrix}.$$

Solutions: The four FOTF blocks can be entered first into the MATLAB workspace, then the normal matrix input statement can be used to enter the multivariable FOTF matrix. Since there is no semicolon at the end of the final statement, the FOTF matrix is displayed directly.

```
>> g1=fotf([1.5 0.7],[1.2 0],1,0,0.5);
   g2=fotf([1.2 1],[1.1 0],2,0,0.2);
   g3=fotf([0.7 1.5],[1.3 0],3,0);
   g4=fotf([1.3 0.6],[1.1 0],2,0,0.2); G = [g1,g2; g3,g4]
```

4.2 Fractional-order state space models

For better describing fractional-order state space models, especially in the case of improper models, that is, the numerator order is higher than the denominator order, a generalized state space model is defined

$$\begin{cases} \mathbf{E}D^{\alpha}\mathbf{x}(t) = \mathbf{A}\mathbf{x}(t) + \mathbf{B}\mathbf{u}(t-T), \\ \mathbf{y}(t) = \mathbf{C}\mathbf{x}(t) + \mathbf{D}\mathbf{u}(t-T), \end{cases} \tag{11}$$

where if the system is improper, the descriptor matrix **E** is singular. For a proper model, according to MATLAB convention, **E** is an empty matrix.

A class for fractional-order state space models can be assigned as FOSS, with a folder `@foss` be created first. The relevant fields for an FOSS class are matrices **A**, **B**, **C**, **D**, and **E**, with the `ioDelay`. Some of the overload functions may also be modified accordingly in folder `foss`.

An FOSS object can be entered into MATLAB with the syntaxes:

(1) Enter directly with G=`foss`($\mathbf{A},\mathbf{B},\mathbf{C},\mathbf{D},\alpha,T,\mathbf{E}$)
(2) Transform to FOSS with G=`foss`(a), where a can be a constant, a TF or an SS model, or an FOTF object. The model described in a can be converted into an FOSS object.

4.3 Modeling under complicated connections

Models in typical connections, including series, parallel and negative feedback connections, can be handled by the redefined operators `*`, `+`, and `feedback()` function directly. Compatible FOTF matrix objects can also be handled with the functions in the FOTF Toolbox. The connected FOTF objects can be handled and simplified easily. Some other supporting functions are also provided to redefine operators "/" and others. Also `simplify()` is provided into FOTF with collected terms.

Example 12. The typical unity negative feedback control system is

$$G(s) = \frac{0.8s^{1.2} + 2}{1.1s^{1.8} + 0.8s^{1.3} + 1.9s^{0.5} + 0.4}, \quad G_c(s) = \frac{1.2s^{0.72} + 1.5s^{0.33}}{3s^{0.8}}.$$

Find the closed-loop model.
Solutions: The following statements can be used to find the closed-loop model:

```
>> G = fotf([1.1,0.8 1.9 0.4],[1.8 1.3 0.5 0],[0.8 2],[1.2 0]);
   Gc = fotf([3],[0.8],[1.2 1.5],[0.72 0.33]);
   G1 = feedback(G*Gc,1)
```

4.4 Analysis of linear fractional-order systems

With the help of FOTF or FOSS classes, some overload functions are implemented, which make linear fractional-order system very easy. In this section, stability analysis, time and frequency domain analysis, and root locus analysis of linear fractional-order systems are performed.
(1) *Stability assessment*. Consider the FOTF model in (9). If the orders are all multiples of a base order α, that is, $\gamma_i = (m-i)\alpha$, $\eta_i = (n-i)\alpha$, let $\lambda = s^\alpha$, the FOTF model can be converted to an integer-order transfer function of λ, such that

$$G(\lambda) = \frac{b_1\lambda^{m-1} + b_2\lambda^{m-2} + \cdots + b_m}{a_1\lambda^{n-1} + a_2\lambda^{n-2} + \cdots + a_{n-1}\lambda + a_n}. \tag{12}$$

The model is also known as the commensurate-order model.

The poles of λ can easily be found with the overload function `eig()`. If all the poles of λ satisfy $|\arg(p_i)| > \alpha\pi/2$, the system is stable. Another function `key=isstable(G)` is provided to assess the stability of an FOTF object, with `key=1` for stable.

Example 13. Check the stability of FOTF model

$$G(s) = \frac{-2s^{0.63} - 4}{2s^{3.501} + 3.8s^{2.42} + 2.6s^{1.798} + 2.5s^{1.31} + 1.5}.$$

Solutions: The model can be entered first, then `isstable()` can be called to assess the stability of the model.

```
>> a = [2,3.8,2.6,2.5,1.5]; na = [3.501,2.42,1.798,1.31,0];
   nb = [0.63,0]; b = [-2,-4]; G = fotf(a,na,b,nb);
   key = isstable(G)
```

Since the returned key is 1, the system is stable. The function is rather time consuming for this example, because the order of the commensurate-order model is as high as 3501.

If the order of the resulted commensurate-order model in (12) is too high, it may not be possible to assess the stability in this way. Alternative methods should be employed.

If all the poles of s in the denominator pseudo-polynomial can be found somehow, they can be used to assess the stability of the system. A MATLAB function more_sols() is provided in FOTF Toolbox, capable of finding all the solutions of a given algebraic equation.

Example 14. Consider the model with irrational orders

$$G(s) = \frac{1}{s^{\sqrt{5}} + 25s^{\sqrt{3}} + 16s^{\sqrt{2}} - Ks^{0.4} + 7}.$$

How can we assess the stability of the system $G(s)$, when $K = 3$? What is the critical value of K such that the system goes unstable?

To find the poles of the system, the following commands can be used, and only two solutions are obtained, with the poles at $s = -0.0812 \pm 0.2880\mathrm{i}$. Since the all the two poles are located on the left-hand side of the s-plane, the system is stable.

```
>> K = 3; f = @(s)s^sqrt(5)+25*s^sqrt(3)+16*s^sqrt(2)-K*s^0.4+7;
   more_sols(f,zeros(1,1,0),100+100i), X
```

Now let us answer the second question. With the same method, it can be found if $K = 10$, the system is unstable. Bisection method can be used to find the critical gain, with the result $K_0 = 7.8492$.

```
>> a = 3; b = 10;
   while (b-a)>0.001, K = 0.5*(a+b);
      f = @(s)s^sqrt(5)+25*s^sqrt(3)+16*s^sqrt(2)-K*s^0.4+7;
      more_sols(f,zeros(1,1,0),100+100i,eps,3);
      if real(X(1))>0, b = K; else, a = K; end
   end, K0 = K
```

(2) *Time domain analysis*. Similar to the time domain response functions provided in the Control System Toolbox. The overload functions are provided in FOTF Toolbox as follows:

- y=step(G,**t**): step response of system G
- y=impulse(G,**t**): impulse response of the system
- y=lsim(G,**u**,**t**): system response driven by **u**.

It should be noted that if **y** appears in the function call, the data for the response is returned, while if there is no returned argument, the plot is drawn automatically. If **t** is not specified in the function call, a default value of 10 is accepted. The three functions are applicable if G is a single variable or multivariable FOTF and FOSS object.

Example 15. Please draw the time domain response of the FOTF model given in Example 13, and $r(t) = \sin t$.
Solutions: The fractional-order transfer function model can be entered first, then the sinusoidal responses can be obtained.

```
>> b = [-2,-4]; a = [2,3.8,2.6,2.5,1.5];
   nb = [0.63,0]; na = [3.501,2.42,1.798,1.31,0];
   G = fotf(a,na,b,nb); t = 0:0.01:30; u = sin(t); lsim(G,u,t);
```

Example 16. Draw the step response plot for the multivariable system shown in Example 11.
Solutions: The multivariable system model can be entered in MATLAB first, then step() function can be evaluated and drawn in Figure 5.

```
>> g1=fotf([1.5 0.7],[1.2 0],1,0,0.5);
   g2=fotf([1.2 1],[1.1 0],2,0,0.2);
   g3=fotf([0.7 1.5],[1.3 0],3,0);
   g4=fotf([1.3 0.6],[1.1 0],2,0,0.2);
   G = [g1,g2; g3,g4]; step(G,10)
```

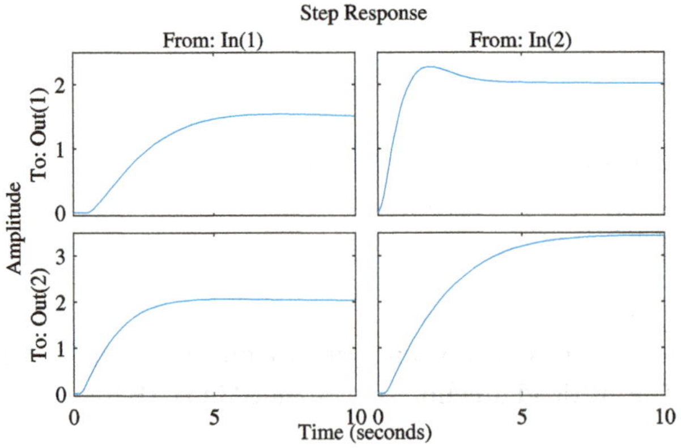

Figure 5: Step responses of multivariable system.

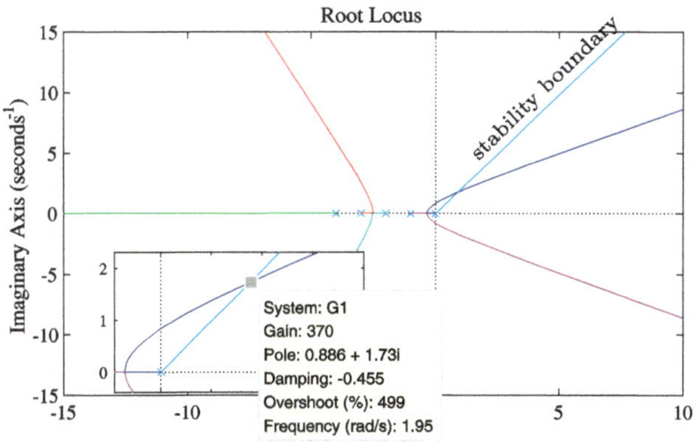

Figure 6: Root locus of the system with critical gain.

(3) *Root locus*. The root locus of a given linear fractional-order object G can be drawn directly with the syntax rlocus(G). To draw root locus of a system, the original FOTF or FOSS object G should be converted first to a commensurate-order one. If the commensurate-order converted is too high, an approximate root locus can be drawn.

Example 17. Assume that the fractional-order transfer function is given by

$$G(s) = \frac{1}{s^{3.5} + 10s^{2.8} + 35s^{2.1} + 50s^{1.4} + 24s^{0.7}},$$

please draw the root locus and find the critical gain.

Solutions: It can be seen that the base order is $\alpha = 0.7$. The root locus of the fractional-order system can be obtained as shown in Figure 6. Zooming the root locus, the critical gain can be read $K = 370$, as shown in the same figure.

```
>> G = fotf([1 10 35 50 24],0.7*[5:-1:1],1,0); rlocus(G)
```

Example 18. Consider the model in Example 13, find the critical gain with root locus method.

Solutions: If $\lambda = s^{0.001}$ is selected, then a 3501th order model can be obtained. It is for sure that the root locus cannot be drawn for such a higher-order system. Approximations should be made first. For instance, by approximating $s^{3.501}$ by $s^{3.5}$, $s^{1.798}$ by $s^{1.8}$, $s^{1.31}$ by $s^{1.3}$, $s^{2.42}$ by $s^{2.4}$, and $s^{0.63}$ by $s^{0.6}$, the root locus can be obtained, and the zoomed root locus is shown, manually, in Figure 7, from which it can be found that the critical gain is $K = 0.320$.

```
>> b = [-2 -4]; nb = [0.6 0]; a = [2 3.8 2.6 2.5 1.5];
   na = [3.5 2.4 1.8 1.3 0]; G1 = fotf(a,na,b,nb); rlocus(G1)
```

Apply a slightly smaller $K = 0.319$ back to the original system, the closed-loop step response can be obtained with oscillation of almost identical magnitudes, indicating the approximate critical gain is correct.

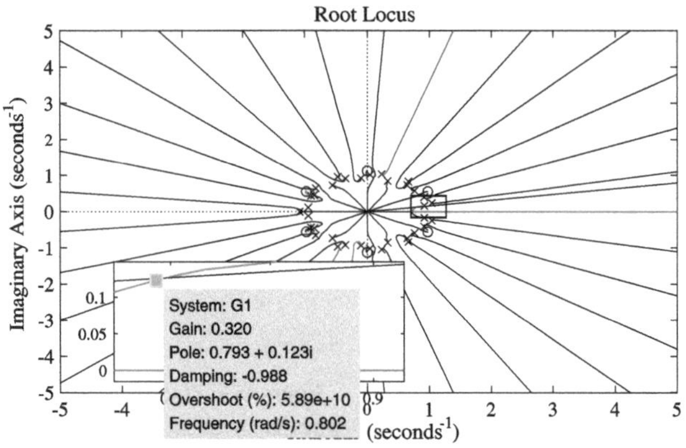

Figure 7: Root locus with critical gain.

```
>> G1.nn = [0.63 0]; G1.nd = [3.501 2.42 1.798 1.31 0];
   K = 0.319; G = feedback(K*G1,1); step(G,200)
```

(4) *Frequency domain analysis*. Consider a fractional-order transfer function $G(s)$. If $j\omega$ is used to substitute s, through simple complex number computation, the exact frequency response data can be obtained directly. The data can be written in the form of the `frd()` function in the Control System Toolbox, so that the frequency domain analysis functions such as `bode()` can be used to draw frequency domain plots. The syntaxes of the functions are

- **H**=`bode(`G`,`**w**`)`: Bode diagram of system G
- **H**=`nyquist(`**G**`,`**w**`)`: Nyquist plot of the system
- **H**=`nichols(`**G**`,`**w**`)`: Nichols chart of the system
- `[`G_1`,`γ`,`ω_1`,`ω_2`]=nichols(`**G**`,`**w**`)`: evaluate the gain margin G_1 and frequency ω_1, and phase margin γ and frequency ω_2.

The returned variable **H** is the frequency response data in the format of the MFD Toolbox [6]. Again, if no arguments are returned in function call, the plots will be drawn automatically.

Example 19. Draw the Bode diagram of the system in Example 13.
Solutions: The fractional-order transfer function model can be entered first, then the Bode diagram of the system can be drawn with the direct used of `bode()` function.

```
>> b = [-2 -4]; nb = [0.6 0]; a = [2 3.8 2.6 2.5 1.5];
   na = [3.5 2.4 1.8 1.3 0];  G = fotf(a,na,b,nb); bode(G)
```

In particular, for multivariable systems, frequency response data can be evaluated with the function `mfrd()`, with the syntax **H**=`mfrd(`**G**`,`**w**`)`, where **G** is the multivariable FOTF or FOSS model, and **w** is the vector of frequency samples. The returned argument **H** is the frequency response data in the MFD Toolbox format. Nyquist plots with Gershgorin bands can be drawn with `gershgorin(H)` function of the FOTF Toolbox. If the Gershgorin bands enclose the origin of the Nyquist plots, the system is said diagonal dominant. If the system is not diagonal dominant, there exist strong interaction, which means that some kind of decoupling needed, before controllers can be designed.

Example 20. Check whether the multivariable system studied in Example 11 is diagonal dominant or not.

Solutions: The multivariable system model can be entered in MATLAB first, then `step()` function can be evaluated and drawn in Figure 8.

```
>> g1=fotf([1.5 0.7],[1.2 0],1,0,0.5);
   g2=fotf([1.2 1],[1.1 0],2,0,0.2);
   g3=fotf([0.7 1.5],[1.3 0],3,0);
   g4=fotf([1.3 0.6],[1.1 0],2,0,0.2); G=[g1,g2; g3,g4];
   w=logspace(-1,1); H=mfrd(G,w); gershgorin(H)
```

(5) *Other system analysis functions.* The norms of the system can be evaluated with `norm(`G`)` and `norm(`G`,inf)`, which returns the $\mathcal{H}_2$ and $\mathcal{H}_\infty$ norms of the system; Function `sigma(`$\mathbf{G}$`)` can be used to draw singular value plot of a multivariable system $\mathbf{G}$; Function `[`α,$\mathbf{r}$,$\mathbf{p}$,K`]=residue(`G`)` can be used to find partial fraction expansion of an FOTF object.

5 Design of fractional-order systems

The controller design facilities are provided also in the FOTF Toolbox. In this section, several controller design methods are presented.

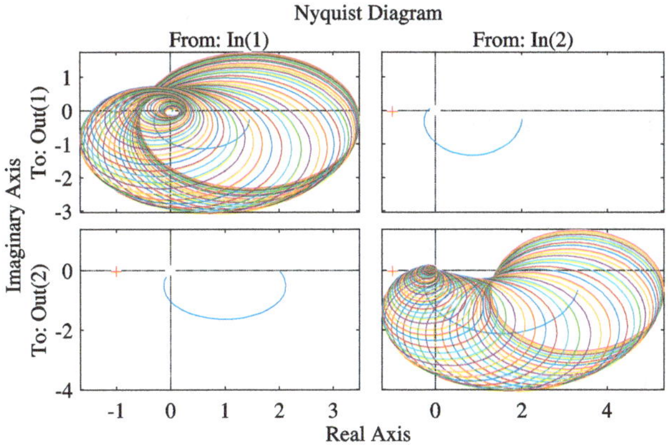

Figure 8: Nyquist plots with Gershgorin bands.

5.1 Fractional-order PID controller design

The typical form of a fractional-order PID controller is

$$G_c(s) = K_p + \frac{K_i}{s^\lambda} + K_d s^\mu. \tag{13}$$

The major difference in the controller and the conventional PID controller is that two extra tuning knobs λ and μ are introduced, such that the controller is more flexible, however, difficulties are increased in tuning the controllers. The fractional-order PID controller is also known as $PI^\lambda D^\mu$controller [10].

An optimal $PI^\lambda D^\mu$controller design function, `fpidtune()`, is proposed in FOTF Toolbox, aiming at designing $PI^\lambda D^\mu$controllers for single variable FOTF plant model. The syntax of the function is G_c=`fpidtune(`$\mathbf{x}_0$`,`$\mathbf{x}_m$`,`$\mathbf{x}_M$`,key)`, where $\mathbf{x}_0$, $\mathbf{x}_m$, $\mathbf{x}_M$ are initial values, lower and upper bounds of the decision variable $\mathbf{x}$ - the parameters of the $PI^\lambda D^\mu$controller. The argument `key` specifies the algorithm used in optimization process, with the default value of 1.

The plant model and other parameters are passed to the function via global variables. Use the command to declare them `global` G $\mathbf{t}$ `key1 key2`, and then set the variables. G is the string expression of the FOTF model, $\mathbf{t}$ is the time vector, `key1` is the type of criterion, with options `'itae'`, `'ise'` `'iae'`, and `'itse'` where `'itae'` is recommended one, while `key2` is the expected controller type, with options `'fpid'` `'fpi'`, `'fpd'`, `'fpidx'` and `'pid'`, with `'fpidx'` for PID^μ controller with integer integral.

Example 21. Consider the plant model

$$G(s) = \frac{1}{0.8s^{2.2} + 0.5s^{0.9} + 1}.$$

Please design an optimum $PI^\lambda D^\mu$ controller for ITAE criterion, and compute the closed-loop step responses.

Solutions: The plant model should first be described, then, for ITAE criterion, the following command can be specified to design optimal $PI^\lambda D^\mu$ controllers:

```
>> clear; global G t key1 key2
   s = fotf('s'); G = 1/(0.8*s^2.2+0.5*s^0.9+1);
   t = 0.01:0.01:20; key2 = 'itae'; key1 = 'fpid';
   xm = zeros(5,1); xM = [20;20;20;2;2];
   x0 = rand(5,1); Gc = fpidtune(x0,xm,xM,1)
```

After the optimization process, an optimum $PI^\lambda D^\mu$ controller is designed

$$G_c(s) = 0.45966 + \frac{0.5761}{s^{0.99627}} + 0.49337s^{1.3792}.$$

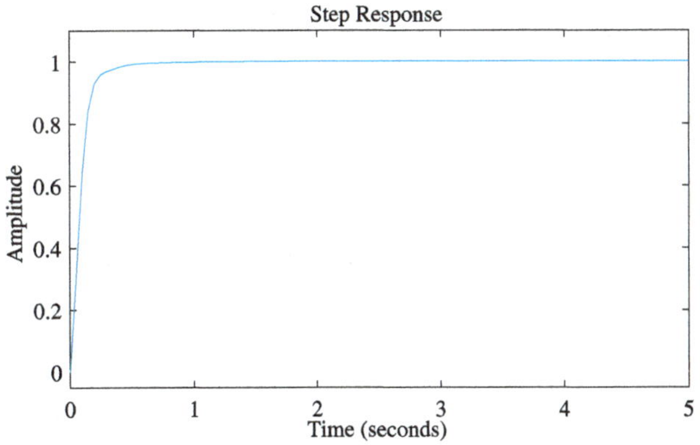

Figure 9: Step response under optimum $PI^{\lambda}D^{\mu}$ controller.

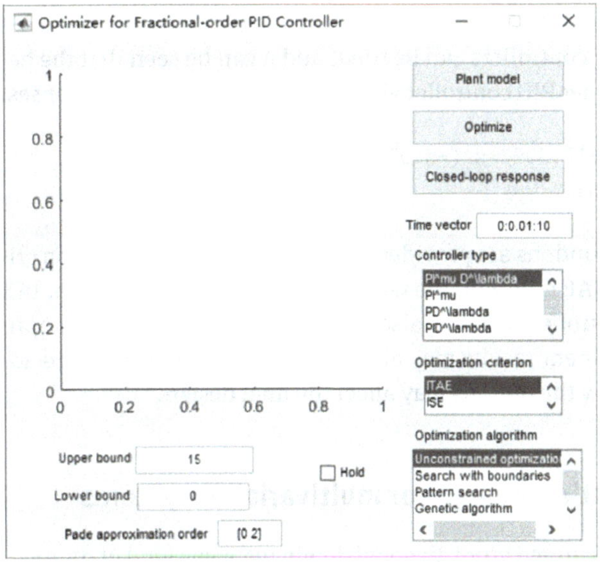

Figure 10: Optimal fractional-order PID controller design interface.

It can be seen that the integral is almost of first order, therefore, a PID^{μ} can also be used, if needed. The closed-loop step response under the controller can be obtained, and the control behavior is shown in Figure 9.

```
>> step(feedback(G*Gc,1),15)
```

5.2 A fractional-order PID controller design interface

A MATLAB interface `optimfopid` [16] is provided as shown in Figure 10. The use of the interface is demonstrated through examples.

Example 22. Consider the plant model in Example 21. Please design optimum fractional-order PID controllers.
Solutions: The model can be input into MATLAB environment as follows:

```
>> G = fotf([0.8 0.5 1],[2.2 0.9 0],1,0)
```

Click "plant model" button in the interface to load the FOTF object, and then specify a time vector. Set then the upper bounds of the controller parameters to 15. It should be noted that the upper bounds of controller parameters may sometimes affect the final search results.

Click the "Optimize" button to initiate the optimization process, and the optimal fractional-order controller can be obtained, and for this example, the optimal vector is $\mathbf{x} = [4.1557, 12.3992, 8.8495, 0.9794, 1.1449]$. The step response of the closed loop under fractional-order PID controller can be obtained and is very close to the result obtained previously. Different controllers can be tried, and it can be seen that the behavior under the fractional-order PID controller yields the best closed-loop responses.

```
>> s=fotf('s'); Gc=4.1557+12.3992*s^(-0.9794)+8.8495*s^1.1449;
   step(feedback(Gc*G,1),5)
```

In the interface, different options are provided. For instance, the optimization criteria can be assigned to the ITAE criterion (the recommended one), ISE criterion, IAE criterion, and so on; PID controller type has also many options; with different optimization algorithms, and different results may also be obtained. The readers are advised to try for themselves how the options may affect the final design.

5.3 Pseudodiagonalization design for multivariable systems

If the transfer function matrix studied is not diagonal dominant, some kind of compensation methods should be introduced, so that it can be converted to diagonal dominant matrices. Then individual loop single-variable design method can be used, regardless of the coupling. The typical block diagram of Nyquist-type methods is shown in Figure 11, where $\mathbf{K}_\mathrm{p}(s)$ is the precompensating matrix such that $\mathbf{G}(s)\mathbf{K}_\mathrm{p}(s)$ is a diagonal dominant matrix. Matrix $\mathbf{K}_\mathrm{d}(s)$ can be used to introduce further dynamic compensation to diagonal dominant matrices.

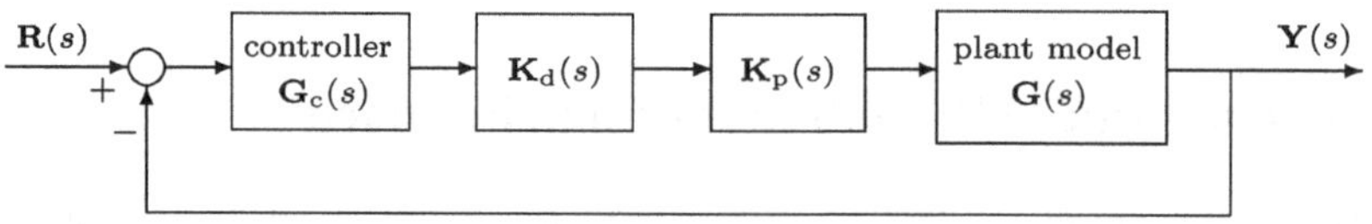

Figure 11: Typical block diagram of multivariable systems.

The pseudodiagonalization method is presented to select the precomposition matrix $\mathbf{K}_p$ [4]. Assume that at frequency $j\omega_0$, the compensation matrix can be design with $\mathbf{K}_p$=`pseudiag(`$\mathbf{H}$`)`, where $\mathbf{H}$ is the samples of frequency response of the plant $\mathbf{G}$. An example is presented to design controllers with the pseudodiagonalization method.

Example 23. For the multivariable plant,

$$\mathbf{G}(s) = \begin{bmatrix} 1/(1.35s^{1.2} + 2.3s^{0.9} + 1) & 2/(4.13s^{0.7} + 1) \\ 1/(0.52s^{1.5} + 2.03s^{0.7} + 1) & -1/(3.8s^{0.8} + 1) \end{bmatrix}$$

Use the pseudodiagonalization method to design a controller and study the behaviors of the matrix.

Solutions: The plant model can be entered first, and with the following statements, the Nyquist plot with Gershgorin bands can be obtained. It can be seen that the plant is not diagonal dominant. in fact, there exists strong coupling in the system.

```
>> s = fotf('s'); g1 = 1/(1.35*s^1.2+2.3*s^0.9+1);
   g2 = 2/(4.13*s^0.7+1); g3 = 1/(0.52*s^1.5+2.03*s^0.7+1);
   g4 = -1/(3.8*s^0.8+1); G = [g1,g2; g3,g4];
   w = logspace(0,1); H = mfrd(G,w); gershgorin(H)
```

Selecting a frequency vector in the range of $(1, 10)$ rad/s, the frequency response of the plant can be obtained. Based on the response, a pre-compensation matrix $\mathbf{K}_p$ can be designed

```
>> w = logspace(0,1); H = mfrd(G,w); Kp = pseudiag(H),
   w1 = logspace(-1,2); H1 = mfrd(G*Kp,w1); gershgorin(H1)
```

The compensation behavior under such a matrix is obtained. It can be seen that by applying such a precompensation matrix, the compensated system is diagonal dominant.

It can be seen that the compensated system is diagonal dominant, however, if open-loop step response of the compensated system is to be drawn, it can be seen that the step responses of the diagonal elements are negative, due to the shapes and directions of the diagonal Nyquist plots. Therefore, dynamic matrix with negative gain should be introduced, for instance,

$$\mathbf{K}_d(s) = \begin{bmatrix} -1/(2.5s + 1) & 0 \\ 0 & -1/(s + 1) \end{bmatrix}.$$

The Nyquist plot with Gershgorin bands under the dynamic matrix is obtained, and the diagonal dominance is further enhanced. This can further be verified by the open-loop step responses.

```
>> s=tf('s'); Kd=[-1/(2.5*s+1), 0; 0, -1/(s+1)];
   G0=G*Kp*Kd; H3=mfrd(G0,w1); gershgorin(H3)
```

Now, two $PI^\lambda D^\mu$ controllers can be designed individually for the two input-output pairs, with the following commands:

```
>> G = G0(1,1); x0 = rand(5,1); global G type key t;
   t = 0:0.02:10; xm = [0 0 0 0 0]; xM = [15 15 15 2 2];
   type = 'fpid'; key = 'itae'; [Gc,x] = fpidtune(x0,xm,xM,1)
```

With the above commands, after certain waiting time, an optimal $PI^\lambda D^\mu$ controller $c_1(s)$ can be designed. Similarly, when $G_0(2,2)$ is used as the equivalent plant model, the above statements may yield another optimal $PI^\lambda D^\mu$controller $c_2(s)$. The two optimal $PI^\lambda D^\mu$ controllers are respectively

$$
\begin{aligned}
c_1(s) &= 10.7003 + 2.9743s^{-0.86736} + 15s^{0.7876},\\
c_2(s) &= 14.848 + 10.1421s^{-0.81932} + 14.6848s^{0.7355}.
\end{aligned}
$$

In order to check the closed-loop behaviors of the two controllers, a Simulink model is established as shown in Figure 12.

The closed-loop step responses can be evaluated with the following statements, and the responses are obtained as shown in Figure 13, and it can be seen that the results are satisfactory. The fractional blocks used in the Simulink block diagram are the blocks provided in the FOTF blockset.

```
>> G=[g1 g2; g3 g4]; s=fotf('s');
   c1=10.7003+2.9743*s^-0.86736+15*s^0.7876;
   c2=14.848+10.1421*s^-0.81932+14.6848*s^0.7355;
   u1=1; u2=0; [t1,~,y1]=sim('c10mpdm2');
   u1=0; u2=1; [t2,~,y2]=sim('c10mpdm2');
   subplot(221), plot(t1,y1(:,1)), ylim([-0.1 1.1])
   subplot(223), plot(t1,y1(:,2)), ylim([-0.1 1.1])
```

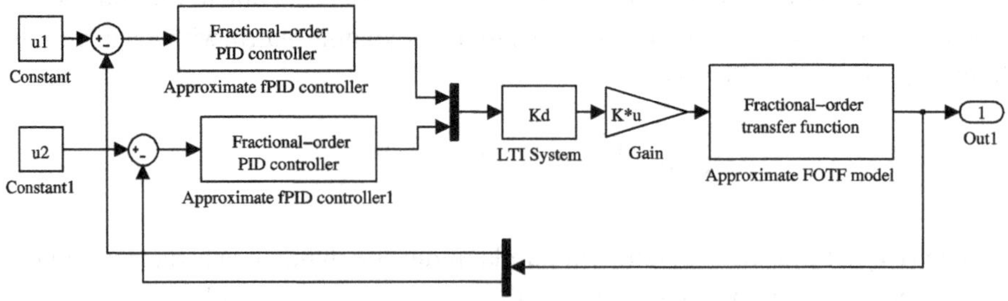

Figure 12: Simulink model of multivariable system (c10mpdm2.slx).

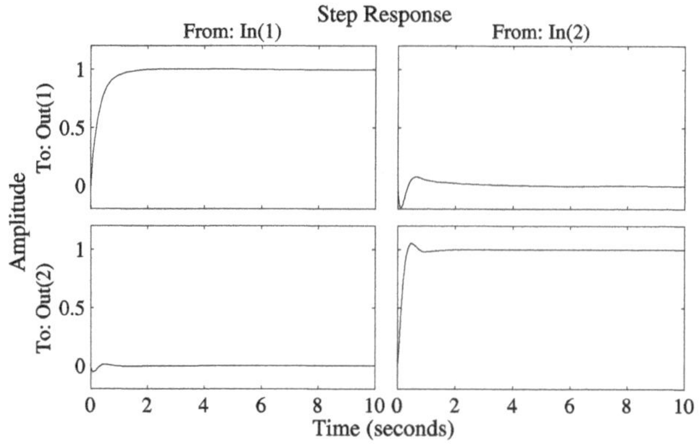

Figure 13: Closed-loop step response of the multivariable system.

```
subplot(222), plot(t2,y2(:,1)), ylim([-0.1 1.1])
subplot(224), plot(t2,y2(:,2)), ylim([-0.1 1.1])
```

5.4 Parameter optimization design for MIMO systems

A parameter optimization algorithm was proposed to design controllers for multivariable systems in [3]. The algorithm can be applied directly to design integer-order controllers. In this section, we shall apply the algorithm to multivariable fractional-order plant models. It will be shown that the integer-order controllers thus designed are robust, therefore, it may not be necessary to design fractional-order controllers.

MFD Toolbox provides a function `fedmunds()` to implement the parameter optimization algorithm. The traditional algorithm is extended, since the common denominator $d(s)$ is no longer needed. Each of the components in a controller matrix can be set independently.

Selecting a target closed-loop system $\mathbf{T}_t(s)$, the target controller $\mathbf{K}_t(s)$ can be obtained from

$$\mathbf{K}_t(s) = \mathbf{G}^{-1}(s)\mathbf{T}_t(s)[\mathbf{I} - \mathbf{T}_t(s)]^{-1}, \tag{14}$$

from which, the pole positions can be read out. Optimization should then be made to find the zeros of the controllers.

The syntax of the function is $\mathbf{N} =$ `fedmunds`$(\mathbf{w}, \mathbf{G}_\omega, \mathbf{T}_\omega, \mathbf{N}_0, \mathbf{D})$, where $\mathbf{w}$ is a vector of selected frequencies, $\mathbf{G}_\omega$ and $\mathbf{T}_\omega$ are the frequency responses of the plant $\mathbf{G}(s)$ and target system $\mathbf{T}_t(s)$, respectively. $\mathbf{D}$ is the polynomial matrix of the denominator, while $\mathbf{N}_0$ represents the structure of the polynomial matrix of numerator. If a component in matrix $\mathbf{N}_0$ is zero, it means that this component needs not be optimized, thus the whole parameter optimization process is simplified. Matrix $\mathbf{N}$ returns the numerator coefficients optimized, as will be demonstrated in the next example.

Example 24. If there exist time delays in the multivariable system studied in the previous example, where

$$\mathbf{G}(s) = \begin{bmatrix} e^{-0.2s}/(1.35s^{1.2} + 2.3s^{0.9} + 1) & 2e^{-0.2s}/(4.13s^{0.7} + 1) \\ 1/(0.52s^{1.5} + 2.03s^{0.7} + 1) & -e^{-0.5s}/(3.8s^{0.8} + 1) \end{bmatrix},$$

please design a multivariable controller with parameter optimization approach and observe the results.

Solutions: A target closed-loop system should be selected as a diagonal matrix, whose diagonal elements should be selected as integer-order transfer functions with good behavior. For instance, the following target model can be selected:

$$\mathbf{T}_t(s) = \begin{bmatrix} 9/(s+3)^2 & 0 \\ 0 & 100/(s+10)^2 \end{bmatrix}.$$

The Bode magnitude plots of the target controller can be obtained with the following statements, as shown in Figure 14.

```
>> s = fotf('s'); g1 = 1/(1.35*s^1.2+2.3*s^0.9+1); g1.ioDelay = 0.2;
   g2 = 2/(4.13*s^0.7+1); g2.ioDelay = 0.2;
   g4 = -1/(3.8*s^0.8+1); g4.ioDelay = 0.5;
   g3 = 1/(0.52*s^1.5+2.03*s^0.7+1); G = [g1,g2; g3,g4];
   s = tf('s'); T = [9/(s+3)^2, 0; 0, 100/(s+10)^2];
   w=logspace(-2,3); Gw=mfrd(G,w); Tw=mfrd(fotf(T),w); I=eye(2);
   h1=finv(w,fadd(w,-Tw,I)); h2=finv(w,Gw); h3=fmulf(w,h2,Tw);
   Kt = fmulf(w,h3,h1); H = mfd2frd(Kt,w); bodemag(H)
```

It can be seen that the Bode magnitude plots all start with decreasing segments; there should be a pole at $s = 0$. Let us see the $(1,1)$ subplot, the second decreasing turning point is about $s = -2$. In fact, in the other subplots, the poles can also be fixed at $s = -2$.

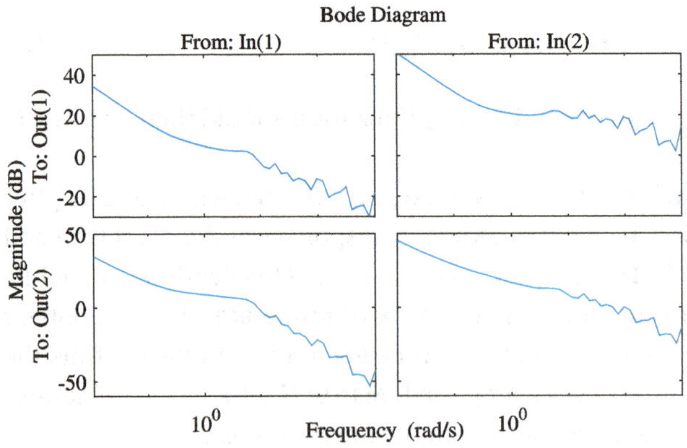

Figure 14: Target Bode magnitude plots.

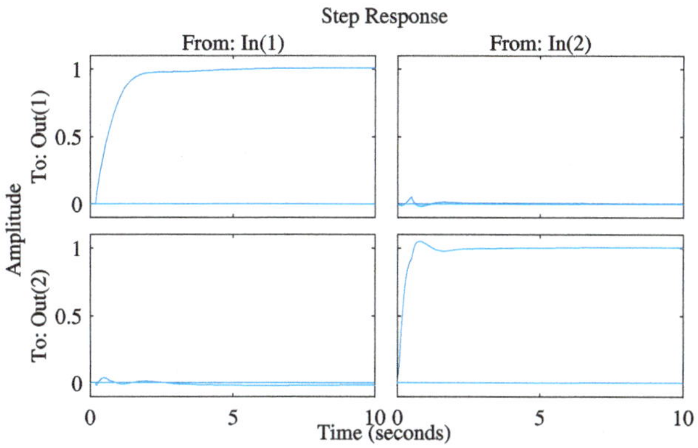

Figure 15: Closed-loop step responses.

Assign the elements in the two matrices

$$\mathbf{N}_0 = \begin{bmatrix} 1 & 1 & 1 & 1 & 1 & 1 \\ 1 & 1 & 1 & 1 & 1 & 1 \end{bmatrix}, \quad \mathbf{D} = \begin{bmatrix} 1 & 2 & 0 & 1 & 2 & 0 \\ 1 & 2 & 0 & 1 & 2 & 0 \end{bmatrix}.$$

The controller can be designed with the following statements:

```
>> N0=[1,1,1, 1,1,1; 1,1,1, 1,1,1]; D=[1,2,0,1,2,0; 1,2,0,1,2,0];
   N=fedmunds(w,Gw,Tw,N0,D), k11=tf(N(1,1:3),D(1,1:3));
   k12=tf(N(1,4:6),D(1,4:6)); k21=tf(N(2,1:3),D(2,1:3));
   k22=tf(N(2,4:6),D(2,4:6)); K=[k11 k12; k21 k22]; zpk(K)
   u1=1; u2=0; [t1,x,y1]=sim('c10mpopt',10);
   subplot(221), plot(t1,y1(:,1)), ylim([-0.1 1.1])
   subplot(223), plot(t1,y1(:,2)), ylim([-0.1 1.1])
   u1=0; u2=1; [t2,x,y2]=sim('c10mpopt',10);
   subplot(222), plot(t2,y2(:,1)), ylim([-0.1 1.1])
   subplot(224), plot(t2,y2(:,2)), ylim([-0.1 1.1])
```

The controller can be designed as

$$\mathbf{K}(s) = \frac{1}{s(s+2)} \begin{bmatrix} -0.57479(s-5.84)(s+0.4024) & 1.5644(s+3.609)(s+0.1528) \\ 9.3963(s+2.513)(s+0.2727) & -1.6243(s+9.073)(s+0.3796) \end{bmatrix},$$

and the closed-loop step response of the system can be obtained as shown in Figure 15. It can be seen that the control result is satisfactory.

6 Conclusions

In this chapter, the major functions in the FOTF Toolbox are demonstrated through examples. Numerical solutions to fractional calculus and fractional-order differential equation problems can be obtained with the high precision solvers. Also modeling, analysis and design facilities for linear fractional-order systems can be carried out through the overload functions in the toolbox, and it can be seen that the analysis and design tasks can be made as easy as the functions in the Control System Toolbox.

Bibliography

[1] Y. Q. Chen, I. Petráš, and D. Xue, Fractional control—a tutorial, in *Proceedings of the American Control Conference*, pp. 1397–1410, 2009.
[2] K. Diethelm, *The Analysis of Fractional Differential Equations: an Application-Oriented Exposition Using Differential Operators of Caputo Type*, Springer, New York, 2010.
[3] J. M. Edmunds, Control system design and analysis using closed-loop Nyquist and bode arrays, *International Journal of Control*, **30**(5) (1979), 773–802.
[4] D. J. Hawkins, Pseudodiagonalisation and the inverse Nyquist array method, *Proceedings of IEE, Part D*, **119**(3) (1972), 337–342.
[5] Z. Li, L. Liu, S. Dehghan, Y. Q. Chen, and D. Y. Xue, A review and evaluation of numerical tools for fractional calculus and fractional order control, *International Journal of Control*, **90**(6) (2015), 1165–1181.
[6] J. M. Maciejowski, *Multivariable Feedback Design*, Addison-Wesley, Wokingham, England, 1989.
[7] A. Oustaloup, *La Commande CRONE*, Hermès, Paris, 1991.
[8] A. Oustaloup, F. Levron, F. Nanot, and B. Mathieu, Frequency band complex non integer differentiator: characterization and synthesis, *IEEE Transactions on Circuits and Systems. I, Fundamental Theory and Applications*, **47**(1) (2000), 25–40.
[9] I. Podlubny, *Fractional Differential Equations*, Academic Press, San Diego, 1999.
[10] I. Podlubny, Fractional-order systems and $PI^{\lambda}D^{\mu}$-controllers, *IEEE Transactions on Automatic Control*, **44**(1) (1999), 208–214.
[11] A. Tepljakov, *Fractional-Order Calculus Based Identification and Control of Linear Dynamic Systems*, Master thesis, Tallinn, University of Technology, 2011.
[12] D. Valério, *Ninteger v. 2.3 Fractional Control Toolbox for MATLAB*, Universudade Téchica de Lisboa, 2006.
[13] D. Y. Xue, *Fractional-Order Control Systems—Fundamentals and Numerical Implementations*, de Gruyter, Berlin, 2017.
[14] D. Y. Xue, *Revisit to Calculus Problems*, Tsinghua University Press, Beijing, 2019, (in Chinese) to appear.
[15] D. Y. Xue and L. Bai, Benchmark problems for Caputo fractional-order ordinary differential equations, *Fractional Calculus & Applied Analysis. An International Journal for Theory and Applications*, **20**(5) (2017), 1305–1312.
[16] D. Y. Xue and Y. Q. Chen, Optimfopid: a MATLAB interface for optimum fractional-order pid controller design for linear fractional-order plants, in *Proceedings of Fractional Derivatives and Its Applications*, Nanjing, China, p. 307, 2012.

Paolo Lino and Guido Maione

Fractional-order controllers for mechatronics and automotive applications

Abstract: This chapter proposes an approach to design fractional-order controllers for systems that are very common in applications. Namely, in many cases, simple models of the controlled systems are used. In particular, a first-order lag plus time delay system is a typically employed model. Then fractional-order controllers can be employed to improve the tradeoff between robustness and performance of the considered control loop. In particular, the controller structure is based on a noninteger-order integration that replaces the classical integer-order one in PI/PID controllers. In this way, control design can take advantage of the order of integration, which is a noninteger number, and develop relatively easy-to-use rules to set the other controller parameters. This chapter surveys a frequency-domain, loop-shaping design approach that allows to set the controller and meet desired robustness and performance specifications. The settings directly relate specifications to the parameters. Moreover, the design is completed by a realization procedure that easily determines the rational transfer function necessary to approximate the irrational compensator. The characteristics of the proposed realization technique is guaranteeing not only stability and minimum-phase properties of the controller but also the interlacing between zeros and poles of the approximating function. The approach is tested on two case-study systems: a DC-servomotor very useful in mechatronics and the injection system of a compressed natural gas engine developed in industry.

Keywords: Fractional-order controllers, fractional-order PI controllers, loop-shaping, DC-servomotors, injection systems of compressed natural gas engines

MSC 2010: 34H05, 93C80, 93C95, 30B70, 41A20

1 Introduction

In the last few decades, noninteger-order controllers, also called Fractional-Order Controllers (FOC), have received an increasing and great attention. The motivation lies in their capacity of achieving high robustness of the control loops with respect to gain and parameter variations of the plant. The misname FOC actually indicates controllers in which an integral or derivative action is of noninteger (fractional) order

Paolo Lino, Guido Maione, Department of Electrical and Information Engineering, Polytechnic University of Bari, Via E. Orabona 4, 70125 Bari, Italy, e-mails: paolo.lino@poliba.it, guido.maione@poliba.it, https://orcid.org/0000-0003-3583-7962, https://orcid.org/0000-0003-0993-8030

https://doi.org/10.1515/9783110571745-012

as opposed to the standard action of integer order that is used in PID controllers. However, despite a plethora of FOC design strategies and tuning methods, until now the literature has offered few methods that can be accepted and widely employed by practitioners. This makes the benefits of FOC still far to be acknowledged and exploited in real applications that usually rely on PID controllers. So systematic and simple rules can help a lot.

In this chapter, the aim is to provide a repeatable method that applies open-loop shaping ideas for designing noninteger-order PI controllers for a specific class of plants, that is, first-order lag systems with or without time delay. Simplicity and satisfaction of requirements are remarkable characteristics of the method. However, it must be completed by a second stage in which irrational differential or integral operators involved in the FOC are approximated by rational transfer functions. The approximation techniques lead to functions enjoying stability, minimum-phase, and zero-pole interlacing properties. As test cases, the described design techniques are applied to two mechatronics and automotive systems. Before going into details, a short overview of the main ideas about fractional-order control is given.

Fractional Calculus (FC) is an old topic with a strong mathematical foundation. It dates back to a famous correspondence between Leibniz and de L'Hôpital, who queried about the meaning of a 1/2-derivative. Leibniz answered that *"it will lead to an apparent paradox, from which one day useful consequences will be drawn."* Later on, Heaviside also said that *"there is a universe of mathematics lying in between the complete differentiations and integrations."* And many other scientists contributed to the field. In control engineering, it is important to profitably use a differential or integral operator $s^{\pm\nu}$. Some difficulties arise because the operator is irrational, but there exist numerous methods to achieve approximations in sufficiently large frequency ranges.

The question is why using FC for control?

In recent years, many researches applied FC for systems modeling and control. As regards modeling, it is widely recognized that FC provides a better tool than integer-order calculus (IC) to represent complex phenomena and dynamical systems for which IC is not sufficient. See, for example, nonexponential models to represent anomalous relaxation behaviors [10]. In general, FC has the ability of better describing real dynamic systems, so that characteristics and behaviors are fractional more than integer in nature [7, 41].

For control, the origin is probably the seminal idea by Bode that a noninteger-order integrator is the ideal open-loop transfer function for feedback amplifiers [4]. Other pioneering works exist, for example, for controlling the dynamics of flexible spacecraft structures [28] or the position of massive objects [42]. In particular, Tustin approximated a fractional integrator of order $\nu = 0.5$ and achieved a constant phase margin of 45° over a wide frequency range around the gain crossover ω_{gc}, that is, between $0.2\,\omega_{gc}$ and $1.4\,\omega_{gc}$. Many other applications were developed more recently. Probably, the most successful are in controlling automotive car suspensions by the CRONE (Commande Robuste d'Ordre Non Entiér) control approach [32].

In synthesis, the success and growing interest of FC can be synthetically explained by the following reasons:

- From the modeling point of view, noninteger/fractional-order systems are very powerful mathematical tools for representing: long-term memory properties of certain phenomena; fractional dynamics of systems that cannot be described by integer-order differential equations; nonlinear chaotic dynamics of structures and circuits; visco-elasticity in some mechanical structures; diffusion and transport phenomena; behavior of neurons; etc.
- From the control point of view, linear FOC are conceived as a generalization and more flexible version of classical PID controllers, because the integral and derivative actions are of noninteger/fractional order; then the notations FOC, Fractional-Order PID (FOPID) controllers, and $PI^{\lambda}D^{\mu}$-controllers, where λ and μ are the noninteger orders of integration and differentiation, respectively [34]. With respect to PID or more complex integer-order controllers, FOC provide benefits in terms of increased closed-loop performance and enhanced robustness, both when the controlled plant is modeled as an integer-order system and especially when the plant is described as a noninteger-order system, in its turn [5, 35].

Although the parameters λ and μ give more design degrees of freedom, there are not widely accepted methods for designing or tuning FOC. There are no rules standing comparison with the popular Ziegler–Nichols settings or other well-established rules for PID. Therefore, if one thinks about all industrial loops in which PID are dominant [1] and the promising mechatronics/automotive applications, there are many open research issues and many potentialities to increase the impact of FOC. In particular, the authors worked on new design procedures and more efficient techniques to approximate the irrational Laplace operators that concur to define FOC [15–17, 22–25, 27].

Here, the authors propose a loop-shaping approach for designing fractional-order PI (for brevity FOPI) controllers that can be used for a wide class of plants, namely those described by a first-order lag plus time delay system. In particular, a mechatronic system and an automotive application are considered. In the first case, a typical DC-servomotor is controlled to achieve desired speed. In the second case, a Compressed Natural Gas (CNG) engine is examined. The developed rules to set controller parameters guarantee a good tracking performance in a significant frequency range and robust stability to parameter changes, with an almost constant phase margin in a sufficiently wide range around the gain crossover frequency. These requirements are satisfied by appropriately shaping the open-loop frequency response.

Section 2 describes the loop-shaping design method developed for the FOPI controllers. Then FOPI can be compared to integer-order PI controllers. The last are instead tuned by other techniques that are typically employed in the considered applications. Since the FOPI controllers are based on irrational Laplace operators, they require a rational approximation. Section 3 describes some approaches based on Continued Fraction Expansions (CFEs) to approximate the FOPI function [21, 24]. In this context, an important result (see Subsection 3.3) proves how to formally guarantee stability and minimum-phase properties for certain classes of approximations [23]. Section 4 shows the application of the method to a mechatronic system based on a DC-servomotor. Section 5 shows the application to an automotive system. Section 6 provides some final remarks.

2 Design of FOPI controllers by loop-shaping

The method here described modifies what has been proposed by [17], and improves and extends preliminary results that were provided by [26]. Let us refer to a classical unitary feedback control system (see Figure 1), where the plant is a first-order system

$$G_p(s) = \frac{K_E}{1 + T_E\, s} \tag{1}$$

and the fractional-order controller takes the form

$$G_c(s) = K_P + \frac{K_I}{s^\nu} = \frac{K_I}{s^\nu}\,(1 + T_C\, s^\nu) \tag{2}$$

where $T_C = K_P/K_I$ is the controller integral time constant and $1 < \nu < 2$ is the non-integer order of integration. The specified range for values of ν allows to include an integer-order integrator that is necessary to obtain zero steady-state error and reject plant disturbances. Hence, the open-loop frequency response associated with $G(s) = G_c(s)\, G_p(s)$ is

$$G(j\omega) = \frac{K_E\, K_I\, \{1 + \omega^\nu\, T_C\, [\cos(\hat{\theta}) + j\sin(\hat{\theta})]\}}{\omega^\nu\, [\cos(\hat{\theta}) + j\sin(\hat{\theta})]\, (1 + j\,\omega\, T_E)} \tag{3}$$

where $\hat{\theta} = \nu\,\pi/2$. Moreover, introducing the nondimensional frequency $u = \omega\, T_E$ leads to

$$G(ju) = \frac{K_E\, K_I\, \{1 + (\frac{u}{T_E})^\nu\, T_C\, [\cos(\hat{\theta}) + j\sin(\hat{\theta})]\}}{(\frac{u}{T_E})^\nu\, [\cos(\hat{\theta}) + j\sin(\hat{\theta})]\, (1 + ju)}. \tag{4}$$

Since the closed-loop transfer function is

$$F(ju) = \frac{1}{1 + G^{-1}(ju)} \tag{5}$$

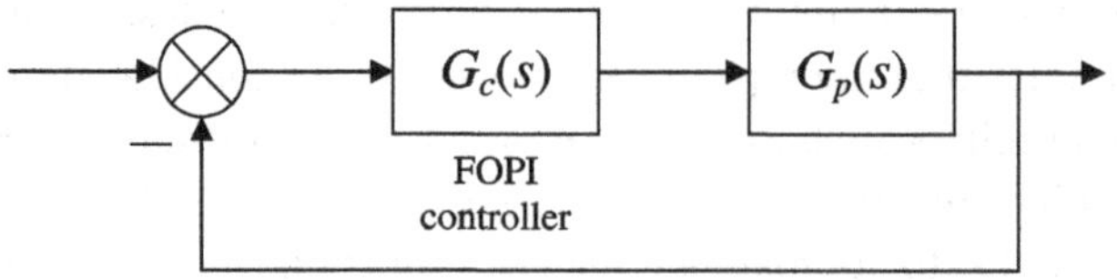

Figure 1: Block diagram with a FOPI controller.

applying the requirement $|F(ju)| \equiv 1$ for all frequencies u states a perfect input-output tracking. Namely, as R. E. Kalman said: *"A feedback system is optimal if and only if the absolute value of the return difference is at least one at all frequencies"* [13]. Of course, physical systems can only approximate this condition, in a limited bandwidth that is specified by $u_B = \omega_B T_E$. For a good tracking performance, $|F(ju)| \approx 1$ should hold. This condition is transformed into an equivalent constraint on the open-loop gain, that is, $|G^{-1}(ju)|^2 \approx 1$.

To obtain a stable performance despite changes in parameters, stability margins are ensured by appropriately shaping. the open-loop frequency response around the crossover frequency. To this aim, the fractional integrator is profitably used. It shows a "flat" phase diagram in a sufficiently wide frequency interval and a magnitude diagram with fractional slope of -20ν dB/decade. More specifically, the values of T_C and ν are selected to obtain a desired and specified phase margin PM_s, which is held nearly constant in a sufficiently wide interval around the 0-dB crossover frequency $u_C = \omega_C T_E$.

The tuning procedure first considers the demand of a bandwidth u_B ensuring a good tracking response of the system. By u_B, a good estimation of the 0-dB crossover frequency u_C is a value belonging to the range $[\frac{u_B}{1.7}, \frac{u_B}{1.3}]$ (see [18, 19]). Hence, assuming, for instance, $u_C = \frac{u_B}{1.7}$, gives the value of the frequency around which to achieve the specified phase margin, say PM_s, in a sufficiently wide range. The phase of the open-loop transfer function (4) is

$$\angle G(ju) = \tan^{-1}\left(\frac{(\frac{u}{T_E})^\nu T_C S}{1 + (\frac{u}{T_E})^\nu T_C C}\right) - \tan^{-1}(u) - \hat{\theta} \tag{6}$$

with $S = \sin(\hat{\theta})$ and $C = \cos(\hat{\theta})$. Then, by using $u = u_C$, the definition of phase margin gives

$$PM = \pi + \angle G(ju_C) = \varphi_1(u_C) - \varphi_2(u_C) + (2 - \nu)\pi/2 \tag{7}$$

where $\varphi_1(u)$ and $\varphi_2(u)$ are the first and second arguments in (6), respectively. Now, by putting $\varphi_1(u_C) = \varphi_2(u_C)$, it holds

$$PM = PM_s = (2 - \nu)\pi/2 \Leftrightarrow \nu = 2 - PM_s/(\pi/2). \tag{8}$$

If there exists an appropriate parameter value $T_C = \overline{T}_C$ leading to $\varphi_1(u_C) = \varphi_2(u_C)$, then (8) directly relates the fractional order, ν, and the requested phase margin, PM_s. One degree of freedom is lost to obtain $\varphi_1(u_C) = \varphi_2(u_C)$; however, an easy-to-use relation between ν and PM_s can be used: the required ν to get the specified margin PM_s is given by the right side of (8), or the achieved phase margin when fixing ν is given by the left side of (8). For $T_C = \overline{T}_C$, it holds

$$u_C = \frac{(\frac{u_C}{T_E})^\nu \overline{T}_C S}{1 + (\frac{u_C}{T_E})^\nu \overline{T}_C C}. \tag{9}$$

Equation (9) yields

$$\overline{T}_C = \frac{u_C}{(\frac{u_C}{T_E})^\nu (S - u_C C)}. \tag{10}$$

Now, as an example, by putting $u_C = \frac{u_B}{1.7}$ in (10), $\overline{T}_C$ can be expressed in terms of u_B or u_C and in terms of ν as follows:

$$\overline{T}_C = \overline{a}\, u_B^{(1-\nu)}\, T_E^\nu = \overline{b}\, u_C^{(1-\nu)}\, T_E^\nu \tag{11}$$

where

$$\overline{a} = \frac{(1.7)^\nu}{1.7\, S - u_B\, C} \quad \text{and} \quad \overline{b} = \frac{1}{S - u_C\, C}. \tag{12}$$

Obviously, if a different relation between u_C and u_B is used, then a number different from 1.7 appears in (12).

Note that $\overline{a} > 0$ and $\overline{b} > 0$ must hold true to guarantee $\overline{T}_C > 0$. Therefore, values of u_B and ν are limited by this constraint.

At this point, to determine K_I, consider the open-loop transfer function for $u = u_C$. The definition of gain crossover frequency gives $G(ju_C) = 1\, e^{-j\varphi_C}$, where $\varphi_C = \angle G(ju_C)$. Hence, $|G^{-1}(ju_C)|^2 = 1$. Using $u = \omega\, T_E$ yields

$$\left|G^{-1}(ju_C)\right|^2 = A(u_C)\, B(u_C) \tag{13}$$

with

$$A(u_C) = \frac{1}{K_E^2 K_I^2} \left(\frac{u_C}{T_E}\right)^{2\nu} \tag{14}$$

and

$$B(u_C) = \frac{1 + u_C^2}{1 + \overline{b}^2\, u_C^2 + 2\overline{b}\, u_C\, C}. \tag{15}$$

Therefore, the obvious equality $A(u_C)\, B(u_C) = 1$ leads to $K_I = \overline{K}_I$ with

$$\overline{K}_I = \frac{1}{K_E} \left(\frac{u_C}{T_E}\right)^\nu \sqrt{B(u_C)} = \frac{1}{K_E} \left(\frac{u_C}{T_E}\right)^\nu \frac{S - u_C\, C}{S} \tag{16}$$

where the rightmost expression can be obtained after simple algebra.

As an example, Table 1 shows the values of the FOPI controller parameters for a specific plant, for a certain bandwidth specification and for different desired phase margins. If a too low phase margin is chosen, the system is not much robust to parametric changes. Usually, as robustness specification, a good value of PM_s is 35° or higher [37]. Therefore, values of $\nu > 1.6$ are not considered useful.

Table 1: FOPI controller parameters for the plant with $K_E = 0.9779$ and $T_E = 0.0798$ s and for the performance specification $u_B = 0.7$ ($u_C = u_B/1.7$).

PM_s	ν	$\bar{a}$	$\bar{b}$	$\bar{T}_C$	$\bar{K}_I$	$\bar{K}_P$
0.35π (63°)	1.3	1.0878	0.9277	0.0452	10.4439	0.4726
0.3π (54°)	1.4	1.1764	0.9514	0.0394	13.2152	0.5205
π/4 (45°)	1.5	1.3061	1.0017	0.0352	16.9214	0.5955
π/5 (36°)	1.6	1.4930	1.0859	0.0324	22.1278	0.7164

2.1 Design in presence of time delays

The loop-shaping method can be extended to consider a deadtime in the plant model. Therefore, it can be applied also to the class of plants exhibiting delays that may be intrinsic or induced by the propagation of signals in the loop. To this aim, consider the following plant transfer function:

$$G_p(s) = \frac{K_E}{1 + T_E\, s}\, e^{-L_E\, s} \tag{17}$$

where L_E is the plant deadtime. Then the open-loop frequency response becomes

$$G(j\omega) = \frac{K_E\, K_I\, \{1 + \omega^\nu\, T_C\, [\cos(\hat{\theta}) + j\sin(\hat{\theta})]\}}{\omega^\nu\, [\cos(\hat{\theta}) + j\sin(\hat{\theta})]\, (1 + j\,\omega\, T_E)}\, e^{-j\, L_E\, \omega} \tag{18}$$

and, by the nondimensional frequency $u = \omega\, T_E$, it can be expressed as

$$G(ju) = \frac{K_E\, K_I\, \{1 + (\frac{u}{T_E})^\nu\, T_C\, [\cos(\hat{\theta}) + j\sin(\hat{\theta})]\}}{(\frac{u}{T_E})^\nu\, [\cos(\hat{\theta}) + j\sin(\hat{\theta})]\, (1 + ju)}\, e^{-j\,\frac{L_E\, u}{T_E}}\,. \tag{19}$$

Remark 1. It is now recalled that, if a phase margin PM_s is specified, then a common stability condition requires that $L_E < DM$, where $DM = \frac{PM_s}{\omega_C} = \frac{PM_s\, T_E}{u_C}$ is the so-called delay margin, that is, the maximum allowed time delay corresponding to PM_s and to the gain crossover frequency ω_C. For example, if the phase margins in Table 1 are considered, then the delay margins are, respectively, 0.2131, 0.1827, 0.1522, and 0.1218 s.

The deadtime L_E does not affect the amplitude of $G(ju)$, whereas it affects the specification on the stability margin. Namely, the new argument of $G(ju)$ is

$$\angle G(ju) = \tan^{-1}\left(\frac{(\frac{u}{T_E})^\nu\, T_C\, S}{1 + (\frac{u}{T_E})^\nu\, T_C\, C}\right) - \tan^{-1}(u) - \hat{\theta} - \frac{L_E\, u}{T_E} \tag{20}$$

and leads to a new phase margin in the crossover frequency u_C:

$$PM = \pi + \angle G(ju_C) = \varphi_1(u_C) - \varphi_2(u_C) - \frac{L_E\, u_C}{T_E} + (2 - \nu)\,\pi/2 \tag{21}$$

where $\varphi_1(u)$ and $\varphi_2(u)$ represent the first and second phase components in (20). If PM_s is specified and $PM_s = (2 - v)\pi/2$ is set, then the value $T_C = \overline{T}_C$ is found leading to $\varphi_1(u_C) - \frac{L_E u_C}{T_E} = \varphi_2(u_C)$. By this position, it follows

$$u_C = \tan(\alpha - \beta) \tag{22}$$

with

$$\alpha = \arctan\left(\frac{(\frac{u_C}{T_E})^v \overline{T}_C S}{1 + (\frac{u_C}{T_E})^v \overline{T}_C C}\right) \quad \text{and} \quad \beta = \frac{L_E u_C}{T_E}. \tag{23}$$

Then some algebraic manipulations can be performed to find $\overline{T}_C$. It holds

$$\overline{T}_C = \frac{u_C + \tau}{(\frac{u_C}{T_E})^v [S - u_C C - \tau (C + u_C S)]} \tag{24}$$

with $\tau = \tan(\frac{L_E u_C}{T_E})$. Equation (24) takes the same form of (11) by putting

$$\overline{a} = \frac{(1.7)^v (u_B + 1.7\tau)}{u_B [1.7 S - u_B C - \tau (1.7 C + u_B S)]} \tag{25}$$

and

$$\overline{b} = \frac{u_C + \tau}{u_C [S - u_C C - \tau (C + u_C S)]} \tag{26}$$

and it can be clearly verified that, with no delay, $\tau = 0$ and (24)–(26) reduce to (10)–(12).

Hence, with the previous procedure and the new values of $\overline{a}$ or $\overline{b}$ given by (25) or (26), the new value of $\overline{K}_I$ is determined again by the same formula (16) because the delay does not affect the amplitude of the frequency response. Obviously, given the delay L_E and the specifications u_B and PM_s, then the respective values of u_C and v, the controller parameters are different because $\overline{a}$ and $\overline{b}$ are changed: $\overline{T}_C$ and $\overline{K}_I$ have new and different values from the case with no delay. See Table 2 for the example with delay $L_E = 0.0191$ s.

Table 2: FOPI controller parameters for the plant with $K_E = 0.9779$, $T_E = 0.0798$ s, $L_E = 0.0191$ s, and for $u_B = 0.7$ ($u_C = u_B/1.7$).

PM_s	v	$\overline{a}$	$\overline{b}$	$\overline{T}_C$	$\overline{K}_I$	$\overline{K}_P$
0.35π (63°)	1.3	1.3383	1.1413	0.0557	10.4762	0.5832
0.3π (54°)	1.4	1.4248	1.1523	0.0477	13.4662	0.6423
$\pi/4$ (45°)	1.5	1.5556	1.1931	0.0419	17.5331	0.7349
$\pi/5$ (36°)	1.6	1.7452	1.2694	0.0378	23.3609	0.8841

3 Approximation of irrational operators by continued fractions

All the design approaches for FOC are based on using irrational differential (or integral) operators s^ν, with $\nu > 0$ (or $\nu < 0$). These operators require a rational approximation for realization purpose. The problem is to obtain a rational transfer function with stable poles and minimum-phase zeros. Namely, stability and minimum-phase properties are important for control.

Many approximation methods exist; one of the earliest is the Oustaloup's recursive approximation, which is employed in the CRONE approach [32, 33]. The main idea of this technique consists in approximating the 20ν dB/decade slope of s^ν by a number of alternate slopes of ±20 and 0 dB/decade (+20 for $\nu > 0$ and −20 for $\nu < 0$), corresponding to alternate zeros and poles in the transfer function approximating s^ν. After the pioneering CRONE method, other mathematical approaches approximate s^ν by truncated continued fractions and rational function interpolation (see [6, 43] for a short review) or by signal processing techniques [2]. These methods are particularly interesting because many of them lead to rational functions with simple zeros interlacing simple poles on the negative real axis of the s-plane [22, 23]. Namely, controller synthesis requires minimum-phase and stability properties, because right half-plane (RHP) zeros and RHP poles in the open-loop transfer function imply inherent limitations to the benefits of feedback. In particular, RHP zeros put a limit on the maximum achievable bandwidth of the feedback controlled systems [12, 36]. To improve the existing realizations, one should adjust the distances between pairs of interlaced zeros and poles of a rational realization, which should be of a conveniently limited order.

Continued Fraction Expansions (CFEs) allow to obtain fast convergence and wider domain of convergence than other techniques, for example, those based on power series expansions. Basically, the irrational operator is expanded as an infinite continued fraction. Truncation provides a finite continued fraction that defines the so-called *convergent*. The method developed by [21] allows to obtain a stable and minimum-phase approximation that has the added, remarkable property of simple, interlaced zeros and poles located along the negative real half-axis of the s-plane. Moreover, the method is based on formulas that are in closed-form and allow easy computation of the convergent. To synthesize, for $0 < \nu < 1$, it holds

$$s^\nu \approx \frac{\alpha_N(\nu, s)}{\beta_N(\nu, s)} \tag{27}$$

where both the denominator and numerator are N-degree polynomials, with $N \geq 1$, whose coefficients depend on ν:

$$\alpha_N(\nu, s) = \alpha_{N0}(\nu)\, s^N + \alpha_{N1}(\nu)\, s^{N-1} + \cdots + \alpha_{NN}(\nu), \tag{28}$$

$$\beta_N(\nu, s) = \beta_{N0}(\nu)\, s^N + \beta_{N1}(\nu)\, s^{N-1} + \cdots + \beta_{NN}(\nu). \tag{29}$$

More specifically, the coefficients of $\alpha_N(\nu, s)$ are in the following closed form:

$$\alpha_{Nj}(\nu) = (-1)^j B(N,j)\,(\nu + j + 1)_{(N-j)}\,(\nu - N)_{(j)} \tag{30}$$

where $B(N,j) = \frac{N!}{j!(N-j)!}$, for $j = 0, \dots, N$, are binomial coefficients and

$$(\nu + j + 1)_{(N-j)} = (\nu + j + 1)(\nu + j + 2) \cdots (\nu + N), \tag{31}$$

$$(\nu - N)_{(j)} = (\nu - N)(\nu - N + 1) \cdots (\nu - N + j - 1) \tag{32}$$

define the Pochhammer functions, with $(\nu + N + 1)_{(0)} = (\nu - N)_{(0)} = 1$. In addition, it holds: $\alpha_{Nj}(\nu) = \beta_{N,N-j}(\nu)$.

The interlacing property was analyzed in detail and proven in [21] and [23] for two different classes of approximations based on CFEs, as described below.

3.1 The first form of Thiele's continued fraction

The classical Thiele's CFE was first applied to control problems by [29]. It can be used to approximate the operator s^ν with $\nu \in \mathbb{R}$, $0 < \nu < 1$ and $s \in \mathbb{C}$. The approach is also applicable to $s^{-\nu}$. The approximation is developed by specifying agreement with the magnitude ω^ν at the distinct and equally logarithmic spaced sample points ω_k, for $k = 0, \dots, 2N$, that are chosen inside the approximation range. Then the classical first form of Thiele's CFE is

$$F_{T1}(\omega) = d_0 + \cfrac{\omega - \omega_0}{d_1 + \cfrac{\omega - \omega_1}{d_2 + \cfrac{\cdots}{\cdots + \cfrac{\omega - \omega_k}{d_{k+1} + \cdots}}}} \tag{33}$$

where $d_0 = \omega_0^\nu$, $d_1 = \frac{\omega_1 - \omega_0}{\omega_1^\nu - \omega_0^\nu}$, and d_k, with $K \geq 2$, can be computed by the so-called "reciprocal differences" that do not depend on the ordering of the samples $\omega_0, \dots, \omega_{2N}$, and overcome computational difficulties. Now, if in (33) a numerator of one of the constituent partial fractions vanishes, this and the following constituents do not affect the value of the CFE and, therefore, can be ignored. So, if in (33) $\omega = \omega_{2N}$, then the obtained rational function agrees in value with ω^ν at the sample points $\omega_0, \dots, \omega_{2N}$ and gives the well-known Thiele's formula for $(2N + 1)$ interpolation points. In the control community, it is known as the Matsuda's formula for approximating an irrational function by a rational one. If s replaces ω in (33), the convergent of the resulting expansion, say $F_{T1}(s)$, leads to the rational transfer function $G_{T1}(s)$. The approximation to s^ν, for $s = i\omega$, has modulus equal to $(\omega_k)^\nu$ at the sample points ω_k, $k = 0, \dots, 2N$.

The definition of reciprocal differences for ω^ν for every $\omega \in \Omega$ (the approximation interval) is given by: $\rho_0[\omega_0] = \omega_0^\nu$, $\rho_1[\omega_0, \omega_1] = \frac{\omega_1 - \omega_0}{\omega_1^\nu - \omega_0^\nu}$, and

$$\begin{aligned}\rho_k[\omega_0, \dots, \omega_{k-1}, \omega_k] &= \rho_{k-2}[\omega_0, \dots, \omega_{k-2}] \\ &\quad + \frac{\omega_k - \omega_{k-1}}{\rho_{k-1}[\omega_0, \dots, \omega_{k-2}, \omega_k] - \rho_{k-1}[\omega_0, \dots, \omega_{k-2}, \omega_{k-1}]}\end{aligned} \tag{34}$$

for $k \geq 2$ and for every $\omega_0, \dots, \omega_k \in \Omega$. Moreover, it can be also shown that, for $k \geq 2$, and for all $\omega_0, \dots, \omega_k \in \Omega$, it holds:

$$d_k = \rho_k[\omega_0, \dots, \omega_{k-1}, \omega_k] - \rho_{k-2}[\omega_0, \dots, \omega_{k-2}]. \tag{35}$$

Truncating $F_{T1}(s)$ to the $(2N+1)$-th term, that is, to d_{2N}, and using (35) yields the transfer function $G_{T1}(s)$:

$$G_{T1}(s) = \frac{A_{2N}(\nu, s)}{B_{2N}(\nu, s)} = \frac{p_{T1_{N,0}} s^N + p_{T1_{N,1}} s^{N-1} + \dots + p_{T1_{N,N}}}{q_{T1_{N,0}} s^N + q_{T1_{N,1}} s^{N-1} + \dots + q_{T1_{N,N}}}. \tag{36}$$

Remark 2. Note that a physically implementable convergent, with degree of numerator equal to the degree of denominator, is only obtained by using an odd number of samples. This motivates truncation of $F_{T1}(s)$ to d_k, with k even.

The numerator and denominator polynomials can be easily computed by a recurrence:

$$\begin{cases} A_k(\nu, s) = b_k A_{k-1}(\nu, s) + a_k A_{k-2}(\nu, s), \\ B_k(\nu, s) = b_k B_{k-1}(\nu, s) + a_k B_{k-2}(\nu, s) \end{cases} \tag{37}$$

for $k = 1, \dots, 2N$, starting with $A_{-1}(\nu, s) = 1$, $A_0(\nu, s) = d_0$, $B_{-1}(\nu, s) = 0$, $B_0(\nu, s) = 1$, and using $b_k = d_k$, and $a_k = (s - \omega_{k-1})$.

3.2 The second form of Thiele's continued fraction

Another approximation comes from a second form of Thiele's CFE. It is based on reciprocal differences, that are very useful when all $\omega_0, \dots, \omega_{2N}$ become equal. More precisely, the first Thiele's approximation requires that the difference between the magnitude of the convergent and ω^ν vanishes at $(2N+1)$ points. On the contrary, it can be specified that the difference and its first $2N$ derivatives vanish at a single point ω_0. In this case, the CFE becomes

$$F_{T2}(s) = c_0 + \cfrac{s - \omega_0}{c_1 + \cfrac{s-\omega_0}{c_2 + \cfrac{\cdots}{\cdots + \cfrac{s-\omega_0}{c_{n+1} + \cdots}}}} \tag{38}$$

where $c_k = \rho_k(\omega_0) - \rho_{k-2}(\omega_0)$, with

$$\rho_k(\omega) = \lim_{\omega_0, \dots, \omega_k \to \omega} \rho_k[\omega_0, \dots, \omega_k]. \tag{39}$$

If $\varphi_k(\omega) = \rho_k(\omega) - \rho_{k-2}(\omega)$ is defined by the same limit process applied to inverse divided differences and if the relation between reciprocal differences and inverse divided differences is used, then a simple reasoning leads to [24]:

$$\varphi_k(\omega) = \frac{k}{\frac{d\rho_{k-1}(\omega)}{d\omega}} \tag{40}$$

and

$$\rho_k(\omega) = \rho_{k-2}(\omega) + \varphi_k(\omega), \quad \varphi_{k+1}(\omega) = \frac{k+1}{\frac{d\rho_k(\omega)}{d\omega}}. \tag{41}$$

Hence, since $c_k = \rho_k(\omega_0) - \rho_{k-2}(\omega_0)$, the coefficients can be computed by (41), with $\rho_{-2}(\omega) = \rho_{-1}(\omega) = 0$, $\rho_0(\omega) = \omega^\nu$, and $\rho_1(\omega) = \omega^{1-\nu}/\nu$. By induction, it follows

$$\rho_{2m}(\omega) = \rho_0(\omega) \prod_{i=1}^{m} \frac{i+\nu}{i-\nu}, \tag{42}$$

$$\rho_{2m+1}(\omega) = (m+1)\,\rho_1(\omega) \prod_{i=1}^{m} \frac{i+1-\nu}{i+\nu} \tag{43}$$

for $m \geq 1$, that are useful for computing c_k. Then the convergent of (38) provides the following approximant:

$$G_{T2}(s) = \frac{A_{2N}(\nu,s)}{B_{2N}(\nu,s)} = \frac{p_{T2_{N,0}} s^N + p_{T2_{N,1}} s^{N-1} + \cdots + p_{T2_{N,N}}}{q_{T2_{N,0}} s^N + q_{T2_{N,1}} s^{N-1} + \cdots + q_{T2_{N,N}}}. \tag{44}$$

Remark 3. The truncation of (38) to c_k, with k even, always guarantees physical implementation and yields the same degree N for the numerator and denominator. The last can be determined by (37), with $A_{-1}(\nu,s) = 1$, $A_0(\nu,s) = c_0$, $B_{-1}(\nu,s) = 0$, $B_0(\nu,s) = 1$, $b_k = c_k$, and $a_k = (s - \omega_0)$. Putting $x = s - \omega_0$ simplifies the computation and gives

$$G_{T2}(x) = \frac{P_{T2}(x)}{Q_{T2}(x)} = \frac{\tilde{p}_{T2_{N,0}} x^N + \tilde{p}_{T2_{N,1}} x^{N-1} + \cdots + \tilde{p}_{T2_{N,N}}}{\tilde{q}_{T2_{N,0}} x^N + \tilde{q}_{T2_{N,1}} x^{N-1} + \cdots + \tilde{q}_{T2_{N,N}}}. \tag{45}$$

Since the second Thiele's approximation is obtained by imposing the vanishing of both the deviation between s^ν and the convergent and the first $2N$ derivatives of such deviation at ω_0, the approximation is effective close to ω_0. On the contrary, the first Thiele's approximation works in a larger frequency interval because it is built by imposing coincidence of magnitude at $(2N+1)$ points.

3.3 Interlacing: a formal property

The valuable interlacing property was proven for two classes of approximations based on CFEs. They enjoy interlaced negative zeros and poles, for any fractional order ν, with $|\nu| < 1$, and for any degree, N, of the approximating transfer function [23]. The proof was derived from a classical result by [38], which is based on the Stieltjes' form:

$$f^S(s) = 1 + \cfrac{1}{\alpha_1 s + \cfrac{1}{\beta_2 + \cfrac{1}{\alpha_3 s + \cfrac{1}{\beta_4 + \cdots}}}}, \tag{46}$$

where the partial numerators and denominators are respectively equal to 1 and to $\alpha_{2k-1} s$ and β_{2k}, with $k \geq 1$, and $\alpha_{2k-1}, \beta_{2k} \in \mathbb{R}$. Terminating (46) gives the rational approximation

$$f_{2N}^{S}(s) = 1 + \cfrac{1}{\alpha_1 s + \cfrac{1}{\beta_2 + \cfrac{1}{\alpha_3 s + \cfrac{1}{\cdots + \cfrac{1}{\alpha_{2N-1} s + \cfrac{1}{\beta_{2N}}}}}}}, \tag{47}$$

which is called the $(2N)$-th *convergent*, where $N \geq 1$ is a positive integer. Clearly, the convergent (47) can be transformed into a rational transfer function, with numerator and denominator polynomials having degree N. According to Stieltjes, if the zeros and poles are all simple, negative, real, and interlaced, then $\alpha_{2k-1}, \beta_{2k} > 0$. The converse is also true: if $\alpha_{2k-1}, \beta_{2k} > 0$, then the interlacing property follows (see [9]).

Since many nonterminating CFEs approximate s^{ν}, one should convert them into the Stieltjes' form and truncate it to verify the interlacing property. But it is difficult to find simple and direct conversion formulas among the different types of CFEs. However, the convergents of any CFE can be expressed as rational transfer functions, from which repeated division gives the approximation $f_{2N}^{S}(s)$.

However, using convergents of a different CFE for generating the finite Stieltjes' form is usually not convenient. Namely, since the coefficients of the numerator and denominator polynomials of the convergents depend on ν, the usual conversion to the Stieltjes' form can lead to relations $\alpha_{2k-1}(\nu) > 0$ and $\beta_{2k}(\nu) > 0$ that express complex conditions on ν.

To solve this problem, changes of variables and equivalence transformations are employed to establish direct relations between the partial numerators and denominators of two different types of CFEs and the corresponding partial denominators $\alpha_{2k-1}(\nu)\, s$ and $\beta_{2k}(\nu)$ of the Stieltjes' form. The first type of CFE has been given in [20, 21] and is indicated by $f_{2N}^{M}(\nu, s-1)$. The second type is the second Thiele's CFE and is indicated by $f_{2N}^{T2}(\nu, s-\omega_0)$. For both types, their partial numerators and denominators allow to easily analyze the interlacing property for $\nu \in (0, 1)$.

Since the two CFEs $f_{2N}^{M}(\nu, s-1)$ and $f_{2N}^{T2}(\nu, s-\omega_0)$ share the same form, for sake of simplicity, the results were obtained with reference to $f_{2N}^{M}(\nu, x)$, with $x = s - 1$.

Now, let $\hat{A}_{2N}(\nu, s) = K_A \prod_{i=1}^{N}(s + \mu_i^S)$ and $\hat{B}_{2N}(\nu, s) = K_B \prod_{i=1}^{N}(s + \lambda_i^S)$ be, respectively, the resolutions of the numerator and the denominator of a convergent $\hat{G}_N(\nu, s)$ into N factors of first degree. The following definition is recalled.

Definition 1. The polynomials $\hat{A}_{2N}(\nu, s)$ and $\hat{B}_{2N}(\nu, s)$ of the same degree N form a positive pair iff (if and only if) their roots alternate:

$$0 < \mu_1^S < \lambda_1^S < \mu_2^S < \lambda_2^S < \cdots < \mu_N^S < \lambda_N^S \tag{48}$$

and their leading coefficients are both positive.

Equation (48) explicitly expresses the interlacing property between zeros of $\hat{A}_{2N}(v,s)$ and poles of $\hat{B}_{2N}(v,s)$. Moreover, the following result was proved [38].

Theorem 1. *For every positive pair of polynomials of degree $N \geq 1$, there exists one, and only one, expansion of the ratio of the polynomials into a continued fraction of the form* (47). *In this expansion, it always holds: $\alpha_{2k-1} > 0$, $\beta_{2k} > 0$, for $k = 1,\ldots,N$.*

The form $f_{2N}^M(v,x)$ is useful in applying equivalence transformations. Let us put $x = 1/w$, then the following lemma holds true [23].

Lemma 1. *The poles of the convergents of $f^M(v,w)$ are all real, simple, and located in the interval* $(-1,0)$.

The proof of the lemma relies on repeated transformations leading to a Jacobi's expansion $f^J(w)$. On the basis of partial numerators and denominators of $f^J(w)$, a tridiagonal Jacobi's matrix, J, is constructed having all negative real and simple eigenvalues. Moreover, the characteristic polynomials φ_k of the principal submatrices formed by the first k rows and k columns of $(J - wI)$ are the canonical denominators of the convergents of $f^J(w)$, that form a Sturm sequence, $\varphi_N(w), \varphi_{N-1}(w), \ldots, \varphi_0(w)$, for $\varphi_N(w)$. Hence, if $SC(-1)$ and $SC(0)$ are the number of sign changes of the Sturm sequence at $w = -1$ and $w = 0$, respectively, then $[SC(0) - SC(-1)]$ gives the number of real roots in the interval $(-1,0)$. In [23], it is shown that $[SC(0) - SC(-1)] = N$, that is, $\varphi_N(w)$ has N simple real poles in $(-1,0)$. Since the denominator of the N-degree convergent of $f^J(w)$ coincides with $\varphi_N(w)$, and $f^M(v,w) = 1 + f^J(w)$, the result follows.

The final result is the following [23].

Theorem 2. *For any $N \geq 1$ and $0 < v < 1$, $\hat{A}_{2N}(v,s)$ and $\hat{B}_{2N}(v,s)$ form a positive pair.*

The proof of this theorem can be given by using equivalence transformations, the Stieltjes' theorem, and the previous lemma.

4 Fractional-order PI control of a mechatronic system

The developed design method can be easily applied to a wide class of mechatronic systems, for example, to DC-servomotors or other servosystems, which are very popular in applications and industry. This section briefly recalls the characteristics, properties, and operation of a DC-servomotor and represents a sufficiently accurate mathematical model.

The DC-servomotor with ball bearings and graphite brushes has both a static friction producing an inner initial resistance to motion and a dynamic resistance, which is more evident when the motion is reversed and the brushes introduce a nonlinear distortion in the speed dynamics. The brushes cause friction, losses, and thermal variations. Thus, a high starting torque is necessary and the motor static I/O characteristic is asymmetric. The consequent behavior of the system is nonlinear and time-variant. However, a nonlinear dynamical model would require more sophisticated controllers and make the tuning harder to carry out. On the contrary, the linear model is sufficient to capture the important dynamics and allows us to compare the FOPI with the PI controllers that are very popular in industry. Moreover, the robustness of the FOPI controllers can compensate the nonlinearities without explicitly modeling them.

If a DC-servomotor with independent field excitation winding is considered and the field flux is kept constant, or if the motor is excited by permanent magnets, then the armature voltage equation is given by

$$v_a(t) = R_a\, i_a(t) + L_a \frac{di_a(t)}{dt} + K_v\, \Phi\, \omega_r(t) \tag{49}$$

where R_a and L_a are the armature winding resistance and inductance, respectively, and $T_a = L_a/R_a$ is the time constant; K_v is the voltage constant; Φ is the constant excitation flux due to permanent magnets or independent field excitation winding; $v_a(t)$, $i_a(t)$, and $e_a(t) = K_v\, \Phi\, \omega_r(t)$ are the armature voltage, current, and the back-electromotive force, respectively; finally, $\omega_r(t)$ is the rotor speed, which is the controlled variable. The mechanical equations are

$$J \frac{d\omega_r(t)}{dt} = c_e(t) - B\, \omega_r(t) - c_L(t) \quad \text{and} \quad \frac{d\theta_r(t)}{dt} = \omega_r(t) \tag{50}$$

where $c_e(t) = K_t\, \Phi\, i_a(t)$ is the electromagnetic torque developed by the motor, K_t is the torque constant, J is the inertia moment of the rotor and connected load, B is the viscous friction coefficient, $c_L(t)$ is the external load torque, and $\theta_r(t)$ is the rotor position. In the Laplace-transform s-domain, the previous equations become

$$I_a(s) = \frac{1/R_a}{1 + T_a\, s}\left[V_a(s) - K_v\, \Phi\, \Omega_r(s)\right], \tag{51}$$

$$\Omega_r(s) = \frac{1/B}{1 + T_m\, s}\left[C_e(s) - C_L(s)\right] \tag{52}$$

where $T_m = J/B$ and the Laplace-transforms of time variables have been indicated by uppercase letters: $I_a(s) = \mathscr{L}\{i_a(t)\}$, $V_a(s) = \mathscr{L}\{v_a(t)\}$, $\Omega_r(s) = \mathscr{L}\{\omega_r(t)\}$, $C_e(s) = \mathscr{L}\{c_e(t)\}$, $C_L(s) = \mathscr{L}\{c_L(t)\}$.

Since $C_e(s)$ and $C_L(s)$ are the manipulated and disturbance inputs, respectively, the model (1) is obtained, in which $K_E = 1/B$ and $T_E = T_m$. Taking into account the static friction, the mechanical model of the motor can be extended by including a pure time delay L_E. Then, for the purpose of this chapter and as usually done by research studies, the DC-servomotor speed behavior is well described by a first-order lag plus time delay system, as shown by (17).

As test-case, the FOPI design method was applied for controlling the 370W DC-servomotor in Figure 2. The experimental set-up includes a PC equipped with a floating point 250 Mhz Motorola PPC DS1104 dSPACE board (with 16 bit AD/DA Converters,

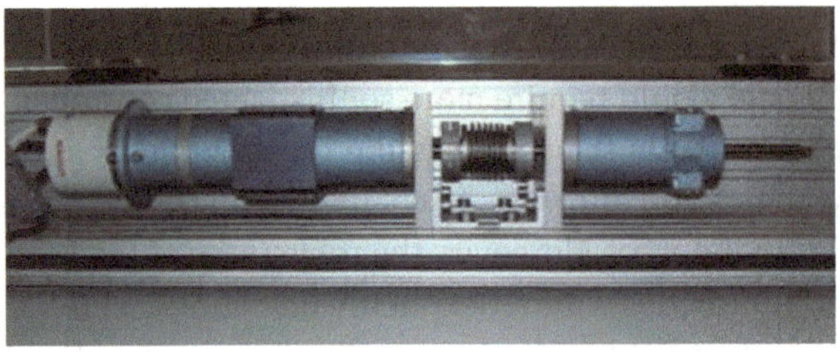

Figure 2: DC-servomotor for the FOPI control experiments.

Table 3: FOPI controller parameters for the DC-servomotor.

PM_s	ν	$\overline{K}_I$	$\overline{K}_P$
$0.3\pi\,(54^\circ)$	1.4	148.38	2.58
$\pi/4\,(45^\circ)$	1.5	289.88	2.96
$\pi/5\,(36^\circ)$	1.6	563.38	3.56

Control Desk to run control routines with 1 ms sampling period in MATLAB/Simulink), a power amplifier to apply the control action, a 1024 pulses incremental encoder to measure the rotor position. As previously remarked, the motor is highly nonlinear, then an identification was necessary to properly determine the parameters of the linear plant model. The interested reader is referred to [16, 17] for similar simulation and experimental analysis, but performed in the case of DC-servomotor position control.

The control system employing a FOPI controller is compared with the standard one using an integer-order PI controller tuned by the *absolute value optimum criterion* [30]. Moreover, in both cases a pre-filter on the set-point is used to smooth the response. This filter is tuned by trial-and-error. The controllers and the filters are implemented through a Simulink block diagram, which is used by the dSPACE board to control the DC-servomotor inside the control loop.

Frequency-domain identification provides the plant parameters: $K_E = 0.9843$, $T_E = 0.0651\,\mathrm{s}$, $L_E = 0.02\,\mathrm{s}$. The formulas (24) and (16) are used to obtain the FOPI controller parameters (Table 3). The fractional orders 1.4, 1.5, 1.6 are selected to provide good phase margins, as robustness specification, and $u_C = 1.8$ is selected for obtaining performance in speed tracking. Realization of the FOPI controllers is obtained by an approximant of order $N = 5$.

Figure 3 shows the closed-loop response to a reference step that is initially applied with no load. Then a load is applied after a certain period of time and maintained during the rest of the time (full load motor operation). In this way, it is possible to test the controllers capability to reject disturbance. As it is shown, the FOPI controllers allow a great reduction of overshoot in the response to the step reference, whereas the PI controller exhibits a 10 % overshoot. The settling times also diminish by using the FOPI controllers. Finally, the load disturbance is better rejected by FOPI controllers in terms of the lower initial undershoot. Moreover, if $\nu = 1.4$ is considered, the settling is

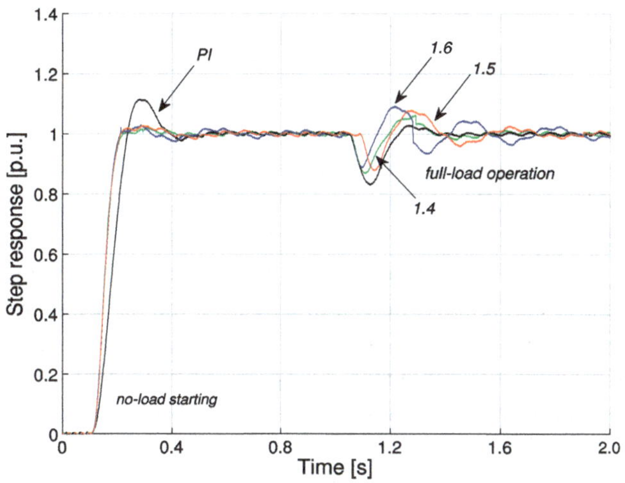

Figure 3: Control of the speed of the DC-servomotor for $\nu = \{1.4, 1.5, 1.6\}$.

similar to the PI controller. To synthesize, a proper choice of ν, also slightly different from the provided fractional values, leads to better performance combined with an excellent phase margin and, above all, a superior robustness due to the flatness of the Bode phase diagram.

5 Fractional-order PI control for an automotive system

FOPI controllers can be effectively applied also to the injection system of compressed natural gas (CNG) engines. For internal combustion engines, it is known that performance strictly depends on the fuel injection dynamics and on the metering of the air-fuel mixture [11]. Owing to a better control on the air-fuel ratio, the Common Rail (CR) technology allows the injection system to remarkably reduce noxious emissions, noise and fuel consumption in Diesel engines, while improving efficiency and available power [40]. These goals are typically achieved in two contemporaneous ways: (1) by setting the injection pressure in the common rail volume to a fixed value; (2) by electronically controlling, for each operating condition, the start and duration of fuel injection that is determined by the injectors. In fact, the injection flow mainly depends on rail pressure and injection timings, which are precisely driven by an Electronic Control Unit (ECU), whereas the large volume of the common rail helps in damping the oscillations due to the operation of pressure controller and injectors. It is also well known that, if compared to liquid fuels, the CNG reduces polluting gaseous emissions of CO, NOx, HC, and of particulate matter and guarantees a better efficiency thanks to its good antiknock properties [44]. However, greater difficulties in metering make the use of CNG less worthwhile. Namely, the gas compressibility makes the working conditions of the injection system vary with driver power requests, engine speed and load to a large extent. Besides, large parametric variations and nonlinearities affect the behavior of the system. Then it is hard to control the process with a high level of robustness.

Recently, to overcome these difficulties, an innovative CNG CR injection system was experimented [14], in which an ECU controls the opening/closing rate of the electro-valve feeding the rail. This allows to achieve an effective regulation of the rail pressure, and hence an accurate metering of the fuel by electronically driving the injectors. In addition, the ECU can modify the working point of the rail pressure for adaptation to the changes of load, to the request of acceleration, etc. However, improving the control design requires a suitable model for predicting the system behavior. Since models are application oriented, none of them has absolute validity. Models that differ for complexity and accuracy can be defined to take into account the main physical phenomena at various accuracy levels [3, 8, 31].

To tradeoff between model accuracy and simplicity of control design, a reliable and accurate state-space nonlinear model of a CNG CR injection system is derived. The nonlinear model can be then linearized at different equilibrium points, depending on the working conditions. Then a transfer function representation derived from the linearized model equations and taking into account the dominant dynamics, is used to design a FOPI controller for every operating condition. The FOPIs used for the different conditions have an inherent robustness to parametric variations and model uncertainties. Finally, gain scheduling is used to cope with the changes in working conditions and has the effect of improving robustness to changes in working conditions.

5.1 Description of the CNG injection system

The main components of the CNG injection system are a tank, a pressure controller, a solenoid valve, a common rail, and the electro-injectors (see Figure 4).

The fuel stored in the tank at a pressure between 20 and 200 bar is delivered to the pressure controller. Fuel pressure is reduced to a value slightly above the pressure inside the intake manifold, then fuel is sent to the common rail. Finally, the fuel flows through the electronically controlled injectors from the common rail to the intake manifold, where the air-fuel mixture is obtained. The aim of the pressure controller is to ensure a constant pressure for a correct metering of the injected fuel, depending on the engine load and speed. The fuel pressures inside the tank and in the common rail are measured to give the driver information about the remaining fuel and to close the control loop, respectively.

The pressure controller consists of a main chamber and a control chamber, both fed by the fuel coming from the tank. Their pressures are regulated by controlling the variable inflow sections. More in details, the control chamber inflow section is

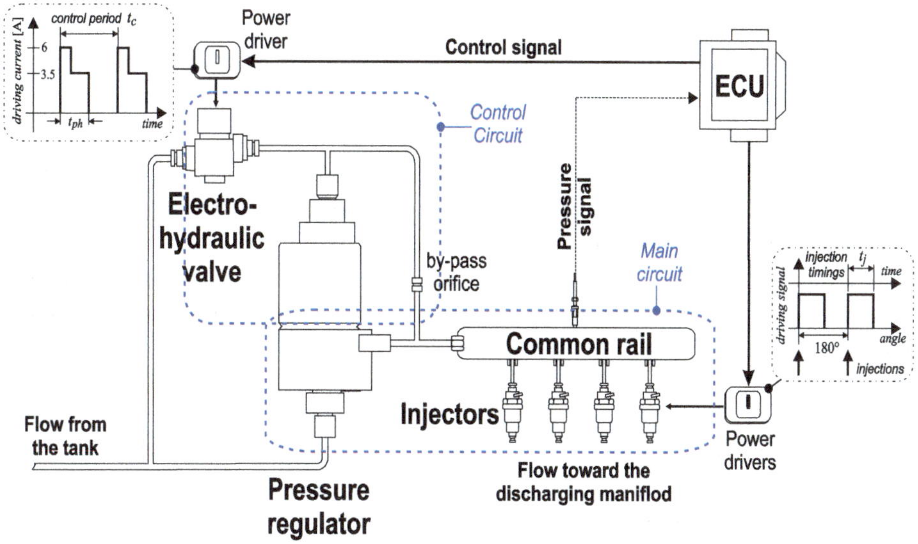

Figure 4: Scheme of the CNG injection system.

changed by varying the driving current of the solenoid valve. The main chamber inflow section depends on the axial displacement of a spherical shutter over a conical seat. A piston placed between the two chambers is provided with a dynamic sealing and operates as actuator for the main chamber shutter. The equilibrium of the applied forces, including the control and main chamber pressures, determines piston and shutter displacements. Therefore, adjusting the pressure in the control circuit by driving the solenoid valve enables the regulation of the main chamber inflow and pressure. Moreover, since the main chamber and the common rail have almost equal pressures, the rail pressure can be indirectly controlled by operating on the valve. Then the ECU processes the signal coming from the sensors and drives the solenoid valve to track the rail reference pressure according to the control algorithm.

5.2 The nonlinear model

The steps to derive the model and the relevant simplifying assumptions are now described. The most significant assumption consists in representing the system as the composition of two subsystems, that is, the control circuit, including the control chamber, and the main circuit, including the main chamber and the common rail. These subsystems are considered as control volumes, in which the fluid thermodynamic properties are spatially constant and time variant, while a constant temperature is assumed.

The pressures dynamics in each control volume is described by the following state equation, derived from the ideal gas law and the continuity equation [39]:

$$\dot{p}_i = \frac{R\Theta}{V}(\dot{m}_{\text{in}} - \dot{m}_{\text{out}}) \tag{53}$$

where p_i is the pressure inside the control volume i, R is the gas constant, Θ is the temperature, V is the instantaneous volume, $\dot{m}_{\text{in}}$ and $\dot{m}_{\text{out}}$ are the input and output mass flows, respectively. The effect of volume variations on pressure dynamics is negligible. Mass flows in (53) are given by the momentum equation, considering different situations. For stationary flows, as for the flow between the control chamber and the main circuit, the following equation is used [39]:

$$\dot{m}_{\text{in}} = c_d A \sqrt{\rho_{\text{out}} (p_{\text{in}} - p_{\text{out}})} \tag{54}$$

where c_d accounts for nonuniformity of the mass flow-rate and for the effect of kinetic energy losses in the nozzle section A, ρ_{out} is the outtake gas density. If the output/input pressure ratio across the flow section satisfies the condition $p_{\text{out}}/p_{\text{in}} < 0.5444$ ("critical flow condition" in presence of sonic speed flows), the resulting nonstationary flow is described by the following equation [45]:

$$\dot{m}_{\text{in}} = c_d A \rho_{\text{in}} \sqrt{k R \Theta \left(\frac{2}{k+1}\right)^{\frac{k+1}{k-1}}} \tag{55}$$

where k is the gas elastic constant and ρ_{in} is the intake gas density. The previous equation applies for the flows from the tank toward the control chamber and the main chamber, and through injectors.

The axial displacement h_s of the coupling between shutter and piston in the main chamber determines the main chamber inflow section A_m. The displacement h_s is computed by applying the equilibrium of the forces acting on the piston and shutter. In addition to each pressure force p_i acting on the surface S_i, the balance depends on the coulomb friction F_c, which is assumed to be constant, and on the force $F_k = -k_s \cdot h_s + F_{so}$ applied by a preloaded spring, where k_s is the spring constant and F_{so} the spring preload. The spring takes the shutter closed against the pressure forces, and prevents the gas from flowing into the main chamber when the solenoid valve is not energized. Viscous friction and inertias of the moving elements are neglected with respect to the large hydraulic forces. Summing up, the force balance gives

$$\sum_i p_i S_i + F_k - F_c = 0. \tag{56}$$

The flow section A_m, perpendicular to the flow direction, is the lateral surface of a truncated cone (see [14]). If β_s is the slope of the conical seat and d_0 is the shutter diameter, the flow section can be approximated by [14]:

$$A_m = 0.5 \pi\, d_0\, h_s \sin(2\beta_s). \tag{57}$$

Hence, (56) provides h_s, and then (57) gives A_m.

To reduce the model order and the computational effort, further simplifications are used for the other components of the injection system, without a sensible degradation of model accuracy:

- the tank pressure is considered as an input, as it is approximately constant within a large time interval and its measurement is always available;
- pressure dynamics inside pipes are not modeled, due to their short length;
- the intake manifold is assumed to have an infinite capacity;
- the solenoid valve and the injectors are represented as simple control valves, as the opening and closing transients are very fast; basically, it is assumed that the flow sections can be completely opened or closed, depending on the actual driving currents of the valve and of the injectors (energized/not-energized circuits).

The solenoid valve regulates the flux incoming in the control chamber. By the previous assumptions, the inflow section can be calculated as $A_s = ET_{\text{sv}} \cdot A_{s,\max}$, where ET_{sv} is a square signal defining the energizing phase of the solenoid valve. The signal ET_{sv} takes values in $\{0; 1\}$ depending on the valve energizing conditions. Similarly, the injectors flow section can be calculated as $A_{\text{inj}} = ET_{\text{inj}} \cdot A_{\text{inj}}$, where ET_{inj} is a square signal of amplitude $\{1, 2, \dots, n\}$, depending on the number of simultaneously opened injectors. The signal period varies with the engine speed, which determines the number of injection cycles per second.

By assuming the control chamber pressure ($x_1 = p_1$) and the rail pressure ($x_2 = p_2$) as state variables, and the signals driving the solenoid valve ($u_1 = ET_{\text{sv}}$) and the injectors ($u_2 = ET_{\text{inj}}$) as inputs, it is possible to write equations (53)–(56) in a state space form:

$$\begin{cases} \dot{x}_1(t) = c_{11} p_{tk}(t) u_1(t) - c_{12} \sqrt{x_2(t)[x_1(t) - x_2(t)]}, \\ \dot{x}_2(t) = c_{21} p_{tk}(t)[c_{24} x_1(t) - c_{25} x_2(t) - c_{26} p_{tk}(t) - c_{27}] \\ \qquad\qquad - c_{22} x_2(t) u_2(t) + c_{23} \sqrt{x_2(t)[x_1(t) - x_2(t)]} \end{cases} \tag{58}$$

where c_{ii} are constant coefficients that depend on the fluid-dynamics gas characteristics and on the system geometrical parameters. The nonlinear system (58) completely describes the dynamics of control volume pressures.

For control design purpose, the state-space equations are linearized around different equilibrium points, so that each tuning of the controller parameters refers to a current working point. By putting (58) in the form $\dot{\mathbf{x}}(t) = f[\mathbf{x}(t), \mathbf{u}(t)]$, the linearization with respect to $(\bar{\mathbf{x}}, \bar{\mathbf{u}})$ yields

$$\delta\dot{\mathbf{x}}(t) = \left[\frac{\partial f}{\partial \mathbf{x}}\right]_{\bar{\mathbf{x}},\bar{\mathbf{u}}} \delta\mathbf{x}(t) + \left[\frac{\partial f}{\partial \mathbf{u}}\right]_{\bar{\mathbf{x}},\bar{\mathbf{u}}} \delta\mathbf{u}(t) = \mathbf{A}\delta\mathbf{x}(t) + \mathbf{B}\delta\mathbf{u}(t) \tag{59}$$

where $\delta\mathbf{x}(t) = \mathbf{x}(t) - \bar{\mathbf{x}}$, $\delta\mathbf{u}(t) = \mathbf{u}(t) - \bar{\mathbf{u}}$, $\mathbf{A} = [a_{i,j}]$, $\mathbf{B} = [b_{i,j}]$. Choosing $\delta x_2(t)$ as output gives $\delta y(t) = \mathbf{C}\delta\mathbf{x}(t)$, where $\mathbf{C} = [0 \;\; 1]$ is the output matrix. The input–output transfer function relating $U_1(s)$ and $X_2(s)$ can be determined by applying the Laplace transform to the linearized state equation (59). By dropping symbol δ with some notational abuse, and considering that $Y(s) = [C(sI - A)^{-1}B] \cdot U(s)$, the transfer function between $U_1(s)$ and $X_2(s)$ is

$$X_2(s) = \frac{a_{21}b_{11}}{s^2 - (a_{11} + a_{22})s + a_{11}a_{22} - a_{12}a_{21}} U_1(s) \tag{60}$$

in which the constant coefficients depend on the working point. The transfer function poles are $s_{1/2} = 0.5[a_{11} + a_{22} \pm \sqrt{(a_{11} - a_{22})^2 + 4a_{12}a_{21}}]$. Since $\{a_{21}, |a_{22}|\} \gg \{|a_{11}|, a_{12}\}$ and $(a_{11} - a_{22})^2 \gg 4a_{12}a_{21}$, the pole $s_1 = 0.5[a_{11} + a_{22} + \sqrt{(a_{11} - a_{22})^2 + 4a_{12}a_{21}}]$ dominates the transient response, independently from the working point, whereas $s_2 \approx a_{22}$. Then, if the smaller time constant is neglected and a time delay τ is introduced to take into account the pressure propagation from the main chamber to the common rail (that can be experimentally determined), a family of first-order lag plus time delay systems is obtained:

$$\frac{X_2(s)}{U_1(s)} \approx G_p(s) = \frac{K}{1 + Ts} e^{-\tau s} \tag{61}$$

where $K = a_{21} \cdot b_{11}/(s_1 \cdot s_2)$ and $T = -1/s_1$ are functions of the current working point. The equilibrium pressures defining the working points can be determined from (58). In particular, the working points are defined by imposing the tank pressure $\overline{p}_{tk}$, and the mean values within an injection cycle of the input couples $(\overline{u}_1, \overline{u}_2)$, as set by the ECU depending on the engine load.

5.3 Performance of the FOPI controller

The efficiency and robustness of the control is tested by simulation of the nonlinear system model. To this aim, the FOPI controllers, with $\nu = 1.3, 1.4, 1.5, 1.6$, are compared to a PI controller that is typically used for injection control and tuned by the open-loop Ziegler–Nichols rules. Values of $\nu < 1.3$ are not considered because they lead to too high phase margins and overshoots, and values of $\nu > 1.6$ neither because they provide too low phase margins, that is, below 36°.

With reference to the typical working conditions, two different cases are simulated. In the first case, there is a small reference pressure variation from 4 to 5 bar. Moreover, the injection timing is 5 ms, the engine speed is 2500 rpm, and the tank pressure is 50 bar. A single FOPI or PI controller is used and tuned with reference to the (K_E, T_E) plant parameters related to the final pressure to achieve. This means that the FOPI or PI controller parameters are set by using the values of K_E and T_E that are determined by using the working condition associated to 5 bar. Figure 5 shows the closed-loop step response.

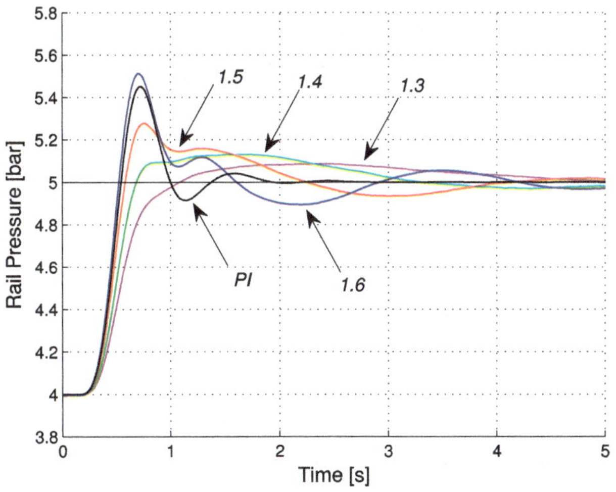

Figure 5: Rail pressure dynamics for constant speed and load: FOPI and PI controllers response to a variation in the reference pressure from 4 to 5 bar.

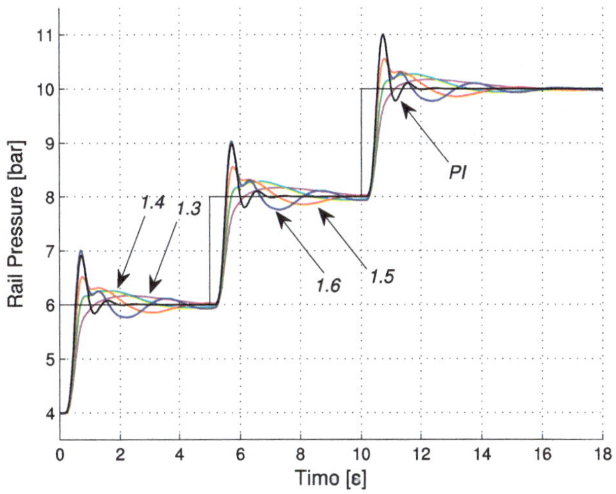

Figure 6: Rail pressure dynamics for constant speed and load: FOPI and PI controllers response to large variations in the reference pressure.

Note that the FOPI controllers allow an improvement with respect to the standard PI controller: overshoot is much reduced with $\nu = 1.5$, then a better accuracy in injection is achieved. The settling times are comparable. The rise time indicates that promptness of the PI is equivalent to that obtained with $\nu = 1.5$.

The second case simulates a large variation in the reference pressure from 4 to 10 bar. Hence, an increment of 6 bar is considered. Therefore, as scheduling strategy, a switch between several controllers is realized. In particular, the pressure variation is divided in 3 steps of 2 bar each, then 3 different FOPI or PI controllers are used, one for each step. The parameters of each different controller are set according to the final value of pressure that must be achieved in each of the successive steps. Then the settings will refer to the plant parameters associated to 6, 8, and 10 bar. In Figure 6, it can be verified that FOPIs reduce the overshoot significantly and that responses are similar to the first case. It can be concluded that FOPIs may help improving pressure regulation also for these large perturbations.

6 Final remarks

This chapter described how to design and realize fractional-order proportional-integral controllers for a specific class of plant models, namely first-order lag plus time delay systems. The design methodology combines the loop-shaping of the frequency response and the pursuing of an optimal feedback system for tracking capability. Then specifications are established in terms of robustness of the control loop and performance of the closed-loop system. Moreover, the irrational transfer function of the FOPI controller is realized by an approximation technique, which is based on peculiar continued fractions allowing the approximant to verify properties of stability, minimum-phase, and interlacing between zeros and poles. The mathematical developments allow to set the controller parameters and to obtain the realization of the controller transfer function by using closed-form expressions that are suitable for real implementation.

The application of the FOPI controllers and the design method was illustrated by two examples taken from the mechatronics and automotive engineering field. The first is the speed control of a DC-servomotor. The second is the control of the common rail pressure in a prototype gas injection system. In both cases, the mentioned simple plant model can be used to derive the FOPI parameters. However, the approach can be extended also to other classes of plant models, for example, the integrating plants with a first-order lag and a time delay or second-order systems.

Bibliography

[1] K.-J. Åström and T. Hägglund, *PID Controllers: Theory, Design, and Tuning*, 2nd ed., Instrument Society of America, Research Triangle Park, NC, 1995.
[2] R.-S. Barbosa, J.-A. Tenreiro Machado, and M.-F. Silva, Time domain design of fractional differintegrators using least squares, *Signal Process.*, **86**(10) (2006), 2567–2581.
[3] T. Bertram, F. Bekes, R. Greul, O. Hanke, C. Haß, J. Hilgert, M. Hiller, O. Ottgen, P. Opgen-Rhein, M. Torlo, and D. Ward, Modelling and simulation for mechatronic design in automotive systems, *Control Eng. Pract.*, **11** (2003), 179–190.
[4] H.-W. Bode, *Network Analysis and Feedback Amplifier Design*, Van Nostrand, New York, 1945.
[5] Y.-Q. Chen, Ubiquitous fractional order controls, in *Proc. of the Second IFAC Symposium on Fractional Derivatives and its Applications (IFAC FDA '06)*, Porto, Portugal, July 19–21, vol. 2, part 1, pp. 168–173, 2006.

[6] Y.-Q. Chen, B. M. Vinagre, and I. Podlubny, Continued fraction expansion approaches to discretizing fractional order derivatives. An expository review, *Nonlinear Dyn., Special Issue on Fract. Derivatives and Their Appl.*, **38**(1–2) (2004), 155–170.

[7] Y. Q. Chen, I. Petráš, and D. Xue, Fractional order control – a tutorial, in *Proc. 2009 American Contr. Conf.*, St. Louis, MO, USA, June 10–12, pp. 1397–1411, 2009.

[8] G. Dellino, P. Lino, C. Meloni, and A. Rizzo, Kriging metamodel management in the design optimization of a CNG injection system, *Math. Comput. Simul.*, **79**(8) (2009), 2345–2360.

[9] F.-R. Gantmacher and M.-G. Krein, *Oscillation Matrices and Kernels and Small Vibrations of Mechanical Systems*, revised ed., AMS Chelsea Publishing, American Mathematical Society, Providence, RI, 2002.

[10] R. Garrappa, F. Mainardi, and G. Maione, Models of dielectric relaxation based on completely monotone functions, *Fract. Calc. Appl. Anal.*, **19**(5) (2016), 1105–1160.

[11] J. Heywood, *Internal Combustion Engine Fundamentals*, McGraw-Hill, New York, 1988.

[12] I. Horowitz, *Synthesis of Feedback Systems*, Academic Press, New York, 1963.

[13] R.-E. Kalman, When is a linear control system optimal?, *Trans. of the ASME, J. Basic Eng.* 86(series D) (1964), 84–90.

[14] P. Lino, B. Maione, and C. Amorese, Modeling and predictive control of a new injection system for compressed natural gas engines, *Control Eng. Pract.*, **16**(10) (2008), 1216–1230.

[15] P. Lino and G. Maione, Tuning PI^{ν} fractional order controllers for position control of DC-servomotors, in *Proc. 2010 IEEE Int. Symp. on Industrial Electronics (ISIE 2010)*, Bari, Italy, July 4–7, pp. 359–363, 2010.

[16] P. Lino and G. Maione, Fractional order PI tuning for integrating plants with time delay, in R. Vilanova and A. Visioli (eds.) *2nd IFAC Conf. Advances in PID Control (PID '12)*, Brescia, Italy, March 28–30, IFAC Proceedings on line, vol. 2, part I, pp. 649–654, 2012.

[17] P. Lino and G. Maione, Loop-shaping and easy tuning of fractional-order proportional integral controllers for position servo systems, *Asian J. Control*, **15**(3) (2013), 796–805.

[18] B.-J. Lurie and P.-J. Enright, *Classical Feedback Control with Matlab*, Control Engineering Series, Marcel Dekker, Inc., New York, 2000.

[19] J.-M. Maciejowski, *Multivariable Feedback Design*, Addison-Wesley, Wokingham, UK, 1989.

[20] G. Maione, Concerning continued fractions representations of noninteger order digital differentiators, *IEEE Signal Process. Lett.*, **13**(12) (2006), 725–728.

[21] G. Maione, Continued fractions approximation of the impulse response of fractional order dynamic systems, *IET Control Theory Appl.*, **2**(7) (2008), 564–572.

[22] G. Maione, High-speed digital realizations of fractional operators in the delta domain, *IEEE Trans. Autom. Control*, **56**(3) (2011), 697–702.

[23] G. Maione, Conditions for a class of rational approximants of fractional differentiators/integrators to enjoy the interlacing property, in S. Bittanti, A. Cenedese, and S. Zampieri (eds.) *Proc. 18th IFAC World Congr.*, Milan, Italy, August 28–September 2, IFAC Proceedings on line, vol. 18, part 1, pp. 13984–13989, 2011.

[24] G. Maione, Thiele's continued fractions in digital implementation of noninteger differintegrators, *Signal Image Video Process., Special Issue: "Fractional signals and systems" (C. M. Ionescu, J. Sabatier, and J. A. Tenreiro Machado (eds.))*, **6**(3) (2012), 401–410.

[25] G. Maione, R. Caponetto, and A. Pisano, Optimization of zero-pole interlacing in discrete approximations of fractional order operators s^{ν}, *Comput. Math. Appl., Special Issue on Fractional Differentiation and its Applications (D. Baleanu, J. A. Tenreiro Machado, and W. Chen (eds.))*, **66**(5) (2013), 746–754.

[26] G. Maione and P. Lino, New tuning rules for fractional PI^{α} controllers, *Nonlinear Dyn.*, **49**(1–2) (2007), 251–257.

[27] G. Maione and A. Punzi, Combining differential evolution and particle swarm optimization to tune and realize fractional order controllers, *Math. Comput. Model. Dyn. Syst.*, **19**(3) (2013), 277–299.
[28] S. Manabe, The non-integer integral and its application to control systems, *Jpn. Inst. Electr. Eng. J.*, **6**(3–4) (1961), 83–87.
[29] K. Matsuda and H. Fujii, H_∞ optimized wave-absorbing control: analytical and experimental results, *J. Guid. Control Dyn.*, **16**(6) (1993), 1146–1153.
[30] R. C. Oldenbourg and H. Sartorius, A uniform approach to the optimum adjustments of control loops, in R. Oldenburger (ed.) *Frequency Response*, The Macmillan Co., New York, 1956.
[31] A. Ollero, S. Boverie, R. Goodall, J. Sasiadek, H. Erbe, and D. Zuehlke, Mechatronics, robotics and components for automation and control, *Annu. Rev. Control*, **30** (2006), 41–54.
[32] A. Oustaloup, *La commande CRONE. Commande Robuste d'Ordre Non Entiér*, Hermés, Paris, 1991.
[33] A. Oustaloup, F. Levron, B. Mathieu, and F. M. Nanot, Frequency-band complex noninteger differentiator: characterization and synthesis, *IEEE Trans. Circuits Syst. I, Fundam. Theory Appl.*, **47**(1) (2000), 25–39.
[34] I. Podlubny, Fractional-order systems and $PI^\lambda D^\mu$-controllers, *IEEE Trans. Autom. Control*, **44**(1) (1999), 208–214.
[35] I. Podlubny, L. Dorcák, and I. Kostial, On fractional derivatives, fractional-order dynamic systems and $PI^\lambda D^\mu$ controllers, in *Proc. 36th Conf. on Dec. and Control*, San Diego, CA, USA, December 10–12, pp. 4985–4990, 1997.
[36] M. J. Sidi, *Design of Robust Control Systems: From Classical to Modern Practical Approaches*, Krieger Publishing Company, Malbar (FLA), USA, 2001.
[37] S. Skogestad and I. Postlethwaite, *Multivariable Feedback Control: Analysis and Design*, 2nd ed., Wiley & Sons, Ltd., Chichester, England, 2005.
[38] T.-J. Stieltjes, *Œuvres Completes de Thomas Jan Stieltjes, Tome II*, P. Noordhoff, Pub. par le soins de la Société Mathématique D'Amsterdam, Groningen, 1918.
[39] V. Streeter, K. Wylie, and E. Bedford, *Fluid Mechanics*, 9th ed., McGraw-Hill, New York, 1998.
[40] G. Stumpp and M. Ricco, Common rail – an attractive fuel injection system for passenger car DI diesel engine, SAE Technical Paper 960870, 1996.
[41] J. A. Tenreiro Machado, V. Kiryakova, and F. Mainardi, Recent history of fractional calculus, *Commun. Nonlinear Sci. Numer. Simul.*, **16**(3) (2011), 1140–1153.
[42] A. Tustin, The design of systems for automatic control of the position of massive objects, *Proc. Inst. Electr. Eng.*, 105(part C, suppl. no. 1) (1958), 1–57.
[43] B. M. Vinagre, I. Podlubny, A. Hernandez, and V. Feliu, Some approximations of fractional order operators used in control theory and applications, *J. Fract. Calc. Appl. Anal.*, **13**(3) (2000), 231–248.
[44] C. S. Weaver, *Natural gas vehicles – a review of the state of the art*, in *Gaseous Fuels: Technology, Performance and Emissions*, Society of Automotive Engineers, Inc., Warrendale, PA, 1989.
[45] M. Zucrow and J. Hoffman, *Gas Dynamics*, John Wiley & Sons, New York, 1976.

Andrzej Dzieliński, Dominik Sierociuk, and Grzegorz Sarwas

Fractional-order modeling and control of selected physical systems

Abstract: In this chapter, an overview of some recent results of modeling and identification of systems with ultracapacitors are discussed. These models are necessary for many algorithms to control the process of its charging and discharging. Presented models are of noninteger order which is the consequence of the internal structure of the ultracapacitor. Theoretical considerations on modeling methods and new models derivations are given. Models properties are discussed and are verified by simulations and real live experiments. The models presented justify some well-known behaviors of this elements and circuits where they are applied.

Keywords: Ultracapacitor, fractional order modeling, ultracapacitors' parameter identification

MSC 2010: 93C20, 93C95, 93A30, 35Q93

1 Introduction

The recent decade has been characterized by increased emphasis on renewable forms of energy and their economical and ecological uses. Devices which can support this kind of systems are required to be relatively small in size and weight in comparison to the energy portion which they can store. Also, the ability of lossless and fast energy transfer-like charging and discharging has become an important factor of new technology. In the industrial applications, the common use of ultracapacitors as elements can support existing energy storage systems such as: electrical storage devices for solar and wind farm, electrical-cars, elevators, etc. has begun in recent decades [12]. To improve the accuracy and to understand the behavior of ultracapacitors, scientists tried to build mathematical and physical models with similar dynamics [5, 24]. It became immediately clear that these devices do not have simple dynamics as traditional capacitors. It is noticeable that in the case of ultracapacitors, energy is stored not in the ions polarization, as in the electrolytic capacitor, but in the ions movement. This feature implies that the behavior in these kinds of systems should be more similar to the dynamics of the diffusion process. In the solution of the diffusion equation [14], which is modeled like the heat transfer process [21], there exists Laplace's complex op-

Andrzej Dzieliński, Dominik Sierociuk, Grzegorz Sarwas, Institute of Control and Industrial Electronics, Warsaw University of Technology, Koszykowa 75, 00-662 Warszawa, Poland, e-mails: andrzej.dzielinski@ee.pw.edu.pl, dominik.sierociuk@ee.pw.edu.pl, grzegorz.sarwas@ee.pw.edu.pl

https://doi.org/10.1515/9783110571745-013

erator s in the form $\sqrt{s}$, which can be interpreted as a derivation of the order 0.5. This is the purpose of use of the fractional-order calculus in the diffusion or heat transfer processes description [1, 11, 15, 16, 18, 21, 23] and also in the case of ultracapacitors' modeling [5, 17, 20, 24].

Utilization of fractional-order calculus for modeling the diffusion process suggests that a lot of electrochemical devices, which construction or mode of action is based on this phenomenon, can be more accurately modeled and controlled using fractional-order models and algorithms. In works [5, 17, 20, 24] were shown very efficient ways of ultracapacitor modeling. These works show the different models but some of them are only useful in a small range of frequency. For the precise control of ultracapacitors, one needs a broad knowledge of its dynamics, hence a new model was necessary. Also, the detailed research about existing models was desired.

Ultracapacitor modeling can be found in the papers [9–11].

In this article, the internal structure of an ultracapacitor is discussed. Based on its description, it is explained why ultracapacitors have the noninteger dynamics. Next, the model based on the traditional normal diffusion and subdiffusion (proposed and derived in [19]) is presented. These two models are compared with other fractional order models. In Section 3 properties of theoretical fractional-order ultracapacitor models are discussed. In this section, the differences between presented models are shown. Also, the ultracapacitor behavior, well known by the producers and engineers, like capacity decreasing with the frequency is discussed and explained. This phenomenon confirms the need to use the fractional-order model in case of modeling the ultracapacitor dynamics. In this section are presented the relations for step responses of frequency theoretical models in time domain what allows to validate achieved model. Next section the results of identification using fractional order ultracapacitor models are shown and discussed. Also, validations of achieved parameters through the comparison with ultracapacitors parameters measured directly for the devices are presented. Further confirmation of the correctness of the application and identification of the studied models by the comparison of theoretical and measured step responses is shown. In the last section, all results are summarized and concluded.

2 Ultracapacitors' models

An ultracapacitor can be shown (Figure 1) as two nonreactive porous plates, or collectors, suspended within an electrolyte, with a voltage potential applied across the collectors. These plates are typically constructed with activated carbon which is characterized by a high degree of microporosity (1 gram of activated carbon has surface area in excess of 500 m^2). In an ultracapacitor, the potential applied to the positive

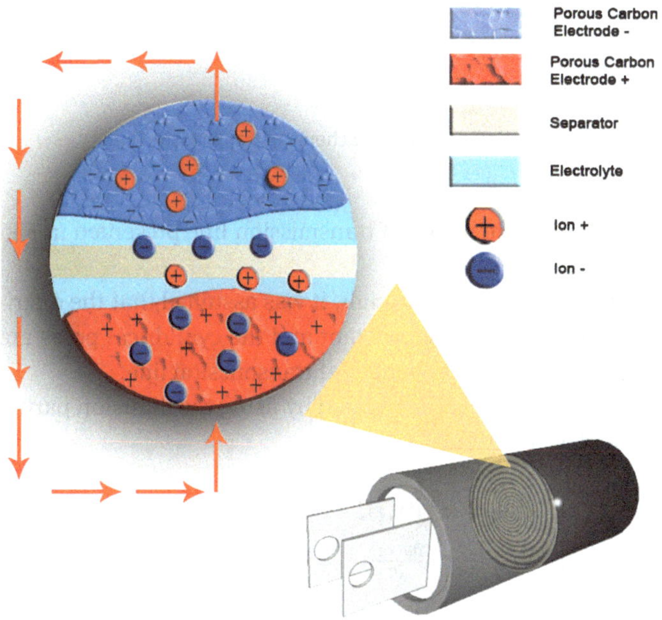

Figure 1: Ultracapacitor scheme.

electrode attracts the negative ions in the electrolyte, while the potential to the negative electrode attracts the positive ions. A dielectric separator between the two electrodes prevents the charge from moving between the two electrodes. Electrical energy storage devices, such as capacitors, store an electrical charge on an electrode. In other devices like electrochemical cells or batteries, the energy is created by a chemical reaction of electrodes. In both of these, the ability to store or create an electrical charge is a function of the surface area of the electrode. In an ultracapacitor, the energy is also the function of the electrodes surface but this energy is collected in an ions movement, differently from an electrolyte capacitor where energy is collected in the ions polarization.

A complicated structure of the supercapacitor and utilization of ions movement to store the energy have a large influence on its dynamics. To derive an analytical model of a double-layer capacitor, it is necessary to focus on the porous structure of electrodes and the electrode-electrolyte interface. The current in supercapacitor is composed of an ionic current through the electrolyte, an electronic current through the electrode material, and a displacement current at the electrode-electrolyte interface [4]. Since the two order of magnitude is higher conductivity of an electrode than the insulator layer, the electrode resistance can be neglected. However, in a full model of an ultracapacitor, this resistance should be taken into consideration because of high current work conditions. For the double-layer capacitor impedance, the main influence has the ions diffusion at the electrode-electrolyte interface layer. The diffusion

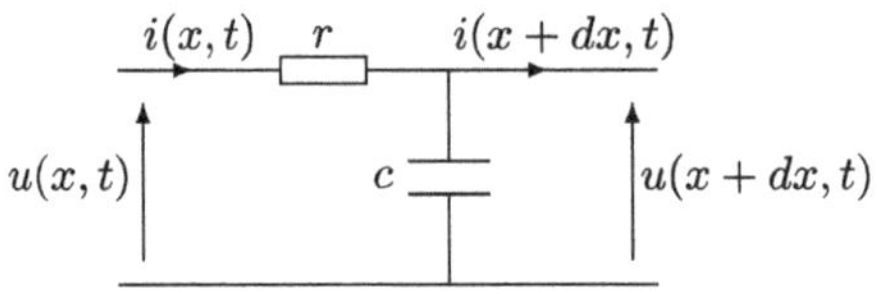

Figure 2: Scheme of transmission line.

process can be modeled using the finite model of transmission line presented in Figure 2 [2].

To derive a simplified model of an ultracapacitor, it was assumed that the distributions of resistance r and capacitance c are constant and the cylindrical pores are uniformly filled with an electrolyte, and d is a length of transmission line.

The current $i(x, t)$ and the voltage $u(x, t)$ are related by using the diffusion partial differential equation presented as follows:

$$\begin{cases} \frac{\partial u(x,t)}{\partial x} = -r \cdot i(x, t), \\ \frac{\partial i(x,t)}{\partial x} = -c \cdot \frac{\partial u(x,t)}{\partial t}. \end{cases} \tag{1}$$

Parameters r and c being constant, equation (1) can be rewritten as

$$\begin{cases} \frac{\partial^2 u(x,t)}{\partial x^2} = r \cdot c \cdot \frac{\partial u(x,t)}{\partial t}, \\ \frac{\partial^2 i(x,t)}{\partial x^2} = r \cdot c \cdot \frac{\partial i(x,t)}{\partial t}. \end{cases} \tag{2}$$

The above relation can be solved like a one-dimensional transient heat conduction equation using the Laplace domain. The Laplace transform derivative with respect to time of the first equation from (2) is presented as follows:

$$\frac{\partial^2 U(x, s)}{\partial x^2} = r \cdot c \cdot s \cdot U(x, s) - r \cdot c \cdot U(x, 0). \tag{3}$$

For the zero initial condition, the above equation can be rewritten as

$$\frac{\partial^2 U(x, s)}{\partial x^2} - r \cdot c \cdot s \cdot U(x, s) = 0. \tag{4}$$

The roots of the characteristic equation of (3) are as follows:

$$\omega_{1,2} = \pm\sqrt{r \cdot c \cdot s}$$

and the solution of the equation (4) is

$$U(x, s) = C_1(s) \sinh(x\sqrt{r \cdot c \cdot s}) + C_2(s) \cosh(x\sqrt{r \cdot c \cdot s}), \tag{5}$$

where $C_1(s)$ and $C_2(s)$ are the constants of the equation.

To appoint constants of above equation, it is necessary to use the boundary condition. For $x = 0$, we have

$$U(0,s) = C_1(s)\sinh(0) + C_2(s)\cosh(0), \tag{6}$$

what gave the relation for the second constant $C_2(s)$ presented as

$$C_2(s) = U(0,s). \tag{7}$$

The constant $C_1(s)$ can be calculated from the first equation from the set of equation (1):

$$-r \cdot i(x,t) = \frac{\partial u(x,t)}{\partial x}. \tag{8}$$

Using the Laplace transform with respect to time and equation (5), we obtained

$$\begin{aligned} -r \cdot I(x,s) = {} & C_1(s)\sqrt{r \cdot c \cdot s} \cdot \cosh(x\sqrt{r \cdot c \cdot s}) \\ & + U(0,s)\sqrt{r \cdot c \cdot s} \cdot \sinh(x\sqrt{r \cdot c \cdot s}). \end{aligned} \tag{9}$$

In the case of ultracapacitors, the limit condition in $x = d$ is $i(d,t) = 0$, which implies $I(d,s) = 0$, so the equation (9) becomes

$$0 = C_1(s)\sqrt{r \cdot c \cdot s} \cdot \cosh(d\sqrt{r \cdot c \cdot s}) + U(0,s)\sqrt{r \cdot c \cdot s} \cdot \sinh(d\sqrt{r \cdot c \cdot s}), \tag{10}$$

what can be rewritten as

$$C_1(s) = -\tanh(d\sqrt{r \cdot c \cdot s})U(0,s). \tag{11}$$

Let us go back to equation (9). For $x = 0$, we obtain

$$-r \cdot I(0,s) = -\tanh(d\sqrt{r \cdot c \cdot s})U(0,s)\sqrt{r \cdot c \cdot s} \cdot 1 + 0, \tag{12}$$

what can be rewritten as

$$I(0,s) = \frac{\sqrt{r \cdot c \cdot s}}{r}\tanh(d\sqrt{r \cdot c \cdot s})U(0,s). \tag{13}$$

Finally, we obtained the transfer function of the transmission line

$$Z_{\mathrm{imp}}(s) = \frac{U(0,s)}{I(0,s)} = \frac{r}{\sqrt{r \cdot c \cdot s}}\coth(d\sqrt{r \cdot c \cdot s}). \tag{14}$$

This allows us to introduce a total capacity of the line C where $C = c \cdot d$ and a total resistance $R_l = r \cdot d$. The pulse impedance of transmission line can be rewritten as

$$\bar{Z}_{\mathrm{imp}}(s) = \frac{R_l}{\sqrt{R_l \cdot C \cdot s}} \cdot \coth(\sqrt{R_l \cdot C \cdot s}). \tag{15}$$

Using a Taylor series, it is possible to approximate coth(x) through the following equation [18]:

$$\coth(x) = \frac{\cosh(x)}{\sinh(x)} \cong \frac{1 + \frac{x^2}{2}}{x}. \tag{16}$$

The expression in numerator $1+\frac{x^2}{2}$ is the approximation of $\sqrt{1+x^2}$. For a low frequency equation, (16) becomes

$$\coth(x) \underset{x\to 0}{\cong} \frac{\sqrt{1+x^2}}{x}. \tag{17}$$

This development is also available for high frequencies as

$$\coth(x) \underset{\infty}{\cong} \frac{e^{\frac{x}{2}}}{e^{\frac{x}{2}}} = 1 \underset{\infty}{\cong} \frac{\sqrt{1+x^2}}{x}. \tag{18}$$

Putting the above relation to equation (15), we achieved a simplified fractional model of ultracapacitor porous impedance:

$$Z_{\text{imp}}(s) = \frac{\sqrt{1 + R \cdot C \cdot s}}{Cs} = \frac{\sqrt{1+Ts}}{Cs}. \tag{19}$$

Model (19) was derived using the idealistic assumptions; however, the real electrodes should be modeled using a more complicated method. The diffusion of liquids [3, 13], suspension in the fractional structures like granular substances or a transport of electrons, holes, ions, spine, etc., in disordered solids should be described by anomalous (fractional) diffusion equations [14].

In normal diffusion, the particle average movement is proportional to $t^{\frac{1}{2}}$, however, in the case of subdiffusion it is smaller, proportional to the $t^{\frac{\alpha}{2}}$, where $\alpha \in (0, 1)$. Based on this assumption in [19] was proposed and proofed in the following lemma.

Lemma 2.1. *The new model of ultracapacitor impedance based on subdiffusion equation is*

$$Z_{\text{imp}}(s) = \frac{\sqrt{1+Ts^{\alpha}}}{C_{\alpha}s^{\alpha}} \tag{20}$$

where C_α is the ultracapacitor fractional capacity and T is the parameter responsible for the changes capacity with frequency.

The full model of supercapacitor including the resistance of the electrodes is presented in Figure 3. The transfer function of the modeled system in Figure 3 is defined as

$$G_{\text{uc}}(s) = \frac{U_{\text{uc}}(s)}{I(s)},$$

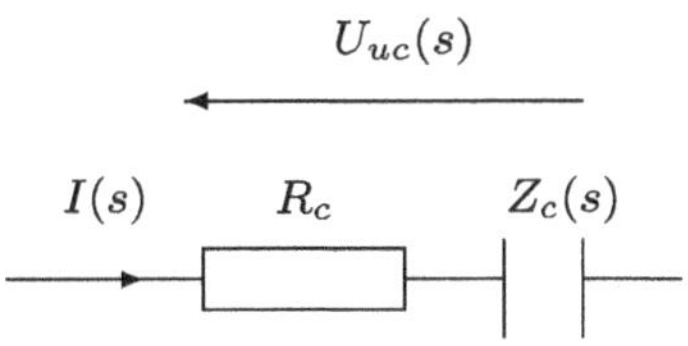

Figure 3: Ultracapacitor equivalent model.

where $U_{\text{uc}}(s)$ is a Laplace transform of the capacitor voltage and $I(s)$ is a Laplace transform of the capacitor current. The whole transfer function of an ultracapacitor presented in Figure 3 is

$$G_{\text{uc}}(s) = \frac{U_{\text{uc}}(s)}{I(s)} = R_c + Z_c(s), \tag{21}$$

where R_c is the resistance of the ultracapacitor and $Z_c(s)$ can be any part of the ultracapacitor capacity model. Using the impedance model from (20), the whole double-layer capacitor model is presented as follows:

$$G_{\text{uc}}(s) = R_c + \frac{\sqrt{1 + Ts^{\alpha}}}{Cs^{\alpha}}, \tag{22}$$

where $R_c \in \mathbb{R}_+$, $T \in \mathbb{R}_+$, and $\alpha \in (0, 1)$.

In [17], J. J. Quintana presented one of the well-known fractional-order capacity model presented below:

$$Z_{c_q}(s) = \frac{(1 + Ts)^{\alpha}}{Cs^{\beta}}. \tag{23}$$

Actually, the analytical derivation of this model has not been presented. It was developed from a realistic capacity model $Z_c = \frac{1}{Cs^{\alpha}}$ presented in [24]. The Quintana's model is the conclusion of results from impedance spectroscopy experiments on the ultracapacitor and can very precisely describe the dynamics of supercapacitor, hence it was meticulously examined and described in the following chapters. The whole Quintana's model of ultracapacitor is

$$G_{uc_q}(s) = R_c + \frac{(1 + Ts)^{\alpha}}{Cs^{\beta}}, \tag{24}$$

where $R_c \in \mathbb{R}_+$, $T \in \mathbb{R}_+$, and $\alpha \in (0, 1)$.

Neither model (22) nor model (24) have capacity which can be compared with the capacity of traditional capacitor. In the denominator is the Cs^{ν}, where $\nu \in \mathbb{R}_+$, which can be called fractional capacity. To create a more physically model capable of comparison with the traditional model of the capacitor, the Quintana capacitance model with the $\beta = 1$ is used:

$$G_{\text{DC}}(s) = R_c + \frac{(1 + Ts)^{\alpha}}{Cs}. \tag{25}$$

Historically, this model was proposed by Davidson and Cole in 1951 like a dielectric model to fit the complex dielectric constant data observed for glycerine [6, 7].

In the next section, the properties of the presented model are discussed. Also, the experiments of ultracapacitor identification and advisability of presented models are described.

3 Properties of theoretical models

In the prior section, the fractional-order models for modeling of ultracapacitors were presented. This section is devoted to the main properties of these models. In the Subsection 3.1 discussion about the new derived model is given. Subsection 3.2 presents most physical ultracapacitor model, which is the special case of the Quintana model for $\beta = 1$. For this model, the derivation of ultracapacitor and quadripole with ultracapacitor step response in the time domain is shown.

3.1 Description of the new model

Let focus on the following fractional-order capacitance model:

$$Z_c(s) = \frac{\sqrt{1 + Ts^\alpha}}{C_\alpha s^\alpha},$$

where $T \in \mathbb{R}_+$ and $\alpha \in (0, 1)$.

To underline that C_α in this model is not a capacity of capacitor in a traditional way, we use extra index C_α. The units of this parameter are the following units $\frac{\text{Farads}}{(\text{sec})^{1-\alpha}}$. It is worth to notice that in fact C_α is not a value of ultracapacitor capacity but it is an ultracapacitor impedance dependent of frequency and α parameter.

Let us take this model into consideration. The behavior at the ends of ranges can be described using following relation:

$$G_c(s) = \begin{cases} \frac{\sqrt{T}}{C_\alpha s^{\alpha/2}} & \text{for } s \gg \frac{1}{T}, \\ \frac{1}{C_\alpha s^\alpha} & \text{for } s \ll \frac{1}{T}. \end{cases}$$

For low frequency, this model tends to the fractional integrator model presented by Westerlund [24]. This model can be rewritten in following form:

$$\frac{1}{C_\alpha s^\alpha} = \frac{1}{C_\alpha \omega^\alpha (\sin \frac{\pi}{2}\alpha + j \cos \frac{\pi}{2}\alpha)} \tag{26}$$

what means that for the $\omega \to 0$ the ultracapacitor capacity is equal the capacity of traditional capacitor.

Even though, this model is the most appropriate to describe the dynamics of ultracapacitor in the frequency domain, it is very hard to calculate the inverse Laplace transform to find the step response of this model in the time domain, therefore in this book, for ultracapacitor modeling, the Davidson–Cole model is used.

3.2 Davidson–Cole Model Properties

Let us take Davidson–Cole model (25), which can be also achieved from Quintana model for $\beta = 1$.

$$G_{\mathrm{DC}} = \frac{(1+Ts)^{\alpha}}{Cs},$$

where C is a capacity of ultracapacitor, $\alpha \in (0,1)$ is a model order and T is a parameter which meaning is presented in this subsection. In this model, parameter C is a traditional capacity with the unit Farads, therefore, this model is more physically-oriented. The behavior of this model on the ends of range is the following:

$$G_{\mathrm{DC}}(s) = \begin{cases} \frac{T^{\alpha}}{Cs^{1-\alpha}} & \text{for } s \gg \frac{1}{T} \\ \frac{1}{Cs} & \text{for } s \ll \frac{1}{T}. \end{cases}$$

It can be noticed that this model for low frequency ($\omega \to 0$) behaves like model of traditional capacitor with decrease of amplitude characteristic 20 dB per decade. For high values of frequency, this model reminds a fractional-order integrator with order equally $1 - \alpha$. The meaning of the T parameter was discussed in [9]. It was explained using the spectral transfer function of Davidson–Cole model given as follows:

$$G_{\mathrm{DC}}(j\omega) = \frac{(Tj\omega + 1)^{\alpha}}{Cj\omega}. \tag{27}$$

The magnitude of this spectral transfer function is

$$A_{\mathrm{DC}}(\omega) = \frac{((T\omega)^2 + 1)^{\frac{\alpha}{2}}}{C\omega}. \tag{28}$$

This magnitude is compared to the magnitude of a traditional capacitor of capacity C' is

$$\frac{((T\omega)^2 + 1)^{\frac{\alpha}{2}}}{C\omega} = \frac{1}{C'\omega}, \tag{29}$$

which yields

$$C' = \frac{C}{((T\omega)^2 + 1)^{\frac{\alpha}{2}}}, \tag{30}$$

where C' is the capacity equivalent of the ultracapacitor for given frequency ω. This equivalent capacity illustrates what capacity the traditional capacitor should have in order to have the same magnitude for desired value of frequency. For $\alpha = 0.5$, we have

$$C' = \begin{cases} C & \text{for } \omega \ll \frac{1}{T}, \\ \frac{C}{\sqrt{T\omega}} & \text{for } \omega \gg \frac{1}{T}. \end{cases} \tag{31}$$

The frequency f_c for which the capacity equivalent decreases 2 times is given as follows:

$$f_c = \frac{\sqrt{2^{\frac{2}{\alpha}} - 1}}{2\pi T}. \tag{32}$$

A fact of decrease of ultracapacitor capacity with frequency is known by the producers but only using of fractional-order model of presented form for ultracapacitor modeling explains this behavior.

Now let us take a full ultracapacitor model based on the Davidson–Cole capacitance model:

$$G_c(s) = R_c + \frac{(1 + Ts)^{\alpha}}{Cs}.$$

The step response of this model in the time domain is discussed below.

Lemma 3.1. *The time domain step response of the ultracapacitor itself is as follows:*

$$u_c(t) = \mathcal{L}^{-1}\left[G_c(s)\frac{I}{s}\right] = \left(R_c + \frac{T^{\alpha}}{C}e^{-\frac{t}{T}}\frac{t^{1-\alpha}}{\Gamma(2-\alpha)}{}_1F_1\left(2; 2-\alpha; \frac{t}{T}\right)\right)I, \tag{33}$$

where the ${}_1F_1(2; 2-\alpha; \frac{t}{T})$ is a confluent hypergeometric function.

Proof. [9, 10] To calculate the of $G_c(s)$, the ultracapacitor capacity model form is necessary to find the inverse Laplace transformation presented below:

$$u_c(t) = \mathcal{L}^{-1}\left[(R_c + G_{DC}(s))\frac{I}{s}\right]. \tag{34}$$

The hardest task in this calculus is finding the inverse Laplace transformation of unit step of the capacity part:

$$G_{DC}(s)\frac{1}{s} = \frac{(Ts + 1)^{\alpha}}{Cs}\frac{1}{s}, \tag{35}$$

which can be rewritten as

$$\mathcal{L}^{-1}\left[G_{DC}(s)\frac{1}{s}\right] = \mathcal{L}^{-1}\left[\frac{T^{\alpha}(s + \frac{1}{T})^{\alpha}}{Cs^2}\right]. \tag{36}$$

Using the theorem of the complex shift of the Laplace transformation, we obtain

$$\mathcal{L}^{-1}\left[G_{DC}(s)\frac{1}{s}\right] = \frac{T^{\alpha}}{C}e^{-\frac{t}{T}}\mathcal{L}^{-1}\left[\frac{s^{\alpha}}{(s-\frac{1}{T})^2}\right]. \tag{37}$$

To solve the inverse Laplace transformation from (37), the following formula is used:

$$\mathcal{L}^{-1}\left[\frac{s^{\gamma-\beta}}{(s-a)^{\gamma}}\right] = \frac{t^{\beta-1}}{\Gamma(\beta)}{}_1F_1(\gamma;\beta;at), \tag{38}$$

where $\gamma, \beta \in \mathbb{R}_+$, and the ${}_1F_1(\gamma;\beta;at)$ is a confluent hypergeometric function define as

$${}_1F_1(a;c;x) = \sum_{n=0}^{\infty}\frac{(a)_n}{(c)_n}\frac{x^n}{n!}, \quad -\infty < x < \infty, \tag{39}$$

where $(a)_n$ and $(c)_n$ are the Pochhammer symbols defined by

$$(g)_n = \frac{\Gamma(g+n)}{\Gamma(g)}, \quad n = 0, 1, 2, \ldots. \tag{40}$$

Using (38) to (37), the step response of the ultracapacitor capacitance in the time domain can be calculated:

$$\mathcal{L}^{-1}\left[G_{DC}(s)\frac{I}{s}\right] = \frac{T^{\alpha}}{C}e^{-\frac{t}{T}}\frac{t^{1-\alpha}}{\Gamma(2-\alpha)}{}_1F_1\left(2;2-\alpha;\frac{t}{T}\right)I. \tag{41}$$

Finally, adding the resistance part the Lemma thesis is achieved:

$$u_c(t) = \mathcal{L}^{-1}\left[G_c(s)\frac{I}{s}\right] = \left(R_c + \frac{T^{\alpha}}{C}e^{-\frac{t}{T}}\frac{t^{1-\alpha}}{\Gamma(2-\alpha)}{}_1F_1\left(2;2-\alpha;\frac{t}{T}\right)\right)I. \tag{42}$$

□

Analytically, the response of the ultracapacitor with the linear function of current ($i(t) = It$, where I = const) as an input signal can be also calculated:

$$\mathcal{L}^{-1}\left[\frac{G_c(s)I}{s^2}\right] = \left(R_cI + \frac{T^{\alpha}}{C}e^{-\frac{t}{T}}\frac{t^{2-\alpha}}{\Gamma(3-\alpha)}{}_1F_1\left(3;3-\alpha;\frac{t}{T}\right)\right)I. \tag{43}$$

Let us take it to consideration the RC quadripole presented in Figure 4.

The transfer function of this system is defined as follows:

$$G_{RC}(s) = \frac{U_{uc}(s)}{U(s)} = \frac{G_{uc}(s)}{G_{uc}(s)+R}.$$

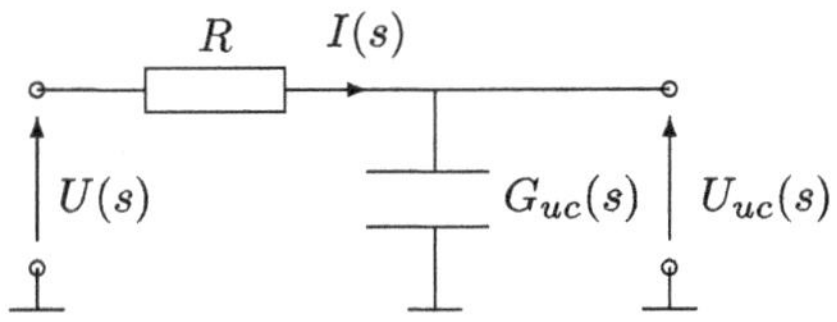

Figure 4: RC quadripole model.

In our case, using $G_c(s)$ like ultracapacitor model $G_{uc}(s)$, this model has the following form:

$$G_{RC}(s) = \frac{G_c(s)}{G_c(s)+R} = \frac{(Ts+1)^{\alpha}+R_cCs}{(Ts+1)^{\alpha}+(R+R_c)Cs},$$

where C, R_c is the ultracapacitor capacity and resistance, respectively, T is a parameter of capacity decreasing with frequency, and R is a system resistance (matching resistance in experimental setup).

Calculation of the step response of RC quadripole with the ultracapacitor model $G_c(s)$ for arbitrary α is not easy. Therefore, the step response of this model for $\alpha = 0.5$ is presented. This order can describe the dynamics of the low-capacity ultracapacitor what was confirmed in experiments.

Lemma 3.2. *The unit step response of the system with the ultracapacitor for $\alpha = 0.5$ is given by the following equation:*

$$u_{uc}(t) = \mathcal{L}^{-1}\left[G_{RC}(s)\frac{1}{s}\right] = e^{-\frac{t}{T}}\sum_{i=1}^{4}\left(\frac{F_i}{t^{0.5}}\cdot E_{1,0.5}(H_i^2t)+F_iH_ie^{H_i^2t}\right), \tag{44}$$

where

$$F_1 = \frac{\sqrt{T}}{2}, \quad F_2 = -\frac{\sqrt{T}}{2}, \quad F_3 = \frac{(B-A)T}{\sqrt{T(T^2+4A^2)}}, \quad F_4 = \frac{(A-B)T}{\sqrt{T(T^2+4A^2)}},$$

$$H_1 = \sqrt{\frac{1}{T}}, \quad H_2 = -\sqrt{\frac{1}{T}}, \quad H_3 = \frac{-T^{1.5}+\sqrt{T^3+4A^2T}}{2AT},$$

$$H_4 = \frac{-T^{1.5}-\sqrt{T^3+4A^2T}}{2AT}, \qquad A = (R+R_c)C, \quad B = R_cC$$

and $E_{1,0.5}(H_i^2t)$ is a two-parameter Mittag-Leffler function. Parameters C, R_c, T are the parameters of the ultracapacitor $G_{uc}(s)$ model and R is the resistance of RC quadripole's resistor.

Proof. [9, 10] To calculate the step response of the system with the ultracapacitor in the time domain for $\alpha = 0.5$, it is necessary to find the inverse Laplace transformation of the following transfer function:

$$\mathcal{L}^{-1}\left[G_{RC}(s)\frac{1}{s}\right] = \mathcal{L}^{-1}\left[\frac{\sqrt{Ts+1}+R_cCs}{\sqrt{Ts+1}+(R+R_c)Cs}\frac{1}{s}\right]. \tag{45}$$

The above equation can be rewritten as

$$\mathcal{L}^{-1}\left[G_{RC}(s)\frac{1}{s}\right] = \mathcal{L}^{-1}\left[\frac{\sqrt{Ts+1}+Bs}{\sqrt{Ts+1}+As}\frac{1}{s}\right], \tag{46}$$

where $B = R_cC$ and $A = (R+R_c)C$.

As a result of the applied theorem of the complex shift of the Laplace transformation, we obtain

$$\mathcal{L}^{-1}\left[G_{RC}(s)\frac{1}{s}\right] = e^{-\frac{t}{T}}\mathcal{L}^{-1}\left[\frac{\sqrt{T}\sqrt{s} + B(s - \frac{1}{T})}{(\sqrt{T}\sqrt{s} + A(s - \frac{1}{T}))(s - \frac{1}{T})}\right]. \tag{47}$$

Let us define the auxiliary variable:

$$G(s) = \frac{\sqrt{T}\sqrt{s} + B(s - \frac{1}{T})}{(\sqrt{T}\sqrt{s} + A(s - \frac{1}{T}))(s - \frac{1}{T})}. \tag{48}$$

Using the auxiliary complex variable $w = s^{0.5}$, we have

$$G(w) = \frac{(BTw^2 + T^{1.5}w - B)T}{(Tw^2 - 1)(ATw^2 + T^{1.5}w - A)}, \tag{49}$$

After decomposition of $G(w)$ into the partial fractions and inverse Laplace transform, back in the complex variable s, we find

$$\begin{aligned} G(s) = {} & \frac{1}{2\sqrt{\frac{1}{T}}(s^{0.5} - \sqrt{\frac{1}{T}})} - \frac{1}{2\sqrt{\frac{1}{T}}(s^{0.5} + \sqrt{\frac{1}{T}})} \\ & - \frac{(-B + A)T}{\sqrt{T(T^2 + 4A^2)}(s^{0.5} - \frac{-T^{1.5} + \sqrt{T^3 + 4A^2T}}{2AT})} \\ & + \frac{(-B + A)T}{\sqrt{T(T^2 + 4A^2)}(s^{0.5} - \frac{-T^{1.5} - \sqrt{T^3 + 4A^2T}}{2AT})}. \end{aligned}$$

By linearity of the Laplace transformation, the goal of finding the inverse Laplace transformation can be decomposed to four identical problems,

$$\mathcal{L}^{-1}\left[G_{RC_3}(s)\frac{1}{s}\right] = e^{-\frac{t}{T}}\mathcal{L}^{-1}[G_1(s) + G_2(s) + G_3(s) + G_4(s)], \tag{50}$$

where $G_1(s)$, $G_2(s)$, $G_3(s)$, and $G_4(s)$ are the individual partial fraction. All of them can be transformed using new parameters to the presented form:

$$G_i(s) = \frac{F_i}{s^{0.5} - H_i}.$$

Using the following equation for decomposition of rational function to partial fraction,

$$\frac{1}{s^{\nu} - a} = \frac{1}{a}\sum_{k=1}^{n} \frac{a^k}{s^{(\nu k - 1)}(s - a^{\frac{1}{\nu}})}, \tag{51}$$

where $\nu = \frac{1}{n} \in \mathbb{R}_+$, we obtain

$$G_i(s) = \frac{F_i}{s^{0.5} - H_i} = F_i \cdot \left(\frac{s^{0.5}}{s - H_i^2} + \frac{H_i}{s - H_i^2}\right).$$

The inverse Laplace transformation can be calculated using the following relation:

$$\mathcal{L}^{-1}\left[\frac{1}{s^{\kappa}(s \pm a^2)}\right] = t^{\kappa} E_{1,\kappa+1}(\mp at), \tag{52}$$

where function $E_{1,\kappa+1}(\mp at)$ is a two-parameter Mittag-Leffler function, defined as

$$E_{\alpha,\beta}(z) = \sum_{k=0}^{\infty} \frac{z^k}{\Gamma(\beta + \alpha k)}, \quad \Re(\alpha), \Re(\beta) > 0,\ z \in \mathbb{C}. \tag{53}$$

Finally, from (50), (51), and (52) we obtain the unit step response of the system with a ultracapacitor:

$$u_{\mathrm{uc}}(t) = \mathcal{L}^{-1}\left[G_{\mathrm{RC}}(s)\frac{1}{s}\right] = e^{-\frac{t}{T}} \sum_{i=1}^{4} \left(\frac{F_i}{t^{0.5}} \cdot E_{1,0.5}(H_i^2 t) + F_i H_i e^{H_i^2 t}\right), \tag{54}$$

where

$$F_1 = \frac{\sqrt{T}}{2}, \quad F_2 = -\frac{\sqrt{T}}{2}, \quad F_3 = \frac{(B-A)T}{\sqrt{T(T^2+4A^2)}}, \quad F_4 = \frac{(A-B)T}{\sqrt{T(T^2+4A^2)}},$$
$$H_1 = \sqrt{\frac{1}{T}}, \quad H_2 = -\sqrt{\frac{1}{T}}, \quad H_3 = \frac{-T^{1.5} + \sqrt{T^3 + 4A^2 T}}{2AT},$$
$$H_4 = \frac{-T^{1.5} - \sqrt{T^3 + 4A^2 T}}{2AT}, \quad A = (R + R_c)C, \quad B = R_c C.$$

□

In the next section, the results of ultracapacitors' modeling are presented. This chapter shows fractional calculus like a proper tool for ultracapacitor modeling.

4 Identification experiments

In the research, two types of an ultracapacitor were used. First, three ultracapacitors of nominal capacity 0.047 F, 0.1 F, 0.33 F, produced by Panasonic®, are called in this paper low capacity ultracapacitors. Their maximum operation voltage, provided by the manufacturer, is 5.5 V. Another type of ultracapacitors contained two devices produced by Maxwell® of the nominal capacity 1500 F/2.7 V (BCAP1500) and 3000 F/2.7 V (BCAP3000). These ultracapacitors are named high capacity ultracapacitors.

4.1 Results of experiments

First experiments were focused on the ultracapacitors' modeling using the fractional-order model. Values of the parameters were obtained as a result of Bode diagram matching. The model parameters were adjusted using the Matlab function *fmincon* by minimizing a sum of Root Mean Square Error (RMSE) from amplitude and phase identification. It is very important to remember that ultracapacitors are electrolytic capacitors and they can accept only positive voltages, consequently in the case of measuring low capacity ultracapacitor, using the voltage follower, the input signal with a constant component was

$$u(t) = 2V + \sin(\omega t).$$

Capacitor voltage (in steady state) in this case was equal to

$$u_{\mathrm{uc}}(t) = A_c(\omega)\sin(\omega t + \varphi_u)$$

and the capacitor current was

$$i(t) = A_i(\omega)\sin(\omega t + \varphi_u).$$

In case of using the type of setup with a current converter, to model high capacity ultracapacitor, input signal was a current sine wave:

$$i(t) = A_i(\omega)\sin(\omega t + \varphi_u).$$

Capacitor voltage in this case contained initial value of voltage (u_0) depending on signal frequency, thus the output voltage signal was equal to

$$u_{\mathrm{uc}}(t) = u_0 + A_c(\omega)\sin(\omega t + \varphi_u).$$

The Bode diagram was obtained from the following relations:

$$M(\omega) = 20\log\left(\frac{A_c(\omega)}{A_i(\omega)}\right), \quad \varphi(\omega) = \varphi_i(\omega) - \varphi_u(\omega).$$

In effect of measurements, two diagrams were obtained. First, one represents the magnitude of the system examined and the second one expresses the phase shift frequency response.

4.2 System identification using Davidson–Cole model

Let us recall the Davidson–Cole model:

$$G_c = R_c + \frac{(1 + Ts)^{\alpha}}{Cs},$$

where R_c is the ultracapacitor resistance, C is the ultracapacitor capacity in the traditional definition, and T is a parameter of capacity decrease with frequency, which is explained in [9].

Table 1: Identified ultracapacitors' parameters.

Capacitor	T	C (F)	R_c (Ω)	α
0.047 F	5.2261	0.06	32	0.6
0.1 F	14.7979	0.094	52	0.6
0.33 F	56.9669	0.27	29	0.57
1500 F	1.006	1348	0.25 m	0.62
3000 F	0.6369	2410	0.13 m	0.7

Table 2: RMSE of Davidson–Cole model identification.

Capacitor	RMSE magnitude	RMSE phase
0.047 F	0.0985	0.6152
0.1 F	0.1521	0.6366
0.33 F	1.3146	0.7857
1500 F	0.9932	1.6518
3000 F	0.8154	1.5344

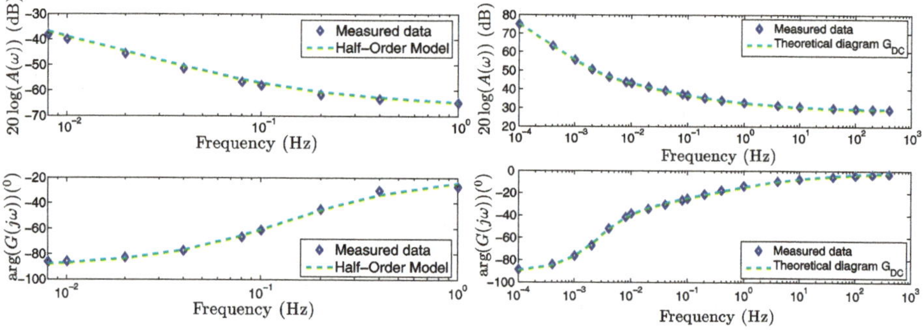

(a) Measured and D-C Model theoretical Bode diagrams of ultracapacitor 1500F.

(b) Measured and D-C Model theoretical Bode diagrams of ultracapacitor 0.33F.

Figure 5: Measured and D-C Model theoretical Bode diagrams.

Model parameters achieved as a result of identification by diagrams matching are presented in the Table 1.

RMSE of identification for this model is shown in the Table 2. Values presented in this table highlighted good matching of identified model. Mean error for magnitude and phase is lower than 1.5 dB and 1.66°, respectively. The examples of ultracapacitor identification in frequency domain in Figures 5(a), 5(b) are presented. These diagrams confirm that the Davidson–Cole model can describe ultracapacitor dynamics in the fairly wide range of frequencies. For high capacity ultracapacitors, the identification absolute error of magnitude is lower than 1.5 dB for all samples and phase absolute error is lower than 4 degrees almost for all samples. In the case of identification, low capacity ultracapacitor diagrams matching has great accuracy in the whole range of measured frequencies. Also, it is worth to note that $\alpha = 0.6$ is constant for all identified ultracapacitors, which means that its value is close to the model derived from the normal diffusion equation, therefore, acceptable results of modeling (especially for the low capacity ultracapacitor) can be achieved using $\alpha = 0.5$. Results of ultracapacitors' modeling using the Half-Order ultracapacitor model are also presented in this section.

Now, let us focus on the T parameter. This parameter is responsible for ultracapacitor capacity decreasing with the frequency. The f_c, for which the equivalent capacity decreases two times is given as follows (30):

$$f_c = \frac{\sqrt{2^{\frac{2}{\alpha}} - 1}}{2\pi T}$$

and the values for tested ultracapacitors are summarized in Table 3.

Following the Table 3 the same dependencies can be noted. For high capacity ultracapacitors produced by Maxwell, the value of f_c frequency increases with capacity, which means that a bigger ultracapacitor has more stable capacity as a function of frequency. In the case of a low capacity, higher capacity causes less stationarity of the capacitance parameter.

Subsequently identified parameters with parameters measured directly from ultracapacitors were compared. Use of the Davidson–Cole model allows for this validation, because unlike other fractional-order models presented in the Section 2, this model contains resistance R_c (Ω) and capacity C (F) in traditional form. To measure the real values of an ultracapacitor's resistance and capacity, the step response of the circuit was examined. The value of real resistance R_{real} (Ω) was measured like an ultracapacitor voltage step for the state current immediately after starting charging process and the value of real capacity C_{real} (F) was measured like a charge stored on a ultracapacitor divided by its voltage. The parameters comparison is presented in the Table 4.

Table 4 confirms that using the fractional-order model G_c one can very precisely identify real parameters of ultracapacitors. Identification results are equivalent to these obtained from real ultracapacitor by different methods, for example, the step response. Differences between the C parameter achieved in the identification process

Table 3: Values of frequency for which equivalent decreases capacity two times.

Capacitor	T	α	f_c (mHz)
0.047 F	5.2261	0.6	91.8
0.1 F	14.7979	0.6	32.4
0.33 F	56.9669	0.57	9
1500 F	1.006	0.62	457
3000 F	0.6369	0.7	625

Table 4: Parameters comparison.

Capacitor	C (F)	R_c (Ω)	C_{real} (F)	R_{real} (Ω)
0.047 F	0.06	32	0.06	32
0.1 F	0.094	52	0.1	42
0.33 F	0.27	29	0.27	28
1500 F	1348	0.25 m	1189	0.26 m
3000 F	2446	0.13 m	2435.4	0.15 m

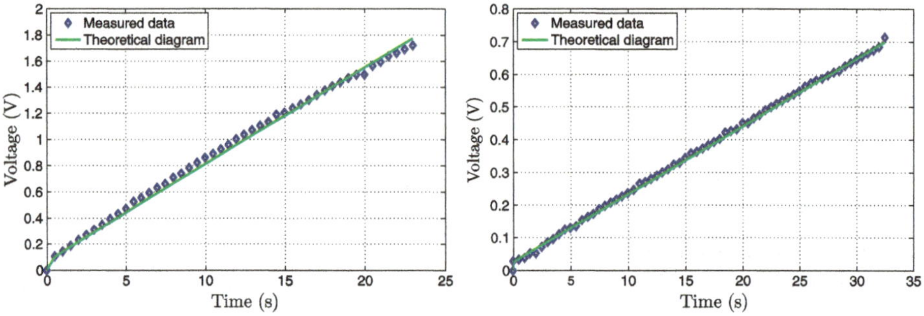

(a) Step response of model $G_c(s)$ for 1500F and current I(t)=100A

(b) Step response of model $G_c(s)$ for 3000F and current I(t)=50A

Figure 6: Step responses of model $G_c(s)$.

and the real values in the case of modeling high capacity ultracapacitors are the effect of not precise enough measurements of fast changing signals. Lack of high frequency samples also had some influence on the identification process.

Having identified models for each measured ultracapacitor, we can validate them in the time domain through the comparison of theoretical step responses of achieved models (33) with step responses of a circuit. The result of this validation is presented in Figures 6(a) and 6(b). The first of these figures shows the step response of an ultracapacitor with nominal capacity 1500 F for a current input signal equal to 100 A and the second presents the step response of an ultracapacitor with nominal capacity 3000 F for a current equal to 50 A.

4.3 System identification using New Model

New Model is an ultracapacitor model derived using the anomalous diffusion partial differential equation:

$$G_{C_\alpha} = R_c + \frac{\sqrt{1 + T_\alpha s^\alpha}}{C_\alpha s^\alpha},$$

Table 5: Identified ultracapacitors parameters for new model.

Capacitor	T	C_α ($\frac{F}{s^{1-\alpha}}$)	R_c (Ω)	α	C_{real} (F)	R_{real} (Ω)
0.047 F	5.9257	0.0533	35	0.9887	0.06	32
0.1 F	18.0459	0.0941	58	0.9969	0.1	42
0.33 F	64.4422	0.2512	32	0.9870	0.27	28
1500 F	1.0132	1490	0.3 m	0.9945	1189	0.26 m
3000 F	0.5	2497	0.18 m	0.9810	2435	0.15 m

Table 6: Error of New Model identification.

Capacitor	Magnitude RMSE	Phase RMSE
0.047 F	0.3121	1.4608
0.1 F	0.3513	1.4979
0.33 F	0.4190	1.7736
1500 F	0.3555	1.8486
3000 F	0.3729	0.7570

where R_c is the ultracapacitor resistance, C_α is the ultracapacitor fractional capacity with units $\frac{\text{Farads}}{(\text{sec})^{1-\alpha}}$, and T_α is the parameter of capacity decrease with frequency.

Comparison of the model parameters achieved as a result of identification by diagrams matching with real ultracapacitors' parameters are presented in Table 5.

In fact, we cannot compare the decrease of the ultracapacitor capacity with frequency, because parameter C_α should be interpreted rather as ultracapacitor impedance than capacity like in the case of parameter C for traditional capacitor (see [9]), but we have to notice that the values of the α parameters in Table 5 demonstrate that for this model the best results of ultracapacitor modeling can be achieved for α slightly less than 1, which means that also for $\alpha = 1$ we should achieve similar results. Although the parameters C and C_α have different units, for α approximately equal to 1, we can try to put these parameters together to see if their values are identified correctly. The presented table confirms that achieved results are proper, because their values are quite close to the real values of ultracapacitor parameters.

The RMSE of identification is presented in the Table 6.

Achieved values confirm good results of the model identification. In the case of using new model, we can see that Magnitude diagram has better accuracy than the Phase one. Examples of the results of ultracapacitor modeling for the new model are presented in Figures 7(a) and 7(b). The maximum identification error of magnitude and phase is less than 0.8 db and 3^0, respectively. One of the identification confirmations is what we can see in Figure 8. These plots present the measured and theoretical diagram for the quadripole system with ultracapacitor of an approximately capacity equal to 0.047 F.

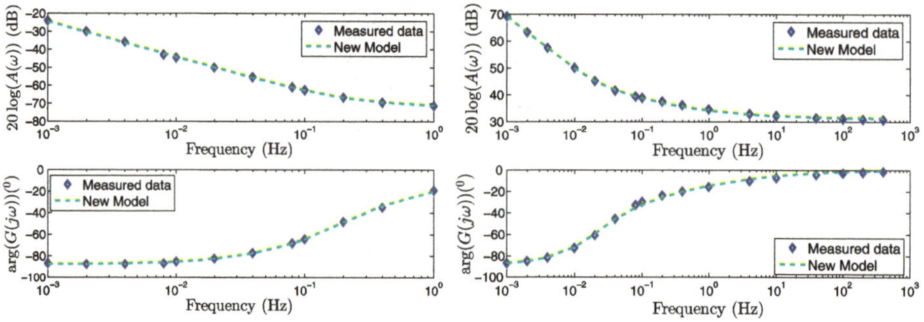

(a) Measured and New Model theoretical Bode's diagrams of ultracapacitor 3000F.

(b) Measured and New Model theoretical Bode's diagrams of ultracapacitor 0.047F.

Figure 7: Measured and New Model theoretical Bode's diagrams.

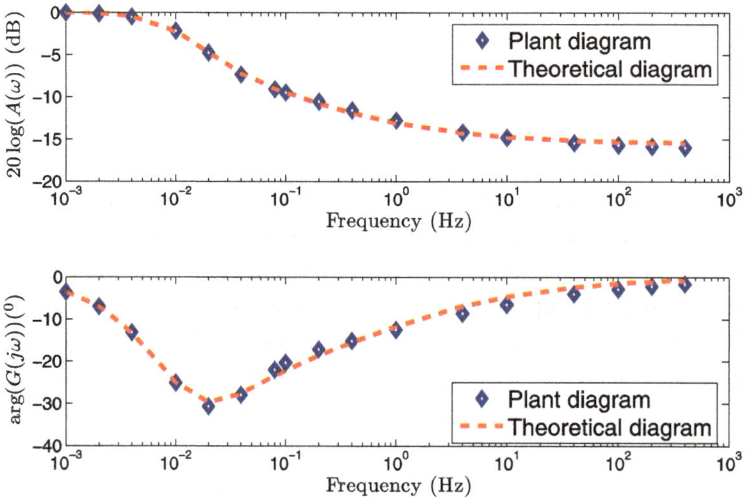

Figure 8: Measured and New Model theoretical Bode's diagrams of system for ultracapacitor 0.047 F.

Having a confirmation of good ability of a new model for ultracapacitor modeling, we can examine the Half Order model ($\alpha = 0.5$). This model has more useful properties because fractional capacity parameters is equivalent to standard capacity.

4.4 System identification using the Half-Order model

The Half-Order model is the ultracapacitor model derived using the diffusion partial differential equation:

$$G_{0.5} = R_c + \frac{\sqrt{1 + Ts}}{Cs},$$

Table 7: Identified ultracapacitors parameters for half–order model.

Capacitor	T	C (F)	R_c (Ω)	C_{real} (F)	R_{real} (Ω)
0.047 F	6.5231	0.056	35	0.06	32
0.1 F	18.5672	0.1	58	0.1	42
0.33 F	73.2900	0.27	32	0.27	28
1500 F	1.3059	1371	0.25 m	1189	0.26 m
3000 F	0.9668	2446	0.13 m	2435	0.13 m

Table 8: Error of half order model identification.

Capacitor	Magnitude RMSE	Phase RMSE
0.047 F	0.3175	1.5169
0.1 F	0.3546	1.4971
0.33 F	0.5053	2.0026
1500 F	0.9533	2.7004
3000 F	0.6875	2.8670

where R_c is the ultracapacitor resistance, C is the ultracapacitor capacity in the traditional definition, and T is a parameter of capacity decrease with frequency. This model can be also achieved as a particular case of the Davidson–Cole model for $\alpha = 0.5$. However, this model has the simplest structure, its construction allows for model validation in the time domain not only for a high capacity ultracapacitor but also for a low capacity one using experimental setups presented at the beginning of this chapter.

Comparison of the model parameters achieved as a result of identification by diagrams matching with real ultracapacitors' parameters are presented in Table 7.

The summary of the identification process is presented in Table 8. It can be noticed that the root mean square error of this experiment is also small which means that the Half-Order model is precise enough for ultracapacitor modeling. The examples of achieved Bode diagrams are presented in Figures 9(a) and 9(b).

Since in the Half-Order model parameter C describes the device capacity in the traditional form, then it can be compared with the traditional capacity. The f_c dependent on T, for which the equivalent capacity decreases two times for the Half-Order model is given as follows:

$$f_c = \frac{\sqrt{2^4 - 1}}{2\pi T}$$

and these values for tested ultracapacitors are summarized in Table 9.

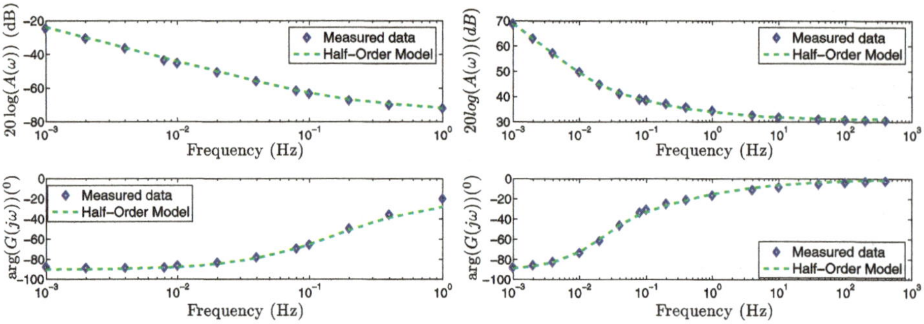

(a) Measured and theoretical Bode's diagrams of ultracapacitor 3000F.

(b) Measured and theoretical Bode's diagrams of ultracapacitor 0.047F.

Figure 9: Measured and theoretical Bode's diagrams.

Table 9: Values of frequency for which equivalent capacity decreases two times.

Capacitor	T	f_c (mHz)
0.047 F	6.5231	94.5
0.1 F	18.5672	33.2
0.33 F	73.29	8.41
1500 F	1.3059	472
3000 F	0.9668	637.6

Achieved values are very close to the values obtained for the Davidson–Cole model. Also, this model shows that for a low capacity ultracapacitor frequency of two time capacity decrease is smaller for higher capacity, and in the case of a high capacity supercapacitor, the value of this parameter is increasing with a capacity of the device.

Because of physical model parameters and order $\alpha = 0.5$ it is possible to validate an achieved model for low capacity ultracapacitors using the relation presented in Lemma 3.2. In Figures 10(a), 10(b), the step responses of an ultracapacitor of a nominal capacity 0.33 F and 0.047 F and for the voltage input signal $v(t) = 5\,\mathrm{V}$ are presented. Presented validation also confirms that the achieved model is proper for ultracapacitor modeling.

Presented results confirm good accuracy of the model achieved from a normal diffusion equation with the dynamics of physical systems; however, they are a little worse than results achieved for Davidson–Cole's model. The greatest advantage of the Half-Order model is the integer system order in the denominator which implies the use of physical system parameters able to compare with parameters of a traditional capacitor.

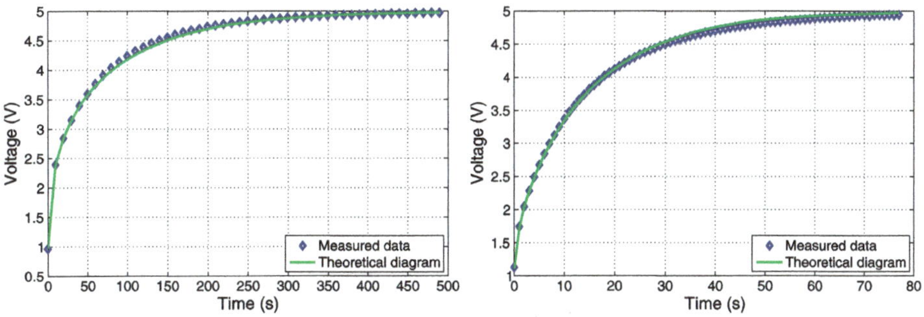

(a) Step response of model $G_{0.5}$ for system with ultracapacitor 0.33F.

(b) Step response of model $G_{0.5}$ for system with ultracapacitor 0.047F.

Figure 10: Step responses of model $G_{0.5}$.

5 Resonance in circuit with ultracapacitors

In analyzing the resonance phenomena, many authors treat ultracapacitors as ordinary classical capacitors. They neglect the fact that ultracapacitors have different dynamics which influences the every phenomena related to its operation and circuits containing it. In [22], it was shown that it is necessary to redefine the equation for the resonance frequencies of the electric circuit with ultracapacitors. In following the authors, we should focus on the spectral transfer function corresponding to anomalous diffusion model 22:

$$Z_{\mathrm{uc}}(j\omega) = R_c + \frac{\sqrt{1+T(j\omega)^{\alpha}}}{C_{\alpha}(j\omega)^{\alpha}}, \tag{55}$$

which can be approximated in the following way:

$$Z_{\mathrm{uc}}(j\omega) \approx \begin{cases} R_c + \frac{\sqrt{T}}{C_{\alpha}(j\omega)^{\alpha/2}} & \text{for } \omega \gg \frac{1}{T}, \\ R_c + \frac{1}{C_{\alpha} j\omega^{\alpha}} & \text{for } \omega \ll \frac{1}{T}, \end{cases} \tag{56}$$

where $R_c > 0$, $T > 0$, C_{α}, and $\alpha \in (0,1)$. The resonance phenomenon of a serial RLC connection in traditional form takes place when an imaginary part of impedance

$$Z = R + j\omega L + \frac{1}{j\omega C} \tag{57}$$

is equal 0. What is equivalent to say is that resonance occurs when the magnitude of the impedance is minimal (as a function of ω), which means that $\omega_{\mathrm{LC}} = \frac{1}{\sqrt{LC}}$. This role is changed when the fractional model of the ultracapacitor has been used. For a model of circuit with an ultracapacitor, we can define the resonance as a frequency for which the magnitude of impedance takes minimal value, or the frequency for which the imaginary part of the impedance is equal 0.

5.1 Simplified model for high frequencies

When $\omega \gg \frac{1}{T}$ can be used, the capacitive reactant of a simplified ultracapacitor model (56) is given by

$$Z_c(\omega) = \mathrm{Im}[G_c(j\omega)] = \mathrm{Im}\left[R + \frac{\sqrt{T}}{C_\alpha (j\omega)^{\frac{\alpha}{2}}}\right]$$
$$= \mathrm{Im}\left[R + \frac{\sqrt{T}}{C_\alpha \omega^{\frac{\alpha}{2}}(\cos\frac{\pi}{2}\frac{\alpha}{2} + j\sin\frac{\pi}{2}\frac{\alpha}{2})}\right]$$
$$= \mathrm{Im}\left[R + \frac{\sqrt{T}(\cos\frac{\pi\alpha}{4} - j\sin\frac{\pi\alpha}{4})}{C_\alpha \omega^{\frac{\alpha}{2}}}\right] = -\frac{\sqrt{T}}{C_\alpha \omega^{\frac{\alpha}{2}}}\sin\frac{\pi\alpha}{4}.$$

The real part of this reactant is

$$Z_r(\omega) = R + \frac{\sqrt{T}}{C_\alpha \omega^{\frac{\alpha}{2}}}\cos\frac{\pi\alpha}{4}. \tag{58}$$

In serial connection of ultracapacitor and inductance, the voltage resonance occurs when $\omega L + Z_c(\omega) = 0$, that is,

$$L\omega = \frac{\sqrt{T}}{C_\alpha \omega^{\frac{\alpha}{2}}}\sin\frac{\pi\alpha}{4}. \tag{59}$$

The last equation yields

$$\omega_{\mathrm{lowLC}_\alpha} = \left(\frac{\sqrt{T}\sin\frac{\pi\alpha}{4}}{LC_\alpha}\right)^{\frac{1}{1+\frac{\alpha}{2}}}. \tag{60}$$

Recall that for an integer order model

$$\omega_{\mathrm{LC}} = \frac{1}{\sqrt{LC}}, \tag{61}$$

thus, setting $C_\alpha = C$, one obtains

$$\omega_{\mathrm{lowLC}_\alpha} = \left(\omega_{\mathrm{LC}}^2 \sqrt{T}\sin\frac{\pi\alpha}{4}\right)^{\frac{1}{1+\frac{\alpha}{2}}}. \tag{62}$$

Additionally, following [22], the approximated formula for the resonance frequency which maximized the current on RLC circle is

$$\omega = \frac{1}{2}\sqrt[3]{\frac{T}{L^2 C_\alpha^2}}. \tag{63}$$

5.2 Simplified model for low frequencies

For $\omega \ll \frac{1}{T}$, one has

$$\begin{aligned} Z_{\mathrm{uc}}(\omega) &= R_{\mathrm{c}} + j\omega L + \frac{1}{C_\alpha (j\omega)^\alpha} \\ &= R_{\mathrm{c}} + j\omega L + \frac{1}{C_\alpha \omega^\alpha e^{j\frac{\pi}{2}\alpha}} \\ &= R_{\mathrm{c}} + \frac{\cos\frac{\pi\alpha}{2}}{C_\alpha \omega^\alpha} + j\left(\omega L - \frac{\sin\frac{\pi\alpha}{2}}{C_\alpha \omega^\alpha}\right), \end{aligned}$$

hence the resonance frequency is

$$\omega_{\mathrm{hiLC}_\alpha} = \left(\frac{\sin\frac{\pi\alpha}{2}}{C_\alpha \omega^\alpha}\right)^{\frac{1}{\alpha+1}}.$$

5.3 Complete model

The spectral function of a RLC serial connection with an ultracapacitor is

$$G(j\omega) = R + \frac{(1 + Tj\omega)^\alpha}{Cj\omega} + j\omega L = R + \frac{(1 + T^2\omega^2)^{\frac{\alpha}{2}} e^{j\varphi\alpha}}{Cj\omega} + j\omega L, \tag{64}$$

where $\varphi = \arctan(T\omega)$.

Explicit separation of the real and imaginary parts yields

$$G(j\omega) = R + \frac{\sin(\varphi\alpha)(1 + T^2\omega^2)^{\frac{\alpha}{2}}}{C\omega} + j\left(\omega L - \frac{\cos(\varphi\alpha)(1 + T^2\omega^2)^{\frac{\alpha}{2}}}{C\omega}\right),$$

where $\varphi = \arctan(T\omega)$.

For $T\omega \gg 1$, one has $\varphi \approx \frac{\pi}{2}$, $1 + T^2\omega^2 \approx T^2\omega^2$. Resonance frequency (zero reactant)

$$\omega L = \frac{\cos(\varphi\alpha)(1 + T^2\omega^2)^{\frac{\alpha}{2}}}{C\omega},$$

for $\varphi \approx \frac{\pi}{2}$, after simple algebraic transformations, one obtains the following final formula for approximation value of resonance frequency:

$$\omega_{LC_\alpha} = \left(\frac{T^\alpha \cos\frac{\pi\alpha}{2}}{LC}\right)^{\frac{1}{2-\alpha}}. \tag{65}$$

For a complete model, the approximated formula for the resonance frequency which maximized the current on the RLC circle can be derived using the following assumption:

$$T\omega \gg 1, \quad 1 + T^2\omega^2 \approx T^2\omega^2.$$

One obtains

$$\left[R + \frac{T^{\alpha}\sin(\varphi\alpha)}{C}\omega^{\alpha-1}\right](\alpha - 1)\frac{T^{\alpha}\sin(\varphi\alpha)}{C}\omega^{\alpha-2} + \left[\omega L - \frac{T^{\alpha}\cos(\varphi\alpha)}{C}\omega^{\alpha-1}\right] \times \left[L - (\alpha - 1)\frac{T^{\alpha}\cos(\varphi\alpha)}{C}\omega^{\alpha-2}\right] = 0. \qquad (66)$$

Equation (66) is highly nonlinear and difficult to solve analytically. The simplifications and assumptions one has to make in order to find an approximative solution yield the result (65), that is, the same as resonance frequency. Nevertheless, this result can be a good starting point for various numerical methods which was confirmed in [22].

6 Conclusion

In this chapter, the theoretical considerations and the experiments of ultracapacitors' modeling and identification using fractional-order models are presented. This results in high accuracy models which are essential for many control methods (like MPC; see, e. g., [8]). Results obtained confirm that the fractional-order model can very precisely represent the dynamics of an ultracapacitor in the frequency domain and in the time domain. Achieved results are crucial in the process of control of a system with an ultracapacitor during charging/discharging. The model which can describe ultracapacitor dynamics in the best way and for a wide range of frequencies is the model derived from the subdiffusion equation (New Model), but for this model it is impossible to compare achieved parameters with the parameters of a traditional capacitor because of nonphysical parameters in the denominator of the model transfer function. However, using this model, we proved that the value of the order for which the best dynamics accuracy was achieved is slightly smaller than 0.5 which could have been expected. The subdiffusion, which is a basic phenomenon used in the ultracapacitor for energy storage is a process very similar to the normal diffusion, so the order 0.5 is good enough for approximating the ultracapacitor dynamics. The Half-Order model, which was derived from the traditional diffusion equation, is not as precise, but it has one very important advantage; all parameters of the used model are in the traditional, well-known form, which allows us to compare the parameters and properties of an ultracapacitor with those of a traditional capacitor. The simulation results of ultracapacitor modeling were also achieved using the Davidson–Cole model, which can be also be derived from Quintana's model using the denominator order equal to 1. This model combines the physical parameters of ultracapacitors with standard capacitor dynamics, which is an effect of anomalous diffusion of ions movement. The full model presented by Quintana has not been considered in this paper, because it does not have capacity which can be represented in Farads, and what is also important, in the literature there is no formal derivation which suggests the justification for using the presented model form. The structure of Quintan's model is only the result of the characteristics matching for the enlarged Westerlund's model in the spectroscopy impedance experiments.

Bibliography

[1] J. L. Battaglia, O. Cois, J. C. Batsale, and A. Oustaloup, La dérivation non entière: Outil de modélisation en thermique, in A. Oustaloup and M. Guglielmi (eds.) *Les Sysémes á Dérivées Non Entières: Théorie et Applications*, vol. 7, 2001.

[2] F. Belhachemi, S. Raël, and B. Davat, A physical based model of power electric double-layer supercapacitors, in *Proceedings of IEEE Industry Applications Conference*, vol. 5, October 2000, pp. 3069–3076.

[3] J. P. Bouchaud and A. Georges, Anomalous diffusion in disordered media: statistical mechanisms, models and physical applications, *Physics Reports*, **195** (1990), 127–293.

[4] H. Brouji, J. M. Vinassa, O. Briat, N. Bertrand, and E. Woirgard, Ultracapacitors self-discharge modelling using a physical description of porous electrode impedance, in *Proceedings of IEEE Vehicle Power and Propulsion Conference (VPPC)*, September 2008.

[5] G. E. Carlson and C. A. Halijak, Approximation of fractional capacitors (1/s)1/n by a regular newton process, *IEEE Transactions on Circuit Theory*, **7** (1964), 210–213.

[6] D. W. Davidson and R. H. Cole, *Dielectric Relaxation in Glycerine*, *Journal of Chemical Physics*, **18** (1951), 1417.

[7] D. W. Davidson and R. H. Cole, Dielectric relaxation in glycerine, propylene glycol, and n-propanol, *Journal of Chemical Physics*, **19** (1951), 1484–1490.

[8] S. Domek, *Fractional-Order Differential Calculus in Model Predictive Control*, West Pomeranian University of Technology Academic Press, 2013.

[9] A. Dzieliński, G. Sarwas, and D. Sierociuk, Time domain validation of ultracapacitor fractional order model, in *Proceedings of the IEEE Conference on Decision and Control*, pp. 3730–3735, 2010.

[10] A. Dzieliński, G. Sarwas, and D. Sierociuk, Comparison and validation of integer and fractional order ultracapacitor models, *Advances in Difference Equations*, **2011** (2011) 11.

[11] A. Dzieliński, D. Sierociuk, and G. Sarwas, Some applications of fractional order calculus, *Bulletin of The Polish Academy of Sciences—Technical Sciences*, **58**(4) (2010), 583–592.

[12] P. Górecki, 2700 faradów, czyli super(ultra)kondensatory, *Elektronika dla Wszystkich*, **6** (2001), 21–24.

[13] J. W. Haus and K. W. Kehr, Diffusionnext term in previous termregular and disordered lattices next term, *Physics Reports*, **150** (1987), 263–406.

[14] T. Kosztołowicz, Subdiffusion in a system with a thick membrane, *Journal of Membrane Science*, **320** (2008), 492–499.

[15] J. S. Lai, S. Levy, and M. F. Rose, High energy density double-layer capacitors for energy storage applications, *IEEE Aerospace and Electronic Systems Magazine*, **7**, (1992), 14–19.

[16] I. Podlubny, L. Dorčák, and J. Misanek, Application of fractional-order derivatives to calculation of heat load intensity change in blast furnace walls, *Transactions of the Technical University of Košice*, **5**(5) (1995), 137–144.

[17] J. J. Quintana, A. Ramos, and I. Nuez, Identification of the fractional impedance of ultracapacitors, in *Proceedings of the 2nd IFAC Workshop on Fractional Differentiation and its Applications. IFAC FDA'06*, Porto, Portugal, 19–21 July 2006.

[18] D. Riu, N. Retière, and D. Linzen, Half-order modelling of supercapacitors, in *Proceedings of IEEE Industry Applications Conference, 39th IAS Annual Meeting, vol. 4*, pp. 2550–2554, 2004.

[19] G. Sarwas, *Modelling and control of systems with ultracapacitors using fractional order calculus*, PhD thesis, Warsaw University of Technology (Poland), 2012.
[20] D. Sierociuk, *Estymacja i sterowanie dyskretnych układów dynamicznych ułamkowego rzędu opisanych w przestrzeni stanu*, PhD thesis, Warsaw University of Technology (Poland), 2007.
[21] D. Sierociuk and I. Petráš, Modeling of heat transfer process by using discrete fractional-order neural networks, in *Proceedings of 16th International Conference on Methods and Models in Automation and Robotics, MMAR'11 Międzyzdroje*, Aug. 2011.
[22] D. Sierociuk, G. Sarwas, and M. Twardy, Resonance phenomena in circuits with ultracapacitors, in *2013 12th International Conference on Environment and Electrical Engineering*, pp. 197–202, May 2013.
[23] B. M. Vinagre, *Modelado y Control de Sistemas Dinámicos caracterizados por Ecuaciones Integro-Diferenciales de Orden Fraccional*, PhD thesis, Escuela Técnica Superior de Ingenieros Industriales, Universidad Nacional de Educación a Distancia, 2001.
[24] S. Westerlund and L. Ekstam, Capacitor theory, *IEEE Transactions on Dielectrics and Electrical Insulation*, **1** (1994), 826–839.

Concepción A. Monje, Bastian Deutschmann, Christian Ott, and Carlos Balaguer

Control of a soft robotic link using a fractional-order controller

Abstract: The purpose of this work is to present a novel control approach for a tendon-driven elastic continuum mechanism using a Fractional-Order (FO) controller. The mechanism is composed of a silicon continuum, tendons and antagonistic actuation which yields a highly complex mechanical model. This model is heavily simplified to that of a linear time invariant second-order system and a robust control approach is applied to cope with the neglected dynamic effects. Apart from the specifications for the speed and the overshoot of the system response, the controller must be extensively robust to model parameter mismatches and uncertainties. The FO controller is used to meet the control specifications, taking advantage of the introduction of its fractional order α. Simulation and experimental data are presented to validate the approach in comparison to results from two standard integer-order controllers that are designed using the same specifications.

Keywords: Fractional-order control, model-based control, soft robotics, robust control

MSC 2010: 93C80

1 Introduction

In recent years, inherently soft robots got in focus of research. These systems ensure that collisions with humans or the environment can be handled by the hardware of the robot. The soft robotic system examined in this chapter is a tendon-driven elastic continuum mechanism (ECM) and the softness is gained by the use of silicon. Its compliance allows the structure to deform as a whole, no deflection of joints can be measured directly for feedback control.

The majority of continuum mechanisms are long, thin tubes used in medical applications as a steerable needle and the flexibility is mainly used to bend continuously whereas axial or shear deformations can be neglected. The present system, however, is able to deform considerably in all directions due to a large cross-section and the

Concepción A. Monje, Carlos Balaguer, Systems Engineering and Automation Department, University Carlos III of Madrid, Av. de la Universidad 30, 28911, Leganés, Madrid, Spain, e-mails: cmonje@ing.uc3m.es, balaguer@ing.uc3m.es
Bastian Deutschmann, Christian Ott, Institute of Robotics and Mechatronics, German Aerospace Center (DLR), Oberpfaffenhofen, 82234, Wessling, Germany, e-mails: Bastian.Deutschmann@dlr.de, Christian.Ott@dlr.de

https://doi.org/10.1515/9783110571745-014

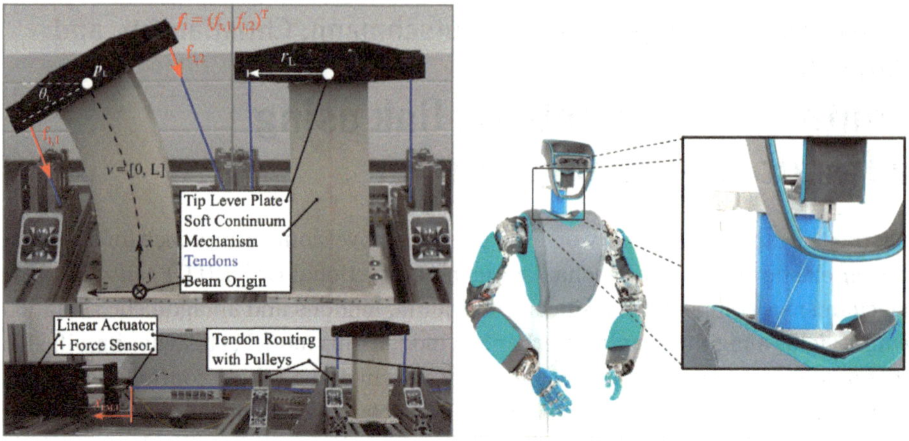

Figure 1: Left: Test setup including two linear actuators which are connected via tendons (blue) to the tip lever plate of the soft continuum mechanism. top: Close up in straight and deflected configurations. bottom: Left side of the antagonistic setup. Right: Tendon-driven soft continuum mechanism as a neck in a humanoid robot [12].

use of silicon [3]. It is a planar prototype for a spine and neck for a future humanoid robot [12] (Figure 1). To control the motion of such a system, model-based control approaches are proposed in the literature. For example, a linearized model is used in [6] to design a vibration damping set point controller and a feedback-linearization based on a nonlinear dynamic model is applied in [2].

A big challenge for model-based approaches are parameter mismatches or changing loading conditions, as they cannot be handled by these approaches instantaneously. Fractional-Order (FO) control can be used for robot motion control with considerable structural flexibility, as successfully demonstrated in [8]. The introduction of the fractional order of the controller provides a performance enhancement of the system even when its structural flexibility incorporates complex dynamics and demands high robustness requirements.

This chapter will provide a more detailed discussion of the control approach established by [11] including the recent passivity investigation from [4]. The approach is a model-based approach in the task space of the robot and the complex system is heavily simplified to that of a second-order system. With that, a FO controller [9] is designed in the frequency domain and modeling errors, changing loading conditions, or external disturbances are covered by the inherent robustness properties of the approach. Two well-established integer-order controllers are designed with the same specifications in the frequency domain and are compared to the FO controller with respect to (w.r.t.) two established performance measures in experiments. These experiments cover the desired nominal behavior and investigate the robustness of each controller by varying model parameters and applying high-frequency noise through external disturbances.

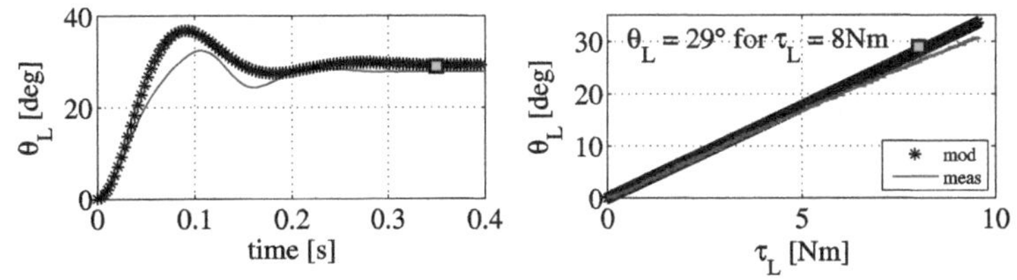

Figure 2: Comparison of the model (mod) and the real system (meas). Left: Dynamic characteristic with an input step of $\tau_L = 8$ Nm. Right: Static characteristic while the mechanism is slowly moved through the workspace.

2 Problem statement and system model

The problem at hand is to control the tip angle $\theta_{\mathrm{L}} \in \mathbb{R}$ of the planar soft continuum mechanism (see Figure 1), whereas the control action is the external torque $\tau_{\mathrm{L}} \in \mathbb{R}$ at the tip generated by two, antagonistically acting tendon forces $\boldsymbol{f}_{\mathrm{t}} \in \mathbb{R}^2$ of each actuator while measuring tendon forces and tendon positions $\boldsymbol{x}_{\mathrm{LM}} \in \mathbb{R}^2$ only. As the full dynamic model of this mechanism includes partial differential equations and a nonlinear dynamic coupling between the tip and the actuator motion, it is not suitable for the design of a linear controller. Instead, a linear time invariant transfer function $G(s) \in \mathbb{R}$ is used as an approximation for the dynamics of the tip position $\Theta_{\mathrm{L}}(s)$ w.r.t. the torque input $\tau_{\mathrm{L}}(s)$,

$$G(s) = \frac{K\omega_{\mathrm{n}}^2}{s^2 + 2\delta\omega_{\mathrm{n}}s + \omega_{\mathrm{n}}^2} = \frac{\Theta_{\mathrm{L}}(s)}{\tau_{\mathrm{L}}(s)}, \tag{1}$$

with the steady-state gain $K = 0.0631$, the eigenfrequency of the system $\omega_n = 37.8021$, and the linear damping $\delta = 0.4$. The parameters have been estimated experimentally by measuring step responses. A comparison between the model and the real system can be found in Figure 2 for the dynamic and the static case.

3 Control strategy

The model of the soft robotic system has been approximated to that of a second- order system with the transfer function (1) and the parameters in Table 1. The input to the system is the torque applied by the tendons, τ_{L} [Nm], and the output is the angular position of the tip, θ_{L} [rad]. Three controllers will be designed for the system: a FO controller, a PID controller, and a lag compensator. A discussion on the required specifications as well as the constraints that need to be fulfilled are presented in the following.

Table 1: Parameters of the simplified model.

steady-state gain	$K = 0.0631$
eigenfrequency	$\omega_n = 37.8021$
damping	$\delta = 0.4$
poles of the system	$p_{1,2} = -15.1208 \pm 34.6462i$
magnitude at $\omega_{cg} = 3$ rad/s	$\lvert G(j\omega_{cg})\rvert_{dB} = -22.6$ dB
phase at $\omega_{cg} = 3$ rad/s	$\arg(G(j\omega_{cg})) = -3.6^\circ$

3.1 Control specifications

The objective of this control approach is to design a controller $C(j\omega)$ in the frequency domain so that the system is robust to system uncertainties and load disturbances. For this reason, specifications related to gain crossover frequency ω_{cg}, phase margin φ_m, and a robustness constraint are going to be considered, due to their important significance regarding performance and stability [10]. The design problem is formulated as follows:

- *Phase margin and gain crossover frequency specifications.* Gain and phase margins are important measures of robustness. It is known that the phase margin is related to the damping of the system and, therefore, can also serve as a performance measure [5]. The equations that define the gain crossover frequency and the phase margin are

$$\left|C(j\omega_{cg})G(j\omega_{cg})\right|_{dB} = 0\ \text{dB}, \tag{2}$$

$$\arg(C(j\omega_{cg})G(j\omega_{cg})) = -\pi + \varphi_m. \tag{3}$$

- *Robustness to variations in the gain of the plant.* According to [1], this requirement can be addressed by demanding

$$\left.\frac{d(\arg(C(j\omega)G(j\omega)))}{d\omega}\right|_{\omega=\omega_{cg}} = 0\ \text{s}. \tag{4}$$

The second specification forces the phase of the open loop system $F(j\omega) = C(j\omega)G(j\omega)$ to be flat at $\omega = \omega_{cg}$ and so, to be almost constant within an interval around ω_{cg}. It means that the system is more robust to gain changes and the overshoot of the response is almost constant within a gain range (iso-damping property of the time response). As for this robustness specification, there are two important considerations to make:

1. In our system, the gain of the model is mainly affected by the load of the device (top masses or pending masses in the experimental results section). Besides, a change in the pulley location can also result in a different gain value. Therefore, with this specification the controller ensures the robustness of the system to changes in its physical parameters, guaranteeing a good performance for a range of uncertainty around the nominal configuration.

2. The design method proposed here for the FO controller will be based on a graphical approach, avoiding the analytical resolution of (4) for the obtaining of the controller parameters; see Section 3.2.

For the case study presented in this work, the following constraints are used in order to compare the different controllers:
- The workspace of the system is $\theta_L = \pm 0.35\,\mathrm{rad}$ ($\approx 20°$).
- The pretension force is $f_{pre} = 10$ N.
- The maximum control torque is $\tau_L = 8$ Nm as the maximum tendon force is $f_{t,max} = 110$ N.
- It is desirable to have a negligible steady-state error and a step response with a settling time of ≈ 1 s, which corresponds to about $\omega_{cg} = 3\,\mathrm{rad/s}$.
- To achieve a better performance, it is desirable to have a slight overshoot. Thus $\varphi_m = 80°$ is demanded.

The different controllers proposed next will be designed ensuring that these specifications are fulfilled. For the sake of comparison, two integer-order controllers including a Lag compensator and a PID controller are designed with the specifications above. The latter ensures no steady-state error as it includes a pure integral term. However, this integral action could brake the system, for example, a tendon, when the mechanism is in contact with the environment.

Furthermore, a FO PD$^\alpha$ controller is designed as it is able to fulfill the control specifications and ensures good robustness properties. A deeper discussion on the advantages of using this controller over the previous mentioned integer-order ones will be presented in Section 4 of experimental results.

3.2 FO controller

As a novel control approach for the soft continuum mechanism, we propose the use of the generalized FO PD$^\alpha$ controller, formulated as [9]:

$$PD^\alpha(s) = \left(\frac{\lambda s + 1}{x\lambda s + 1}\right)^\alpha, \tag{5}$$

which corresponds to a FO lead compensator that can be identified as a PD$^\alpha$ controller plus a noise filter. The parameters $\alpha, \lambda, x \in \mathbb{R}$ are now obtained by the subsequent method.

The design method proposed here for the tuning of the controller is based on the work previously conducted by the authors in [10]. As an improvement, considerations on the dual use of the controller as a lead-lag compensator are introduced. For a specified phase margin φ_m and gain crossover frequency ω_{cg}, the following relationship for the open-loop system is given in the complex plane:

$$G(j\omega_{cg}) \cdot \left(\frac{j\lambda\omega_{cg} + 1}{jx\lambda\omega_{cg} + 1} \right)^{\alpha} = e^{j(-\pi+\varphi_m)},$$
$$\left(\frac{j\lambda\omega_{cg} + 1}{jx\lambda\omega_{cg} + 1} \right)^{\alpha} = \frac{e^{j(-\pi+\varphi_m)}}{G(j\omega_{cg})} = a_1 + jb_1,$$
$$\left(\frac{j\lambda\omega_{cg} + 1}{jx\lambda\omega_{cg} + 1} \right) = (a_1 + jb_1)^{1/\alpha} = a + jb, \tag{6}$$

where $G(s)$ is the plant to be controlled and (a_1, b_1) is called the "*design point,*" which is directly obtained from the system (1) with Table 1 and the values of φ_m and ω_{cg}. After solving (6) for x and λ, we have

$$x = \frac{a - 1}{a(a-1) + b^2}, \tag{7}$$
$$\lambda = \frac{a(a-1) + b^2}{b\omega_{cg}}. \tag{8}$$

Studying the conditions for (a) and (b) to find a solution, it can be concluded that a lead compensator is obtained when $a > 1$ and $b > 0$ [9]. According to this condition and considering (7) and (8), the lead compensation regions in the complex plane for different positive values of α are obtained, as shown in Figure 3, left. The zone to the right of each curve is the lead region, and any design point in this zone can be fulfilled with a FO compensator having a value of α equal or bigger than the one defining the curve which passes through the design point α_{min}. For instance, for the design point $\alpha_{min} = 0.48$ in Figure 3, left, Figure 3, right shows the pairs (x, λ) obtained for each value of α in the range $\alpha_{min} = 0.48 \leq \alpha \leq 2$. Therefore, once the design point is computed, the selection of the controller parameters is direct and flexible through the use of the curves in Figure 3.

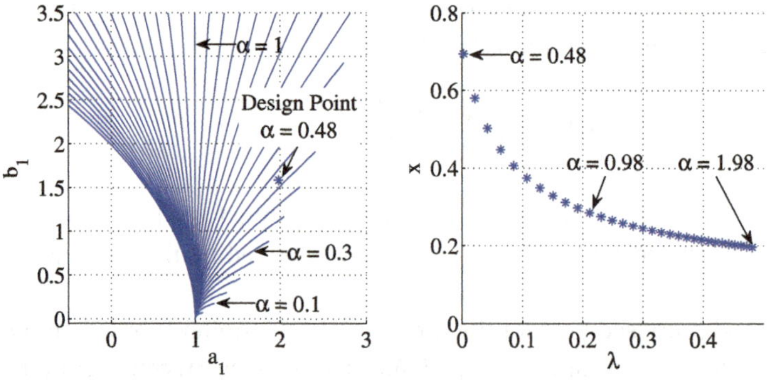

Figure 3: Left: Lead regions for the controller $C(j\omega_{cg})$ for $0 \leq \alpha \leq 2$. Right: Parameter pairs (x, λ) for $\alpha_{min} \leq \alpha \leq 2$.

For the sake of robustness to gain variations, and according to the discussion in Section 3.1, the phase curve of the open-loop system should present a flat shape (zero slope) around the frequency of interest ω_{cg}. It is then desirable for the FO controller to contribute with a phase curve that both guarantees the amount of phase required for the fulfillment of the phase margin specification at that frequency and ensures that the flat phase constraint for the open-loop system is achieved.

Taking into account the phase characteristic of the system $G(s)$ (see (1) with Table 1), which presents a slope close to zero at $\omega_{cg} = 3\,\text{rad/s}$, the condition for the phase curve of the FO controller is also to have a flat behavior around that frequency, ensuring this way that the open-loop system phase is as flat as possible at ω_{cg}. It can be observed from Figure 3 that the maximum value of x is obtained for α_{min}. In other words, by choosing the minimum value α_{min}, the distance between the zero and the pole of the compensator will be the maximum possible, which ensures that the phase curve of the compensator is the flattest possible and variations in a frequency range centered at ω_{cg} will not produce a significant phase change as in other cases. That improves the robustness of the system. This will be the criterion to be followed here and the controller will be tuned according to this α_{min} condition.

In case a phase lag (φ_{lag}) is required for the compensator in order to fulfill the design specifications, it will be designed as a lead compensator giving a phase $|\varphi_{lead}| = |\varphi_{lag}|$, and later the sign of α will be changed so that the phase contribution is negative. Besides, it has to be taken into account that a change in the sign of α for the lag compensation leads to an inverted magnitude of the designed compensator. So, in order to keep the gain unchanged (fulfilling already the specification of crossover frequency), the lag compensator should be multiplied by a gain $k_{lag} = 1/|G(j\omega_{cg})|^2$, as can be reached from (6). Therefore, the fractional lag compensator will be given by

$$C_{FOC}(s) = k_{lag}\left(\frac{\lambda s + 1}{x\lambda s + 1}\right)^{-\alpha}, \tag{9}$$

with α a positive real number.

Now a FO controller of the form in (9) is designed so that the specifications $\varphi_m = 80^\circ$ at $\omega_{cg} = 3\,\text{rad/s}$ are achieved, together with the requirement of robustness to gain variations. After the design process, the resulting controller is

$$C_{FOC}(s) = 905.7688\left(\frac{6.97s + 1}{0.0069s + 1}\right)^{-1.12}, \tag{10}$$

with $\lambda = 6.97$, $x = 9.9291 * 10^{-04}$, $\alpha = 1.12$. The Bode plots of the open-loop system with this FO controller are shown in Figure 4, where it can be observed that the phase margin, gain crossover frequency, and robustness constraints (flat phase) are fulfilled.

In order to evaluate the robustness of the closed-loop system to gain variations, Figure 5, left presents the step responses for varying gains K (1), showing that the overshoot keeps almost constant.

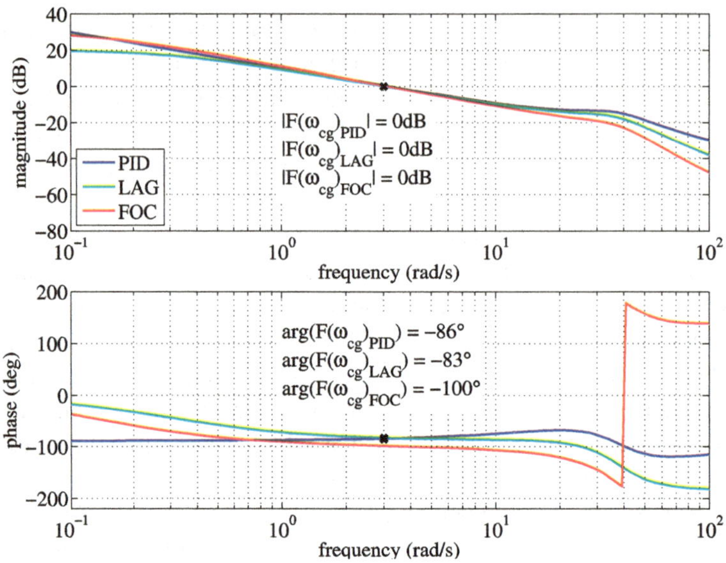

Figure 4: Bode plots of the open-loop system of the three designed controller.

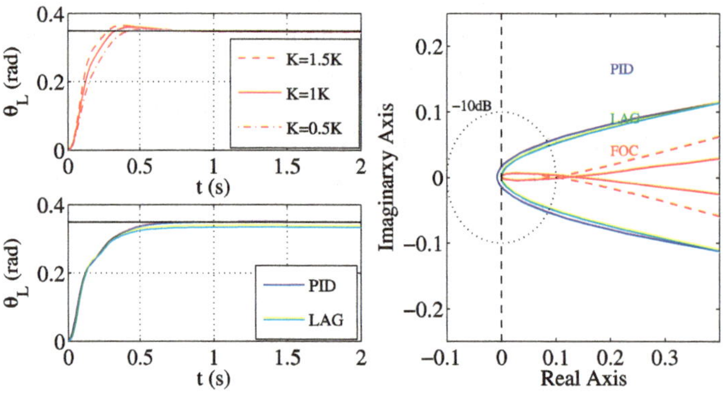

Figure 5: Left: Nominal step responses of the closed-loop system. Top: $C_{FOC}(s)$ controller with different values of the gain (Table 1): K, $0.5K$, and $1.5K$. Bottom: $C_{PID}(s)$ and $C_{LAG}(s)$. $\theta_{L,d}$ in black. Right: Nyquist plots of the $P(s)$ functions corresponding to the three controllers. The dashed, red curve corresponds to the finite approximation of the FO controller.

Integer-order approximation

When FO controllers have to be implemented, fractional transfer functions are usually replaced by integer transfer functions with a behavior close enough to the one desired, but much easier to handle. There are many different ways of finding such approximations but unfortunately it is not possible to say that one of them is the best, because even though some of them are better than others in regard to certain characteristics, the relative merits of each approximation depend on the differentiation order, on whether one is more interested in an accurate frequency behavior or in accurate time responses, or on how large admissible transfer functions may be, among others. A good review of these approximations can be found in [9].

In our particular case, to implement the FO controller in the real setup, the transfer function (10) has been approximated to that of an integer-order controller using a frequency identification method performed by the MATLAB function *invfreqs*. An integer-order transfer function is obtained which fits the frequency response of the FO controller in a frequency of two decades around the gain crossover frequency, with two poles/zeros:

$$\tilde{C}_{FOC}(s) = \frac{0.3908s^2 + 67.3s + 774.4}{s^2 + 7.465s + 0.8204}. \tag{11}$$

This method has been chosen due to its accuracy in the frequency range of interest, as successfully applied for real implementations in [8, 10], but any other of the techniques in [9] could also be suitable for this purpose.

3.3 Integer-order PID controller

A PID controller is designed with the graphical user interface of the *Automated PID Tuning* method incorporated in the *Control Systems Toolbox* in MATLAB. The tuning program uses the *Robust Response Time* method which allows the specification of the bandwidth and the phase margin to be $\omega_{cg} = 3$ rad/s and $\varphi_m = 90^\circ$, respectively. Note that the maximum contribution in phase of this controller is -90°, which makes it impossible to fulfill the phase margin specification of 80° and, in general, a wider range of values for the phase margin. The PID controller designed following these instructions is

$$C_{PID}(s) = 41.17 \cdot \frac{(1 + 0.024s)^2}{s}, \tag{12}$$

whose open-loop Bode plots are presented in Figure 4. The nominal step responses of the closed-loop system is shown in Figure 5, bottom left.

A physical interpretation of this PID controller is that by choosing both zeros of the controller to be at $z_{1,2} \approx 41.6$ rad/s, the controller approximately cancels the poles of the system (Table 1), which extends the inherently flat phase to the system (see Figure 4). As the chosen design method maximizes bandwidth and optimizes phase margin, it is a compromise between robustness and performance.

3.4 Integer-order lag compensator

An integer-order lag compensator is designed for the system. However, the integer order limits the fulfillment of the phase margin specification, since the maximum phase contribution is -90°. Besides, the flat phase robustness constraint cannot be guaranteed due to the lack of an extra parameter.

For the case study presented in this work, an integer-order lag compensator is designed fulfilling the frequency specifications $\omega_{cg} = 3$ rad/s and $\varphi_m = 90^\circ$. The general transfer function of an integer-order lag compensator is

$$C_{LAG}(s) = \frac{s+z}{s+p}, \quad \text{with } z > p. \tag{13}$$

The pole p and the zero z are determined using the specifications that at $\omega_{cg} = 3$ rad/s the magnitude of compensator needs to be 22.6 dB with a phase of -90°. By solving (2) and (3), we find the integer-order lag compensator to be

$$C_{LAG}(s) = \frac{s + 39.74929}{s + 0.3007329}. \tag{14}$$

Figure 4 and Figure 5 show the Bode plots and the step response of the system with this lag compensator, respectively. In these simulation results, it can be seen that the steady-state error is higher in comparison with the FO and PID controllers.

3.5 Passivity analysis

The controller design in the previous section was based on a linear system model. It was shown that the proposed FO controller has advantageous robustness properties, for example, w.r.t. gain variations. The experimental results in Section 4 will confirm that the inaccuracy in the linear approximation of the system dynamics can be handled by these robustness properties. In this section, we want to discuss to which extent one can make analytical stability statements for the application of the (linear) FO controller to the nonlinear system dynamics. In particular, we look at the passivity properties of the closed-loop system. As a purely mechanical structure, the continuum mechanism clearly is a passive system with the physical energy as a storage function, considering the input τ_L and the output $\dot{\theta}_L$. This input–output pair represents the physical power between the controller and the mechanism. In the following, we will show that the designed FO controller represents a passive system w.r.t. the input $\dot{\theta}_L$ and the output $-\tau_L$, that is, in feedback interconnection with the mechanism. Since passivity is preserved by feedback interconnection of passive subsystems [14], we can then conclude passivity of the closed-loop system. This even extends to physical interaction of the mechanism with its environment. As long as the environment is passive, the same holds for the overall system.

For showing the passivity of the controller, we have to consider the transfer function $P(s) = -C_{FOC}(s)\frac{1}{s}$. Passivity requires $\mathrm{Re}(P(j\omega)) \geq 0$ (positive realness) for all frequencies ω, which can be checked with a Nyquist plot [13]. Figure 5, right shows the Nyquist plots of the P(s) functions corresponding to the three controllers discussed in the previous sections. One can easily observe that the FO controller as well as its finite-order approximation are positive real, while this is not the case for the PID controller.

Notice that this analysis assumes that the electrical dynamics of the force-controlled actuators is sufficiently fast so that it can be neglected, and thus the desired control torque τ_L can be realized instantaneously. While unmodeled dynamics at the level of the continuum mechanism can be handled in this way, unmodeled dynamics at the actuator level would need a different approach.

4 Experimental results

4.1 Test setup

The setup used for the experiments is depicted in Figure 1. Two linear actuators from Linmot© with incorporated position sensors are equipped with an axial force sensor from Omega© to measure position and external force at each slider. The designed motion controllers and the force control loops are implemented in MATLAB/Simulink© using real time workshop on a QNX-neutrino 6.5 target. An EtherCAT bus sends the generated control signals to the current controller of the linear motors and receives sensor data within a control cycle of 1 kHz. By applying EtherCAT as an industrial standardized real time communication protocol, a high degree of determinism is ensured. The sensor data are the actuator positions $\boldsymbol{x}_{LM}$ and tendon forces $\boldsymbol{f}_t$. The error level of the force sensors is ±0.1 N with 0.1% of linearity. The mean error of the position encoders is ±0.01 mm.

The soft continuum mechanism is molded out of silicon from Dragonskin© and connected to a 3D-printed bottom plate and a tip lever plate with an anchored area to ensure a firm connection. The polyethylene tendons are looped around a ball bearing, and are mounted to the soft continuum mechanism at a bearing seat on the top lever; see Figure 1. The tendons are routed with pulleys at each side toward the linear actuators.

4.2 Evaluation

As introduced earlier, the performance of each controller is evaluated using the ISE-index,

$$\mathrm{ISE} = \int_0^{t_{\mathrm{end}}} e(t)^2 dt, \tag{15}$$

with $e(t) = \theta_{L,d}(t) - \theta_L(t)$ being the error between the desired and the actual control output. The ISE measure is associated to the error energy and is a well-known performance index in control design [7]. A low ISE-value indicates a small control error, which is desirable. Furthermore, we want to assess the control action by

$$\mathrm{CA} = \frac{1}{t_{\mathrm{end}}} \int_0^{t_{\mathrm{end}}} (\tau_{\mathrm{L}}(t=0) - \tau_{\mathrm{L}}(t))^2 dt \qquad (16)$$

to indicate the amount of energy created by the control law. A low CA-value indicates less control action, which is also desirable.

4.3 Nominal responses

Figure 6, left shows the experimental step responses of the system in closed loop with the FO controller, the PID controller, and the integer-order lag compensator, together with their corresponding control laws. As can be seen in the figure, the FO controller presents a slightly underdamped response (peak value of 20.44°), as expected, and a soft control law under the saturation limit of 8 Nm. The final value of the response is 19.65°, presenting a small but negligible steady-state error due to the absence of an integral component in the controller. The PID controller presents zero steady-state error due to the effect of the integral action. However, the integrator may cause the drifting of the control action when an external force blocks the device, which may eventually break the tendon. The lag compensator presents a higher steady-state error (final value of 17.77°). Though the three controlled systems present a transient with similar settling time, the rise time for the FO controller is lower and its steady-state error negligible, providing a better response than the integer-order controllers. The assessment of the three controllers w.r.t. the performance measures is shown in Table 2. For the nominal response (first row), one can observe that the FO controller has the lowest ISE value, meaning the least error over time to the desired motion. However, the FO controller needs the highest control action, indicated in a higher CA value.

Table 2: ISE and CA of the three different controllers.

Controller		ISE m^2s			CA 10^4 $(Nm)^2$		
		PID	LAG	*FOC*	PID	LAG	*FOC*
Nominal		12.34	17.05	*7.63*	2.52	2.26	*2.94*
	1 Mot.	9.56	13.23	*5.94*	1.88	1.71	*2.19*
	097	20.91	27.34	*13.09*	1.87	1.75	*2.96*
Mass	194	22.07	27.65	*12.54*	1.82	1.73	*2.93*
	380	20.51	26.84	*12.90*	1.82	1.71	*2.84*
	566	20.72	26.46	*12.87*	1.73	1.63	*2.75*
Disturb.	0°	0.90	1.19	*0.69*	0.28	0.23	*0.12*
	20°	0.80	0.90	*0.40*	0.32	0.30	*0.25*

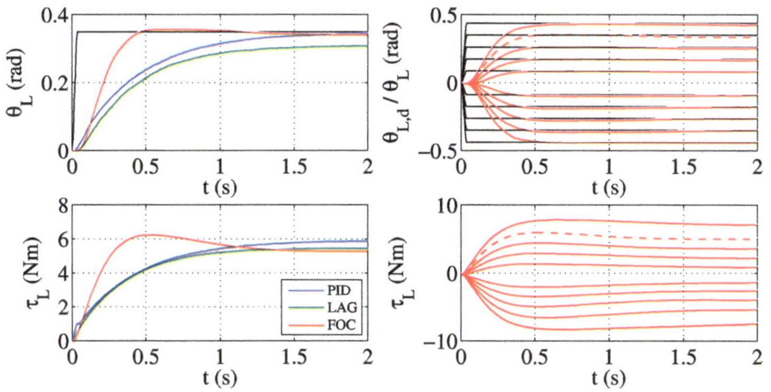

Figure 6: Left: Experimental step responses θ_L of the system in closed-loop and corresponding control inputs τ_L of the three controllers. Right: Experimental step responses of the system with the FO controller (red) for different step inputs with amplitudes in the range $(5^\circ, 25^\circ)$.

In order to test that the device behaves experimentally as a linear system around the nominal working point, the FO control system has been tested for different step inputs with amplitudes in the range $(5^\circ, 25^\circ)$. The results are presented in Figure 6, right, showing that only slight variations are appreciated w.r.t. the nominal performance.

4.4 Mass variation

To evaluate the robustness of the closed-loop system, step responses of the three controllers are measured when one of the tendons is disconnected. This means that mass and frictional damping of the real system decrease w.r.t. the model and its parameters. The results are shown in Figure 7, left for the three controllers. All the controllers result in a stable system.

Another, more critical robustness test of the closed-loop system is the placement of additional masses on the top of the mechanism. This mass implies a change in the total impedance of the system. Particularly, the assumption that the dynamics of the soft continuum mechanism are neglected is jeopardized. The results are shown in Figure 7, right for the FO controller and in Figure 8 for the PID and LAG controller. The additional masses on the top are $m_{ext} = \{97, 194, 380, 566, 663, 849\}$ g. The FO controller presents a robust performance to this variation of the overall impedance, while the PID controller and the lag compensator become unstable for $m_{ext} = 566$ g. For a mass variation, the performance measures were also evaluated and yield the same results as in the nominal response, being a low ISE value and a higher CA value for the FO controller (rows 2–6 in Table 2). The more mass loaded on top of the system, the less control action is necessary to deflect the mechanism, which is displayed in the course of the CA values.

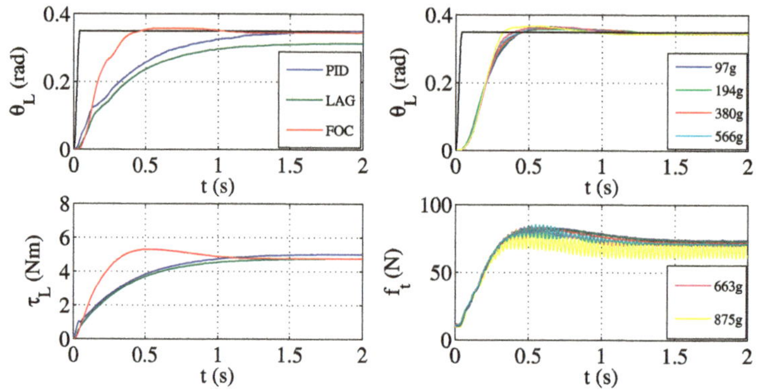

Figure 7: Left: Experimental step responses and tendon forces of the system in closed loop with only one connected motor. Right: Experimental step responses and tendon forces for different external masses placed on the top of the mechanism with the FO controller.

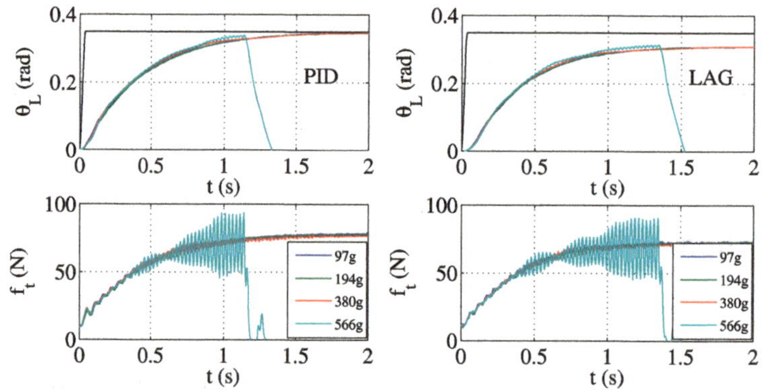

Figure 8: Experimental step responses and control laws of the system in closed loop for different external masses placed on the top of the mechanism. Left: PID controller. Right: LAG compensator.

4.5 External disturbances

A requirement for the present mechanism is to handle external disturbances to a certain extent. To investigate the behavior of the control system while it is subjected to external disturbances, the following experiments are carried out. In our approach, θ_L is controlled at static positions and an external mass of 849 g disturbs the control system in the following way:

1. A disk is attached to the tip lever plate with a cord; the cord is guided with an additional pulley to ensure a disturbance in parallel to the actuation.
2. The disk is dropped from a fixed distance, causing a high impact on the tip. The distance traveled by the disk is about 15 cm.

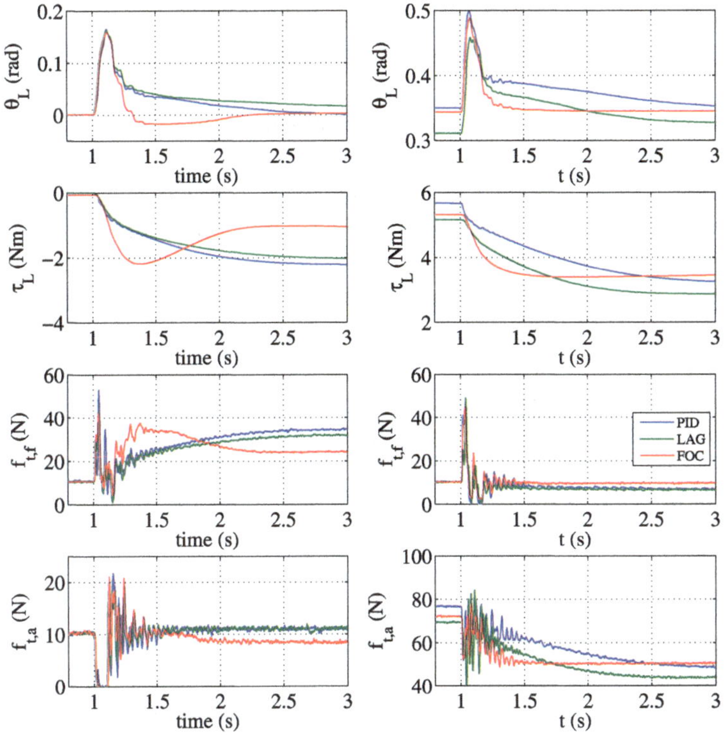

Figure 9: Experimental responses with the FO controller, the PID controller, and the integer-order lag compensator in the presence of disturbances. Left: 0° tip deflection. Right: 20° tip deflection.

This experiment has been done for two static positions, 0° and 20° tip deflection. The results are shown in Figure 9. A robust performance is obtained in all the cases. However, the disturbance rejection property of the FO controller is superior to the integer-order controllers. This is further substantiated by the two performance measures in the last two lines of Table 2. Here, a very interesting result can be reported as the FO controller still impresses with a low ISE value but also with a small control action with lower CA-values than both integer-order controllers.

The reported behavior is sustained by a video that shows the experiments; see https://www.youtube.com/watch?v=ivR-3bN0LVA&feature=youtu.be.

5 Conclusions and future works

This chapter reports on a model-based approach to control the position of a tendon-driven system. The complex mechanical system is heavily simplified to that of a linear time invariant second-order system. A fractional-order lag compensator is designed with this model and is proven to be passive. The required specifications for the closed-loop system are met with this controller and the incorporated robustness is able to cope with unmodeled dynamic effects.

In order to compare the FO controller, two standard linear controllers are designed with the same specifications whereas three different experiments are conducted: a PID controller, which is typically not passive, and a lag compensator, which is passive. The nominal motion corresponds to the system step response of the link angle $\theta_L = 20^\circ$. The three controllers prove an experimentally stable response. However, the rise time of the FO controller is superior to the others while having a negligible steady-state error of 0.35°, which is also reflected in a low ISE value compared to the integer-order controllers.

In order to evaluate the robustness w.r.t. additional parameter variations, several different masses are mounted on the tip of the soft continuum mechanism and again step responses are conducted. Again, the FO controller presented a superior performance, as it showed a stable behavior with higher additional masses than both integer-order controllers, which become unstable for masses higher than 566 g.

Since external collisions are also of interest, the controlled soft robotic system is investigated under these conditions as well. The controllers show a stable behavior in the testbed. However, the superior disturbance rejection of the FO controller needs to be emphasized. This property is also reflected in the performance indices that were investigated: a lower ISE value for the FO controller combined with a lower CA value, meaning that less control energy is necessary in order to track the desired position while being externally disturbed.

Bibliography

[1] Y. Chen, C. Hu, and K. L. Moore, Relay feedback tuning of robust pid controllers with iso-damping property, in *Decision and Control, 2003. Proceedings. 42nd IEEE Conference on*, vol. 3 pp. 2180–2185, IEEE, 2003.

[2] B. Deutschmann, A. Dietrich, and C. Ott, Position control of an underactuated continuum mechanism using a reduced nonlinear model, in *Decision and Control (CDC), IEEE Conference on*, pp. 5223–5230, 2017.

[3] B. Deutschmann, T. Liu, A. Dietrich, C. Ott, and D. Lee, A method to identify the nonlinear stiffness characteristics of an elastic continuum mechanism, *IEEE Robotics and Automation Letters*, **3** (2018), 1450–1457.

[4] B. Deutschmann, C. Ott, C. A. Monje, and C. Balaguer, Robust motion control of a soft robotic system using fractional order control, in *Conference on Robotics in Alpe-Adria Danube Region*, Springer, 2017.

[5] G. F. Franklin, J. D. Powell, and A. Emami-Naeini, *Feedback Control of Dynamics Systems*, Addison-Wesley, Reading, MA, 1994.

[6] I. A. Gravagne, C. D. Rahn, and I. D. Walker, Large deflection dynamics and control for planar continuum robots, *IEEE/ASME Transactions on Mechatronics*, **8**(2) (2003), 299–307.

[7] W. S. Levine, *The Control Handbook*, CRC press, 1996.

[8] C. Monje, F. Ramos, V. Feliu, and B. Vinagre, Tip position control of a lightweight flexible manipulator using a fractional order controller, *IET Control Theory & Applications*, **1**(5) (2007), 1451–1460.

[9] C. A. Monje, Y. Chen, B. M. Vinagre, D. Xue, and V. Feliu-Batlle, *Fractional-Order Systems and Controls: Fundamentals and Applications*, Springer Science & Business Media, 2010.

[10] C. A. Monje, B. M. Vinagre, V. Feliu, and Y. Chen, Tuning and auto-tuning of fractional order controllers for industry applications, *Control Engineering Practice*, **16**(7) (2008), 798–812.

[11] C. A. Monje, C. Balaguer, B. Deutschmann, and C. Ott, Fractional order control of a soft robotic system, in *Advances in Cooperative Robotics*, pp. 771–779, World Scientific, 2017.

[12] J. Reinecke, B. Deutschmann, and D. Fehrenbach, A structurally flexible humanoid spine based on a tendon-driven elastic continuum, in *Robotics and Automation (ICRA), IEEE Conference on*, pp. 4714–4721, 2016.

[13] R. Sepulchre, M. Jankovic, and P. Kokotovic, *Constructive Nonlinear Control*, Springer-Verlag, 1997.

[14] A. van der Schaft, *L_2-Gain and Passivity Techniques in Nonlinear Control*, 2nd edition, Springer-Verlag, 2000.

S. Hassan HosseinNia and Niranjan Saikumar

Fractional-order precision motion control for mechatronic applications

Abstract: This chapter deals with the design of both integer and fractional-order controllers for precision mechatronic systems. System performance and robustness analysis in the frequency domain is studied using closed-loop sensitivity functions. Controller design along with rules of thumb is introduced followed by an analysis of the advantages of using fractional-order controllers and their design. A practical example is considered to highlight the advantages of using fractional-order controllers in precision mechatronics.

Keywords: Mechatronic system design, motion control, precision control, fractional order control, PID

MSC 2010: 34H05, 37N35, 70Q05, 93B52

1 Introduction

Precision motion control design is highly competitive and challenging due to the ever increasing demands on speed and precision. Applications include photolithography systems, atomic force microscopes, nano-positioning systems among others [4, 7]. These require controller design to meet bandwidth specifications in 100 s of Hz, precision in nanometer range, good rejection performance against external vibrations and robustness to system changes to be met simultaneously. Most of the industry today continues to design classical PID using standard loop-shaping method because this method provides a clear insight into the performance aspects of the closed-loop system.

From the loop-shaping perspective, we have to fulfill two requirements. One is high open-loop gain at low frequencies to ensure good tracking and disturbance rejection, and low gain at high frequencies to avoid noise amplification. The other requirement is the phase margin at cross-over frequency (bandwidth) to ensure stability. These two requirements directly conflict with each other. This translates to a trade-off between precision and bandwidth on the one hand versus stability and robustness on the other hand.

Recently, fractional-order calculus (FOC) has gained interest in control theory and entails the use of noninteger orders of derivative and integral action [2–6, 8, 9]. This

S. Hassan HosseinNia, Niranjan Saikumar, Department of Precision and Microsystem Engineering, Faculty of Mechanical, Material and Maritime Engineering, Delft University of Technology, Delft, The Netherlands, e-mails: s.h.hosseinniakani@tudelft.nl, N.Saikumar@tudelft.nl

https://doi.org/10.1515/9783110571745-015

allows for more flexible controller design, especially with derivative part of controller. This increased freedom helps in achieving a less conservative design for the given system specifications reducing the effect of the precision-stability trade-off.

In this chapter, we deal with the design of fractional-order controllers specifically for precision motion systems. General guidelines are provided for tuning fractional-order PID. A toolbox (FLOreS) is introduced for designing fractional-order controllers using the industry standard loop-shaping technique. Finally, a practical example is given to illustrate the advantage of fractional PID for precision motion control.

2 Dynamics of motion in mechatronic systems

Deriving the dynamics of motion is the first step in precision control of a mechatronic system. The simplest mechatronic system is a frictionless moving mass. Its dynamics of motion can be represented as a double integrator, that is, $\frac{1}{ms^2}$ where m is the mass of the system and s is the Laplace variable. The optical lens in CD drive is an example of such a mass system. In practice, the masses are usally suspended by a spring. Servo motors, loud speakers and voice coil actuators are some examples for such system. These systems can be represented by mass-spring system. A linear stage or translation stage is a component of a precise motion system used to restrict an object to a single axis of motion. This system can be represented as a double mass-spring system shown in Figure 1. Depending on where the sensor is located, two different dynamics can be derived. If the sensor and actuator are located on the same mass, then the system is called "collocated," otherwise it is called "non-collocated."

Equations of motion for the double mass-spring system shown in Figure 1 can be derived as

$$\begin{aligned} m_1\ddot{x}_1 + (k_1 + k_2)x_1 - k_2x_2 &= F, \\ m_2\ddot{x}_2 + k_2x_2 - k_2x_1 &= 0 \end{aligned} \tag{1}$$

Converting (1) to Laplace domain, transfer function from sensor to actuator can be found for both collocated and noncollocated systems.

1. *Collocated sensor and actuator*:

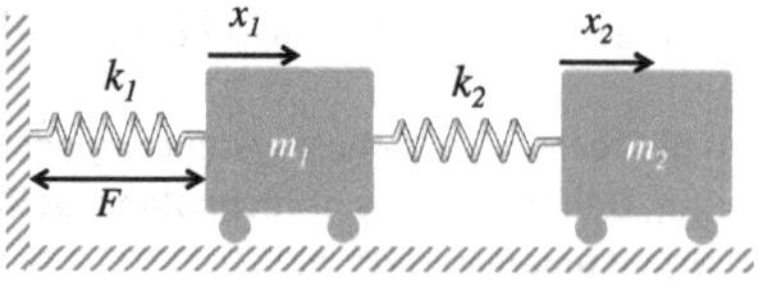

Figure 1: Free body diagram of a double mass spring system.

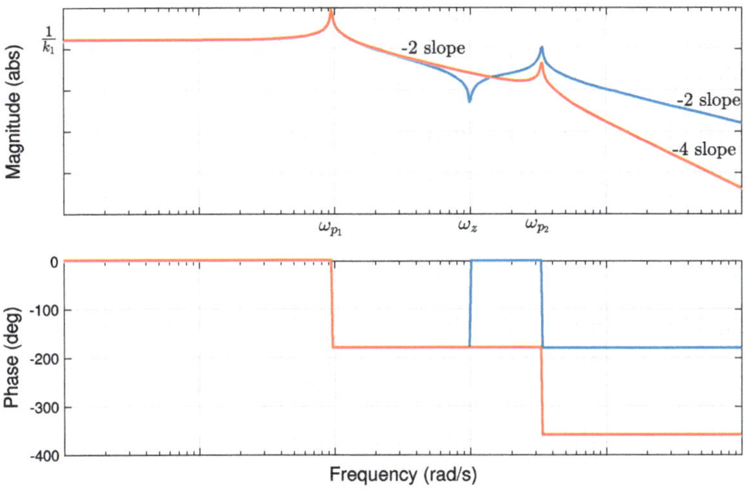

Figure 2: Frequency response of the double mass spring system.

In this case, actuator and sensor are located at m_1. Thus, the transfer function is found as

$$G_{\mathrm{col}} = \frac{X_1}{F} = \frac{m_2 s^2 + k_2}{m_1 m_2 s^4 + (k_2 m_1 + (k_1 + k_2) m_2) s^2 + k_1 k_2}. \tag{2}$$

2. *Noncollocated sensor and actuator*:
   In this case, the sensor is located at m_2 while the actuator is kept at m_1. Hence, the transfer function is obtained as

$$G_{\mathrm{ncol}} = \frac{X_2}{F} = \frac{k_2}{m_1 m_2 s^4 + (k_2 m_1 + (k_1 + k_2) m_2) s^2 + k_1 k_2}. \tag{3}$$

Figure 2 shows the frequency response in both cases. While both systems have similar behavior at low frequency, collocated system has two pole pairs ($\omega_{p_1} \approx \pm j\sqrt{\frac{k_1}{m_1+m_2}}$ and $\omega_{p_2} \approx \pm j\sqrt{\frac{k_2(m_1+m_2)}{m_1 m_2}}$ if $k_1 \ll k_2$) and a zero pair ($\omega_z = \pm j\sqrt{\frac{k_2}{m_2}}$), whereas the non-collocated system does not have any zeros. Pairs of poles and zeros can be seen as a peak and valley, respectively. In addition, the collocated system descends with a –2 slope (–40 dB/decade) while the non-collocated system descends with a –4 slope (–80 dB/decade) at high frequencies.

As mentioned above, most of the precision mechatronic systems can be simplified to a collocated or noncollocated double mass spring system. However, high frequency modes and delay have to be considered as a limiting condition when designing motion control. In the next section, we explain how to design a linear controller for a motion system using loop-shaping method.

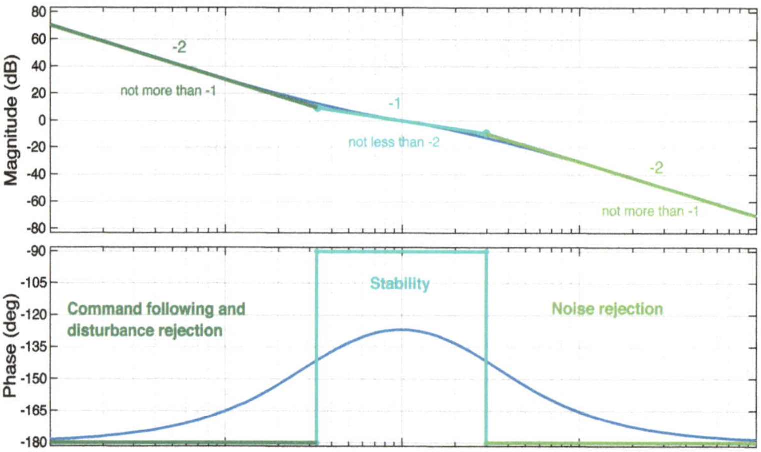

Figure 3: The desired shape of the open loop in loop-shaping.

3 Loop-shaping

Loop-shaping is a method used to design the controller by shaping the open-loop, defined as $L(s) = C(s)G(s)$, as close as possible to the one shown in Figure 3. In large range of frequencies (low and high frequency), it resembles a mass system (double integrator), while in middle frequency range (crossover frequency region) it behaves as a damper system (integrator). In other words, loop-shaping can be divided into three regions,

1. Low frequency region for command following and disturbance rejection shown in dark green: The higher is the gain of open-loop L, the better is command following and disturbance rejection.
2. Crossover frequency region to provide stability shown in light blue: in order to provide sufficient phase margin, the slope of the open loop at this region has to be larger than −2 since at a slope of −2 (less than −2) the system is marginally stable (unstable).
3. High frequency region for noise rejection shown in light green: the lower is the gain of open-loop at this region, the better is the noise rejection.

Based on the above mentioned requirements, three elements are necessary to build a linear controller. The first element is lag element (also known as proportional integrator i. e. PI) defined as

$$C_{\mathrm{PI}} = k_p\left(1 + \frac{\omega_i}{s}\right) \tag{4}$$

where k_p is proportional gain and ω_i is the frequency where the integration stops. Lag element provides the high gain at low frequency which is required to increase the gain in command following and disturbance rejection region.

Lead element is the second element of controller necessary to provide stability at crossover region defined as

$$C_{\text{lead}} = \frac{1 + \frac{s}{\omega_d}}{1 + \frac{s}{\omega_t}} \tag{5}$$

where, ω_d and ω_t is where the differentiation starts and is tamed, respectively.

Finally, the last element which is often used in linear controllers is low pass filter (LPF). It can be a first-order filter defined as

$$C_{\text{lpf}^1} = \frac{1}{1 + \frac{s}{\omega_l}} \tag{6}$$

or second order defined as

$$C_{\text{lpf}^2} = \frac{1}{(\frac{s}{\omega_l})^2 + 2\zeta\frac{s}{\omega_l} + 1} \tag{7}$$

depending on system requirements and noise level. In these equations, ζ is damping factor and ω_l is the cut-off frequency. Combining all three elements, a series PID can be defined as follows:

$$C = \underbrace{k_p\left(1 + \frac{\omega_i}{s}\right)}_{\text{lag}}\underbrace{\left(\frac{1 + \frac{s}{\omega_d}}{1 + \frac{s}{\omega_t}}\right)}_{\text{lead}}\underbrace{\left(\frac{1}{1 + \frac{s}{\omega_l}}\right)}_{\text{LPF}}. \tag{8}$$

For simplicity, we consider the first-order LPF, but one can also use the second-order LPF if needed.

3.1 Closed loops

In order to tune the parameters of the controller, which are, k_p, ω_i, ω_d, ω_t, and ω_l, we need to understand the closed-loop transfer functions. Looking at Figure 4, one can define following four transfer functions:

- Complementary sensitivity function:

$$T = \frac{y}{r} = \frac{L(s)}{1 + L(s)}, \tag{9}$$

- Sensitivity function:

$$S = \frac{y}{n} = \frac{1}{1 + L(s)}, \tag{10}$$

- Process sensitivity function:

$$PS = \frac{y}{d} = \frac{P(s)}{1 + L(s)}, \tag{11}$$

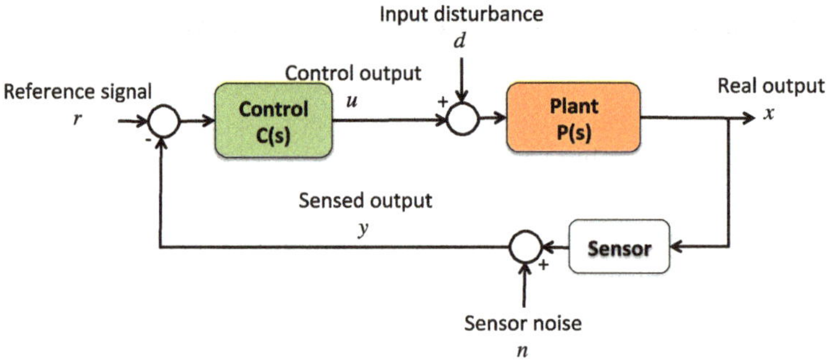

Figure 4: The block diagram of the closed loop.

- Control sensitivity function:

$$CS = \frac{u}{r} = \frac{C(s)}{1 + L(s)}. \tag{12}$$

The real error of the closed loop system can be derived as follows:

$$e_{\text{real}} = r - x = S(s)r - PS(s)d + T(s)n. \tag{13}$$

It should be noted that $T(s) + S(s) = 1$, which is why $T(s)$ is called complementary sensitivity function. In absence of noise and disturbance, one could design a controller such that $T = 1$, and $S = 0$. But in real practice, sensor noise, input and output disturbances limit the performance. For this reason, at low frequency (region 1) $T \approx 1$ and $S \approx 0$ is desired to follow the command and reject the disturbances. In this case, real error will be simplified to

$$e_{\text{real}} \approx n|_{\omega \ll \omega_c}. \tag{14}$$

Since in practice, noise occurs at high frequencies, the above equation goes to zero at low frequencies. At high frequencies, $T \approx 0$ and $S \approx 1$, and hence real error will be

$$e_{\text{real}} \approx r - P(s)d|_{\omega \gg \omega_c}. \tag{15}$$

Since in practice, disturbance and commands are at low frequencies, the above equation tends to zero. To summarize:

- Complementary sensitivity function has to be as close as possible to 1 ($T \approx 1$) at low frequencies (frequencies below crossover frequency also known as bandwidth frequency). In this region, open loop gain $L(s)$ has to be large which can be provided by lag element (integrator).
- Sensitivity function has to be as close as possible to 1 ($S \approx 1$) at high frequencies (frequencies above crossover frequency). In this region, open loop gain $L(s)$ has to be small which can be provided by an LPF.
- In order to provide the stability, a differentiator may be added around the crossover frequency.

3.2 PID and rule of thumb

In literature, one may find optimal ways to tune PID parameters. But here we would like to introduce a rule of thumb:

- *Integrator frequency*: This frequency is set to be 1/10th of the control bandwidth i. e. $\omega_i = \omega_c/10$, where ω_c is cross-over frequency (bandwidth).
- *Differentiation band*: The frequency where differentiation starts and stops are set to be $\omega_d = \omega_c/a$ and $\omega_t = a\omega_c$ where $a = 3$ and it is called differentiation band.
- *Proportional gain*: Assume $k_t = 1/P(\omega_c)$, then

$$k_p = \begin{cases} k_t & \text{if there is no differentiator,} \\ \approx k_t/a & \text{if there is a differentiator.} \end{cases}$$

- *LPF cutoff frequency*: In the case of using LPF, it is set to a frequency 10 times higher than bandwidth, that is, $\omega_l = 10\omega_c$.

With this rule of thumb, all of the control parameters are related to bandwidth ω_c, which makes the rule to be very simple and yet reliable for a wide range of applications. It should be noted that higher is the bandwidth, faster is the motion (command tracking) and better is the disturbance rejection. Therefore, for a motion system, it is important to choose the maximum possible value for bandwidth. This rule of thumb can be considered as a starting point for students and engineers to tune the PID controller. However, it does not necessarily provide robustness. An extensive explanation of the rule of thumb is give in [7].

3.3 Robustness criteria

The next step after applying rule of thumb is to fine-tune the parameters such that the robustness and stability conditions are satisfied. Following are the most important robustness conditions:

- *Phase margin*: The phase margin is amount of phase decrease required to make loop phase go below -180° at bandwidth: $\phi_m = 180^\circ + \measuredangle L(\omega_c) > \mathcal{L}^\circ$.
  If the open-loop $L(s)$ is shaped such that

$$\begin{cases} |L(j\omega)| \gg 1, & \text{for } \omega < \omega_c, \\ |L(j\omega)| \approx -20\,\text{dB/dec}, & \text{for } \omega \approx \omega_c, \\ |L(j\omega)| \ll 1, & \text{for } \omega > \omega_c \end{cases} \tag{16}$$

  following relations hold between frequency and time domains:

$$t_r \approx \frac{1.5}{\omega_c}, \qquad o\,\% \approx 70 - \phi_m \tag{17}$$

  where t_r is the rise time and $o\,\%$ is the percent overshoot.

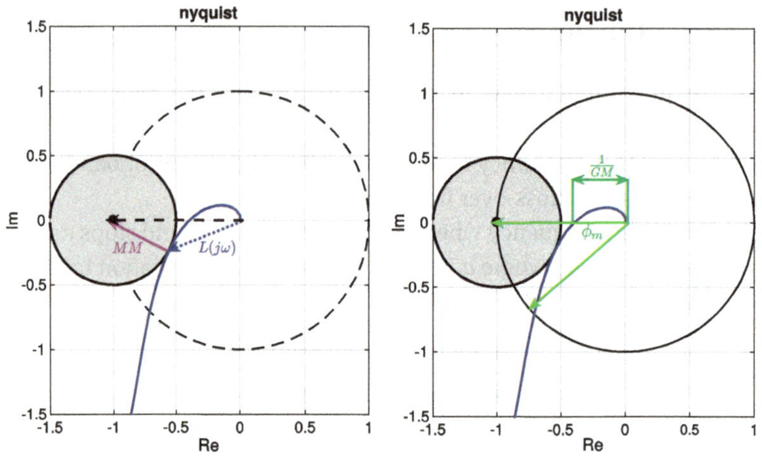

Figure 5: The robust margins.

- *Gain margin*: The gain margin is the amount of gain increase or decrease required to make the loop gain unity at the frequency where the phase angle is −180°. In practice, it is required to be more than 6 dB, that is, $GM = |1/L(\omega)| > 6\,\text{dB}$ for all ω where $\measuredangle L(\omega) = -180°$.
- *Modulus margin*: A large gain margin is not sufficient to have a stable system in presence of gain uncertainty since gain margin only guarantees the stability against the gain variation at frequencies where the phase is crossing −180°. Therefore, there is a need to ensure the robustness against the gain variation in the system. This can be ensured if modulus margin (MM) defined in (18) is greater than 6 dB (peak of sensitivity less than 6 dB).

$$\frac{1}{MM} = \max_{\omega}|S(\omega)| = \max_{\omega} \frac{1}{|1 + L(\omega)|} < 6\,\text{dB}. \tag{18}$$

All these robustness margins are shown graphically in Figure 5.

Following the loop-shaping method, providing high gain at low frequency and low gain at high frequency are desired for higher precision in motion but undesired for stability since they decrease the phase margin. In addition, increasing the phase margin using lead (increasing the gain around the bandwidth frequency by increasing the number of differentiator or band of differentiation (a)), decreases the gain at low frequencies and increases the gain at high frequencies which in result reduces the precision. In addition, increasing the bandwidth increases the gain at low frequencies which is desired for precision but it also increases the gain at high frequencies which is undesired for precision since it amplifies the noises. Therefore, a trade-off is needed but classical PID controller is too limited to provide that. In the next section, we will show that using fractional-order control, we can find a trade-off between precision, stability, and bandwidth.

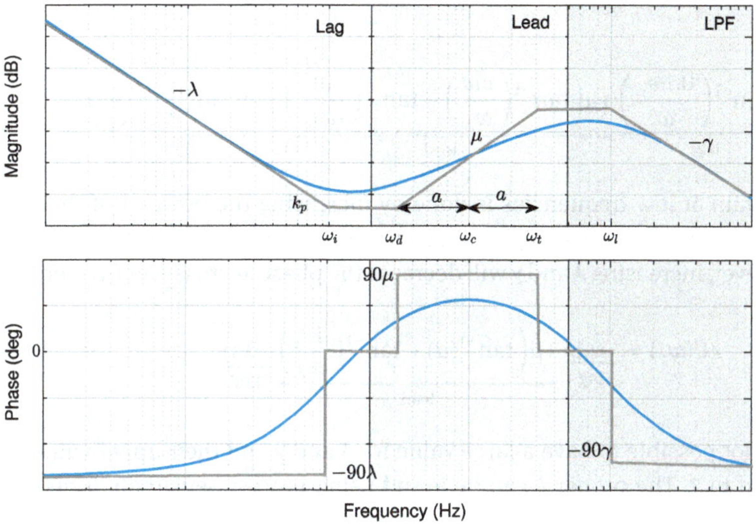

Figure 6: Fractional-order PID control.

4 Fractional-order PID for loop-shaping

Let us define a fractional-order PID controller as follows:

$$C = \underbrace{k_p\left(1+\frac{\omega_i}{s}\right)^{\lambda}}_{\text{lag}} \underbrace{\left(\frac{1+\frac{s}{\omega_d}}{1+\frac{s}{\omega_t}}\right)^{\mu}}_{\text{lead}} \underbrace{\left(\frac{1}{1+\frac{s}{\omega_l}}\right)^{\gamma}}_{\text{LPF}} \tag{19}$$

where λ, μ, and γ are the fractional-order parameters. This definition of fractional-order PID consists of lag, lead, and LPF element as shown in Figure 6. As mentioned before, what interests us from lag and LPF elements is their gain behavior which is useful for tracking, disturbance, and noise rejection and what interests us from lead element is its phase for stability. Therefore, it is important that we analyze the gain and phase behavior of fractional-order PID. For this, we apply the rule of thumb discussed in previous section. Hence, (19) can be rewritten as

$$C = \underbrace{k_p\left(1+\frac{0.1\omega_c}{s}\right)^{\lambda}}_{\text{lag}} \underbrace{\left(\frac{1+\frac{s}{\omega_c/a}}{1+\frac{s}{a\omega_c}}\right)^{\mu}}_{\text{lead}} \underbrace{\left(\frac{1}{1+\frac{s}{10\omega_c}}\right)^{\gamma}}_{\text{LPF}} \tag{20}$$

where proportional gain should be set to $k_p = \frac{k_t}{a^{\mu}}$ in order to satisfy the gain crossover frequency condition. The magnitude and phase of (20) can respectively be found as follows:

$$|C(\omega)| = \underbrace{k_p\left(1+\left(\frac{0.1\omega_c}{\omega}\right)^2\right)^{0.5\lambda}}_{\text{lag}} \underbrace{\left(\frac{1+(\frac{\omega}{\omega_c/a})^2}{1+(\frac{\omega}{a\omega_c})^2}\right)^{0.5\mu}}_{\text{lead}} \underbrace{\left(\frac{1}{1+(\frac{\omega}{10\omega_c})^2}\right)^{0.5\gamma}}_{\text{LPF}} \tag{21}$$

and

$$\measuredangle C(\omega) = \underbrace{-\lambda \tan^{-1}\left(\frac{0.1\omega_c}{\omega}\right)}_{\text{lag}} \underbrace{+\mu\left(\tan^{-1}\left(\frac{a\omega}{\omega_c}\right) - \tan^{-1}\left(\frac{\omega}{a\omega_c}\right)\right)}_{\text{lead}} \underbrace{-\gamma \tan^{-1}\left(\frac{\omega}{10\omega_c}\right)}_{\text{LPF}}. \tag{22}$$

Increasing the gain at low frequencies is done by increasing the power λ of the lag element and decreasing the gain at high frequencies is done by increasing the power γ of the LPF. However, increasing λ and γ will decrease the phase at crossover frequency:

$$\measuredangle C(\omega_c) \approx \underbrace{-0.1\lambda}_{\text{lag}} \underbrace{+\mu\left(\tan^{-1}(a) - \tan^{-1}\left(\frac{1}{a}\right)\right)}_{\text{lead}} \underbrace{-0.1\gamma}_{\text{LPF}}. \tag{23}$$

Therefore, it is not possible to have a large value for λ and γ and their upper value is usually bounded to 2. The proper λ can be found using process sensitivity function (11). This function shows the relation between input disturbance d and output y. Depending on the amplitude and frequency of the disturbance, different values of λ can be obtained. Likewise, γ can be found using sensitivity function (10) depending on requirement for noise attenuation.

Phase margin can be achieved by tuning the power μ of lead element and its differentiation band a. In order to ensure that this does not overlap with the integrator part (otherwise the differentiator cancels the integrator which results in a proportional controller in the overlap region), a is limited to 10. Since the cross-over frequency ω_c is a geometric mean of the corner frequencies ($a\omega_c$ and ω_c/a) of the lead element, the gain at low and high frequencies can be simplified to $\frac{k_t w_i}{a^\mu s^\lambda}$ and $\frac{k_t w_l a^\mu}{s^\gamma}$, respectively. Therefore, the gain at low and high frequencies are altered by factor a^μ with respect to 0 dB line.

Figure 7(A) depicts a^μ versus a with a changing from 1 to 10 and μ from 0 to 2. Figure 7(B) shows the phase lead versus a for all μ. For a given phase lead obtained as

$$\phi_l = \mu\left(\tan^{-1}(a) - \tan^{-1}\left(\frac{1}{a}\right)\right) \tag{24}$$

it is desired to have highest value for a (it results in better robustness) but the lowest value for a^μ (it results in higher precision, that is, better tracking, disturbance, and noise rejection). It is obvious that both cannot be achieved at the same time, since increasing a increases a^μ and decreasing a^μ decreases the a value. Therefore, there has to be a trade-off between these two values.

As an example, suppose a phase lead of $\phi_l = 80^\circ$ is required. There are infinite solutions for (24). The red line with circle in Figure 7(A) shows the solution i.e a^μ for $1 < a < 10$ where $\mu = \frac{\phi_l = 80^\circ}{(\tan^{-1}(a) - \tan^{-1}(\frac{1}{a}))^\circ}$. There is no solution for $\mu \leq 1$ since it cannot provide the required phase for a differentiation band $a < 10$. Among the feasible solutions ($1 < \mu < 2$), solutions with $\mu > 1.5$ result in a very narrow differentiation band and

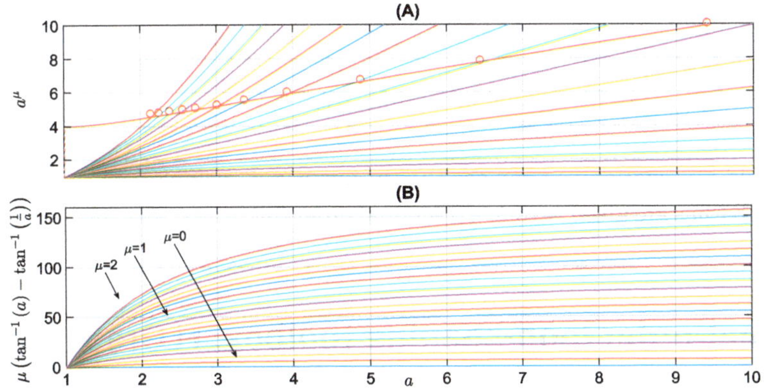

Figure 7: Effect of a and μ on phase and gain of the controller.

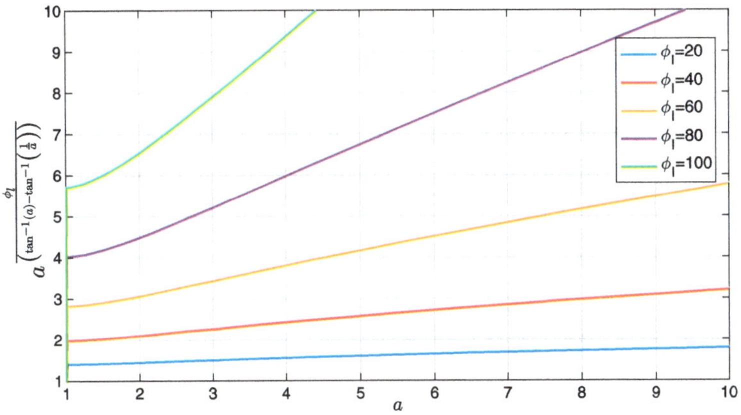

Figure 8: a^{μ} where $\mu = \frac{\phi_l}{(\tan^{-1}(a) - \tan^{-1}(\frac{1}{a}))}$ for different phase margin.

solutions with $\mu < 1.2$ result in a high a^{μ} value. Therefore, solutions with $1.2 \leq \mu \leq 1.5$ are trade-off between minimum a^{μ} and maximum a. This can be done for any other desired ϕ_l. Figure 8 shows the feasible solution for $\phi_l = [20, 40, 60, 80, 100]$. We can conclude that for a small phase lead, smaller μ and larger a is desired while for large phase lead, larger μ but smaller a is desired. It is recommended not to choose the value of a to be less than 3.

To summarize, a guideline to tune fractional-order PID (20) is given as follows. For a given cross-over frequency ω_c (bandwidth) and phase margin ϕ_m:

1. Find proper λ and γ using process sensitivity and sensitivity functions in order to satisfy the real error requirement.
2. Find required phase lead: $\phi_l = \phi_m + 0.1(\gamma + \lambda) - \pi - \angle G$.
3. Find a^{μ} where $\mu = \frac{\phi_l}{(\tan^{-1}(a) - \tan^{-1}(\frac{1}{a}))}$ for each a in interval of $[1, 10]$ (Figure 8).

4. Among all solutions for a, choose a solution that is a trade off between maximum a and minimum a^{μ}.
5. Find $\mu = \frac{\phi_l}{(\tan^{-1}(a)-\tan^{-1}(\frac{1}{a}))}$ for a found in previous step.

For more detailed information you can see [10]. In this paper, a similar approach is taken and thumb rules are given to tune a fractional order PID.

5 Fractional-order loop-shaping toolbox: FLOreS

Loop-shaping is popularly used in the industry for designing and tuning controllers since it is intuitive. With loop-shaping, the control engineer can use prior knowledge and experience to better design control. The control specifications including bandwidth; stability and robustness in terms of gain, phase, and modulus margins; disturbance rejection and noise attenuation are represented graphically using Bode, Nyquist, and Nichols plots.

Loop-shaping with fractional-order controllers is enabled with a user-friendly graphical interface toolbox—FLOreS [1]. The screenshot is shown in Figure 9. The various features of FLOreS which helps the control engineer during the design process are described.

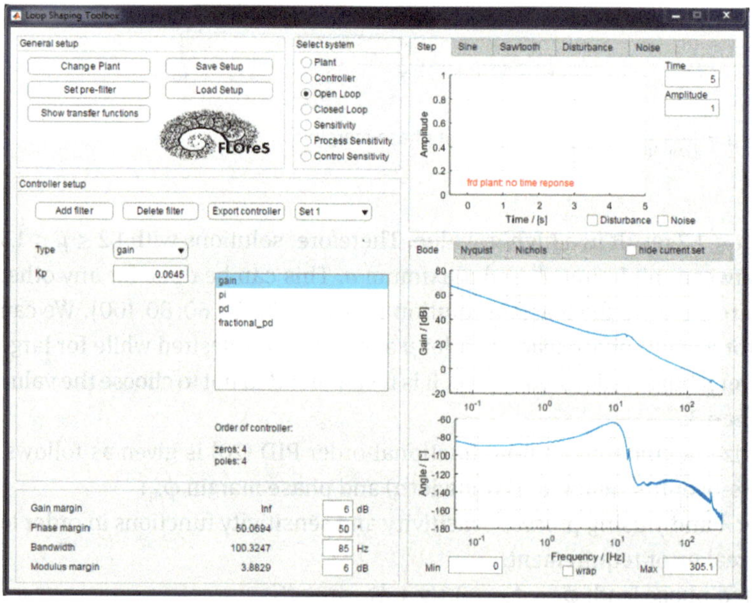

Figure 9: Screenshot of FLOreS.

Plant: FLOreS can be used for designing both integer-order and fractional-order controllers for SISO plants. The plant can be imported generally as a transfer function from workspace. However, it is quite common in the industry for the obtained plant frequency response data to be used for tuning. Hence, FLOreS is designed to directly import this data. Example plants like mass-spring, double mass-spring systems are added to be used for educational purposes.

Controller: This is the most important feature of FLOreS which enables the design and tuning of both integer- and fractional-order controllers. While integer-order filters like *pd, pi, pid, lead lag, notch* are essential and are available as part of toolbox, fractional-order variants of all these filters provide the engineer with greater flexibility in tuning and design. In loop-shaping, limiting filters to the frequency band in which their action is necessary is critical. Hence, CRONE approximation is used as default for all fractional-order filters, since this approximation allows for frequency band to be defined accurately.

FLOreS also allows for multiple controllers to be designed for the same plant for comparison of performance aspects.

Frequency response: The popularity of loop-shaping in industry stems from the fact that the closed-loop performance aspects can be visualized graphically. For this purpose, all sensitivity functions and open-loop frequency response can be observed through *Bode* (with wrapping of phase being an option), *Nyquist* or *Nichols* plots.

Performance: The performance panel is used to provide the engineer with stability margin and bandwidth values to allow for quicker analysis and design. The requirements can be defined beforehand and corresponding performance values are highlighted when they do not satisfy requirements.

Time response: Although loop-shaping involves designing controllers using the frequency domain approach, the visualization of time response further aids the design process. This panel displays the closed-loop system response to *step, sine*, or *sawtooth* reference signal. Additionally, the system response to disturbance and noise can also visualized. However, this function is not available when the plant information is loaded using frequency response data.

Additional features: The controller transfer function can be exported to MATLAB workspace. Also, the complete session on FLOreS can be saved and restarted later or shared with others.

6 A practical example

Let us use a practical setup to design a fractional-order controller on FLOreS. A precision planar positioning stage shown in Figure 10 is used for this purpose. The setup consists of 3 voice-coil actuators (1A, 1B and 1C) which are rigidly attached to 3 masses

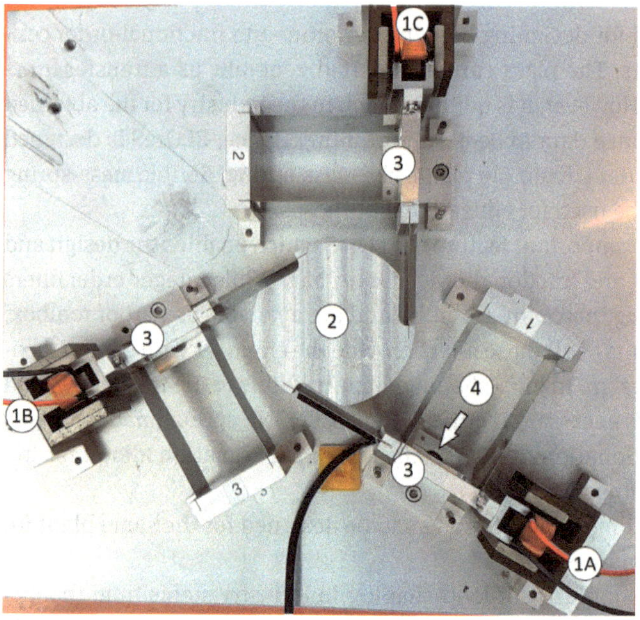

Figure 10: 3 DOF planar precision positioning stage.

numbered '3'. These masses are suspended with respect to the real-world through a pair of leaf flexures each of which behaves as a spring. All these 3 masses are connected to the central mass numbered "2" through "3" individual leaf flexures. Linear encoders placed under masses "3" provide position feedback with a resolution of 0.1 μm. With this system, actuation allows for two planar translations of the central mass in X and Y directions and also allows for in-plane rotation θ.

Since this forms a MIMO system, only one of the actuators (1A) is considered for control of mass "3" rigidly attached to it resulting in a SISO system. From the point of view of actuator 1A, it can be seen that the force is applied on mass "3" which is suspended by springs. Further, this mass is then attached to the central mass through another spring. If we neglect the other actuators and masses, it is clear that this is a double mass-spring system. Since the sensor and actuation take place on the same mass, this is a collocated system. Although the dynamics of this system can be derived analytically, industry standard procedure is followed and frequency response data of this system is obtained and shown in Figure 11.

The controller is to be designed to meet the following specifications:

1. $\phi_l \geq 45^\circ$
2. $GM \geq 6$ dB
3. $MM \geq 6$ dB
4. robust to gain variations of up-to 10 % (not considering delay)

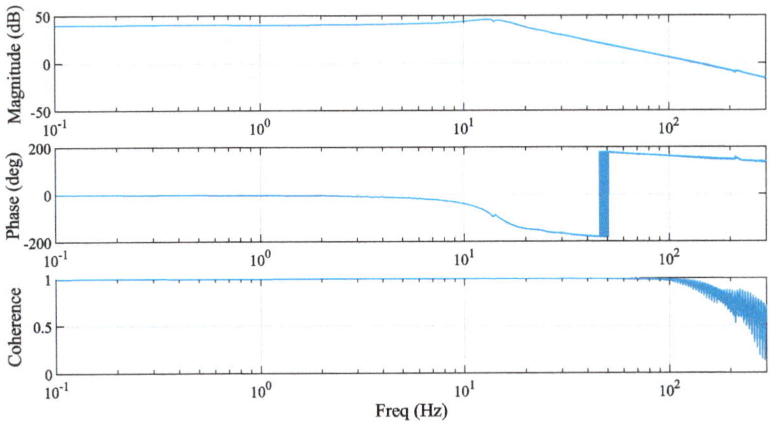

Figure 11: Frequency response data of the precision stage.

One of the advantages of fractional-order PID is that flat phase behavior can be achieved in the region of bandwidth to account for gain variations. When the open-loop phase in the bandwidth region is close to flat, then the change in PM due to variations of plant gain is minimized resulting in a robust system. A fractional-order PID controller is designed on FLOreS for this system and the parameters of the controller corresponding to equation (20) are given below:

1. Bandwidth $\omega_c = 100$ Hz
2. $k_p = 0.0645$
3. $a = 5$
4. $\mu = 1.2$
5. $\lambda = 1$
6. $\gamma = 1$

The guidelines provided earlier were used for this purpose.

- Since no strict requirements are placed on disturbance rejection or noise attenuation of closed-loop system, integer order PI and LPF are used as part of the controller.
- The required phase lead is found to be $\phi_l = 80^\circ$ since the phase of system at bandwidth ($\angle G$) is -205°.
- From Figure 7, multiple solutions exist which are shown along the red circle line.
- We need to find the value of a and μ such that a is maximized and a^μ is minimized. However, we also need the system to be robust to gain variations. Hence, a trade-off is made and $a \approx 5$ is chosen.
- The corresponding value of $\mu = 1.2$ required to provide ϕ_l is used.

The step response obtained from the closed-loop system is shown in Figure 12. The gain of the system is varied by 10 % (both increase and decrease) by changing the

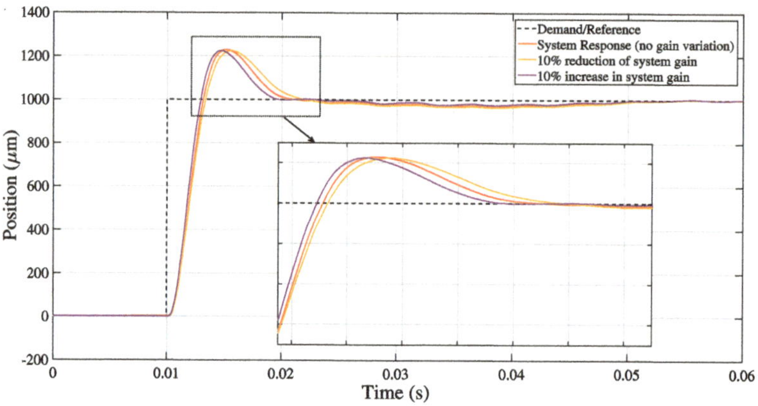

Figure 12: Step response of practical setup with designed controller showing robustness against gain variations.

gain of the actuator power amplifier. In the traditional case with integer-order PID, where the system gain changes, the change in PM results in a significant difference in step response of system especially in the overshoot. However, since the fractional-order PID was designed to be robust to such variations, the change in overshoot due to change in system is negligible.

Bibliography

[1] L. van Duist, G. van der Gugten, D. Toten, N. Saikumar, and S. H. HosseinNia, FLOreS – fractional order loop shaping Matlab toolbox, in *3rd IFAC Conference on Advances in Proportional-Integral-Derivative Control*, 2018.

[2] S. H. Hosseinnia, I. Tejado, V. Milanés, J. Villagrá, and B. M. Vinagre, Experimental application of hybrid fractional-order adaptive cruise control at low speed, *IEEE Transactions on Control Systems Technology*, **22**(6) (2014), 2329–2336.

[3] S. H. HosseinNia, I. Tejado, B. M. Vinagre, and Y. Chen, Iterative learning and fractional reset control, in *ASME 2015 International Design Engineering Technical Conferences and Computers and Information in Engineering Conference*, p. V009T07A041, American Society of Mechanical Engineers, 2015.

[4] M. E. Krijnen, R. A. van Ostayen, and S. H. HosseinNia, The application of fractional order control for an air-based contactless actuation system, *ISA Transactions* (2017).

[5] L. Marinangeli, F. Alijani, and S. H. HosseinNia, Fractional-order positive position feedback compensator for active vibration control of a smart composite plate, *Journal of Sound and Vibration*, **412** (2018), 1–16.

[6] N. Saikumar and S. H. HosseinNia, Generalized fractional order reset element (GFrORE), in *9th European Nonlinear Dynamics Conference (ENOC 2017) in Budapest, Hungary*, 2017.

[7] R. M. Schmidt, G. Schitter, and A. Rankers, *The Design of High Performance Mechatronics: High-Tech Functionality by Multidisciplinary System Integration*, IOS Press, 2014.

[8] I. Tejado, S. H. Hosseinnia, and B. Vinagre, Adaptive gain-order fractional control for network-based applications, *Fractional Calculus and Applied Analysis*, **17**(2) (2014), 462–482.
[9] M. Zarghami and S. H. HosseinNia, Fractional order set point regulator using reset control: application to EGR systems, in *Proceedings of the 2017 The 5th International Conference on Control, Mechatronics and Automation*, pp. 35–41, ACM, 2017.
[10] A. Ahmadi Dastjerdi, N. Saikumar and S. H. HosseinNia, Tuning guidelines for fractional order PID controllers: Rules of thumb, *Mechatronics*, **56** (2018), 26–36.

Costas Psychalinos

Development of fractional-order analog integrated controllers – application examples

Abstract: A systematic design procedure for designing fractional-order controllers, which offers resistorless realizations, with electronic adjustment of their characteristics, as well as capability of on-chip implementation is presented in this chapter. This is achieved through the utilization of appropriate integer-order transfer functions for approximating the required fractional-order integration/differentiation stages. The derived multi-feedback structures are implemented using operational transconductance amplifiers as active elements in order to perform a voltage-mode realization. Also, possible realization in current-mode has been discussed. The provided application examples about the implementation of dc motor controller and, also, about a brake/throttle controller confirm the correct operation of the derived topologies.

Keywords: Fractional-order circuits, fractional-order controllers, fractional-order integrators, fractional-order differentiators, analog integrated circuits

PACS: 85.40.-e, 84.30.-r, 84.30.Bv, 84.30.Vn

1 Introduction

Fractional-order calculus is a generalization of the conventional (i. e., integer) calculus with interdisciplinary applicability. Therefore, in the last two decades, fractional calculus has been rediscovered by scientists and engineers, applied in an increasing number of fields, including electrical and electronic engineering, biology, biomedicine, and control theory [19, 22, 45].

Fractional-order proportional-integral-derivative ($\mathrm{PI}^{\lambda}\mathrm{D}^{\mu}$) controllers, where $0 < \lambda < 1$ and $0 < \mu < 1$ are the orders of integration and differentiation, respectively, have received a considerable attention both from academic and industrial points of view [2–4, 6, 8, 21, 31, 35, 37, 42, 43, 48, 49]. This originates from the fact that integer-order PID controllers have three parameters to select, while $\mathrm{PI}^{\lambda}\mathrm{D}^{\mu}$ controllers have five parameters, offering additional degrees of freedom [7, 12, 30].

The functional block diagram of a $\mathrm{PI}^{\lambda}\mathrm{D}^{\mu}$ controller is depicted in Figure 1, and its transfer function is

$$C(s) = K_p + \frac{K_i}{s^{\lambda}} + K_d s^{\mu}, \tag{1}$$

Costas Psychalinos, University of Patras, Physics Department, Electronics Laboratory, GR-26504, Rio Patras, Greece, e-mail: cpsychal@upatras.gr

https://doi.org/10.1515/9783110571745-016

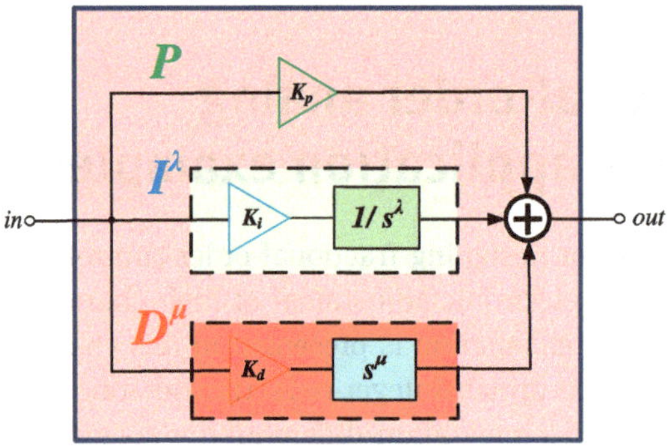

Figure 1: Functional block diagram of a $PI^\lambda D^\mu$ controller.

where K_p is the proportional gain, K_i is the integration constant, K_d is the differentiation constant, and λ, μ are the orders of integration/differentiation, respectively.

The main building blocks for implementing fractional-order controllers are (fractional-order) integrators and differentiators. The implementation of fractional-order differentiators and integrators can be performed through the following procedures:

- By substituting the integer-order capacitors, in the corresponding conventional (i. e., integer-order) differentiation/integration topologies, with their fractional-order counterparts. Although significant research effort toward the physical fabrication of fractional-order capacitors has been performed in the literature [1, 9, 26, 27], they are not yet commercially available. Thus, only approximation of the behavior of these elements is obtainable and this can be performed using appropriately configured RC networks [10, 23, 47]. This procedure has been followed for the implementation of the fractional-order controllers in [11, 18, 38]. Although this solution is easy for implementation, it suffers from the absence of the on-the-fly adjustment of the characteristics of the elements, in the sense that the whole RC network must be re-designed in order to alter the characteristics (such as value and/or order) of the fractional-order capacitors. From the integration point of view, the internal time-constants, which have the form: $\tau = RC$, cannot be accurately realized on silicon, leading to implementations with significant deviations with regards to the theoretically predicted values. In addition, in the case that a low-frequency band is desired, as in control applications, extremely large values of time-constants are required, making them nonpractical for on-chip implementation.
- By employing integer-order transfer functions that approximate the prototype transfer functions of fractional-order differentiators/integrators. The derived multi-feedback structures can be implemented using various types of active elements. Thus, Operational Amplifiers (op-amps), second generation Current Conveyors (CCIIs), Current Feedback Operational Amplifiers (CFOAs), Operational Transconductance Amplifiers (OTAs), and Current-Mirrors (CMs) have

been utilized for implementing fractional-order differentiation/integration stages [5, 17, 20, 25, 28, 29, 44, 46]. The implementations based on the utilization of OTAs and CMs offer the following attractive characteristics: (a) they are resistorless because of the employment of the small-signal transconductance parameter of MOS transistors, (b) they are electronically tunable, and this is originated from the fact that the transconductance parameter can be adjusted through an appropriate dc bias current or voltage, and (c) they are capable of monolithic implementation.

Therefore, this procedure is the best option for implementing fractional-order controllers, which will offer the aforementioned benefits and will be followed in this chapter for designing these topologies. The work presented in this chapter is organized as follows: a review of the approximation of the integrodifferential operator in the Laplace domain (s^q) is presented in Section 2. The design procedure for realizing integratable fractional-order controllers will be also presented in this section. The application examples will be demonstrated in Section 3 and the operation of the presented schemes will be verified using the Analog Design Environment of the Cadence software and the Design Kit provided by the Austria Mikro Systeme (AMS) 0.35 μm CMOS process. In Section 4, a discussion about the implementation issues will be performed and the main conclusions derived through the material presented in this chapter will be given in Section 5.

2 Approximation of fractional-order integrators and differentiators

2.1 Preliminaries

As it has been previously mentioned, fractional-order differentiator and integrator circuits are very useful building blocks for implementing fractional-order controllers. The transfer function of these circuits is described in general as

$$H(s) = (\tau s)^q, \tag{2}$$

where $q = \{-\lambda, +\mu\}$, with (λ) being the order of integration and (μ) being the order of differentiation, respectively. Also, the variable τ represents the time-constant of the differentiator/integrator, related to its unity gain frequency (ω_0) through the formula: $\tau = 1/\omega_0$.

The accuracy level of the fractional integrodifferential operator in the Laplace domain (s^q) is very critical for the approximation of the corresponding stage. The nth-order approximation of the operator around a center frequency $\omega_0 = 1$ rad/sec, is expressed by a rational function defined by the quotient of two polynomials of the variable. Employing the Oustaloup's approximation method [34], the corresponding expression, for geometrically distributed frequencies over the band $[\omega_b, \omega_h]$, is the following:

$$s^q \cong C \cdot \prod_{k=-N}^{N} \frac{s + \omega_k^{/}}{s + \omega_k} = \frac{B_n s^n + B_{n-1} s^{n-1} + \cdots + B_1 s + B_o}{s^n + A_{n-1} s^{n-1} + \cdots + A_1 s + A_o}. \quad (3)$$

Here, the variables $\omega_k^{/}$, ω_k, and C in equation (3) are defined by equation (4):

$$\omega_k^{/} = \omega_b \cdot \left(\frac{\omega_h}{\omega_b}\right)^{\frac{k+N+0.5\cdot(1-q)}{2N+1}}, \quad \omega_k = \omega_b \cdot \left(\frac{\omega_h}{\omega_b}\right)^{\frac{k+N+0.5\cdot(1+q)}{2N+1}}, \quad C = \omega_h^r. \quad (4)$$

Owing to the geometrical distribution of frequencies, the unity-gain frequency (ω_0) is calculated according to the formula: $\omega_0 = \sqrt{\omega_b \cdot \omega_h}$. It must also be mentioned that the order of the transfer function is $n = 2N + 1$ and, therefore, only odd-order approximations are possible through the Oustaloup's method.

Using the *fomcon* toolbox of MATLAB [41], the obtained magnitude and phase responses in frequency domain of the fifth-order approximation (i. e., $N = 2$) of the variable $s^{0.5}$ using the Oustaloup's method for $[\omega_b, \omega_h] = [10^{-3}, 10^{+3}]$ rad/sec, are given in Figure 2. In order to be evident the achieved level of accuracy, the corresponding error plots are provided in Figure 3, where it is derived that an error in gain less than 10 % is achieved in the range $[10^{-3}$ rad/sec, 10^{+3} rad/sec], while for the same level of phase accuracy, the range is $[16 \cdot 10^{-3}, 0.058 \cdot 10^{+3}]$ rad/sec. It is obvious that the accuracy of the phase response imposes limitation into the range where the approximation is valid; in the presented case this range is about $[\omega_0/60, 60\,\omega_0]$, where ($\omega_0$) is the unity gain frequency and this is reasonable in most control applications.

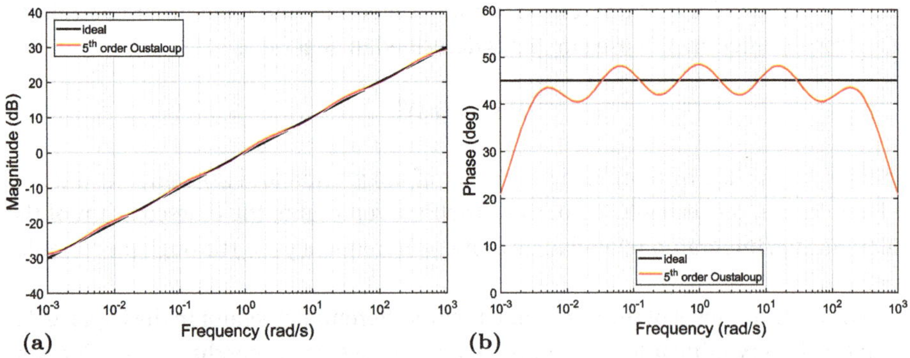

Figure 2: Frequency responses of the $s^{0.5}$ operator using the fifth-order Oustaloup's approximation (a) magnitude, and (b) phase.

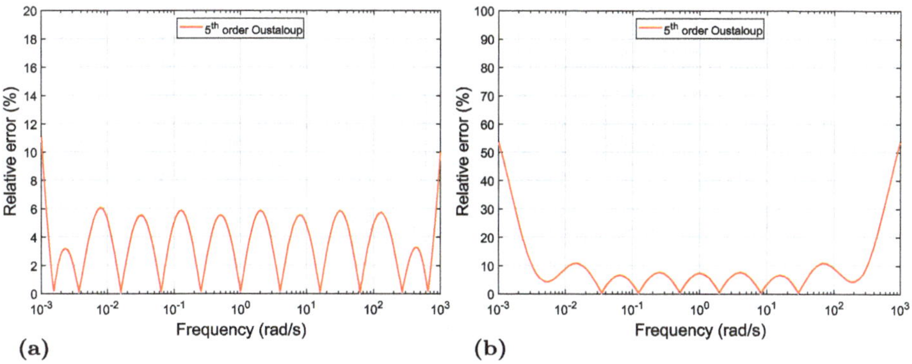

Figure 3: Error plots of the responses in Figure 2 about the (a) magnitude, and (b) phase.

2.2 Multi-feedback topologies for approximating the integrodifferential operator

Inspecting equation (3), it is obtained that the form for both the fractional-order integrator and differentiator is the same and, consequently, it can be implemented by the same topology. Thus, the Follow-the-Leader Feedback (FLF) and Inverse-Follow-the-Leader-Feedback (IFLF) structures [14], represented by the FBDs in Figures 4(a) and 4(b), respectively, can be utilized for performing the fifth-order Oustaloup's approximation. The transfer function realized by both schemes is given by equation (5)

$$H(s) = \frac{G_5 s^5 + (\frac{G_4}{\tau_1})s^4 + \cdots + (\frac{G_0}{\tau_1 \cdot \tau_2 \cdots \tau_5})}{s^5 + (\frac{1}{\tau_1})s^4 + \cdots + (\frac{1}{\tau_1 \cdot \tau_2 \cdots \tau_5})}. \tag{5}$$

The calculation of the time-constants τ_i $(i = 1, \ldots, 5)$ and gain-factors G_j $(j = 0, \ldots, 5)$ is performed by equating the corresponding coefficients of the polynomials in equation (3) with those in equation (5). The resulted design equations are the following:

$$\tau_i = \frac{A_{n+1-i}}{A_{n-i}}, \quad i = 1, 2, \ldots, 5, \tag{6}$$

$$G_j = B_j \times \prod_{i=1}^{n-j} \tau_i, \quad j = 0, 1, \ldots, 5. \tag{7}$$

It should be mentioned at this point that both integrator and differentiator can be implemented by the same structure, just by appropriately selecting the values of time-constants and gain factors, a fact that is very attractive from the design flexibility point of view.

The implementation of the FBD in Figure 4(a), using operational amplifiers (op-amps) as active elements for realizing the lossless integration and summation stages [14], is depicted in Figure 5. The realized time-constants are given by the expression

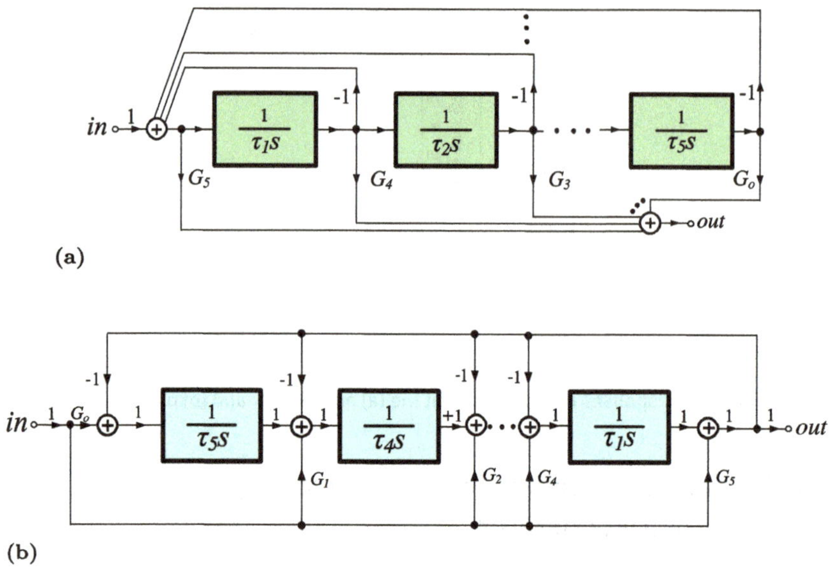

Figure 4: Multi-feedback structures (a) FLF, and (b) IFLF, for the fifth-order approximation of fractional-order differentiators/integrators.

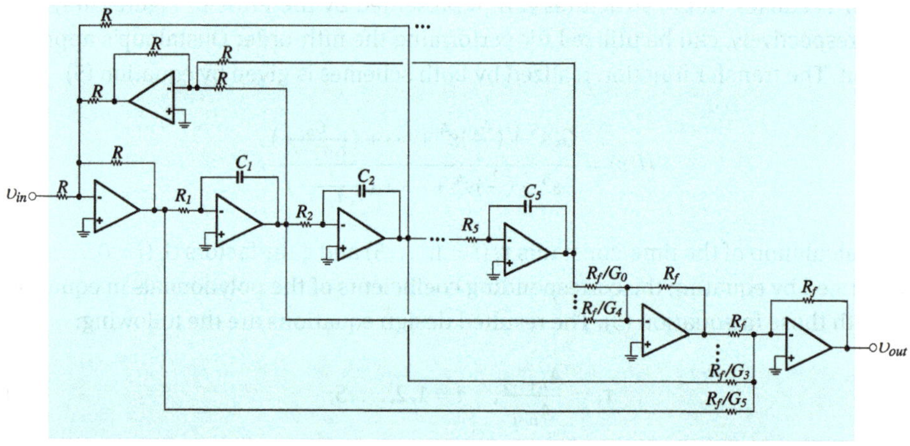

Figure 5: Fifth-order approximation of fractional-order integrators/differentiators using op-amps as active elements.

in equation (8):

$$\tau_i = R_i C_i \quad (i = 1, \dots, 5). \tag{8}$$

The required values of gain-factors are implemented by choosing appropriate values of the resistors in the summation stage at the output of the circuit, as is depicted in Figure 5. The corresponding implementation using second generation Current Conveyors

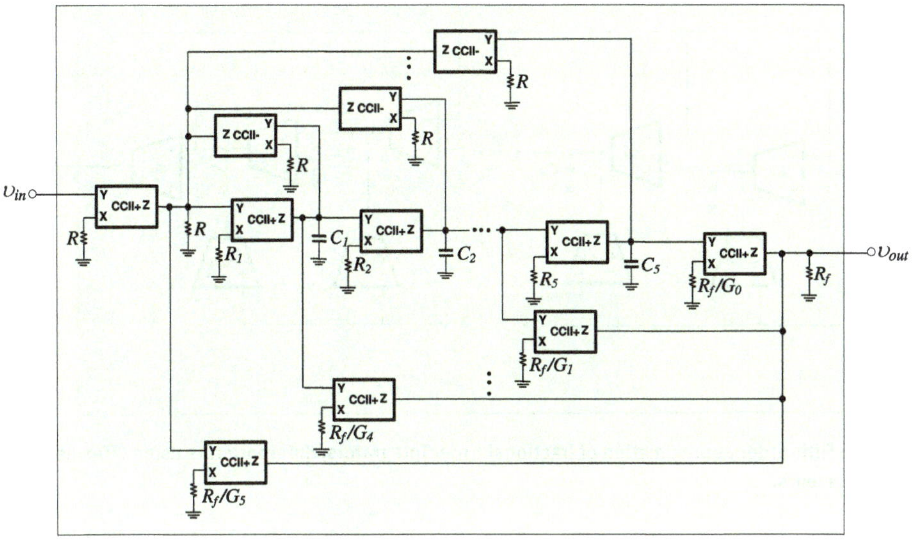

Figure 6: Fifth-order approximation of fractional-order integrators/differentiators using CCIIs as active elements.

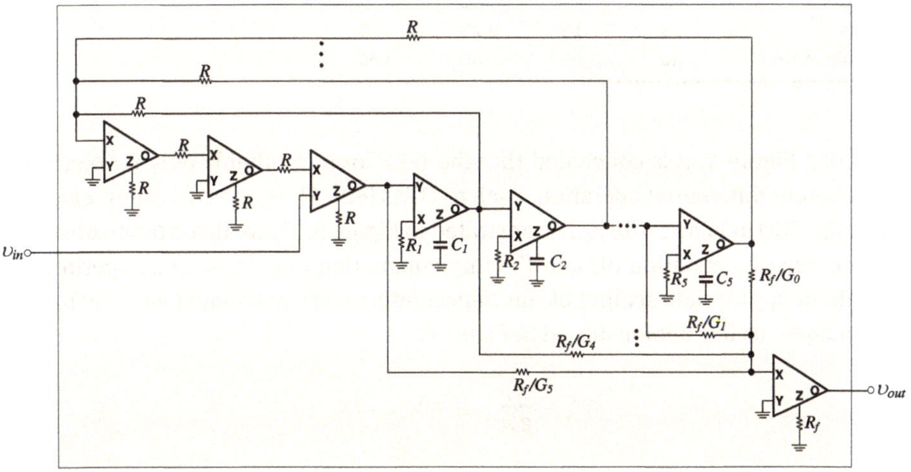

Figure 7: Fifth-order approximation of fractional-order integrators/differentiators using CFOAs active elements.

(CCIIs) or Current Feedback Operational Amplifiers (CFOAs) as active elements [32, 39], are demonstrated in Figures 6–7, respectively. The expressions of time-constants for both implementations are still given by equation (8), while the gain-factors are implemented through the same way as in the case of the op-amp based topology.

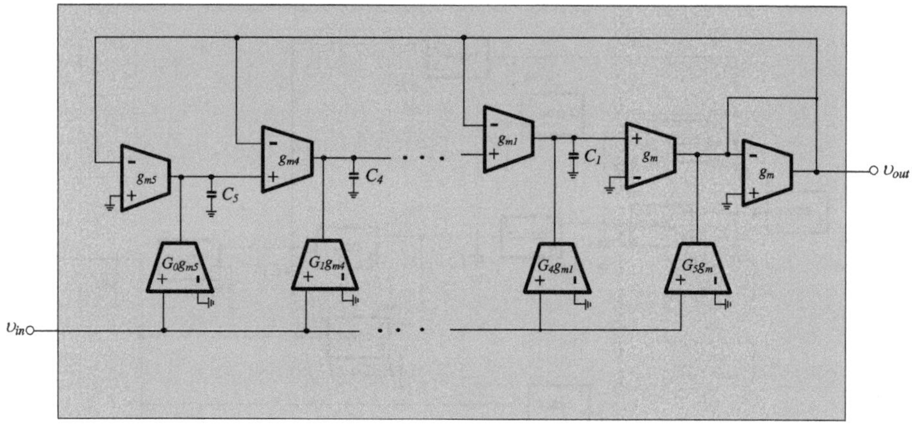

Figure 8: Fifth-order approximation of fractional-order integrators/differentiators using OTAs as active elements.

Table 1: Performance comparison results for the implementations in Figures 5–8.

Factor	Figure 5	Figure 6	Figure 7	Figure 8
active elements	10	18	10	14
capacitors	5	5	5	5
resistors	23	15	22	0
electronic control	no	no	no	yes

Inspecting Figure 4, it is concluded that the IFLF form is suitable only for elements with inherent differential operation, such as OTAs [40]. Thus, the OTA-C implementation of the FBD in Figure 4(b) is demonstrated in Figure 8. The realized time-constants are expressed by equation (9) and the implementation of gain-factors is performed through an appropriate scaling of the transconductances associated with each integration node, as it is demonstrated in Figure 8:

$$\tau_i = \frac{C_i}{g_{mi}} \quad (i = 1, \dots, 5). \tag{9}$$

The comparison results about the circuit complexity of the structures in Figures 5–8 are summarized in Table 1.

According to the provided results, the best option in terms of active component count is the FLF structure implemented using op-amps or CFOAs as active elements. On the other hand 23 or 22 resistors are required leading into an increase of the silicon area, especially in low-frequency applications where large values of resistors are required. Another problem is originated from the limited accuracy of the on-chip implementation of the time-constants, due to their form: τ = RC, making essential the employment of an automatic tuning system. All the aforementioned obstacles can be overcome using the OTA-C realization in Figure 8. Although the number of active elements is increased by a factor 40 %, passive resistors are not required and, most important, the time-constants and gain-factors can be controlled through the small-signal transconductance parameter (g_m). Taking also into account that the implementation of differentiators and integrators can be performed by the same structure, just by changing the values of time-constants and gain-factors, it is obvious that by adjusting the bias currents/voltages, which control the transconductance, fractional-order integrators/differentiators of variable orders and unity-gain frequency can be implemented. This is very important from the design versatility point of view as the derived controller structures will be modular and, consequently, the same core can be used for implementing the integrating and differentiating parts of the controller.

3 Designs of integratable $PI^{\lambda}D^{\mu}$ controllers

3.1 DC motor system controller

3.1.1 Specifications of the DC motor controller

The DC motor is a power actuator, which converts direct electrical energy into rotational mechanical energy. DC motors are still often used in industry and in numerous control applications, robotic manipulators, and commercial applications such as disk drives, tape motors, etc. In this application example, a mini- DC motor with model number PPN13KA12C which is suitable for robots, remote control applications, CD/DVD mechanics, etc. [36] is presented. The design goal is the implementation of a controller, which provides a step response of feedback control with overshoot independent of payload changes (i. e., iso-damping). From the frequency domain point of view, this means that the phase margin is independent of the payload changes. Phase margin of a controlled system is given by equation (10)

$$\Phi_m = \arg\left[C(j\omega_{gc})\, G_{DCM}(j\omega_{gc})\right] + \pi, \tag{10}$$

where $C(s)$ is the transfer function of the controller, $G_{DCM}(s)$ is the transfer function of the DC motor given by: $G_{DCM}(s) = 0.08/s\,(0.05s + 1)$, and $j\omega_{gc}$ is the gain-crossover frequency. Independent phase margin means in other words constant phase. This can be accomplished by controller of the form:

$$C(s) = 12.5\,\frac{0.05\,s + 1}{s^{\gamma}}. \tag{11}$$

Such controller gives a constant phase margin and obtained phase margin is

$$\Phi_m = \pi - (1 + \gamma)\,\frac{\pi}{2}. \tag{12}$$

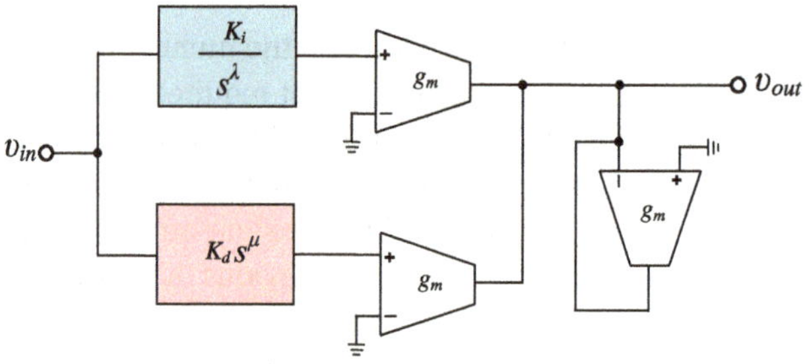

Figure 9: Block diagram of the DC motor system fractional-order controller.

Using the expression in equation (12) and setting $\Phi_m = 45°$, it is obtained that $\gamma = 0.5$. Consequently, the transfer function of the controller in equation (11) can be alternatively written as

$$C(s) = 12.5\,s^{-0.5} + 0.625\,s^{0.5}. \tag{13}$$

According to equation (1), it is a fractional $\mathrm{I}^\lambda\mathrm{D}^\mu$ controller with $K_i = 12.5$, $K_d = 0.625$, and $\lambda = \mu = 0.5$. Comparing each term of the sum in equation (13) with the general expression of the fractional-order integrator/differentiator given by equation (2), it is readily obtained that the corresponding time-constants will be $\tau_{\text{int}} = 6.4$ msec for the integrator, and $\tau_{\text{diff}} = 0.391$ sec for the differentiator. Consequently, the unity-gain frequency of the integrator will be 24.9 Hz while for the differentiator will be 410 mHz. The block diagram of the controller is depicted in Figure 9, and is constructed from fractional-order integrator and differentiator blocks which will be implemented using the topology in Figure 8, and their outputs will be summed up by the employment of the 3 OTAs structure at the output of the controller [15].

3.1.2 Simulation results of the DC motor controller

The implementation of the controller will be performed using the improved linearity OTA given in Figure 10, where the transconductance is given by equation (14) [13]

$$g_m = \frac{5}{9} \cdot \frac{I_B}{nV_T}, \tag{14}$$

where n is the slope factor of a MOS transistor in sub-threshold region ($1 < n < 2$) and V_T is the thermal voltage (26 mV@27°C).

Using equations (9) and (14) the realized time-constants will be given by the expression

$$\tau_i = \frac{9}{5} \cdot \frac{nC_iV_T}{I_{Bi}} \quad (i = 1, \ldots, 5). \tag{15}$$

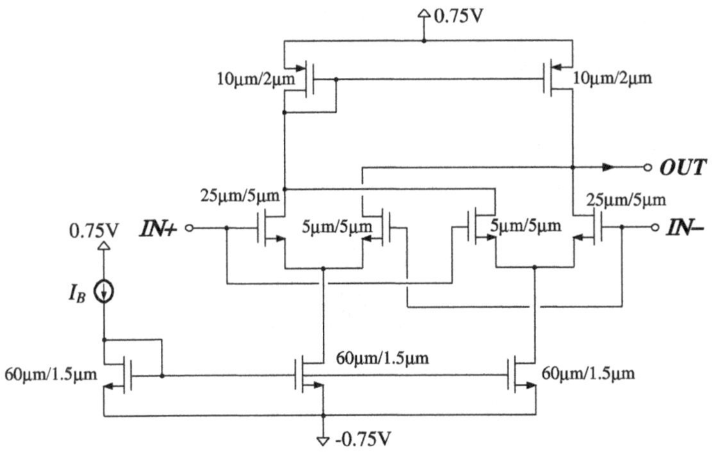

Figure 10: MOS transistor implementation of the OTA.

Table 2: Values of gain factors and time-constants for fractional-order integrator and differentiator in DC motor controller.

	Integrator		Differentiator	
index *i*	G_i	τ_i (sec)	G_i	τ_i (sec)
0	31.63	–	0.032	–
1	7.943	47.6 μ	0.126	0.73 m
2	1.996	803 μ	0.501	12.3 m
3	0.501	12.8 m	1.996	196 m
4	0.126	203 m	7.943	3.12
5	0.032	3.42	31.63	52.5

Using the *fomcon* toolbox of the MATLAB, the derived expressions for approximating the terms of the controller are the following:

$$12.5\,s^{-0.5} = \frac{0.3953\,s^5 + 211.5\,s^4 + 6713\,s^3 + 1.339 \cdot 10^4\,s^2 + 1680\,s + 12.5}{s^5 + 134.4s^4 + 1072s^3 + 537.1s^2 + 16.92s + 0.03162}, \tag{16}$$

$$0.625\,s^{0.5} = \frac{19.76\,s^5 + 2656\,s^4 + 2.118 \cdot 10^4\,s^3 + 1.061 \cdot 10^4\,s^2 + 334.3\,s + 0.625}{s^5 + 534.9\,s^4 + 1.698 \cdot 10^4\,s^3 + 3.389 \cdot 10^4\,s^2 + 4249\,s + 31.62}. \tag{17}$$

Comparing the coefficients of equations (16) and (17) with those of equations (3), and using equations (6)–(7), the calculated values of time-constants and gain-factors are summarized in Table 2.

With regards to the implementation of time-constants, the values of capacitors or dc bias currents are calculated using equation (15), while the implementation of the required gain-factors is performed by scaling G_j times ($j = 0, 1, \ldots, 5$) the dc bias current of the corresponding OTA. In the simulations, it was considered that g_{mi} =

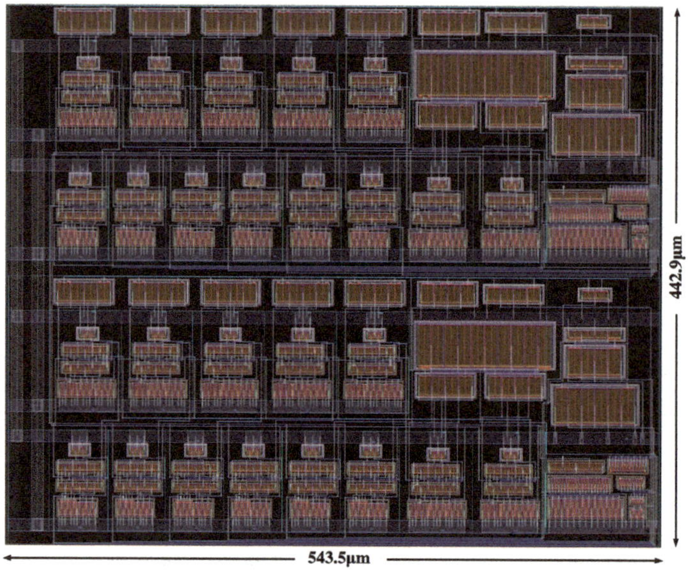

Figure 11: Layout design of the DC motor controller.

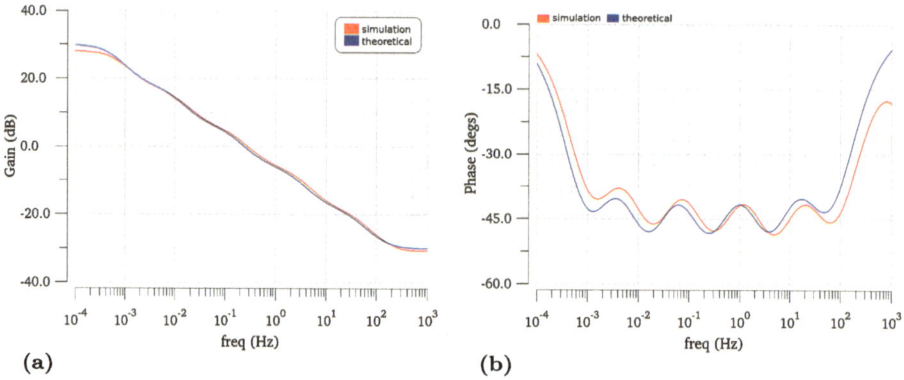

Figure 12: Frequency responses of the fractional-order integrator of the DC motor controller (a) magnitude, and (b) phase.

35.61 nS ($i = 1, \ldots, 5$) and, according to equation (14), the corresponding bias currents have been chosen equal to 2.5 nA.

The layout design of the controller is demonstrated in Figure 11. The obtained post-layout responses of the magnitude and phase of the fractional-order integrator and differentiator are demonstrated in Figures 12–13, where the unity-gain frequencies of integrator and differentiator were 31.26 Hz and 465 mHz, with the corresponding theoretical values being 24.9 Hz and 410 mHz, respectively.

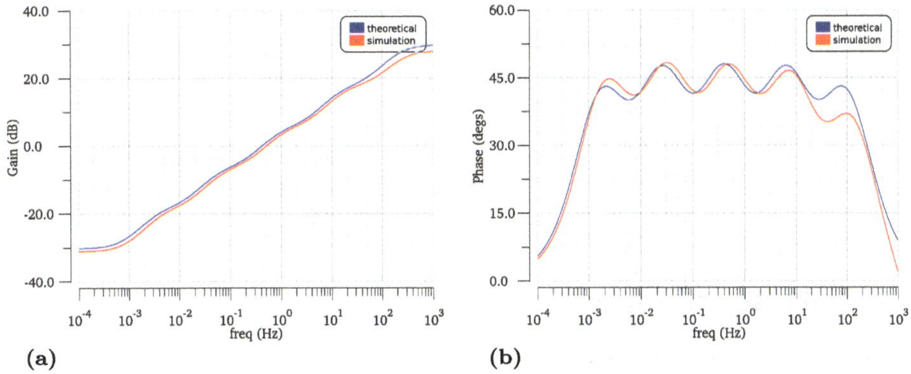

Figure 13: Frequency responses of the fractional-order differentiator of the DC motor controller (a) magnitude, and (b) phase.

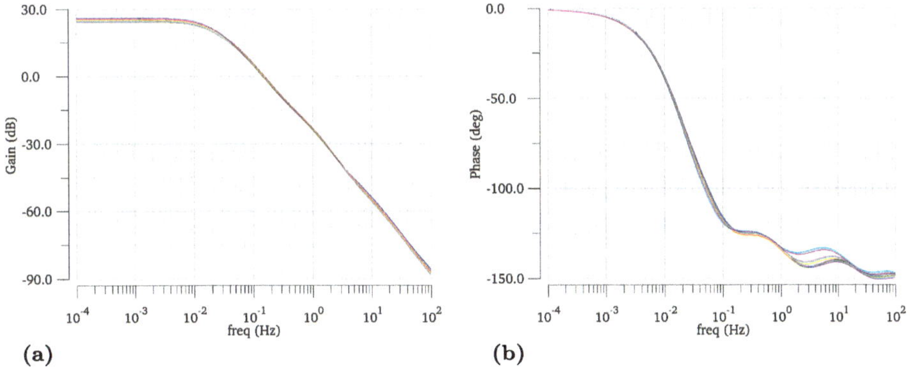

Figure 14: PVT corner analysis for the open-loop frequency response of the DC motor controller (a) magnitude, and (b) phase.

The behavior of the controller at process, power supply voltage, and temperature (PVT) corners has been evaluated over 32 different PVT corners. Thus, worst case power (fast–fast), worst case speed (slow–slow), worst case one (fast–slow), and worst zero (slow–fast) MOS transistors model parameters, 0°C, 27°C, 60°C temperature values, and ±5 % supply voltages variations have been considered. The derived open-loop responses are depicted in the plots of Figure 14, where the measured phase-margin was 53° close to the theoretically predicted value of 45°. Also, the unity-gain frequency was 160 mHz, while the theoretical value was 158 mHz. The corners of the gain at unity-gain frequency (i. e., 158 mHz) were −0.2 dB and +0.1 dB. The corresponding corners of the phase margin were 57° and 51°, respectively. As the corresponding nominal values were 0 dB and 53°, respectively, the designed controller offers robustness.

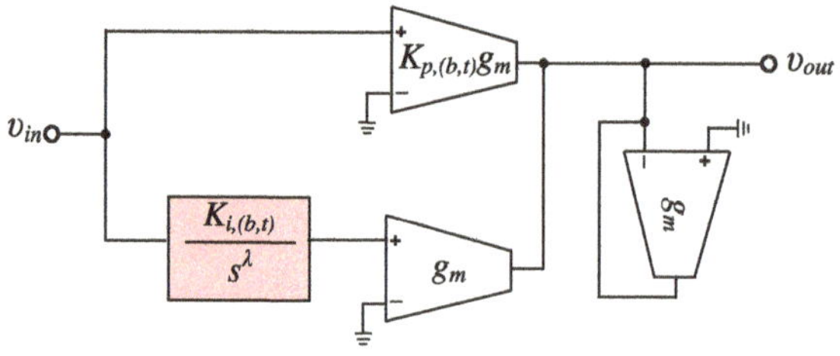

Figure 15: Block diagram of the brake/throttle fractional-order controller.

3.2 Brake/throttle system controller

3.2.1 Specifications of the controller

A structure constructed from two different PI^{λ} controllers has been proposed to control a commercial Citroën C3 prototype vehicle, which has automatic driving capabilities at low speeds, in [24]. More specifically, the controllers have been designed to act over the throttle and brake pedals and their operation is described by equation (18)

$$C(s) = K_{p(b,t)} + \frac{K_{i(b,t)}}{s^{\lambda}}, \tag{18}$$

where K_{pb} is the proportional gain during brake and K_{pt} is the proportional gain during throttle. In a similar way, K_{ib} and K_{it} are the integration constants during brake and throttle, respectively. The block diagram of this scheme is demonstrated in Figure 15 [16].

The vehicle dynamics during braking are given by equation (19)

$$G_b(s) \cong \frac{1}{2.25\,s + 1}, \tag{19}$$

while the corresponding expression during acceleration is given by equation (20)

$$G_t(s) \cong \frac{4.39}{s + 0.1746}. \tag{20}$$

In order to achieve robustness to time-constant variations of the brake controller, the phase margin (Φ_m) has been set equal to 90°, while the phase (p_m) and gain (M_m) at frequency $\omega_m = 5.7$ rad/sec have been set equal to −100° and −20 dB, respectively. Thus, the parameters of the brake controller were: $K_{p,b} = 0.7$, $K_{i,b} = 1.1$, and $\lambda = 0.45$. In the case of the throttle controller, for phase margin $\Phi_m = 90°$, gain-crossover frequency $\omega_{gc} = 0.45$ rad/sec and output disturbance rejection ≤−20 dB for $\omega \leq 0.035$ rad/sec, and the parameters of the controller were: $K_{p,t} = 0.09$, $K_{i,t} = 0.025$, and $\lambda = 0.8$.

The extremely small value (0.025) of the variable $K_{i,t}$ of the integrator can cause nonreasonable values of capacitors and, in order to avoid such obstacle, the integration term in the case of the throttle controller will be expressed as in equation (21).

$$\frac{0.025}{s^{0.8}} = 0.023 \cdot \frac{1.1}{s^{0.45}} \cdot \frac{1}{s^{0.35}}. \tag{21}$$

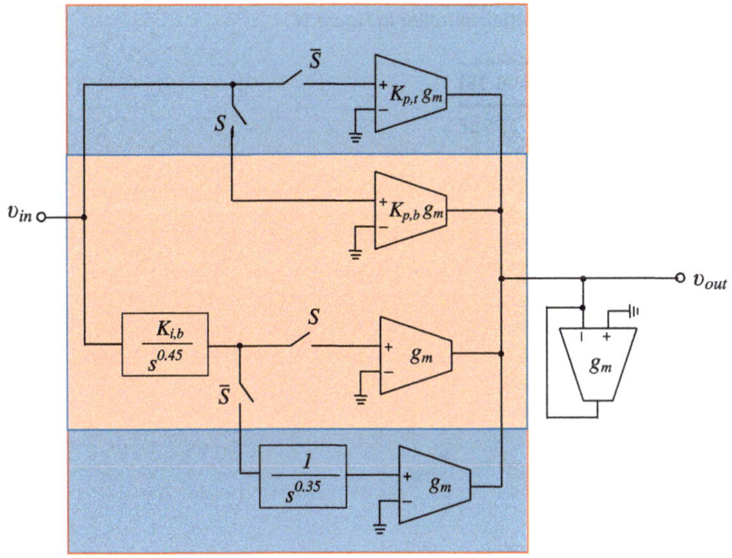

Figure 16: Alternative block diagram of the brake/throttle controller for avoiding large capacitances.

According to equation (21), and taking also into account that for the brake controller $K_{i,b} = 1.1$, and $\lambda = 0.45$, the integrator of the throttle controller will be constructed from: (a) a gain stage, (b) the integrator of the brake controller, and (c) a fractional-order integrator of order equal to 0.35. Owing to the fact that there is nonoverlap between the brake and throttle operations, the block diagram in Figure 15 can be alternatively represented by the block diagram in Figure 16. The activation of the appropriate part of this diagram is performed through the utilization of a nonoverlapping switching scheme (denoted by S and $\bar{S}$) which will be digitally controlled.

3.2.2 Simulation results for the brake/throttle controller

The implementation of the brake/throttle controller has been performed using the OTA cell in Figure 8 with the same bias condition as in the case of the DC motor controller. Using the *fomcon* toolbox of MATLAB [41], the derived rational transfer functions that approximate the operators $s^{-0.45}$ and $s^{-0.35}$ are given by equations (22)–(23):

$$s^{-0.45} \cong \frac{0.0447s^5 + 22.3s^4 + 660.7s^3 + 1230s^2 + 144s + 1}{s^5 + 144s^4 + 1230s^3 + 660.7s^2 + 22.3s + 0.0447}, \tag{22}$$

$$s^{-0.35} \cong \frac{0.0891s^5 + 38.75s^4 + 1000s^3 + 1622s^2 + 165.3s + 1}{s^5 + 165.3s^4 + 1622s^3 + 1000s^2 + 38.75s + 0.0891}. \tag{23}$$

Following a similar procedure as that in Section 3.1, and considering that g_{mi} = 35.61 nS ($i = 1, \ldots, 5$), the calculated values of capacitors are summarized in Table 3.

Table 3: Capacitor values for the brake/throttle controller in Figure 16.

Capacitor	brake (λ = 0.45)	throttle (λ = 0.35)
C_1	200 pF	215 pF
C_2	3.38 nF	3.63 nF
C_3	53.7 nF	57.8 nF
C_4	855 nF	919 nF
C_5	14.4 μF	14.5 μF

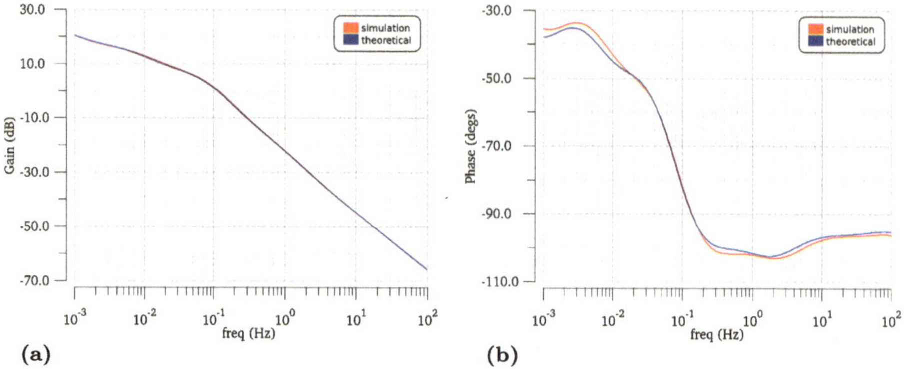

Figure 17: Open loop frequency response of the brake controller (a) magnitude, and (b) phase.

It should be mentioned at this point that in the case that the block diagram in Figure 15 would have been employed the resulted maximum capacitance value would be equal to 2.9 mF, a fact that verifies the improvement offered by the block diagram in Figure 16.

The open-loop gain and phase responses of the brake controller are given in Figure 17, where the measured phase margin was $\Phi_m = 94^\circ$ close to the desired value of 90°. Also the phase (p_m) and gain (M_m) at frequency $\omega_m = 5.7$ rad/sec were equal to –101° and –20.78 dB, respectively, while the corresponding specifications were set equal to –100° and –20 dB. The response of the throttle controller is demonstrated in Figure 18 where the phase margin was $\phi_m = 88^\circ$, and the gain-crossover frequency $\omega_{gc} = 0.502$ rad/sec. The plot of the output disturbance rejection is depicted in Figure 19, where values less than –21 dB are achieved for $\omega \leq 0.035$ rad/sec. The corresponding specifications were $\phi_m = 90^\circ$, $\omega_{cg} = 0.45$ rad/sec, and disturbance rejection $\leq$–20 dB and, therefore, all of them are fulfilled by the designed controller.

The sensitivity behavior of both controllers has been studied using the Monte-Carlo analysis tool offered by the Analog Design Environment of the Cadence software. The values of standard deviation of the unity-gain frequency and phase margin, for $N = 100$ runs, were 1 mHz and 1° for the brake controller while they were 10 mHz and

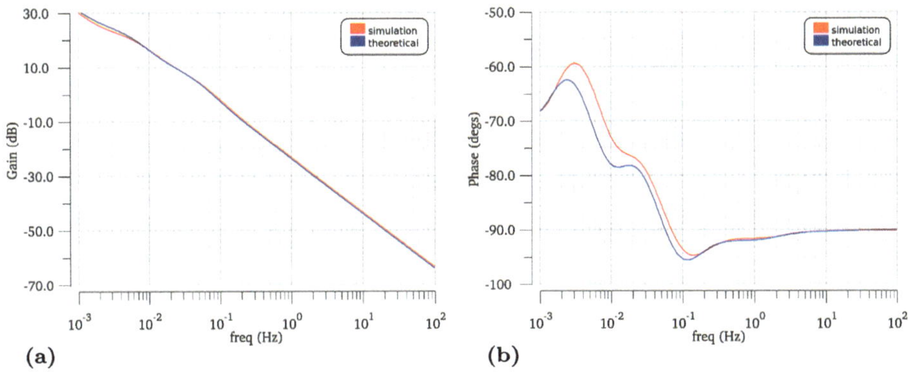

Figure 18: Open-loop frequency response for the throttle controller (a) magnitude, and (b) phase.

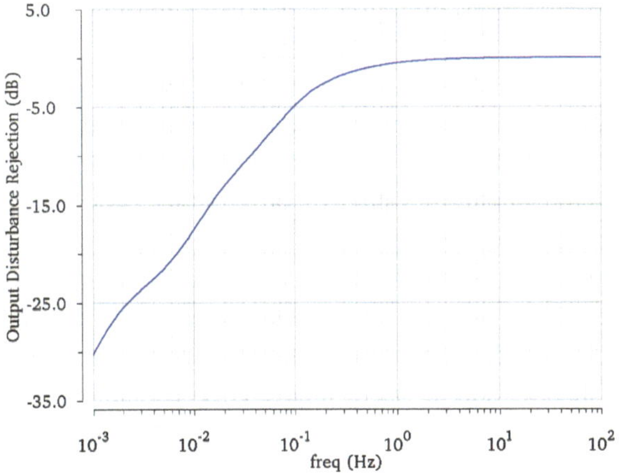

Figure 19: Output disturbance rejection of the throttle controller.

12° for the throttle controller. The increased sensitivity of the throttle controller phase is mainly caused by the cascade connection used for implementing this controller.

4 Discussion

Inspecting the provided values of the gain-factors (G_j) in Table 2, it is readily obtained that their spread (i. e., the ratio between the maximum and minimum values) is equal to 988. Taking also into account that the gain factors are implemented through a scaling of the aspect ratio of the MOS transistors which feed the OTA (see Figure 10), it is obvious that nonrealistic values of aspect ratio will be derived. More specifically, as the aspect ratio of the nMOS diode connected transistor is $60\,\mu m/1.5\,\mu m$, the corresponding minimum aspect ratio must be equal to $1.92\,\mu m/1.5\,\mu m$, which is reasonable,

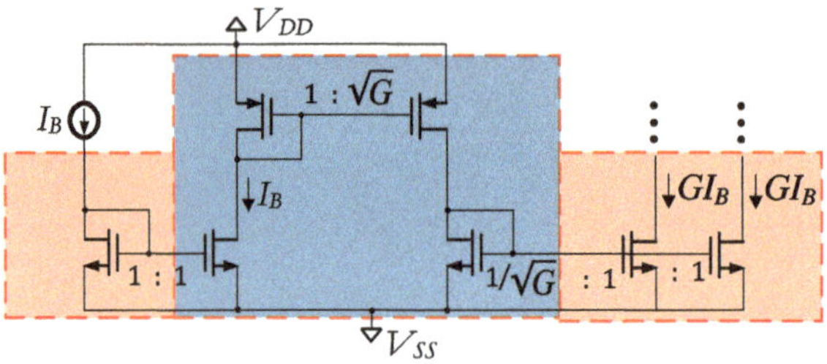

Figure 20: Modified bias scheme of the OTA in order to achieve reasonable values of MOS transistors aspect ratios.

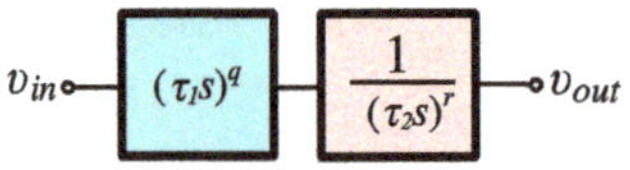

Figure 21: Block diagram for simultaneously realizing fractional-order integrator/differentiator of order (α) and fractional-order differentiator/integrator of order ($1-\alpha$).

but the maximum required value will be 1898 μm/1.5 μm, which is not reasonable for implementation on silicon.

In order to overcome this problem, the solution demonstrated in Figure 20 is proposed. Performing a routine analysis it is easily verified that a bias current equal to GI_B is produced at the output of the topology. In this way, the minimum aspect ratio will be equal to 11 μm/1.5 μm, while the maximum value will be 337 μm/1.5 μm, which is, obviously, reasonable.

As it was previously mentioned, the scheme in Figure 8 is capable for implementing both fractional-order integrators/differentiators through the adjustment of the corresponding dc bias currents. An additional feature is that the topology could be further enhanced in order to simultaneously offer both types of stages. This is demonstrated in the block diagram in Figure 21, where the realized transfer function is the following:

$$H(s) \equiv \frac{v_{\mathrm{out}}}{v_{\mathrm{in}}} = \tau s^{q-r}, \tag{24}$$

where the time-constant of the system is defined as: $\tau \equiv \tau_1^q/\tau_2^r$. According to equation (24), a fractional-order differentiator or integrator of order $(1-\alpha)$ is implemented for $(q,r) = (-\alpha,-1)$ or $(\alpha,1)$, respectively. This could be very useful in cases where fractional-order integration stages of complementary order (i. e., the sum of the order will be equal to one) are required. This was the case of the DC motor controller in Figure 9, where the required fractional-order integrator/differentiator can be implemented using a fractional-order differentiator/integrator with an additional integer-order integrator/differentiator. In this way, the circuit complexity of the controller will be almost halved.

The current-mode approach could be a powerful tool for implementing integratable fractional-order controller structures, due to simple structures of current-mode integrators constructed from current-mirrors [5]. Therefore, the FLF block diagram in Figure 4(a) can be implemented using the lossless integrator in Figure 22. The realized

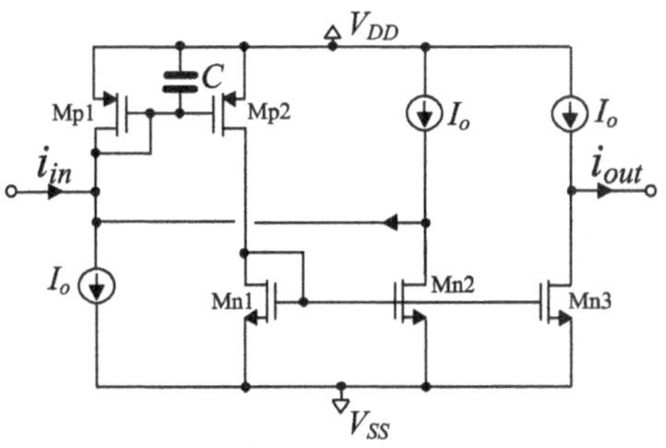

Figure 22: Typical current-mode lossless integrator using current-mirrors.

transfer function is given by equation (25):

$$H(s) = \frac{1}{(\frac{C}{g_m})s}, \tag{25}$$

where g_m is transconductance parameter of the diode-connected nMOS transistor (Mn1) associated with the capacitor C. The required gain-factors can be implemented through a scaling of the aspect ratio of transistor and dc bias current source associated with the output branch. Because of the current-mode nature of the circuit, the output summation of the intermediate currents, required for realizing the total output current, is easily implemented through a direct connection of the corresponding pins.

5 Conclusion

A systematic design procedure for deriving fractional-order controllers with capability of implementation on silicon, is presented in this chapter. The method is based on the utilization of fractional-order integrators and differentiators approximated by multi-feedback structures realized using OTAs as active elements with MOS transistors operated in the subthreshold region. Thus, the main offered benefits are the capability of operation in a low-voltage (±0.75 V) environment, the reduced power consumption, the electronic adjustment of the characteristics of the integrators and differentiators through the tuning of the transconductance (g_m) of the OTAs, and the capability of implementation in monolithic form. The provided design examples confirm the validity of the proposed design method in terms of the performance characteristics as well as of robustness.

Further research effort could be performed for exploiting and other techniques for implementing the controllers, including companding filtering, current-mode filtering, etc. In addition, the development of novel approximation expressions could be a powerful tool for implementing the circuitry of the controllers with reduced active component count making the circuit complexity comparable to that offered by the conventional (i. e., integer-order) controllers [33].

Bibliography

[1] A. Agambayev, S. P. Patole, M. Farhat, A. Elwakil, H. Bagci, and K. N. Salama, Ferroelectric fractional-order capacitors, *ChemElectroChem*, **4**(11) (2017), 2807–2813.

[2] B. B. Alagoz, A. Ates, and C. Yeroglu, Auto-tuning of PID controller according to fractional-order reference model approximation for DC rotor control, *Mechatronics*, **23**(7) (2013), 789–797.

[3] A. Ateş and C. Yeroglu, Optimal fractional order PID design via Tabu Search based algorithm, *ISA Transactions*, **60** (2016), 109–118.

[4] J. Baranowski, W. Bauer, M. Zagórowska, A. Kawala-Janik, T. Dziwiński, and P. Piątek, Adaptive non-integer controller for water tank system, in *Theoretical Developments and Applications of Non-Integer Order Systems*, pp. 271–280, Springer, 2016.

[5] P. Bertsias, C. Psychalinos, A. Elwakil, and B. Maundy, Current-mode capacitorless integrators and differentiators for implementing emulators of fractional-order elements, *AEÜ. International Journal of Electronics and Communications*, **80** (2017), 94–103.

[6] G. W. Bohannan, Analog realization of a fractional control element – revisited, in *Proc. of the 41st IEEE Int. Conf. on Decision and Control, Tutorial Workshop*, vol. 2, pp. 203–208, 2002.

[7] R. Caponetto, G. Dongola, L. Fortuna, and I. Petráš, *Fractional Order Systems: Modeling and Control Applications*, World Scientific, 2010.

[8] R. Caponetto, G. Dongola, G. Maione, and A. Pisano, Integrated technology fractional order proportional-integral-derivative design, *Journal of Vibration and Control*, **20**(7) (2014), 1066–1075.

[9] R. Caponetto, S. Graziani, F. L. Pappalardo, and F. Sapuppo, Experimental characterization of ionic polymer metal composite as a novel fractional order element, *Advances in Mathematical Physics*, **2013** (2013), 953695.

[10] G. Carlson and C. Halijak, Approximation of fractional capacitors $(1/s)^{1/n}$ by a regular Newton process, *IEEE Transactions on Circuit Theory*, **11**(2) (1964), 210–213.

[11] A. Charef, Analogue realisation of fractional-order integrator, differentiator and fractional $\mathrm{PI}^{\lambda}\mathrm{D}^{\mu}$ controller, *IEE Proceedings. Control Theory and Applications*, **153**(6) (2006), 714–720.

[12] Y. Chen, I. Petráš, and D. Xue, Fractional order control – a tutorial, in *American Control Conference, 2009. ACC '09*, pp. 1397–1411, IEEE, 2009.

[13] P. Corbishley and E. Rodriguez-Villegas, A nanopower bandpass filter for detection of an acoustic signal in a wearable breathing detector, *IEEE Transactions on Biomedical Circuits and Systems*, **1**(3) (2007), 163–171.

[14] T. Deliyannis, Y. Sun, and J. K. Fidler, *Continuous-Time Active Filter Design*, CRC Press, 1998.

[15] I. Dimeas, I. Petráš, and C. Psychalinos, New analog implementation technique for fractional-order controller: a DC motor control, *AEÜ. International Journal of Electronics and Communications*, **78** (2017), 192–200.

[16] I. Dimeas, C. Psychalinos, A. Elwakil, and A. Tepljakov, OTA-C realization of PI^{λ} brake and throttle controllers for autonomous vehicles, in *2017 European Conference on Circuit Theory and Design (ECCTD)*, pp. 1–4, IEEE, 2017.

[17] I. Dimeas, G. Tsirimokou, C. Psychalinos, and A. S. Elwakil, Experimental verification of fractional-order filters using a reconfigurable fractional-order impedance emulator, *Journal of Circuits, Systems, and Computers*, **26**(9) (2017), 1750142.

[18] L. Dorčák, J. Valsa, E. Gonzalez, J. Terpák, I. Petráš, and L. Pivka, Analogue realization of fractional-order dynamical systems, *Entropy*, **15**(10) (2013), 4199–4214.

[19] A. S. Elwakil, Fractional-order circuits and systems: an emerging interdisciplinary research area, *IEEE Circuits and Systems Magazine*, **10**(4) (2010), 40–50.

[20] T. J. Freeborn, B. Maundy, and A. S. Elwakil, Field programmable analogue array implementation of fractional step filters, *IET Circuits, Devices & Systems*, **4**(6) (2010), 514–524.

[21] E. Gonzalez, L. Dorčák, C. Monje, J. Valsa, F. Caluyo, and I. Petráš, Conceptual design of a selectable fractional-order differentiator for industrial applications, *Fractional Calculus and Applied Analysis*, **17**(3) (2014), 697–716.

[22] R. E. Gutiérrez, J. M. Rosário, and J. Tenreiro Machado, Fractional order calculus: basic concepts and engineering applications, *Mathematical Problems in Engineering*, **2010** (2010), 375858.

[23] C. Halijak, An RC impedance approximant to $(1/s)^{1/2}$, *IEEE Transactions on Circuit Theory*, **11**(4) (1964), 494–495.

[24] S. H. Hosseinnia, I. Tejado, V. Milanés, J. Villagrá, and B. M. Vinagre, Experimental application of hybrid fractional-order adaptive cruise control at low speed, *IEEE Transactions on Control Systems Technology*, **22**(6) (2014), 2329–2336.

[25] J. Jerabek, R. Sotner, J. Dvorak, J. Polak, D. Kubanek, N. Herencsar, and J. Koton, Reconfigurable fractional-order filter with electronically controllable slope of attenuation, pole frequency and type of approximation, *Journal of Circuits, Systems, and Computers*, **26**(10) (2017), 1750157.

[26] I. S. Jesus and J. T. Machado, Development of fractional order capacitors based on electrolyte processes, *Nonlinear Dynamics*, **56**(1–2) (2009), 45–55.

[27] D. A. John, S. Banerjee, G. W. Bohannan, and K. Biswas, Solid-state fractional capacitor using MWCNT-epoxy nanocomposite, *Applied Physics Letters*, **110**(16) (2017), 163504.

[28] J. Koton, D. Kubanek, K. Vrba, A. Shadrin, and P. Ushakov, Universal voltage conveyors in fractional-order filter design, in *2016 39th International Conference on Telecommunications and Signal Processing (TSP)*, pp. 593–598, IEEE, 2016.

[29] D. Kubanek, J. Koton, J. Jerabek, P. Ushakov, and A. Shadrin, Design and properties of fractional-order multifunction filter with DVCCs, in *2016 39th International Conference on Telecommunications and Signal Processing (TSP)*, pp. 620–624, IEEE, 2016.

[30] Y. Luo and Y. Chen, *Fractional Order Motion Controls*, John Wiley & Sons, 2012.

[31] G. Maione and P. Lino, New tuning rules for fractional PI^{α} controllers, *Nonlinear Dynamics*, **49**(1) (2007), 251–257.

[32] P. A. Mohan, *Current-Mode VLSI Analog Filters: Design and Applications*, Springer Science & Business Media, 2012.

[33] C. Muñiz-Montero, L. V. García-Jiménez, L. A. Sánchez-Gaspariano, C. Sánchez-López, V. R. González-Díaz, and E. Tlelo-Cuautle, New alternatives for analog implementation of fractional-order integrators, differentiators and PID controllers based on integer-order integrators, *Nonlinear Dynamics*, **90**(1) (2017), 241–256.

[34] A. Oustaloup, F. Levron, B. Mathieu, and F. M. Nanot, Frequency-band complex noninteger differentiator: characterization and synthesis, *IEEE Transactions on Circuits and Systems. I, Fundamental Theory and Applications*, **47**(1) (2000), 25–39.

[35] I. Petráš, The fractional-order controllers: methods for their synthesis and application, *Journal of Electrical Engineering*, **50**(9–10) (1999), 284–288.

[36] I. Petráš, Fractional-order feedback control of a DC motor, *Journal of Electrical Engineering*, **60**(3) (2009), 117–128.

[37] I. Podlubny, Fractional-order systems and $PI^{\lambda}D^{\mu}$ controllers, *IEEE Transactions on Automatic Control*, **44**(1) (1999), 208–214.

[38] I. Podlubny, I. Petráš, B. M. Vinagre, P. O'leary, and L. Dorčák, Analogue realizations of fractional-order controllers, *Nonlinear Dynamics*, **29**(1) (2002), 281–296.

[39] R. Senani, D. Bhaskar, A. Singh, and V. Singh, *Current Feedback Operational Amplifiers and Their Applications*, Springer Science & Business Media, 2013.

[40] Y. Sun, *Design of High Frequency Integrated Analogue Filters*, IEE Circuits, Devices and Systems Series, vol. 14, Institution of Electrical Engineers (IEE), 2002.

[41] A. Tepljakov, E. Petlenkov, and J. Belikov, FOMCON: fractional-order modeling and control toolbox for MATLAB, in *2011 Proceedings of the 18th International Conference Mixed Design of Integrated Circuits and Systems (MIXDES)*, pp. 684–689, IEEE, 2011.
[42] A. Tepljakov, E. Petlenkov, and J. Belikov, Efficient analog implementations of fractional-order controllers, in *2013 14th International Carpathian Control Conference (ICCC)*, pp. 377–382, IEEE, 2013.
[43] A. Tepljakov, E. Petlenkov, J. Belikov, and E. A. Gonzalez, Design of retuning fractional PID controllers for a closed-loop magnetic levitation control system, in *2014 13th International Conference on Control Automation Robotics & Vision (ICARCV)*, pp. 1345–1350, IEEE, 2014.
[44] G. Tsirimokou and C. Psychalinos, Ultra-low voltage fractional-order differentiator and integrator topologies: an application for handling noisy ECGs, *Analog Integrated Circuits and Signal Processing*, **81**(2) (2014), 393–405.
[45] G. Tsirimokou, C. Psychalinos, and A. Elwakil, *Design of CMOS Analog Integrated Fractional-Order Circuits: Applications in Medicine and Biology*, Springer, 2017.
[46] G. Tsirimokou, C. Psychalinos, A. S. Elwakil, and K. N. Salama, Electronically tunable fully integrated fractional-order resonator, *IEEE Transactions on Circuits and Systems. II, Express Briefs*, **65**(2) (2018), 166–170.
[47] J. Valsa and J. Vlach, RC models of a constant phase element, *International Journal of Circuit Theory and Applications*, **41**(1) (2013), 59–67.
[48] B. M. Vinagre, C. A. Monje, A. J. Calderón, and J. I. Suárez, Fractional PID controllers for industry application. A brief introduction, *Journal of Vibration and Control*, **13**(9–10) (2007), 1419–1429.
[49] M. Zagórowska, J. Baranowski, P. Bania, W. Bauer, T. Dziwiński, and P. Piątek, Parametric optimization of PD^{α} controller using Laguerre function approximation, in *2015 20th International Conference on Methods and Models in Automation and Robotics (MMAR)*, pp. 104–109, IEEE, 2015.

Changpin Li and Weiyuan Ma

Synchronizations in fractional complex networks

Abstract: Fractional complex networks can excellently describe the memory and hereditary properties of various networks. In virtue of potential applications in secure communication and control processing, synchronization of fractional networks becomes a challenging and interesting topic. Compared to the integer-order models, synchronization of fractional complex networks may be more useful. In this part, pinning control and adaptive control methods are used for the synchronization of fractional complex networks. To demonstrate the validity of the proposed methods, several illustrative examples are presented.

Keywords: Fractional derivative, complex networks, synchronization, numerical simulation

PACS: 05.45.Gg, 05.45.Xt

1 Introduction

Synchronization is an emerging phenomenon of a population of dynamically interacting units. Synchronization processes are ubiquitous in nature and play a very important role in many different contexts such as in biology, ecology, climatology, sociology, technology, and the arts [15, 17]. In 1665, the mathematician and physicist, inventor of the pendulum clock, Huygens, discovered a synchronization phenomenon in two pendulum clocks suspended side-by-side of each other. The pendulum clocks swung with exactly the same frequency and 180 degrees out of phase; when the pendula were disturbed, the antiphase state was restored within half of an hour and persisted indefinitely. Huygens deduced that the crucial interaction for this effect came from "imperceptible movements" of the common frame supporting the two clocks. From that time on, the phenomenon got into the focus of scientists. Among the efforts for the scientific description of synchronization phenomena, there are several typical works that represented a breakthrough in our understanding of these phenomena. Synchronization, as an important and interesting collective behavior of complex networks, has been extensively investigated. It is a fundamental phenomenon that enables coherent

Changpin Li, Department of Mathematics, Shanghai University, Shanghai 200444, China, e-mail: lcp@shu.edu.cn
Weiyuan Ma, School of Mathematics and Computer Science, Northwest Minzu University, Lanzhou 730000, China, e-mail: mwy2004@126.com

https://doi.org/10.1515/9783110571745-017

behaviors in networks as a result of interactions. But the phenomenon of synchronization also have many challenging problems to be solved.

Fractional calculus is as old as the classic calculus. However, fractional calculus has become a hot topic in the recent two decades due to its applications in physics and engineering. As a generalization of ordinary differential equations, the fractional differential equation can capture a nonlocal character in space and time. Thus, the fractional-order models are believed to be more accurate to deal with nonlocal problems. Fractional models have been proven to be an excellent instrument to describe the memory and hereditary properties of various materials and processes, such as dielectric polarization, electrode-electrolyte polarization, electromagnetic waves, viscoelastic systems, quantitative finance, and waves [4–6, 18].

Following these findings, synchronization of fractional complex networks becomes a challenging and interesting realm due to its potential applications in secure communication and control processing. The most fractional complex networks cannot spontaneously reach synchronization. To realize synchronization, many control methods have been proposed, such as pinning control, adaptive control, impulsive control, sliding-mode control, and so on.

Many of complex networks normally have a large number of nodes, therefore, it is usually expensive to control a complex network by designing the controllers for all nodes. To reduce the number of controllers, a pinning control method is proposed. Synchronization via a pinning control of general complex dynamical networks was studied [1, 14, 21, 27, 29, 31]. Due to global dependent property of fractional complex networks, as far as we know, the literature on pinning synchronization of fractional complex networks is still sparse. In [23], based on the eigenvalue analysis and fractional stability theory, local stability properties of pinned fractional networks were derived. The pinning synchronization of new uncertain fractional unified chaotic systems were discussed [16]. A global pinning synchronization for fractional complex dynamical networks was proposed [13].

As is well known, the fractional adaptation law enlarges the parameter adaptation performance by heightening one degree of freedom. The values of adaptive parameters will influence the change speed of the synchronization errors. Then the synchronization errors will converge to the origin more rapidly if we choose the parameters larger. The adaptive synchronization of fractional complex networks have been studied recently. Yang et al. [30] analyzed adaptive synchronization of drive-response fractional complex networks. Wang et al. [25] demonstrated projective outer synchronization of coupled uncertain fractional complex networks. Ma et al. [11] derived adaptive synchronization of fractional neural networks. Liu et al. [9] addressed adaptive synchronization for a class of uncertain fractional neural networks. Ma et al. [12] investigated synchronization of fractional fuzzy cellular neural networks.

Due to impulsive control [10, 22, 24], sliding-mode control [2, 20, 26], and other control methods for fractional complex networks synchronization still on the infant stage, we focused on the pinning control and adaptive control. The rest of this chapter is arranged as follows. In Section 2, some necessary preliminaries are given. In Section 3, the general drive and response fractional complex network models are introduced. Based on the Lyapunov stability theorem, pinning controllers are designed to ensure that the drive and response systems achieve synchronization. The illustrative numerical simulations are also displayed. Section 4 investigates the adaptive synchronization in the drive-response fractional dynamical networks with uncertain parameters. Section 5 concludes this chapter.

Throughout this chapter, let $\|\cdot\|$ the Euclidean norm, I_n the identity matrix with order n. If A is a vector or matrix, its transpose is denoted by A^T. Let $\lambda_{\min}(A)$ and $\lambda_{\max}(A)$ be the smallest and largest eigenvalue of symmetric matrix A, respectively.

2 Preliminaries

Now some lemmas are presented below.

Lemma 1 ([3]). *Let $x(t) = (x_1(t), \ldots, x_n(t))^T \in \mathbb{R}^n$ be a real continuous and differentiable vector function. Then*

$$^C D^\alpha [x^T(t) P x(t)] \le 2x^T(t) P \, ^C D^\alpha x(t),$$

where $0 < \alpha < 1$, $t \ge 0$, P is a symmetric and positive definite matrix.

Lemma 2 ([28]). *Let X and Y be arbitrary n-dimensional real vectors, K a positive definite matrix, and $H \in \mathbb{R}^{n\times n}$. Then the following matrix inequality holds:*

$$2X^T HY \le X^T HK^{-1} X + Y^T KY.$$

Consider the Caputo fractional nonautonomous system

$$^C D^\alpha x(t) = f(t, x(t)) \tag{1}$$

with initial condition $x(t) = x_0(t)$, where $\alpha \in (0,1)$, $f : [0,\infty) \times \Omega \to \mathbb{R}^n$ is piecewise continuous on t and locally Lipschitz with respect to x. And $\Omega \subset \mathbb{R}^n$ is a domain that contains the origin $x = 0$. We always assume that (1) has an equilibrium $x = 0$.

Lemma 3 ([8]). *Let $x = 0$ be an equilibrium point of system (1) and $\Omega \subset \mathbb{R}^n$ be a domain containing the origin. If there exists a Lyapunov-like function $V(t, x(t)) : [0,\infty] \times \Omega \to R$, which is continuously differentiable and locally Lipschitz with respect to x such that*

$$\text{(i)} \quad \alpha_1 \|x(t)\|^a \le V(t, x(t)) \le \alpha_2 \|x(t)\|^{ab}, \tag{2}$$

$$\text{(ii)} \quad ^C D^\alpha V(t, x(t)) \le -\alpha_3 \|x(t)\|^{ab}, \tag{3}$$

where $t \ge 0$, $x \in \Omega$, $\alpha \in (0,1)$, a, b, α_1, α_2, α_3 are positive constants. Then $x = 0$ of system (1) is globally Lyapunov asymptotically stable.

Lemma 4 ([19]). *Assume that* $Q = (q_{ij})_{N\times N}$ *is symmetric. Let* $M^* = \operatorname{diag}(m_1^*, m_2^*, \ldots, m_l^*, \underbrace{0, \ldots, 0}_{N-l})$, $1 \le l \le N$, $m_i^* > 0$ $(i = 1, 2, \ldots, l)$, $Q - M^* = \begin{pmatrix} (E-\widetilde{M}^*)_{l\times l} & S_{l\times(N-l)} \\ S^T_{(N-l)\times l} & \widetilde{Q}_{(N-l)\times(N-l)} \end{pmatrix}$, $\widetilde{M}^* = \operatorname{diag}(m_1^*, \ldots, m_l^*)$, $m^* = \min_{1\le i\le l}\{m_i^*\}$, $\widetilde{Q}$ *is the block matrix of* Q *with order* $N-l$, E *and* S *are matrices with appropriate dimensions. When* $m^* > \lambda_{\max}(E - S\widetilde{Q}^{-1}S^T)$, *then* $Q - M^* < 0$ *is equivalent to* $\widetilde{Q} < 0$.

Lemma 5 ([7]). *If* $x(t) \in C^1[0, b]$ *and* $0 < \alpha < 1$, *then:*

$$\text{(i)}\quad {}^C D^\alpha I^\alpha x(t) = x(t), \quad 0 < t < b; \tag{4}$$

$$\text{(ii)}\quad I^\alpha\, {}^C D^\alpha x(t) = x(t) - x(0), \quad 0 < t < b. \tag{5}$$

Lemma 6 ([8]). *For the real-valued continuous* $f(t, x) : [t_0, \infty) \times \Omega \to \mathbb{R}^n$, *one has*

$$\|I^\alpha f(t, x(t))\| \le I^\alpha \|f(t, x(t))\|, \tag{6}$$

where $\Omega \subset \mathbb{R}^n$, $\alpha \ge 0$ *and* $\|\cdot\|$ *denotes an arbitrary norm.*

Lemma 7 ([8]). *Let* $x = 0$ *be an equilibrium point for the nonautonomous fractional differential system* (1). *Assume that there exists a Lyapunov function* $V(t, x(t))$ *and class-*K *functions* ϕ_i, $i = 1, 2, 3$, *satisfying*

$$\phi_1(\|x\|) \le V(t, x) \le \phi_2(\|x\|), \tag{7}$$

$${}^C D^\alpha V(t, x) \le -\phi_3(\|x\|), \tag{8}$$

then the equilibrium point of system (1) *is asymptotically stable.*

Remark 1 ([8]). A continuous function $g : [0, t) \to [0, \infty)$ is said to belong to class-K if it is strictly increasing and $g(0) = 0$.

3 Pinning synchronization of fractional complex networks

3.1 Fractional complex networks

At present, there are several definitions of fractional differential operators [18], such as Grünwald–Letnikov definition, Riemann–Liouville definition, and Caputo one. Among them, the initial conditions for Caputo derivatives have the same form as those for integer-order ones, and the Caputo derivative not only has a clearly interpretable physical meaning, but also has properly measured initial condition in the simulation. So it may be the most appropriate choice for practical applications.

Consider a general Caputo fractional dynamical complex networks consisting of N nodes, which can be described as follows:

$$ {}^{C}D^{\alpha}x_i(t) = Ax_i(t) + f(x_i(t)) + c\sum_{j=1}^{N} b_{ij}Hx_j(t), \tag{9} $$

where $i = 1, 2, \ldots, N$, $t \geq 0$, $0 < \alpha < 1$ is the fractional order, $x_i(t) = (x_{i1}(t), \ldots, x_{in}(t))^T \in \mathbb{R}^n$ is the state variable of the ith node. Here, $A \in \mathbb{R}^{n \times n}$ is a given linear matrix, and $f(x_i) = [f_1(x_i), f_2(x_i), \ldots, f_n(x_i)]^T : \mathbb{R}^n \to \mathbb{R}^n$ is a smooth function describing the nonlinear dynamics of the node; c is a parameter of coupling strength; $H \in \mathbb{R}^{n \times n}$ is inner coupling matrix; $B = (b_{ij})_{N \times N}$ denotes the coupling configuration matrix of the network; If there is a connection from node i to node j $(i \neq j)$, then $b_{ij} > 0$; otherwise, $b_{ij} = 0$; The diagonal elements of matrix B is given by $b_{ii} = -\sum_{j=1, j \neq i}^{N} b_{ij}$.

If model (9) is referred as the drive system, the response complex network can be chosen as

$$ {}^{C}D^{\alpha}y_i(t) = Ay_i(t) + f(y_i(t)) + c\sum_{j=1}^{N} b_{ij}Hy_j(t) + u_i(t), \tag{10} $$

where $y_i(t) = (y_{i1}(t), y_{i2}(t), \ldots, y_{in}(t))^T \in \mathbb{R}^n$ is the response state vector of the ith node; $u_i(t) \in \mathbb{R}^n$ $(i = 1, 2, \ldots, N)$ are the controllers to be designed; the other parameters have the same meanings as those in (9).

It is not necessary to assume that the inner coupling matrix H and coupling configuration matrix B are symmetric and irreducible. Meanwhile, the corresponding topological graph can be directed or undirected. Throughout the paper, we always assume that nonlinear function $f(x)$ satisfy the uniform Lipschitz conditions,

$$ \|f(x) - f(y)\| \leq L\|x - y\|. \tag{11} $$

According to systems (9) and (10), the error system is described by

$$ {}^{C}D^{\alpha}e_i(t) = Ae_i(t) + f(y_i(t)) - f(x_i(t)) + c\sum_{j=1}^{N} b_{ij}He_j(t) + u_i(t), \tag{12} $$

where $e_i(t) = y_i(t) - x_i(t)$, $i = 1, 2, \ldots, N$. Thus, our objective is to design a suitable controller $u_i(t)$ such that error dynamical system (12) is asymptotically stable, that is,

$$ \lim_{t \to \infty} \|y_i(t; t_0, x_0) - x_i(t; t_0', x_0')\| = 0, \quad i = 1, 2, \ldots, N, $$

which implies the drive system (9) and the response system (10) are synchronized.

3.2 Pinning synchronization

To realize synchronization between (9) and (10), assume that first l $(1 \le l \le N)$ nodes are pinned, the pinning controllers are chosen as

$$\begin{cases} u_i(t) = -p_i e_i(t), & 1 \le i \le l, \\ u_i(t) = 0, & l+1 \le i \le N, \end{cases} \tag{13}$$

where $p_i > 0$ are feedback gains.

Theorem 1. *Suppose that the dynamical function f and nonlinear coupling function g satisfy Lipschitz condition* (11). *If there exists a matrix P satisfying the following conditions*

$$\bar{\mu} = -\lambda_{\max}[(a+L)I_N + ch\hat{B} - P] > 0, \tag{14}$$

where $a = \|A\|$, $h = \|H\|$, then the fractional response network (10) *asymptotically synchronizes to the drive network* (9).

Proof. Construct the following Lyapunov-like function:

$$V(t, e(t)) = \frac{1}{2}\sum_{i=1}^{N} e_i^T(t) e_i(t). \tag{15}$$

Using equation (12) and Lemma 1, the fractional derivative of $V(t, e(t))$ yields

$$\begin{aligned} {}^C D^\alpha V(t, e(t)) &\le \sum_{i=1}^{N} e_i^T(t)\, {}^C D^\alpha e_i(t) \\ &= \sum_{i=1}^{N} e_i^T(t) A e_i(t) + \sum_{i=1}^{N} e_i^T(t)[f(y_i(t)) - f(x_i(t))] \\ &\quad + c\sum_{i=1}^{N}\sum_{j=1}^{N} b_{ij} e_i^T(t) H e_j(t) - \sum_{i=1}^{l} p_i e_i^T(t) e_i(t) \\ &\le (a+L)\sum_{i=1}^{N} e_i^T(t) e_i(t) + c\sum_{i=1}^{N}\sum_{j=1}^{N} b_{ij} e_i^T(t) H e_j(t) - \sum_{i=1}^{l} p_i e_i^T(t) e_i(t). \end{aligned} \tag{16}$$

From Lemma 2, we have

$$\begin{aligned} &\sum_{i=1}^{N}\sum_{j=1}^{N} b_{ij} e_i^T(t) H e_j(t) \\ &\quad = \sum_{i=1}^{N}\sum_{j=1, j\ne i}^{N} b_{ij} e_i^T(t) H e_j(t) + \sum_{i=1}^{N} b_{ii} e_i^T(t)\left(\frac{H + H^T}{2}\right) e_i(t) \\ &\quad \le h\sum_{i=1}^{N}\sum_{j=1, j\ne i}^{N} b_{ij}\|e_i(t)\| \cdot \|e_j(t)\| + \rho_{\min}\sum_{i=1}^{N} b_{ii} e_i^T(t) e_i(t), \end{aligned} \tag{17}$$

where $\rho_{\min}$ is the minimum eigenvalue of the matrix $\frac{H+H^T}{2}$.

Substituting (17) into (16), we obtain that

$$\begin{aligned} {}_C D^{\alpha}_{0,t} V(t, e(t)) &\le \|e(t)\|^T [(a+L)I_N + ch\hat{B} - P] \|e(t)\| \\ &\le -\bar{\mu} \sum_{i=1}^{N} e_i^T e_i(t), \end{aligned} \tag{18}$$

where $\|e(t)\|_V = (\|e_1(t)\|, \|e_2(t)\|, \ldots, \|e_N(t)\|)^T$, $P = \operatorname{diag}(\underbrace{p_1, p_2, \ldots, p_l}_{N-l}, \underbrace{0, 0, \ldots, 0}_{N-l})$, $\hat{B} = \frac{\tilde{B}^T + \tilde{B}}{2}$ and $\tilde{B}$ is a modifying matrix of B via replacing the diagonal elements b_{ii} by $(\rho_{\min}/h) b_{ii}$.

According to Lemma 3, we have $\|e_i(t)\| \to 0$, that is $\|y_i(t) - x_i(t)\| \to 0$ as $t \to \infty$, which means that the asymptotical synchronization between drive system (9) and response system (10) is realized. □

Furthermore, from (18), let $Q = (a+L)I_N + ch\hat{B}$, and $Q - P = \begin{pmatrix} E-P^* & S \\ S^T & \tilde{Q} \end{pmatrix}$, where $1 \le l \le N$, $P^* = \operatorname{diag}(p_1, p_2, \ldots, p_l)$, $\tilde{Q}$ is a block matrix of Q with order $N-l$, E and S are matrices with appropriate dimensions. Based on Lemma 4, and supposing that p_i $(i = 1, \ldots, l)$ are suitably large, $Q - P < 0$ is equivalent to $\tilde{Q} = [(a+L)I_N + ch\hat{B}]_{N-l} < 0$. One has $\lambda_{\max}[(a+L)I_N + ch\hat{B}]_l = (a+L) + ch\lambda_{\max}(\hat{B}_{N-l}) < 0$. So, the following corollary can be immediately obtained.

Corollary 1. *Under assumption* (11), *the fractional response network* (10) *asymptotically synchronizes to the drive one* (9) *using the controller* (13), *where* p_i $(i = 1, \ldots, l)$ *are suitably large, and the following conditions are satisfied:*

$$\bar{\mu} = -[(a+L) + ch\lambda_{\max}(\hat{B}_{N-l})] > 0. \tag{19}$$

3.3 Numerical example

In this section, a numerical example is presented. Consider complex network with 10 nodes, the fractional dynamical equation of each node is described by the following fractional chaotic Lorenz system

$$\begin{cases} {}^C D^{\alpha} x_{i1} = a_L (x_{i2} - x_{i1}), \\ {}^C D^{\alpha x_{i2}} = b_L x_{i1} - x_{i1} x_{i3} - x_{i2}, \\ {}^C D^{\alpha} x_{i3} = x_{i1} x_{i2} - c_L x_{i3}, \end{cases} \tag{20}$$

where $i = 1, 2, \ldots, 10$. The parameters are chosen as $a_L = 10$, $b_L = 28$, $c_L = 8/3$ and $\alpha = 0.995$. From (9) and (10), we known that

$$A = \begin{pmatrix} -a & a & 0 \\ b & -1 & 0 \\ 0 & 0 & -c \end{pmatrix}, \quad f(x_i(t)) = \begin{pmatrix} 0 \\ -x_{i1} x_{i3} \\ x_{i1} x_{i2} \end{pmatrix},$$

and that system (20) is chaotic; see Figure 1.

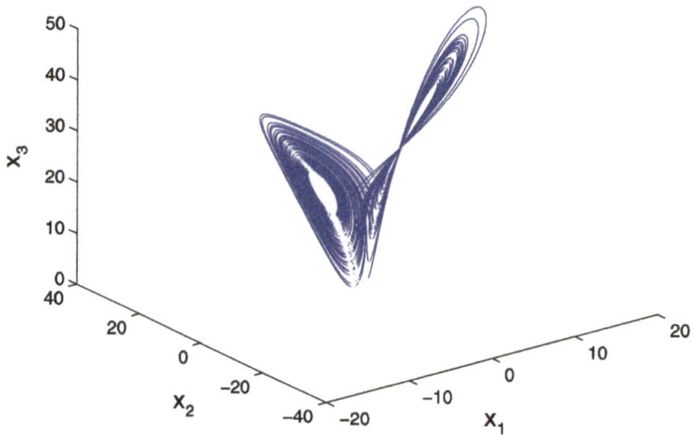

Figure 1: Chaotic attractor of fractional Lorenz system with order $\alpha = 0.995$.

For convenience, let $H = \bar{H} = I$, the coupling configuration matrices B and $\bar{B}$ are given as follows:

$$\begin{pmatrix} -2 & 1 & 0 & 1 & 0 & 0 & 0 & 0 & 0 & 0 \\ 0 & -2 & 1 & 0 & 1 & 0 & 0 & 0 & 0 & 0 \\ 1 & 0 & -3 & 0 & 0 & 1 & 1 & 0 & 0 & 0 \\ 0 & 0 & 1 & -1 & 0 & 0 & 0 & 0 & 0 & 0 \\ 0 & 1 & 0 & 1 & -3 & 0 & 0 & 0 & 1 & 0 \\ 0 & 0 & 1 & 0 & 1 & -2 & 0 & 0 & 0 & 0 \\ 0 & 0 & 0 & 0 & 0 & 1 & -3 & 1 & 0 & 1 \\ 0 & 1 & 0 & 1 & 0 & 0 & 0 & -3 & 1 & 0 \\ 1 & 0 & 1 & 0 & 0 & 0 & 1 & 0 & -4 & 1 \\ 0 & 0 & 1 & 1 & 0 & 0 & 0 & 0 & 0 & -2 \end{pmatrix}.$$

It is known that the Lorenz system is bounded. Actually, $\|x_{i1}\| \le 25$, $\|x_{i2}\| \le 30$, $\|x_{i3}\| \le 60$, $\|y_{i1}\| \le 25$, $\|y_{i2}\| \le 30$, $\|y_{i3}\| \le 60$, $i = 1, 2, \ldots, 10$, and

$$\begin{aligned} &\|f(x_i) - f(y_i)\| \\ &\quad \le \sqrt{(-x_{i1}x_{i3} + y_{i1}y_{i3})^2 + (x_{i1}x_{i2} - y_{i1}y_{i2})^2} \\ &\quad \le 75.83\|e_i\|, \end{aligned}$$

that is $L = 75.83$. Obviously, $\bar{L} = 2$. According to the method proposed in [19], whose out-degrees are bigger than their in-degrees, it should be selected as pinned candidates. The out-degrees of nodes 2, 3, and 4 are bigger than their in-degrees, so we choose them as the pinned nodes. Re-arrange the network nodes and the new order is 4, 3, 2, 1, 6, 10, 5, 7, 8, 9.

Let $p_i = 910$, $i = 1, 2, 3$, when $\alpha = 0.9$, $c = 100$, one has $\bar{\mu} > 0$. From Theorem 1, it is clear that pinning conditions hold. The simulation results are shown in Figure 2,

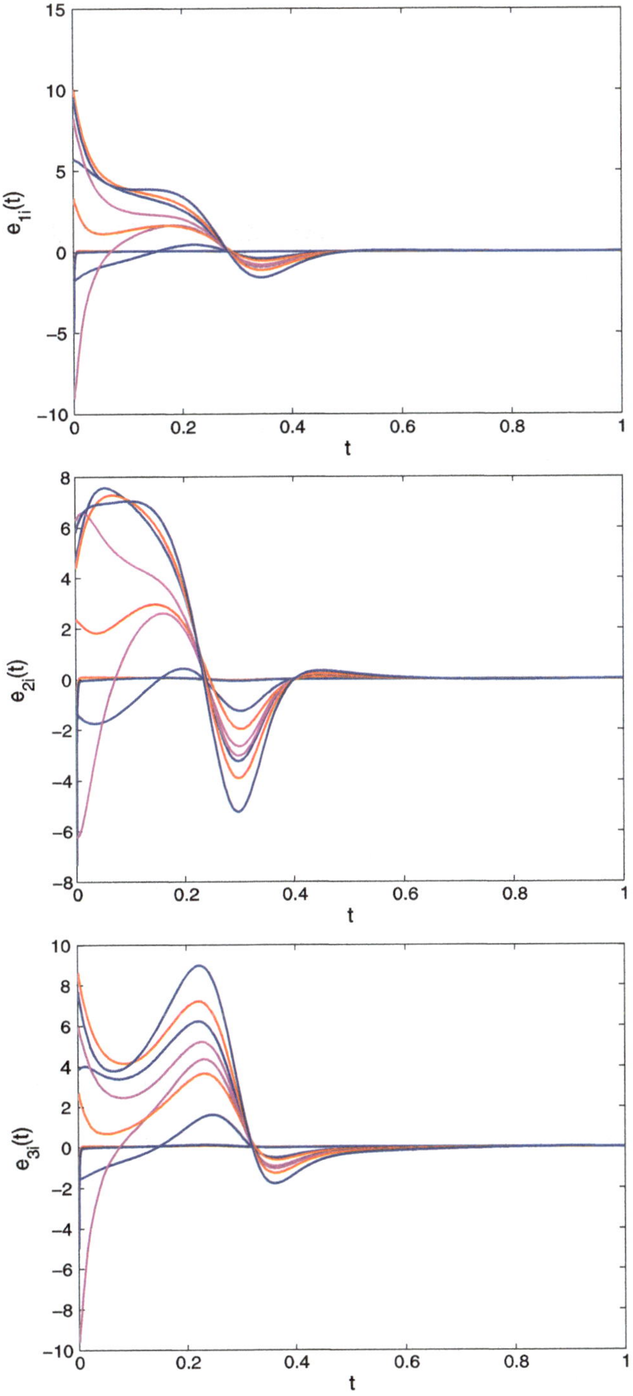

Figure 2: Time evolution of the error states e_{1i}, e_{2i}, and e_{3i}.

which shows the time waveforms of errors $e_{i1}, e_{i2}, e_{i3}, i = 1, 2, \ldots, 10$. From the figures, fractional complex networks (9) and (10) are synchronized, which demonstrate the effectiveness of the proposed method.

4 Adaptive synchronization of uncertain fractional complex networks

4.1 Model description

Considering the drive and response complex networks characterized by N identical fractional system nodes, each of the n-dimensional system governed by fractional differential equations can be described by

$$ {}^{C}D^{\alpha}x_i(t) = f(x_i(t)) + F(x_i(t))\alpha_i + c\sum_{j=1}^{N} b_{ij}Hx_j(t), \tag{21} $$

$$ {}^{C}D^{\alpha}y_i(t) = f(y_i(t)) + F(y_i(t))\hat{\alpha}_i + c\sum_{j=1}^{N} b_{ij}Hy_j(t) + u_i, \tag{22} $$

where $i = 1, 2, \ldots, N$, $x_i = (x_{i1}, x_{i2}, \ldots, x_{in})^T$, $y_i = (y_{i1}, y_{i2}, \ldots, y_{in})^T \in \mathbb{R}^n$ denote the state vectors, $\alpha_i = (\alpha_{i1}, \alpha_{i2}, \ldots, \alpha_{in})^T \in \mathbb{R}^{m_i}$ is the unknown system parameter vector and m_i are nonnegative integers indicating the numbers of unknown parameters, $\hat{\alpha}_i = (\hat{\alpha}_{i1}, \hat{\alpha}_{i2}, \ldots, \hat{\alpha}_{in})^T \in \mathbb{R}^{m_i}$ is the estimation for the unknown system parameter, $f(x_i(t))$, $f(y_i(t))$ are $n \times 1$ continuous differentiable functions, $F(x_i)$ is an $n \times m_i$ function matrix, the other parameters have the same meanings as before.

Remark 2. Many fractional chaotic systems belong to the class individualized by this kind form ${}^{C}D^{\alpha}x_i(t) = f(x_i(t)) + F(x_i(t))\alpha_i$, for example, the fractional Chen system, the fractional Liu model, and the fractional chaotic Rossler system, just to mention a few.

Before starting our proposed solution, we introduce the following useful assumptions.

Assumption 1. Suppose that there exist nonnegative constants L^f such that

$$ \|f(x) - f(y)\| \le L^f \|x - y\|. \tag{23} $$

Assumption 2. Suppose that there exist nonnegative constants L^F such that

$$ \|F(x) - F(y)\| \le L^F \|x - y\|. \tag{24} $$

4.2 Adaptive synchronization

Let $e_i = y_i - x_i$ be the synchronization error. Hence, the error dynamics system takes the form ${}^C D^\alpha e_i = {}^C D^\alpha y_i - {}^C D^\alpha x_i$, where $i \le i \le N$. Then based on the equations (21)–(22), one can get

$$\begin{aligned} {}^C D^\alpha e_i &= \left[f(y_i(t)) + F(y_i(t))\hat{\alpha}_i + c\sum_{j=1}^{N} b_{ij} H y_j(t) + u_i \right] \\ &\quad - \left[f(x_i(t)) + F(x_i(t))\alpha_i + c\sum_{j=1}^{N} b_{ij} H x_j(t) \right] \\ &= f(y_i) - f(x_i) + F(y_i)\hat{\alpha}_i - F(x_i)\alpha_i + c\sum_{j=1}^{N} b_{ij} H e_j + u_i. \end{aligned} \tag{25}$$

The main objective is to realize the identification of unknown parameters in the process of synchronization by using the designed controller and adaptive laws of parameters.

Theorem 2. *If* $\|F^T(x)\| \le M, i = 1, 2, \ldots, N$, *and*

$$v = -\lambda_{\max}[L^f I_N + L^F \Delta + ch\hat{B} - K] > 0,$$

the controllers and the adaptive laws of parameters are taken as

$$\begin{cases} u_i(t) = -k_i(t)e_i(t), \\ {}^C D^\alpha k_i(t) = d_i \|e_i(t)\|^2, \\ {}^C D^\alpha \hat{\alpha}_i(t) = -F^T(y_i)e_i(t), \end{cases} \tag{26}$$

where $d_i > 0$ *is a constant, the drive and response networks* (21) *and* (22) *can achieve synchronization with uncertain parameters.*

Proof. Consider the following Lyapunov function candidate:

$$V(e(t)) = \sum_{i=1}^{N} \left[e_i^{\mathrm{T}}(t)e_i(t) + (\hat{\alpha}_i - \alpha_i)^T(\hat{\alpha}_i - \alpha_i) + \frac{1}{d_i}(k_i(t) - k_i(0))^2 \right]. \tag{27}$$

It is obvious that

$$V(e(t)) \ge \sum_{i=1}^{N} e_i^{\mathrm{T}}(t)e_i(t) = \sum_{i=1}^{N} \|e_i(t)\|^2 = \|e(t)\|^2, \tag{28}$$

where $e(t) = (e_1^T(t), e_2^T(t), \ldots, e_n^T(t))^T \in \mathbb{R}^{nN}$.

From Lemma 5, we get

$$k_i(t) - k_i(0) = d_i \mathrm{I}^\alpha [e_i^{\mathrm{T}}(t)e_i(t)] \ge 0, \tag{29}$$

and

$$\hat{\alpha}_i(t) - \hat{\alpha}_i(0) = \mathrm{I}^\alpha[F^T(y_i(t))e_i(t)]. \tag{30}$$

From (29), (30), and Lemma 6, we have

$$\begin{aligned} V(e) &\le \|e\|^2 + \|\hat{\alpha}(t) - \alpha\|^2 + \sum_{i=1}^{N} \frac{1}{d_i}(d_i \mathrm{I}^\alpha \|e_i\|^2)^2 \\ &\le \|e\|^2 + \|\hat{\alpha}(0) - \alpha + \mathrm{I}^\alpha(F^T(y)e)\|^2 + \sum_{i=1}^{N} d_i(\mathrm{I}^\alpha\|e_i\|^2)^2 \\ &\le \|e\|^2 + [\|\hat{\alpha}(0) - \alpha\| + \|\mathrm{I}^\alpha(F^T(y)e)\|]^2 + \sum_{i=1}^{N} d_i(\mathrm{I}^\alpha\|e_i\|^2)^2 \\ &\le \|e\|^2 + 2\|\hat{\alpha}(0) - \alpha\|^2 + 2(\mathrm{I}^\alpha\|F^T(y)e\|)^2 + d\left(\sum_{i=1}^{N} \mathrm{I}^\alpha\|e_i\|^2\right)^2 \\ &\le \|e\|^2 + 2\|\hat{\alpha}(0) - \alpha\|^2 + 2\left(\mathrm{I}^\alpha\sqrt{\sum_{i=1}^{N}\|F^T(y_i)e_i\|^2}\right)^2 + d(\mathrm{I}^\alpha\|e\|^2)^2 \\ &\le \|e\|^2 + 2\|\hat{\alpha}(0) - \alpha\|^2 + 2\left(\mathrm{I}^\alpha\sqrt{\sum_{i=1}^{N}\|F^T(y_i)\|^2 \cdot \|e_i\|^2}\right)^2 + d(\mathrm{I}^\alpha\|e\|^2)^2 \\ &\le \|e\|^2 + 2\|\hat{\alpha}(0) - \alpha\|^2 + 2M^2(\mathrm{I}^\alpha\|e\|)^2 + d(\mathrm{I}^\alpha\|e\|^2)^2, \end{aligned} \tag{31}$$

where $d = \max_{1\le i\le N}\{d_i\}$. Then, from (28) and (31), condition (4) in Lemma 7 is held.

Calculating the fractional derivative of $V(t)$ along the trajectories of error system (25) and using Lemma 1 yield

$$\begin{aligned} {}^C\mathrm{D}^\alpha V(e) &\le 2\sum_{i=1}^{N}\left[e_i^{\mathrm{T}}\, {}^C\mathrm{D}^\alpha e_i + (\hat{\alpha}_i - \alpha_i)^T\, {}^C\mathrm{D}^\alpha(\hat{\alpha}_i - \alpha_i) + \frac{1}{d_i}(k_i - k_i(0))\, {}^C\mathrm{D}^\alpha k_i\right] \\ &\le 2\sum_{i=1}^{N}\left[e_i^{\mathrm{T}}\left(f(y_i) - f(x_i) + F(y_i)\hat{\alpha}_i - F(x_i)\alpha_i + c\sum_{j=1}^{N} b_{ij}He_j - k_i e_i\right)\right. \\ &\quad \left. - (\hat{\alpha}_i - \alpha_i)^T F^T(y_i)e_i + (k_i(t) - k_i(0))\|e_i\|^2\right] \\ &= 2\sum_{i=1}^{N} e_i^{\mathrm{T}}(f(y_i) - f(x_i)) + 2\sum_{i=1}^{N} e_i^{\mathrm{T}}[F(y_i)\alpha_i - F(x_i)\alpha_i] \\ &\quad + 2c\sum_{i=1}^{N}\sum_{j=1}^{N} b_{ij}e_i^{\mathrm{T}}He_j - 2\sum_{i=1}^{N} k_i(0)\|e_i\|^2 \\ &\le 2L^f\sum_{i=1}^{N} e_i^{\mathrm{T}}e_i + 2\sum_{i=1}^{N} e_i^{\mathrm{T}}[F(y_i)\alpha_i - F(x_i)\alpha_i] \\ &\quad + 2c\sum_{i=1}^{N}\sum_{j=1}^{N} b_{ij}e_i^{\mathrm{T}}He_j - 2\sum_{i=1}^{N} k_i(0)\|e_i\|^2. \end{aligned} \tag{32}$$

From (17), we get

$$
\begin{aligned}
{}^{C}\mathrm{D}^{\alpha}V(e) &\le 2L^{f}\sum_{i=1}^{N} e_i^{\mathrm{T}} e_i + 2L^{F}\sum_{i=1}^{N} e_i^{\mathrm{T}} e_i\|\alpha_i\| \\
&\quad + 2ch\sum_{i=1}^{N}\sum_{j=1,j\neq i}^{N} b_{ij}\|e_i(t)\|\cdot\|e_j(t)\| \\
&\quad + 2c\rho_{\min}\sum_{i=1}^{N} b_{ii}e_i^{T}(t)e_i(t) - 2\sum_{i=1}^{N} k_i(0)\|e_i\|^2 \\
&\le 2\|e(t)\|_v^{T}[L^{f}I_N + L^{F}\Delta + ch\hat{B} - K]\|e(t)\|_v \\
&\le -2v\sum_{i=1}^{N} e_i^{T} e_i(t) = -2v\|e(t)\|,
\end{aligned} \tag{33}
$$

where $\Delta = \operatorname{diag}(\|\alpha_1\|, \|\alpha_2\|, \ldots, \|\alpha_N\|)$, and $K = \operatorname{diag}(k_1(0), k_2(0), \ldots, k_N(0))$. By Lemma 7, the proof is completed. □

4.3 Examples of applications

In this subsection, we further test the performance of the proposed method under the case that the nodes of network are hyperchaotic system. We consider a network with $N = 10$ nodes. The fractional hyperchaotic Lorenz system is used as the node dynamics in the networks. The single hyperchaotic fractional Lorenz system in the ith node can be described as below:

$$
\begin{cases}
{}^{C}D^{\alpha}x_1 = a(x_2 - x_1) + x_4, \\
{}^{C}D^{\alpha}x_2 = bx_1 - x_1x_3 - x_2, \\
{}^{C}D^{\alpha}x_3 = x_1x_2 - cx_3, \\
{}^{C}D^{\alpha}x_4 = -x_2x_3 + rx_4,
\end{cases} \tag{34}
$$

the parameters are chosen as $a = 10$, $b = 28$, $c = 8/3$, $r = -1$, and $\alpha = 0.98$. The fractional Lorenz hyperchaotic attractor is shown in Figure 3. The fractional hyperchaotic Lorenz system can also be rewritten as

$$
\begin{bmatrix} {}^{C}D^{\alpha}x_1 \\ {}^{C}D^{\alpha}x_2 \\ {}^{C}D^{\alpha}x_3 \\ {}^{C}D^{\alpha}x_4 \end{bmatrix}
=
\begin{bmatrix} x_2 - x_1 & 0 & 0 & 0 \\ 0 & x_1 & 0 & 0 \\ 0 & 0 & -x_3 & 0 \\ 0 & 0 & 0 & x_4 \end{bmatrix}
\begin{bmatrix} a \\ b \\ c \\ r \end{bmatrix}
+
\begin{bmatrix} x_4 \\ -x_1x_3 - x_2 \\ x_1x_2 \\ -x_2x_3 \end{bmatrix}. \tag{35}
$$

We consider the drive and response complex network which consists of ten identical fractional hyperchaotic Lorenz systems. For the sake of simplicity, we set the inner-coupling matrix $H = I$, $c = 1$, the coupling configuration matrix B of networks is ran-

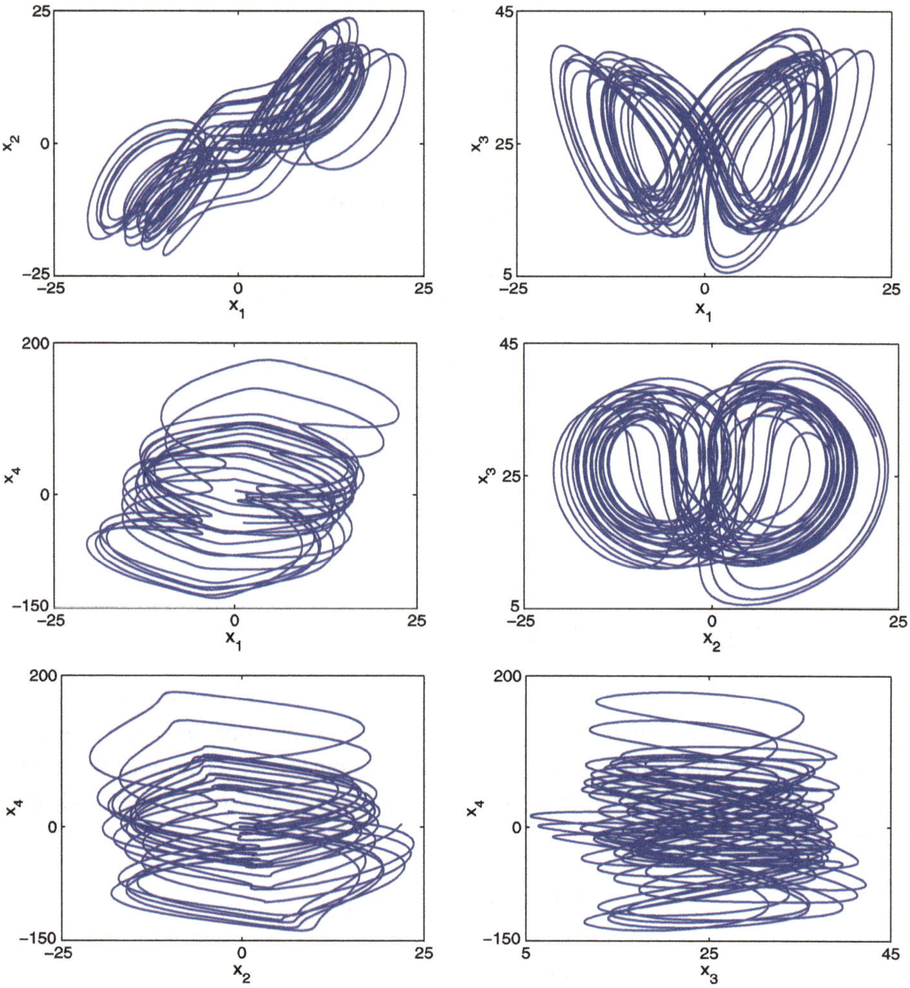

Figure 3: The attractor of fractional Lorenz hyperchaotic system.

domly selected as follows:

$$\begin{pmatrix} -4 & 1 & 0 & 0 & 0 & 0 & 1 & 0 & 1 & 1 \\ 0 & -2 & 0 & 0 & 0 & 0 & 0 & 1 & 1 & 0 \\ 1 & 1 & -4 & 1 & 0 & 0 & 1 & 0 & 0 & 0 \\ 0 & 1 & 0 & -3 & 2 & 0 & 0 & 0 & 0 & 0 \\ 0 & 0 & 0 & 1 & -2 & 1 & 0 & 0 & 0 & 0 \\ 1 & 0 & 0 & 1 & 0 & -4 & 0 & 1 & 1 & 0 \\ 0 & 1 & 0 & 0 & 0 & 1 & -2 & 0 & 0 & 0 \\ 0 & 1 & 0 & 0 & 1 & 0 & 1 & -3 & 0 & 0 \\ 0 & 1 & 1 & 0 & 0 & 1 & 1 & 0 & -5 & 1 \\ 0 & 0 & 1 & 0 & 1 & 0 & 1 & 0 & 0 & -3 \end{pmatrix}.$$

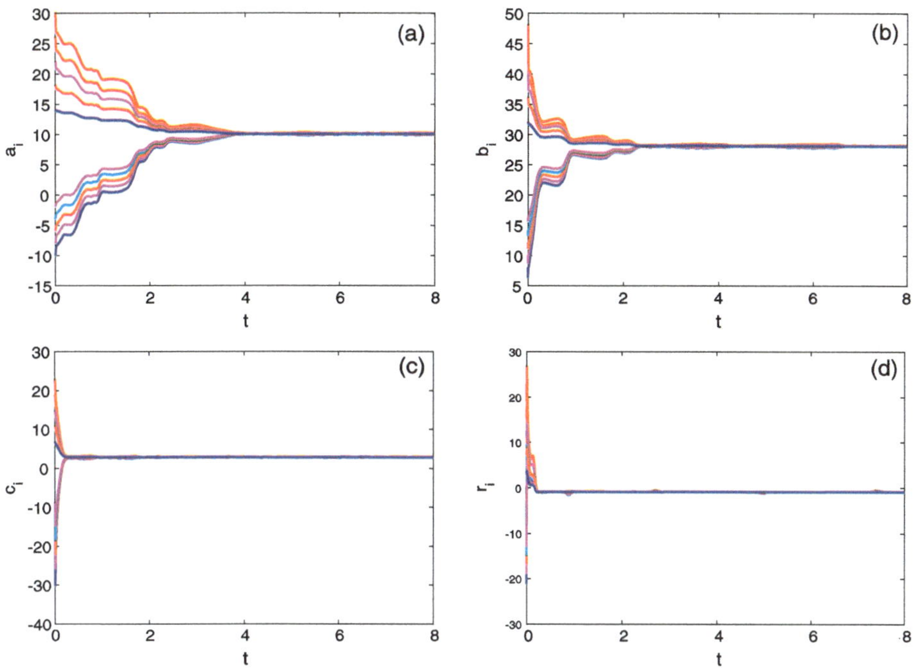

Figure 4: Identification of system parameters.

Analogously, we would also like to give the numerical simulations to verify the effectiveness of the above-designed controller and the updating laws. In these numerical simulations, we take the initial states as

$$\begin{cases} x_i(0) = (1 + 0.5i, 2 + i, 0.5 + 3i, 2 + 5i)^T, \\ y_i(0) = (2 + 3i, 1 + 5i, 1 + 2i, 5 + 2i)^T, \\ k_i(0) = 100. \end{cases} \tag{36}$$

We choose all the positive constants $d_i = 10$ for $i = 1, 2, \ldots, 10$; and the fractional order $\alpha = 0.98$. From Figure 4, it can be seen that all the unknown parameters in the complex network are identified to their true values. Figure 5 depicts the behavior of the synchronization errors which converge to zeros asymptotically.

5 Conclusion

In this part, the pinning synchronization and adaptive synchronization of complex networks with fractional derivatives are discussed. By means of the stability theory of fractional differential system, the pinning control method, and the adaptive control technique, the control strategies are proposed to realize synchronization. Several

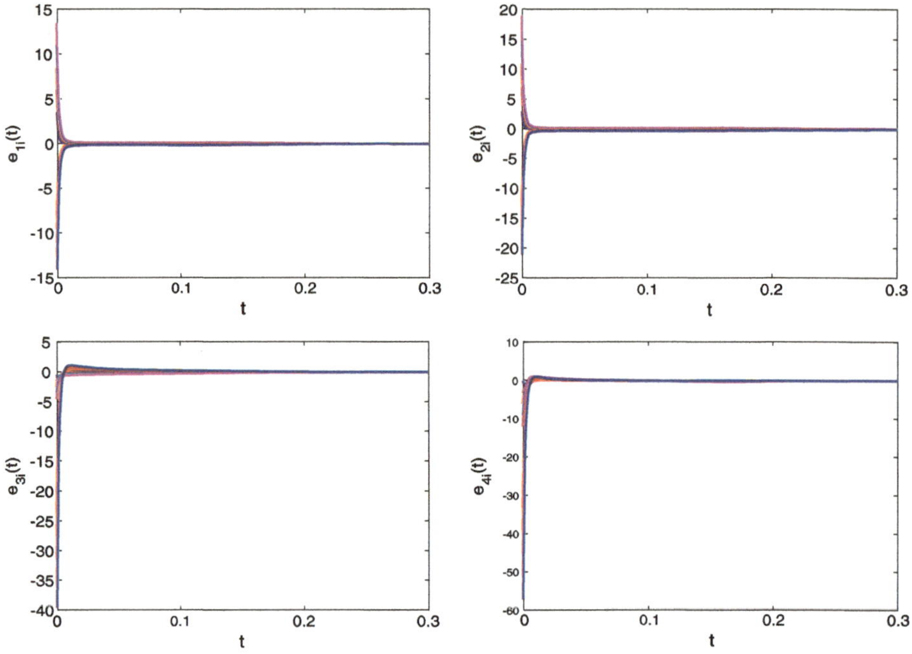

Figure 5: The time evolution of synchronization errors $e_i(t)$ ($i = 1, 2, \ldots, 10$).

criteria of networks synchronization are deduced. Especially, the designed adaptive controllers for network synchronization are rather simple in form. The method can be well applied to other complex networks. The effectiveness of the presented synchronization scheme is validated by the theoretical proofs and some numerical examples.

Bibliography

[1] R. Cheng, M. Peng, and W. Yu, Pinning synchronization of delayed complex dynamical networks with nonlinear coupling, *Physica A: Statistical Mechanics and Its Applications*, **413** (2014), 426–431.

[2] H. Delavari, A novel fractional adaptive active sliding mode controller for synchronization of non-identical chaotic systems with disturbance and uncertainty, *International Journal of Dynamics and Control*, **5**(1) (2017), 102–114.

[3] M. Duarte-Mermoud, N. Aguila-Camacho, J. Gallegos, and R. Castro-Linares, Using general quadratic Lyapunov functions to prove Lyapunov uniform stability for fractional order systems, *Communications in Nonlinear Science and Numerical Simulation*, **22**(1–3) (2015), 650–659.

[4] O. Heaviside, *Electromagnetic Theory*, pp. 1–130, Chelse, New York, 1971.

[5] R. C. Koeller, Application of fractional calculus to the theory of viscoelasticity, *Journal of Applied Mechanics*, **51**(2) (1984), 294–298.

[6] R. C. Koeller, Polynomial operators, Stieltjes convolution, and fractional calculus in hereditary mechanics, *Acta Mechanica*, **58**(3) (1986), 251–264.

[7] C. Li and W. Deng, Remarks on fractional derivative, *Applied Mathematics and Computation*, **187** (2007), 777–784.
[8] Y. Li, Y. Chen, and I. Podlubny, Mittag-Leffler stability of fractional order nonlinear dynamic system, *Automatica*, **45**(8) (2009), 1965–1969.
[9] H. Liu, S. Li, and H. Wang et al., Adaptive synchronization for a class of uncertain fractional-order neural networks, *Entropy*, **17**(10) (2015), 7185–7200.
[10] W. Ma, C. Li, and Y. Wu, Impulsive synchronization of fractional Takagi–Sugeno fuzzy complex networks, *Chaos*, **26**(8) (2016), 084311.
[11] W. Ma, C. Li, and Y. Wu et al., Adaptive synchronization of fractional neural networks with unknown parameters and time delays, *Entropy*, **16**(12) (2014), 6286–6299.
[12] W. Ma, C. Li, and Y. Wu et al., Synchronization of fractional fuzzy cellular neural networks with interactions, *Chaos*, **27**(10) (2017), 103106.
[13] W. Ma, Y. Wu, and C. Li, Pinning synchronization between two general fractional complex dynamical networks with external disturbances, *IEEE/CAA Journal of Automatica Sinica*, **4**(2) (2017), 328–335.
[14] F. Nian and X. Wang, Optimal pinning synchronization on directed complex network, *Chaos: An Interdisciplinary Journal of Nonlinear Science*, **21**(4) (2011), 043131.
[15] G. V. Osipov, J. Kurths, and C. Zhou, *Synchronization in Oscillatory Networks*, Springer, Berlin, Germany, 2007.
[16] L. Pan, W. Zhou, J. Fang, and D. Li, Synchronization and anti-synchronization of new uncertain fractional-order modified unified chaotic systems via novel active pinning control, *Communications in Nonlinear Science and Numerical Simulation*, **15**(12) (2010), 3754–3762.
[17] A. Pikovsky, M. Rosenblum, and J. Kurths, *Synchronization*, Cambridge University Press, Cambridge, UK, 2001.
[18] I. Podlubny, *Fractional Differential Equations*, pp. 201–307, Academic Press, New York, 1999.
[19] Q. Song and J. Cao, On pinning synchronization of directed and undirected complex dynamical networks, *IEEE Transactions on Circuits and Systems. I, Regular Papers*, **57**(3) (2010), 672–680.
[20] X. Song, S. Song, and I. Balsera et al., Synchronization of two fractional-order chaotic systems via nonsingular terminal fuzzy sliding mode control, *Journal of Control Science and Engineering*, **2017** (2017), 9562818.
[21] F. Sorrentino, M. Bernardo, F. Garofalo, and G. Chen, Controllability of complex networks via pinning, *Physical Review E*, **75**(4) (2007), 046103.
[22] W. Sun, Z. Chen, and S. Chen, Synchronization of impulsively coupled complex networks, *Chinese Physics B*, **21**(5) (2012), 126–132.
[23] Y. Tang, Z. Wang, and J. Fang, Pinning control of fractional-order weighted complex networks, *Chaos: An Interdisciplinary Journal of Nonlinear Science*, **19**(1) (2009), 013112.
[24] F. Wang, Y. Yang, and A. Hu et al., Exponential synchronization of fractional-order complex networks via pinning impulsive control, *Nonlinear Dynamics*, **82**(4) (2015), 1979–1987.
[25] J. Wang and Y. Zhang, Robust projective outer synchronization of coupled uncertain fractional-order complex networks, *Central European Journal of Physics*, **11**(6) (2013), 813–823.
[26] L. Wang, Y. Tang, and Y. Chai et al., Generalized projective synchronization of the fractional-order chaotic system using adaptive fuzzy sliding mode control, *Chinese Physics B*, **23**(10) (2014), 64–70.
[27] X. Wang and G. Chen, Pinning control of scale-free dynamical networks, *Physica A: Statistical Mechanics and Its Applications*, **310**(3–4) (2002), 521–531.
[28] J. Wu and L. Jiao, Synchronization in complex delayed dynamical networks with nonsymmetric coupling, *Physica A: Statistical Mechanics and Its Applications*, **386**(1) (2007), 513–530.
[29] Y. Xu, C. Xie, and D. Tong, Adaptive synchronization for dynamical networks of neural type with time-delay, *Optik – International Journal for Light and Electron Optics*, **125**(1) (2014), 380–385.

[30] L. Yang and J. Jiang, Adaptive synchronization of drive-response fractional-order complex dynamical networks with uncertain parameters, *Communications in Nonlinear Science and Numerical Simulation*, **19**(5) (2014), 1496–1506.
[31] W. Yu, G. Chen, J. Lü, and J. Kurths, Synchronization via pinning control on general complex networks, *SIAM Journal on Control and Optimization*, **51**(2) (2013), 1395–1416.

Adel Ouannas and Viet-Thanh Pham

New trends in synchronization of fractional-order chaotic systems

Abstract: This chapter presents different schemes of coexistence of some types of synchronization between fractional-order chaotic and hyperchaotic systems with different orders and dimensions. Numerical examples show the effectiveness of the introduced synchronization schemes.

Keywords: Synchronization, chaos, hyperchaos, fractional order

PACS: 05.45.Ac, 05.45.Gg, 05.45.Xt

1 Introduction

Chaos synchronization was discovered by Pecora and Carroll in dynamical systems described by differential equations [59]. The objective in chaos synchronization is to make the slave system variables synchronized in time with the corresponding chaotic master system variables [4, 18, 62]. By considering the historical timeline of the topic, it can be observed that a large variety of approaches have already been successfully applied to the problem of chaotic synchronization [9, 19–30, 36, 37, 39, 44, 52, 54, 55].

Besides integer-order systems, attention has been recently focused on systems described by fractional-order differential equations [60, 65]. Recently, some efforts have been devoted to the synchronization of dynamical systems described by fractional-order differential equations [3, 35, 38, 41, 50, 57, 58]. In many literatures, synchronization among fractional-order systems is only investigated through numerical simulations that are based on the stability criteria of linear fractional-order systems, such as the work presented in [12, 17], or based on Laplace transform theory, such as the work presented in [68]. Fractional chaos synchronization has great potential applications in secure communication and cryptography [16].

Recently, many inverse types of synchronization have been developed such that inverse hybrid function projective synchronization [8], inverse matrix projective synchronization [34, 47, 51], inverse full state function projective synchronization [46, 49], inverse generalized synchronization [32]. Studying inverse types of chaos synchronization is an attractive and important idea. Obviously, complexity of inverse types of synchronization can have important effect in applications.

Adel Ouannas, University of Larbi Tebessi, Tebessa, 12002, Algeria, e-mail: ouannas_adel@yahoo.fr
Viet-Thanh Pham, School of Electronics and Telecommunications, Hanoi University of Science and Technology, 01 Dai Co Viet, Hanoi, Vietnam, e-mail: pvt3010@gmail.com

https://doi.org/10.1515/9783110571745-018

Very recently, an interesting phenomenon that may occur is the coexistence of several synchronization types. Not long ago, some papers studying the problem of coexistence of different types of chaos synchronization. For example, the coexistence of projective synchronization, full state hybrid projective synchronization and generalized synchronization between discrete-time hyperchaotic systems was studied in [40]. A general control scheme was proposed in [31], to study the coexistence of inverse projective synchronization, inverse generalized synchronization and Q-S synchronization between arbitrary 3D hyperchaotic maps. Also, in discrete time, the coexistence of full state hybrid projective synchronization, generalized synchronization and antiphase synchronization is shown to be achieved with different dimensions [48]. For integer-order chaotic systems, two synchronization schemes of coexistence has been introduced in [43]. Referring to fractional-order differential systems, many schemes of coexistence of various types of synchronization were investigated [2, 33, 42, 45, 53, 56]. It is important to note that the topic of coexistence is important and has many applications in various disciplines. For instance, it can be used to enhance the level of security in wireless communication networks utilizing chaotic encryption.

The outline of the rest of this work is organized as follows. Section 2 provides some preliminaries and basic concepts which are helpful in the proving analysis of the proposed approaches. In Section 3, coexistence of full state hybrid projective synchronization (FSHPS) and inverse full state hybrid projective synchronization (IFSHPS) is studied. Coexistence of generalized (GS) and inverse generalized synchronization (IGS) is proposed in Section 4. Coexistence of complete synchronization (CS), anti-synchronization (AS) and inverse full state hybrid function projective synchronization (IFSHFPS) is investigated in Section 6. Finally, the conclusion is drawn in Section 6.

2 Basic concepts

In this section, we present some basic concepts of fractional derivatives and stability of fractional systems which are helpful in the proving analysis of the proposed approaches. The idea of fractional integrals and derivatives has been known since the development of the regular calculus, with the first reference probably being associated with Leibniz in 1695. There are several definitions of fractional derivatives [10]. The Caputo derivative [5] is a time domain computation method. In real applications, the Caputo derivative is more popular since the unhomogenous initial conditions are permitted if such conditions are necessary. Furthermore, these initial values are prone to measure since they all have idiographic meanings [63]. The Caputo derivative definition is given by

$$D_t^p f(t) = J^{m-p} f^m(t) \quad \text{with } 0 < p \leq 1 \tag{1}$$

where $m = [p]$, that is, m is the first integer which is not less than p, f^m is the m-order derivative in the usual sense, and J^q ($q > 0$) is the q-order Riemann–Liouville integral operator with expression:

$$J^q f(t) = \frac{1}{\Gamma(q)} \int_0^t (t-\tau)^{q-1} f(\tau) d\tau \tag{2}$$

where Γ denotes Gamma function.

Some basic properties and lemmas of fractional derivatives and integrals used in this study are listed as follows.

Property 1. *For $p = n$, where n is an integer, the operation D_t^p is the same result as classical integer order n. Particularly, when $p = n$, the operation D_t^p is the same as the ordinary derivative, that is, $D_t^p f(t) = \frac{df(t)}{dt}$; when $p = 0$, the operation $D_t^p f(t)$ is the identity operation: $D_t^0 f(t) = f(t)$.*

Property 2. *Fractional differentiation (fractional integration) is linear operation:*

$$D_t^p [af(t) + bg(t)] = aD_t^p f(t) + bD_t^p g(t). \tag{3}$$

Property 3. *The fractional differential operator D_t^p is left-inverse (and not right-inverse) to the fractional integral operator J^p, that is,*

$$D_t^p J^p f(t) = D^0 f(t) = f(t). \tag{4}$$

Lemma 1 ([61]). *The Laplace transform of the Caputo fractional derivative rule reads*

$$\mathbf{L}(D_t^p f(t)) = s^p \mathbf{F}(s) - \sum_{k=0}^{n-1} s^{p-k-1} f^{(k)}(0), \quad (p > 0, n-1 < p \le n). \tag{5}$$

Particularly, when $0 < p \le 1$, we have

$$\mathbf{L}(D_t^p f(t)) = s^p \mathbf{F}(s) - s^{p-1} f(0). \tag{6}$$

Lemma 2 ([11]). *The Laplace transform of the Riemann–Liouville fractional integral rule satisfies*

$$\mathbf{L}(J^q f(t)) = s^{-q} \mathbf{F}(s), \quad (q > 0). \tag{7}$$

2.1 Stability of linear fractional systems

Consider the following linear fractional system:

$$D_t^{p_i} x_i(t) = \sum_{j=1}^{n} a_{ij} x_j(t), \quad i = 1, 2, \ldots, n, \tag{8}$$

where p_i is a rational number between 0 and 1 and $D_t^{p_i}$ is the Caputo fractional derivative of order p_i, for $i = 1, 2, \ldots, n$. Assume that $p_i = \frac{\alpha_i}{\beta_i}$, $(\alpha_i, \beta_i) = 1$, $\alpha_i, \beta_i \in \mathbb{N}$, for $i = 1, 2, \ldots, n$. Let d be the least common multiple of the denominators β_i's of p_i's.

Lemma 3 ([15]). *If p_i's are different rational numbers between 0 and 1, then the system (8) is asymptotically stable if all roots λ of the equation*

$$\det(\operatorname{diag}(\lambda^{dp_1}, \lambda^{dp_2}, \ldots, \lambda^{dp_n}) - A) = 0, \tag{9}$$

satisfy $|\arg(\lambda)| > \frac{\pi}{2d}$, where $A = (a_{ij})_{n \times n}$.

2.2 Fractional Lyapunov method

Definition 4. A continuous function γ is said to belong to class-K if it is strictly increasing and $\gamma(0) = 0$.

Theorem 5 ([13]). *Let $X = 0$ be an equilibrium point for the following fractional-order system:*

$$D_t^p X(t) = F(X(t)), \tag{10}$$

where $0 < p \leq 1$. Assume that there exists a Lyapunov function $V(X(t))$ and class-K functions γ_i $(i = 1, 2, 3)$ satisfying

$$\gamma_1(\|X\|) \leq V(X(t)) \leq \gamma_2(\|X\|), \tag{11}$$

$$D_t^p V(X(t)) \leq -\gamma_3(\|X\|). \tag{12}$$

Then the system (10) is asymptotically stable.

Theorem 6 ([6]). *If there exists a positive definite Lyapunov function $V(X(t))$ such that $D_t^p V(X(t)) < 0$, for all $t > 0$, then the trivial solution of system (10) is asymptotically stable.*

In the following, a new lemma for the Caputo fractional derivative is presented.

Lemma 7 ([1]). *Let $X(t) \in \mathbf{R}^n$ be a continuous and derivable function. Then, for any time instant $t \geq t_0$*

$$\frac{1}{2} D_t^p (X^T(t)X(t)) \leq X^T(t) D_t^p (X(t)). \tag{13}$$

3 Master-Slave systems description

Suppose that the master system in the form

$$D_t^{p_i} x_i(t) = \sum_{j=1}^{3} a_{ij} x_j(t) + f_i(X(t)), \quad i = 1, 2, 3, \tag{14}$$

where $X(t) = (x_1(t), x_2(t), x_3(t))^T$ is the state vector of the master system, $(a_{ij})_{3\times3}$, $f_i : \mathbf{R}^3 \to \mathbf{R}$, $i = 1,2,3$, are nonlinear functions, p_i are different rational numbers between 0 and 1, $D_t^{p_i}$ is the Caputo fractional derivative of order p_i, for $i = 1,2,3$.

The slave system is defined by

$$D_t^{q_i} y_i(t) = \sum_{j=1}^{4} b_{ij} y_j(t) + g_i(Y(t)) + u_i, \quad i = 1,2,3,4, \tag{15}$$

where $Y(t) = (y_1(t), y_2(t), y_3(t), y_4(t))^T$ is the state vector the slave system, $(b_{ij})_{4\times4}$, $g_i : \mathbf{R}^4 \to \mathbf{R}$, $i = 1,2,3,4$, are nonlinear functions, q_i are different rational numbers between 0 and 1, $D_t^{q_i}$ is the Caputo fractional derivative of order q_i, for $i = 1,2,3,4$ and u_i, $i = 1,2,3,4$, are controllers.

4 Coexistence of FSHPS and IFSHPS

Definition 8. We say that full state hybrid projective synchronization (FSHPS) and inverse full state hybrid projective synchronization (IFSHPS) coexist in the synchronization of the systems (14) and (15), if there exists controllers u_i, $1 \le i \le 4$, and real numbers $(\alpha_i)_{1\le i\le4}$, $(\beta_i)_{1\le i\le3}$, $(\gamma_i)_{1\le i\le4}$, $(\theta_i)_{1\le i\le3}$ such that the synchronization errors

$$\begin{aligned} e_1(t) &= \sum_{j=1}^{4} \alpha_j y_j(t) - x_1(t), \\ e_2(t) &= y_2(t) - \sum_{j=1}^{3} \beta_j x_j(t), \\ e_3(t) &= \sum_{j=1}^{4} \gamma_j y_j(t) - x_3(t), \\ e_4(t) &= y_4(t) - \sum_{j=1}^{3} \theta_j x_j(t), \end{aligned} \tag{16}$$

satisfy

$$\lim_{t\longrightarrow+\infty} e_i(t) = 0, \quad i = 1,2,3,4. \tag{17}$$

4.1 Analytical result

The error system (16) can be derived as follows;

$$\frac{d^{q_1}}{dt^{q_1}} e_1(t) = \sum_{j=1}^{4} \alpha_j \frac{d^{q_1}}{dt^{q_1}} y_j(t) - \frac{d^{q_1}}{dt^{q_1}} x_1(t), \tag{18}$$

$$\frac{d^{q_3}}{dt^{q_3}}e_2(t) = \frac{d^{q_2}}{dt^{q_2}}y_2(t) - \frac{d^{q_2}}{dt^{q_2}}\left[\sum_{j=1}^{3}\beta_j x_j(t)\right],$$

$$\frac{d^{q_3}}{dt^{q_3}}e_3(t) = \sum_{j=1}^{4}\gamma_j \frac{d^{q_3}}{dt^{q_3}}y_j(t) - \frac{d^{q_3}}{dt^{q_3}}x_3(t),$$

$$\frac{d^{q_4}}{dt^{q_4}}e_4(t) = \frac{d^{q_4}}{dt^{q_4}}y_4(t) - \frac{d^{q_2}}{dt^{q_2}}\left[\sum_{j=1}^{3}\theta_j x_j(t)\right].$$

Furthermore, the error system (18) can be written as

$$\begin{aligned}
\frac{d^{q_1}}{dt^{q_1}}e_1(t) &= \sum_{j=1}^{4}(b_{1j} - c_{1j})e_j(t) + \alpha_1 u_1 + R_1,\\
\frac{d^{q_3}}{dt^{q_3}}e_2(t) &= \sum_{j=1}^{4}(b_{2j} - c_{2j})e_j(t) + u_2 + R_2,\\
\frac{d^{q_3}}{dt^{q_3}}e_3(t) &= \sum_{j=1}^{4}(b_{3j} - c_{3j})e_j(t) + \gamma_3 u_3 + R_3,\\
\frac{d^{q_4}}{dt^{q_4}}e_4(t) &= \sum_{j=1}^{4}(b_{4j} - c_{4j})e_j(t) + u_4 + R_4,
\end{aligned} \tag{19}$$

where $(c_{ij}) \in \mathbb{R}^{4\times 4}$ are control constants and

$$\begin{aligned}
R_1 &= \sum_{j=1}^{4}(c_{1j} - b_{1j})e_j(t) + \sum_{j=2}^{4}\alpha_j \frac{d^{q_1}}{dt^{q_1}}y_j(t)\\
&\quad + \alpha_1\left(\sum_{j=1}^{4} b_{1j}y_j(t) + g_1(Y(t))\right) - \frac{d^{q_1}}{dt^{q_1}}x_1(t),\\
R_2 &= \sum_{j=1}^{4}(c_{2j} - b_{2j})e_j(t) + \sum_{j=1}^{4} b_{2j}y_j(t)\\
&\quad + g_2(Y(t)) - \frac{d^{q_2}}{dt^{q_2}}\left[\sum_{j=1}^{3}\beta_j x_j(t)\right],\\
R_3 &= \sum_{j=1}^{4}(c_{3j} - b_{3j})e_j(t) + \sum_{\substack{j=1\\ j\neq 3}}^{4}\gamma_j \frac{d^{q_3}}{dt^{q_3}}y_j(t)\\
&\quad + \gamma_3\left(\sum_{j=1}^{4} b_{3j}y_j(t) + g_3(Y(t))\right) - \frac{d^{q_3}}{dt^{q_3}}x_3(t),\\
R_4 &= \sum_{j=1}^{4}(c_{4j} - b_{4j})e_j(t) + \sum_{j=1}^{4} b_{4j}y_j(t)\\
&\quad + g_4(Y(t)) - \frac{d^{q_3}}{dt^{q_3}}\left[\sum_{j=1}^{3}\theta_j x_j(t)\right].
\end{aligned} \tag{20}$$

Rewriting the error system described by equation (19) in the compact form

$$\frac{d^q}{dt^q}e(t) = (B - C) + \mathrm{diag}(\alpha_1, 1, \gamma_3, 1)U + R, \tag{21}$$

where $\frac{d^q}{dt^q} = [\frac{d^{q_1}}{dt^{q_1}}, \frac{d^{q_2}}{dt^{q_2}}, \frac{d^{q_3}}{dt^{q_3}}, \frac{d^{q_4}}{dt^{q_4}}]$, $e(t) = (e_i(t))_{1\le i\le 4}$, $B = (b_{ij})_{4\times 4}$, $C = (c_{ij})_{4\times 4}$ is a control constant matrix, $R = (R_i)_{1\le i\le 4}$, and $U = (u_i)_{1\le i\le 4}$ is the vector controller. We assume that $\alpha_1, \gamma_3 \neq 0$.

Theorem 9. *There exists a suitable feedback gain matrix C to realize the co-existence of FSHPS and IFSHPS between the master system* (14) *and the slave system* (15) *under the following control law:*

$$U = -\,\mathrm{diag}\left(\frac{1}{\alpha_1}, 1, \frac{1}{\gamma_3}, 1\right) \times R. \tag{22}$$

Proof. Applying the control law (22) to equation (21) yields the resulting error dynamics as follows:

$$\frac{d^q}{dt^q}e(t) = (B - C)e(t). \tag{23}$$

The control matrix C is chosen such that all roots λ of the equation

$$\det(\mathrm{diag}(\lambda^{Mq_1}, \lambda^{Mq_2}, \lambda^{Mq_3}, \lambda^{Mq_4}) + C - B) = 0 \tag{24}$$

satisfy $|\arg(\lambda_i)| > \frac{\pi}{2M}$, $i = 1, 2, 3, 4$ where M is the least common multiple of the dominators of q_1, q_2, q_3, and q_4. According Lemma 3, we conclude that all solutions of the error system (23) go to zero as $t \longrightarrow +\infty$. Therefore, the two systems (14) and (15) are globally synchronized. □

4.2 Numerical application

We assumed that the incommensurate fractional-order unified system is the master system and the hyperchaotic fractional-order Lorenz system is the slave system. The master system is defined as

$$\begin{aligned}
\frac{d^{p_1}}{dt^{p_1}}x_1 &= (25r + 10)(x_2 - x_1), \\
\frac{d^{p_2}}{dt^{p_2}}x_2 &= (28 - 35r)x_1 + (29r - 1)x_2 + x_1x_3, \\
\frac{d^{p_3}}{dt^{p_3}}x_3 &= \frac{-(r + 8)}{3}x_3 + x_1x_2.
\end{aligned} \tag{25}$$

This system, exhibits chaotic behavior when $(p_1, p_2, p_3) = (0.85, 0.9, 0.95)$ and $r = 1$ [7]. The chaotic attractors of the incommensurate fractional-order unified system are as shown in Figure 1.

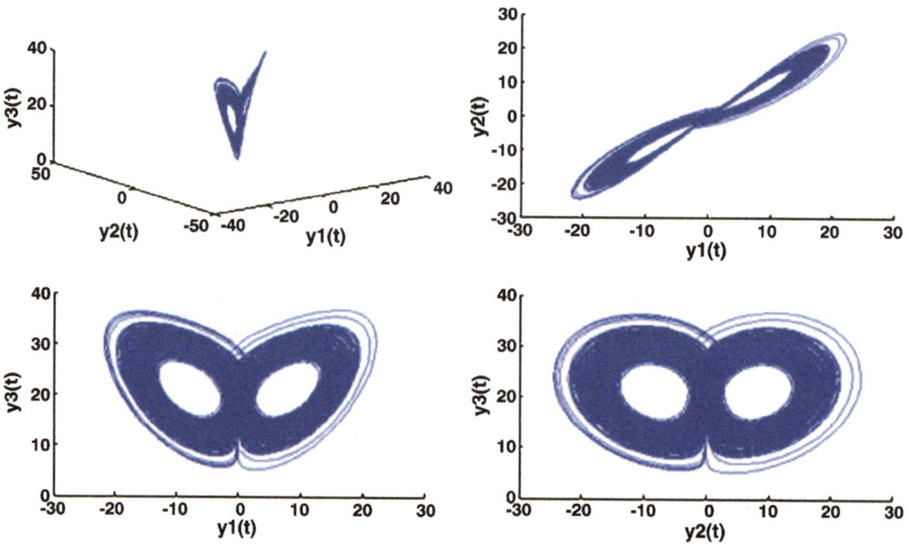

Figure 1: Chaotic attractors of the incommensurate fractional-order unified system.

The slave system is described by

$$\begin{aligned}\frac{d^{q_1}}{dt^{q_1}}y_1 &= a(y_2 - y_1) + y_4 + u_1,\\ \frac{d^{q_2}}{dt^{q_2}}y_2 &= cy_1 - y_2 - y_1y_3 + u_2,\\ \frac{d^{q_3}}{dt^{q_3}}y_3 &= -by_3 + y_1y_2 + u_3,\\ \frac{d^{q_4}}{dt^{q_4}}y_4 &= dy_4 + y_2y_3 + u_4,\end{aligned} \tag{26}$$

where $U = (u_1, u_2, u_3, u_4)^T$, $(q_1, q_2, q_3, q_4) = (0.94, 0.96, 0.97, 0.99)$ and $(a, b, c, d) = (10, \frac{8}{3}, 28, -1)$. The projections of the hyperchaotic Lorenz attractor, when $u_1 = u_2 = u_3 = u_4 = 0$, are as shown in Figure 2 [66].

Compare system (26) with systems (15), one can have

$$B = (b_{ij}) = \begin{pmatrix} -10 & 10 & 0 & 1\\ 28 & -1 & 0 & 0\\ 0 & 0 & -\frac{8}{3} & 0\\ 0 & 0 & 0 & -1 \end{pmatrix}, \quad (g_i)_{1\le i\le 4} = \begin{pmatrix} 0\\ -y_1y_3\\ y_1y_2\\ y_2y_3 \end{pmatrix}.$$

According to our synchronization approach, the errors between the master system (25) and the slave system (26) can be defined as

$$\begin{aligned} e_1 &= \alpha_1y_1 + \alpha_2y_2 + \alpha_3y_3 + \alpha_4y_4 - x_1,\\ e_2 &= y_2 - \beta_1x_1 - \beta_2x_2 - \beta_3x_3,\end{aligned} \tag{27}$$

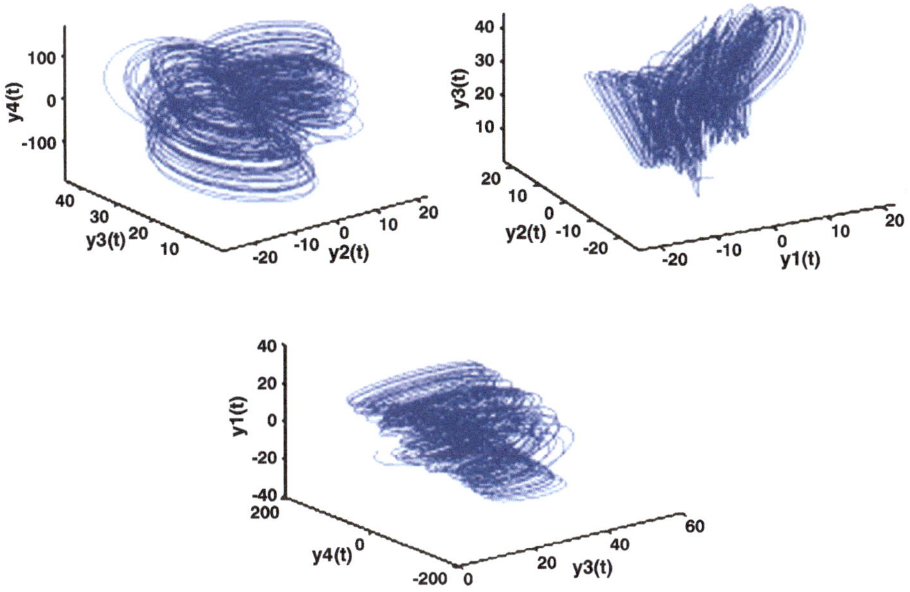

Figure 2: Chaotic attractors of the incommensurate the hyperchaotic fractional-order Lorenz system.

$$e_3 = \gamma_1 y_1 + \gamma_2 y_2 + \gamma_3 y_3 + \gamma_4 y_4 - x_3,$$
$$e_4 = y_4 - \theta_1 x_1 - \theta_2 x_2 - \theta_3 x_3,$$

where $(\alpha_1, \alpha_2, \alpha_3, \alpha_4) = (1, 0, 0, 4)$, $(\beta_1, \beta_2, \beta_3) = (1, 2, 0)$, $(\gamma_1, \gamma_2, \gamma_3, \gamma_4) = (1, 0, 3, 1)$, and $(\theta_1, \theta_2, \theta_3) = (1, 0, 1)$.

Using Theorem 9, there exists a gain matrix C so that systems (25) and (26) realize the synchronization. For example, if the control matrix C can be chosen as

$$C = \begin{pmatrix} 0 & 10 & 0 & 1 \\ 28 & 0 & 0 & 0 \\ 0 & 0 & 0 & 0 \\ 0 & 0 & 0 & 0 \end{pmatrix},$$

then the error system can be written as follows:

$$\begin{aligned} \frac{d^{0.94}}{dt^{0.94}} e_1 &= -10e_1, \\ \frac{d^{0.96}}{dt^{0.96}} e_2 &= -e_2, \\ \frac{d^{0.97}}{dt^{0.97}} e_3 &= -\frac{8}{3} e_3, \\ \frac{d^{0.99}}{dt^{0.99}} e_4 &= -e_4. \end{aligned} \tag{28}$$

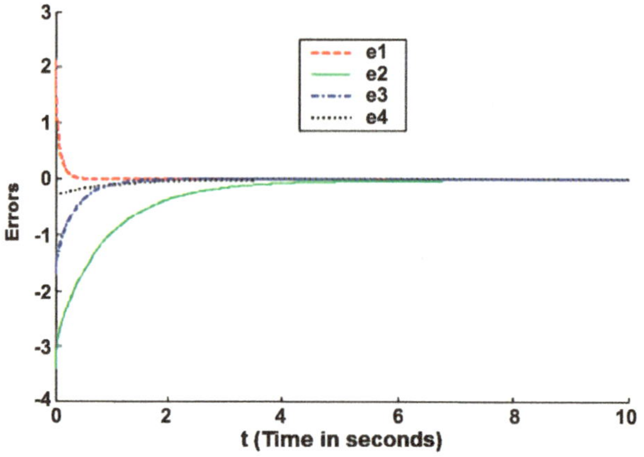

Figure 3: Time evolution of error system between the master system (25) and the slave system (26).

Then the roots of the equation $\det(\mathrm{diag}(\lambda^{M0.94}, \lambda^{M0.96}, \lambda^{0.97}, \lambda^{M0.99}) + C - B) = 0$ are

$$\begin{aligned} \lambda_1 &= 10^{\frac{1}{M0.94}}\left[\cos\left(\frac{\pi}{M0.94}\right) + \mathbf{i}\sin\left(\frac{\pi}{M0.94}\right)\right], \\ \lambda_2 &= \cos\left(\frac{\pi}{M0.96}\right) + \mathbf{i}\sin\left(\frac{\pi}{M0.96}\right), \\ \lambda_3 &= \left(\frac{8}{3}\right)^{\frac{1}{M0.97}}\left[\cos\left(\frac{\pi}{M0.97}\right) + \mathbf{i}\sin\left(\frac{\pi}{M0.97}\right)\right], \\ \lambda_4 &= \cos\left(\frac{\pi}{M0.99}\right) + \mathbf{i}\sin\left(\frac{\pi}{M0.99}\right), \end{aligned} \tag{29}$$

where M is the least common multiple of the denominators of 0.94, 0.96, 0.97, and 0.99. It is easy to see that $|\arg(\lambda_i)| > \frac{\pi}{2M}$, $i = 1, 2, 3, 4$. According to the stability theory of the fractional-order systems, we can get $\lim_{t\to\infty} \|e(t)\| = 0$. Therefore, systems (25) and (26) are globally synchronized. The numerical simulations of the error functions evolution are shown in Figure 3.

5 Coexistence of GS and IGS

Definition 10. We say that generalized synchronization (GS) and inverse generalized synchronization (IGS) coexists in the synchronization of the master system (14) and the slave system (15), if there exists controllers u_i, $1 \le i \le 4$, and differentiable functions $\phi_1, \phi_2 : \mathbf{R}^3 \to \mathbf{R}$, $\varphi_1, \varphi_2 : \mathbf{R}^4 \to \mathbf{R}$, such that the synchronization errors

$$\begin{aligned} e_1(t) &= \varphi_1(Y(t)) - x_1(t), \\ e_2(t) &= y_2(t) - \phi_1(X(t)), \end{aligned} \tag{30}$$

$$
\begin{aligned}
e_3(t) &= \varphi_2(Y(t)) - x_3(t),\\
e_4(t) &= y_4(t) - \phi_2(X(t)),
\end{aligned}
$$

satisfy that $\lim_{t\to\infty} e_i(t) = 0$, $i = 1, 2, 3, 4$.

5.1 Description of the method

The error system (30) can be derived as follows:

$$
\begin{aligned}
\dot{e}_1(t) &= \sum_{j=1}^{4} \frac{\partial \varphi_1}{\partial y_j}\dot{y}_j(t) - \dot{x}_1(t),\\
D_t^{q_2} e_2(t) &= D_t^{q_2} y_2(t) - D_t^{q_2}\phi_1(X(t)),\\
\dot{e}_3(t) &= \sum_{j=1}^{4} \frac{\partial \varphi_2}{\partial y_j}\dot{y}_j(t) - \dot{x}_3(t),\\
D_t^{q_4} e_4(t) &= D_t^{q_4} y_4(t) - D_t^{q_4}\phi_2(X(t)).
\end{aligned}
\tag{31}
$$

We suppose that the controllers u_i, $i = 1, 2, 3, 4$, can be designed in the following form:

$$
\begin{aligned}
u_1 &= -\sum_{j=1}^{4} b_{1j} y_j(t) - g_1(Y(t)) + J^{1-q_1}(v_1),\\
u_2 &= v_2,\\
u_3 &= -\sum_{j=1}^{4} b_{3j} y_j(t) - g_3(Y(t)) + J^{1-q_3}(v_3),\\
u_4 &= v_4
\end{aligned}
\tag{32}
$$

where v_i, $1 \le i \le 4$, are new controllers to be determined later. By substituting equation (32) into equation (15), we can rewrite the slave system as

$$
\begin{aligned}
D_t^{q_i} y_i(t) &= J^{1-q_i}(v_i), \quad i = 1, 3,\\
D_t^{q_i} y_i(t) &= \sum_{j=1}^{4} b_{ij} y_j(t) + g_i(Y(t)) + v_i, \quad i = 2, 4.
\end{aligned}
\tag{33}
$$

Now, applying the Laplace transform to states $y_i(t)$, $i = 1, 3$, and letting $\mathbf{F}_i(s) = \mathbf{L}(y_i(t))$, $i = 2, 4$, we obtain

$$
s^{q_i}\mathbf{F}_i(s) - s^{q_i-1} y_i(0) = s^{q_i-1}\mathbf{L}(v_i), \quad i = 1, 3 \tag{34}
$$

multiplying both the left-hand and right-hand sides of (34) by s^{1-q_i}, $i = 1, 3$, and again applying the inverse Laplace transform to the result, we obtain a new equation for the states $y_i(t)$, $i = 1, 3$,

$$
\dot{y}_i(t) = v_i, \quad i = 1, 3. \tag{35}
$$

Furthermore, the error system (31) can be written as

$$\begin{aligned} \dot{e}_1(t) &= b_{11}e_1(t) + b_{13}e_3(t) + \frac{\partial \varphi_1}{\partial y_1}v_1 + \frac{\partial \varphi_1}{\partial y_3}v_3 + R_1, \\ D_t^{q_2}e_2(t) &= b_{21}e_2(t) + b_{24}e_4(t) + v_2 + R_2, \\ \dot{e}_3(t) &= b_{31}e_1(t) + b_{33}e_3(t) + \frac{\partial \varphi_2}{\partial y_1}v_1 + \frac{\partial \varphi_2}{\partial y_3}v_3 + R_3, \\ D_t^{q_4}e_4(t) &= b_{42}e_2(t) + b_{44}e_4(t) + v_4 + R_4, \end{aligned} \tag{36}$$

where

$$\begin{aligned} R_1 &= -b_{11}e_1(t) - b_{13}e_3(t) + \frac{\partial \varphi_1}{\partial y_2}\dot{y}_2(t) + \frac{\partial \varphi_1}{\partial y_4}\dot{y}_4(t) - \dot{x}_1(t), \\ R_2 &= -b_{21}e_2(t) - b_{24}e_4(t) + \sum_{j=1}^{4} b_{2j}y_j(t) + g_2(Y(t)) - D_t^{q_2}\phi_1(X(t)), \\ R_3 &= -b_{31}e_1(t) - b_{33}e_3(t) + \frac{\partial \varphi_2}{\partial y_2}\dot{y}_2(t) + \frac{\partial \varphi_2}{\partial y_4}\dot{y}_4(t) - \dot{x}_3(t), \\ R_4 &= -b_{42}e_2(t) - b_{44}e_4(t) + \sum_{j=1}^{4} b_{4j}y_j(t) + g_4(Y(t)) - D_t^{q_4}\phi_2(X(t)). \end{aligned} \tag{37}$$

Rewriting the error system (36) in the compact form

$$\dot{e}_I(t) = B_I e_I(t) + \mathbf{J}\times V_I + R_I, \tag{38}$$

and

$$D_t^{q}e(t) = B_{II}e_{II}(t) + V_{II} + R_{II}, \tag{39}$$

where $\dot{e}_I(t) = (\dot{e}_1(t), \dot{e}_3(t))^T$, $D_t^q e_{II}(t) = (D_t^{q_2}e_2(t), D_t^{q_4}e_4(t))^T$, $B_I = \begin{pmatrix} b_{11} & b_{13} \\ b_{31} & b_{33} \end{pmatrix}$, $B_{II} = \begin{pmatrix} b_{22} & b_{24} \\ b_{42} & b_{44} \end{pmatrix}$, $\mathbf{J} = \begin{pmatrix} \frac{\partial \varphi_1}{\partial y_1} & \frac{\partial \varphi_1}{\partial y_3} \\ \frac{\partial \varphi_2}{\partial y_1} & \frac{\partial \varphi_2}{\partial y_3} \end{pmatrix}$, $R_I = (R_1, R_3)^T$, $R_{II} = (R_2, R_4)^T$, $V_I = (v_1, v_3)^T$, and $V_{II} = (v_2, v_4)^T$.

To achieve synchronization between the master system (14) and the slave system (15), we assume that the matrix $\mathbf{J}$ is invertible and it's inverse matrix is $\mathbf{J}^{-1}$. Hence, we have the following result.

Theorem 11. *There exist two control matrices C_I and C_{II} to realize the co-existence of GS and IGS between the master system* (14) *and the slave system* (15) *under the following control laws:*

$$V_I = -\mathbf{J}^{-1} \times (C_I e_I(t) + R_I), \tag{40}$$

and

$$V_{II} = -C_{II}e_{II}(t) - R_{II}, \tag{41}$$

where $C_I = \begin{pmatrix} c_{11} & c_{13} \\ c_{31} & c_{33} \end{pmatrix}$ and $C_{II} = \begin{pmatrix} c_{22} & c_{24} \\ c_{42} & c_{44} \end{pmatrix}$.

Proof. First, applying the control law (40) to equation (38) yields the resulting error dynamics as follows:

$$\dot{e}_I(t) = (B_I - C_I)e_I(t). \tag{42}$$

Construct the candidate Lyapunov function in the form: $V(e_I(t)) = e_I^T(t)e_I(t)$, we obtain

$$\begin{aligned}\dot{V}(e_I(t)) &= \dot{e}_I^T(t)e_I(t) + e_I^T(t)\dot{e}_I(t)\\ &= e_I^T(t)(B_I - C_I)^T e_I(t) + e_I^T(t)(B_I - C_I)e_I(t)\\ &= e_I^T(t)[(B_I - C_I)^T + (B_I - C_I)]e_I(t).\end{aligned}$$

If we select the control matrix C_I such that $(B_I - C_I)^T + (B_I - C_I)$ is a negative matrix, then we get $\dot{V}(e(t)) < 0$. Thus, from the Lyapunov stability theory of integer-order systems, that is,

$$\lim_{t\to\infty} e_1(t) = \lim_{t\to\infty} e_3(t) = 0. \tag{43}$$

Second, by substituting equation (41) into equation (39), one can have

$$D_t^q e_{II}(t) = (B_{II} - C_{II})e_{II}(t). \tag{44}$$

The control matrix C_{II} is chosen such that all roots λ of

$$\det(\operatorname{diag}(\lambda^{Mq_2}, \lambda^{Mq_4}) + C_{II} - B_{II}) = 0, \tag{45}$$

satisfy $|\arg(\lambda)| > \frac{\pi}{2M}$, where M is the least common multiple of the denominators of q_2 and q_4. According to Lemma 3, we conclude that the zero solution of the error system (44) is globally asymptotically stable, that is,

$$\lim_{t\to\infty} e_2(t) = \lim_{t\to\infty} e_4(t) = 0. \tag{46}$$

Finally, from equations (43) and (46), we conclude that the master system (14) and the slave system (15) are globally synchronized. □

5.2 Test of the method

As the master system, we consider the incommensurate fractional-order modified coupled dynamos system and the controlled hyperchaotic fractional-order Lorenz system as the slave system. The master system is defined as

$$D^{p_1}x_1 = -\alpha x_1 + (x_3 + \beta)x_2, \tag{47}$$

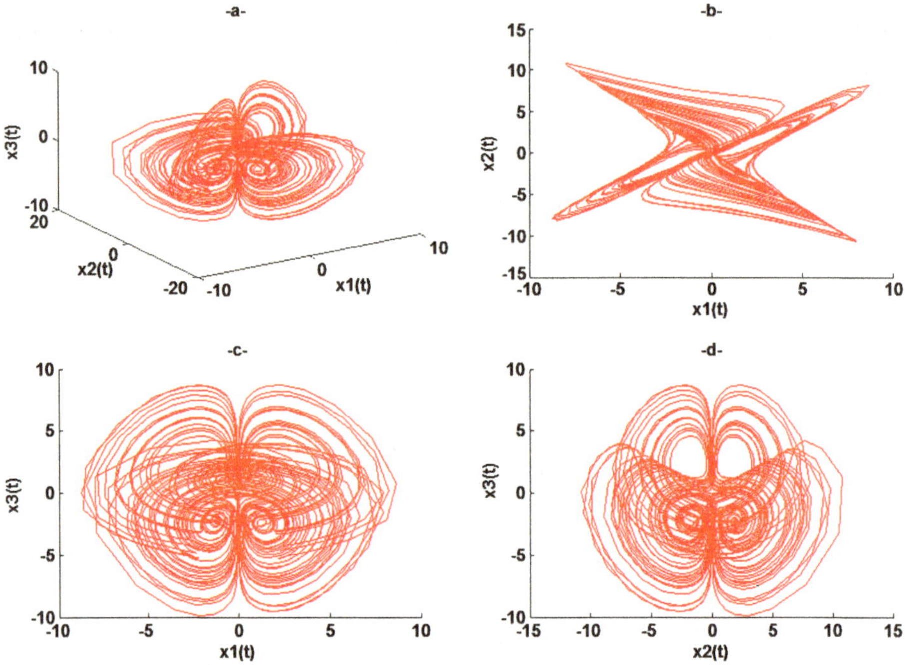

Figure 4: Chaotic attractors of incommensurate fractional-order modified coupled dynamos system.

$$D^{p_2}x_2 = -\alpha x_2 + (x_3 - \beta)x_1,$$
$$D^{p_3}x_3 = x_3 - x_1x_2.$$

This system, as shown in [64], exhibits chaotic behaviors when $(p_1, p_2, p_3) = (0.9, 0.93, 0.96)$ and $(\alpha, \beta) = (2, 1)$. The chaotic attractors of the incommensurate fractional-order modified coupled dynamos system (47) are shown in Figure 4.

The slave system is described by

$$\begin{aligned} D^{q_1}y_1 &= a(y_2 - y_1) + y_4 + u_1, \\ D^{q_1}y_2 &= cy_1 - y_2 - y_1y_3 + u_2, \\ D^{q_1}y_3 &= -by_3 + y_1y_2 + u_3, \\ D^{q_1}y_4 &= dy_4 + y_2y_3 + u_4, \end{aligned} \tag{48}$$

where $(q_1, q_2, q_3, q_4) = (0.94, 0.96, 0.97, 0.99)$ and $(a, b, c, d) = (10, \frac{8}{3}, 28, -1)$.

Compare system (48) with system (15), one can have

$$B = (b_{ij}) = \begin{pmatrix} -10 & 10 & 0 & 1 \\ 28 & -1 & 0 & 0 \\ 0 & 0 & -\frac{8}{3} & 0 \\ 0 & 0 & 0 & -1 \end{pmatrix}, \quad (g_i)_{1\le i\le 4} = \begin{pmatrix} 0 \\ -y_1y_3 \\ y_1y_2 \\ y_2y_3 \end{pmatrix}.$$

The error system between the master system (47) and the slave system (48) can be defined as

$$\begin{aligned}
e_1 &= \varphi_1(y_1, y_2, y_3, y_4) - x_1, \\
e_2 &= y_2 - \phi_1(x_1, x_2, x_3), \\
e_3 &= \varphi_2(y_1, y_2, y_3, y_4) - x_3, \\
e_4 &= y_4 - \phi_2(x_1, x_2, x_3),
\end{aligned} \tag{49}$$

where

$$\begin{aligned}
\varphi_1(y_1, y_2, y_3, y_4) &= y_1 + \frac{y_2 y_4}{2} + y_3, \\
\phi_1(x_1, x_2, x_3) &= x_1 x_2 + x_3^2, \\
\varphi_2(y_1, y_2, y_3, y_4) &= y_1 + y_2 - y_3 - y_4, \\
\phi_2(x_1, x_2, x_3) &= x_1 + x_2 x_3.
\end{aligned} \tag{50}$$

So,

$$\mathbf{J} = \begin{pmatrix} 1 & 1 \\ 1 & -1 \end{pmatrix}, \quad \text{and} \quad \mathbf{J}^{-1} = \begin{pmatrix} \frac{1}{2} & \frac{1}{2} \\ \frac{1}{2} & -\frac{1}{2} \end{pmatrix}.$$

According to Theorem 11, there exists two control matrices C_I and C_{II} so that systems (47) and (48) realize the synchronization. For example, if we select C_I and C_{II} as

$$C_I = \begin{pmatrix} 1 & 0 \\ 0 & \frac{1}{3} \end{pmatrix}, \quad C_{II} = \begin{pmatrix} 1 & 0 \\ 0 & 0 \end{pmatrix},$$

then it is easy to know that $(B_I - C_I)^T + (B_I - C_I)$ is a negative definite matrix where

$$B_I = \begin{pmatrix} -10 & 0 \\ 0 & -\frac{8}{3} \end{pmatrix},$$

also if B_I is given by

$$B_{II} = \begin{pmatrix} -1 & 0 \\ 0 & -1 \end{pmatrix},$$

we can see that the roots of $\det(\operatorname{diag}(\lambda^{M0.96}, \lambda^{M0.99}) + C_{II} - B_{II}) = 0$, satisfy $|\arg(\lambda)| > \frac{\pi}{2M}$, where M is the least common multiple of the denominators of 0.96 and 0.99. The conditions of Theorem 11 are satisfied and the synchronization between systems (47) and (48) is achieved. Hence, the error system can be described as follows:

$$\dot{e}_1 = -11 e_1, \tag{51}$$

$$\dot{e}_3 = -e_3,$$

and

$$\begin{aligned} D^{0.94} e_2 &= -2e_2, \\ D^{0.99} e_4 &= -e_4. \end{aligned} \tag{52}$$

Figures 5 and 6 display the synchronization errors between systems (47) and (48).

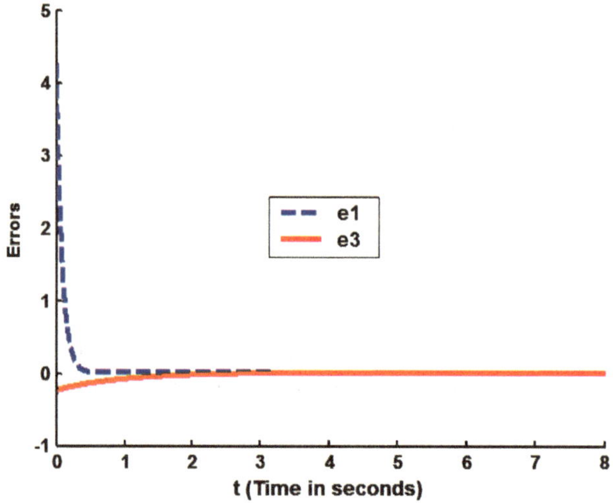

Figure 5: Time evolution of the errors e_1 and e_3.

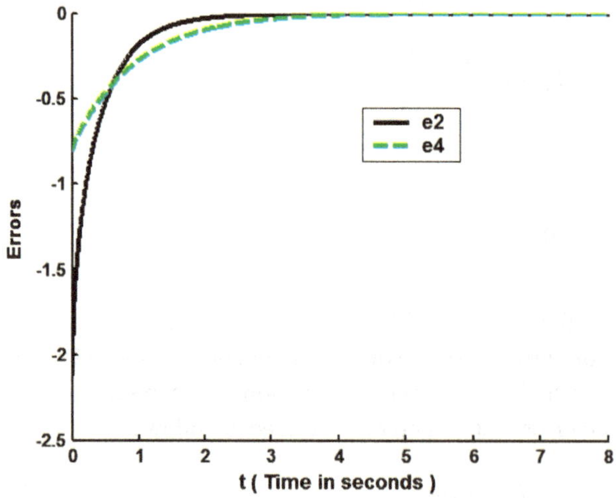

Figure 6: Time evolution of the errors e_2 and e_4.

6 Coexistence of CS, AS, and IFSHFPS

Based on the master-slave synchronizing system described by (14)–(15), the following definition of coexistence of different synchronization types can be given

Definition 12. Complete synchronization (CS), antisynchronization (AS) and inverse full state hybrid function projective synchronization (IFSHFPS) coexist in the synchronization of the master system (14) and the slave system (15), if there exist controllers u_i, $1 \le i \le 4$, and given differentiable functions $\alpha_j(t)$ $(1 \le j \le 4)$, such that the synchronization errors

$$\begin{aligned} e_1(t) &= y_1(t) - x_1(t), \\ e_2(t) &= y_2(t) + x_2(t), \\ \dot{e}_3(t) &= \sum_{j=1}^{4} \alpha_j(t) y_j(t) - x_3(t), \end{aligned} \tag{53}$$

satisfy that $\lim_{t\to\infty} e_i(t) = 0$, $i = 1, 2, 3$.

6.1 Main results

The error system (53) can be derived as

$$\begin{aligned} D_t^{q_1} e_1(t) &= D_t^{q_1} y_1(t) - D_t^{q_1} x_1(t), \\ D_t^{q_2} e_2(t) &= D_t^{q_2} y_2(t) + D_t^{q_3} x_2(t), \\ \dot{e}_3(t) &= \sum_{j=1}^{4} \dot{\alpha}_j(t) y_j(t) + \sum_{j=1}^{4} \alpha_j(t) \dot{y}_j(t) - \dot{x}_3(t). \end{aligned} \tag{54}$$

The error system (54) can be divided in two subsystems as follows:

$$\left(D_t^{q_1} e_1(t), D_t^{q_2} e_2(t)\right)^T = (B - L)\left(e_1(t), e_2(t)\right)^T + (u_1, u_2)^T + (R_1, R_2)^T, \tag{55}$$

and

$$\dot{e}_3(t) = \alpha_3(t) \dot{y}_3(t) + R_3, \tag{56}$$

where $B = (b_{ij})_{1\le i;j\le 2}$, $C = (c_{ij})_{1\le i;j\le 2}$ is a control matrix to be selected and

$$\begin{aligned} R_1 &= (c_{11} - b_{11}) e_1(t) + (c_{12} - b_{12}) e_2(t) \\ &\quad + \sum_{j=1}^{4} b_{1j} y_j(t) + g_1(Y(t)) - D_t^{q_1} x_1(t), \end{aligned} \tag{57}$$

$$R_2 = (c_{21} - b_{21})e_1(t) + (c_{22} - b_{22})e_2(t) + \sum_{j=1}^{4} b_{2j}y_j(t) + g_2(Y(t)) - D_t^{q_2}x_2(t),$$

$$R_3 = \sum_{j=1}^{4} \dot{\alpha}_j(t)y_j(t) + \sum_{\substack{j=1\\ j\neq 3}}^{4} \alpha_j(t)\dot{y}_j(t) - \dot{x}_3(t).$$

To achieve synchronization between the master (14) and the slave system (15), we assume that $\alpha_3(t) \neq 0$ for all $t \geq 0$. Hence, we have the following result.

Theorem 13. *CS, AS, and IFSHFPS coexist between the master system* (14) *and the slave system* (15) *under the following conditions:*

(i) $\binom{u_1}{u_2} = -\binom{R_1}{R_2}$, $u_3 = -\sum_{i=1}^{4} b_{3j}y_j(t) - g_3(Y(t)) + J^{1-q_3}[\frac{1}{\alpha_3(t)}((b_{33} - c)e_3(t) - R_3)]$ *and* $u_4 = 0$.

(ii) *All roots of* $\det(\operatorname{diag}(\lambda^{Mq_1}, \lambda^{Mq_2}) + C - B) = 0$, *satisfy* $|\arg(\lambda)| > \frac{\pi}{2M}$, *where M is the least common multiple of the denominators of q_1 and q_2.*

(iii) *The control constant c is chosen such that* $b_{33} - c < 0$.

Proof. First by using (i), the subsystem of errors (55) can be written as follows:

$$D_t^q \hat{e}(t) = (B - C)\hat{e}(t), \tag{58}$$

where $D_t^q \hat{e}(t) = (D_t^{q_2}e_1(t), D_t^{q_1}e_2(t))^T$. If the feedback gain matrix C is chosen as (ii). Then, according to Lemma 3, we conclude that

$$\lim_{t\longrightarrow+\infty} e_1(t) = \lim_{t\longrightarrow+\infty} e_2(t) = 0. \tag{59}$$

Now, by using the controller u_3, the state $y_3(t)$ can be described as

$$D_t^{q_3}y_3(t) = J^{1-q_3}\left[\frac{1}{\alpha_3(t)}((b_{33} - c)e_3(t) - R_3)\right]. \tag{60}$$

Applying the Laplace transform to (60) and letting $\mathbf{F}(s) = \mathbf{L}(y_3(t))$, we obtain

$$s^{q_3}\mathbf{F}(s) - s^{q_3-1}y_3(0) = s^{q_3-1}\mathbf{L}\left(\frac{1}{\alpha_3(t)}((b_{33} - c)e_3(t) - R_3)\right), \tag{61}$$

multiplying both the left-hand and right-hand sides of (61) by s^{1-q_3} and applying the inverse Laplace transform to the result, we get the following equation:

$$\dot{y}_3(t) = \frac{1}{\alpha_3(t)}((b_{33} - c)e_3(t) - R_3). \tag{62}$$

From equations (62) and (15), the dynamics of error $e_3(t)$ can be written as

$$\dot{e}_3(t) = (b_{33} - c)e_3(t), \tag{63}$$

If c is selected as (ii), then we get

$$\lim_{t\to\infty} e_3(t) = 0. \tag{64}$$

Finally, from equations (59) and (64), we conclude that the master system (14) and the slave system (15) are globally synchronized. □

When $q_1 = q_2 = q$, we can conclude the following result.

Proposition 14. *The condition* (ii) *of Theorem* 13, *can be replaced by the following condition:*
(ii) *The control matrix C is selected such that $B - C$ is a negative definite matrix.*

Proof. If a Lyapunov function candidate is chosen as $V(\hat{e}(t)) = \frac{1}{2}\hat{e}^T(t)\hat{e}(t)$. Then the time Caputo fractional derivative of order q of V along the trajectory of the system (18) is $D_t^q V(\hat{e}(t)) = D_t^q(\frac{1}{2}\hat{e}^T(t)\hat{e}(t))$, and by using Lemma 7 we get

$$\begin{aligned} D_t^q V(e(t)) &\le \hat{e}^T(t) D_t^q \hat{e}(t) \\ &= \hat{e}^T(t)(B - C)\hat{e}(t). \end{aligned}$$

If we select the matrix C such that $B-C$ is a negative matrix, then we get $D_t^q V(\hat{e}(t)) < 0$. From Theorem 6, the zero solution of the system (58) is a globally asymptotically stable, that is,

$$\lim_{t\longrightarrow+\infty} e_1(t) = \lim_{t\longrightarrow+\infty} e_2(t) = 0. \tag{65}$$

□

6.2 Numerical example

In this example, the master system is the fractional-order Liu system:

$$\begin{aligned} D^{p_1}x_1 &= a(x_2 - x_1), \\ D^{p_2}x_2 &= bx_1 - x_1x_3, \\ D^{p_3}x_3 &= -cx_3 + 4x_1^2. \end{aligned} \tag{66}$$

This system exhibits chaotic behavior when $(p_1, p_2, p_3) = (0.93, 0.94, 0.95)$ and $(a, b, c) = (10, 40, 2.5)$ [14]. Attractors of the fractional-order Liu system (66) are shown in Figure 7.

The slave system is the hyperchaotic fractional-order Lorenz system:

$$\begin{aligned} D^{q_1}y_1 &= 0.56y_1 - y_2 + u_1, \\ D^{q_2}y_2 &= y_1 - 0.1y_2y_3^2 + u_2, \end{aligned} \tag{67}$$

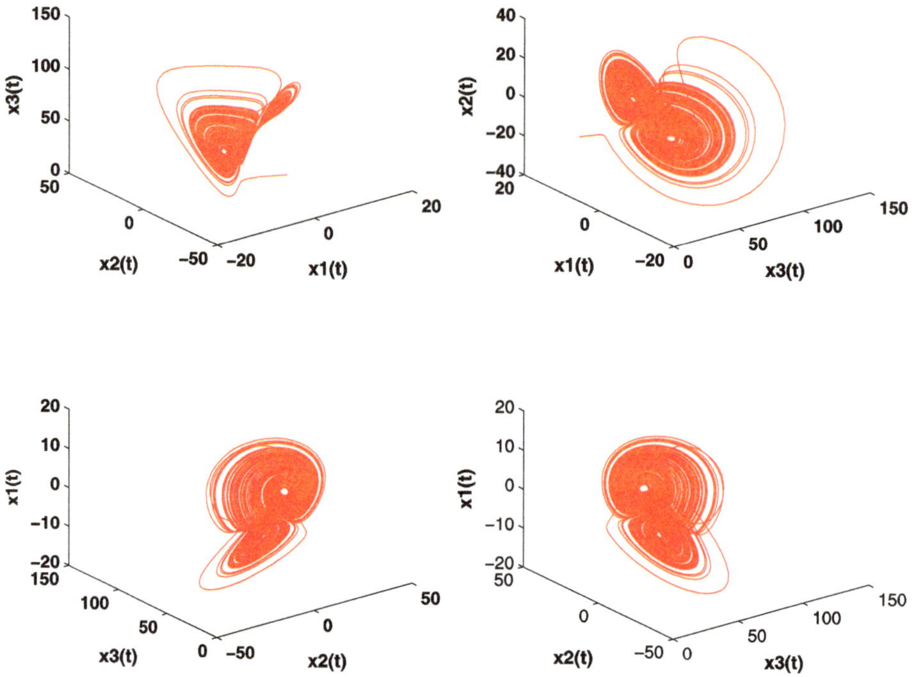

Figure 7: Chaotic attractors of the fractional-order Liu system.

$$D^{q_3}y_3 = 4y_2 - y_3 - 6y_4 + u_3,$$
$$D^{q_4}y_4 = 0.5y_3 + 0.8y_4 + u_4,$$

where u_1, u_2, u_3, u_4 are synchronization controllers. This system, as shown in [67], exhibits hyperchaotic behavior when $(u_1, u_2, u_3, u_4) = (0, 0, 0, 0)$, and $(q_1, q_2, q_3, q_4) = (0.98, 0.98, 0.95, 0.95)$. Attractors of the uncontrolled system (67) are shown in Figure 8.

The error system between the master system (66) and the slave system (67) is

$$\begin{aligned} e_1 &= y_1 - x_1, \\ e_2 &= y_2 + x_2, \\ e_3 &= \sin(t)y_1 + \cos(t)y_2 + \frac{1}{t^2+1}y_3 + y_4 - x_3. \end{aligned} \tag{68}$$

Based on the notations presented in the previous subsection, the matrix B, the control matrix C, the element b_{33} and the control constant c are given as follows:

$$B = \begin{pmatrix} 0.56 & -1 \\ 1 & 0 \end{pmatrix}, \quad C = \begin{pmatrix} 1 & -1 \\ 1 & 1 \end{pmatrix}, \quad b_{33} = -1 \quad \text{and} \quad c = 0.$$

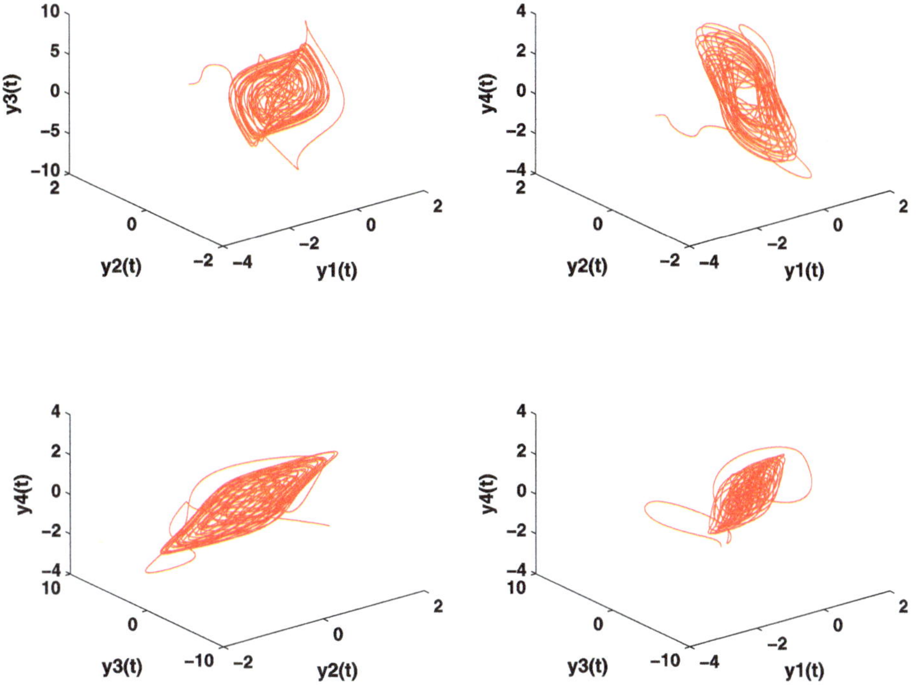

Figure 8: Attractors the hyperchaotic fractional-order Lorenz system.

and the controllers u_1, u_2, u_3 and u_4 are constructed as follows:

$$\begin{aligned} u_1 &= -0.44e_1 - 0.56y_1 + y_2 - D_t^{0.98}x_1, \\ u_2 &= -e_2 + -y_1 + 0.1y_2y_3^2 - D_t^{0.98}x_2, \\ u_3 &= -4y_2 + y_3 + 6y_4 \\ &\quad + J^{0.03}\left[(t^2+1)\left(-e_3 - \cos(t)y_1 + \sin(t)y_2 - \frac{2t}{(t^2+1)^2}y_3\right)\right], \\ &\quad + J^{0.03}[(t^2+1)(-\sin(t)y_1 - \cos(t)y_2 - y_4 + \dot{x}_3)], \\ u_4 &= 0. \end{aligned} \tag{69}$$

We can show that $B - C$ is a negative definite matrix and the condition of Proposition 14 is satisfied. Therefore, systems (66) and (67) are globally synchronized. The error systems can be written as follows:

$$\begin{aligned} D^{0.98}e_1 &= -0.44e_1, \\ D^{0.98}e_2 &= -e_2, \end{aligned} \tag{70}$$

and

$$\dot{e}_3 = -e_3. \tag{71}$$

Time evolution of the errors e_1, e_2, and e_3 are shown in Figures 9 and 10, respectively.

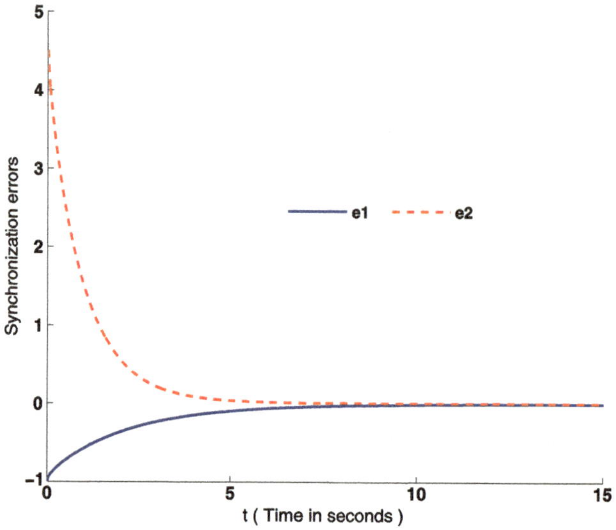

Figure 9: Time evolution of the synchronization errors e_1 and e_2 between the master system (66) and the slave system (67).

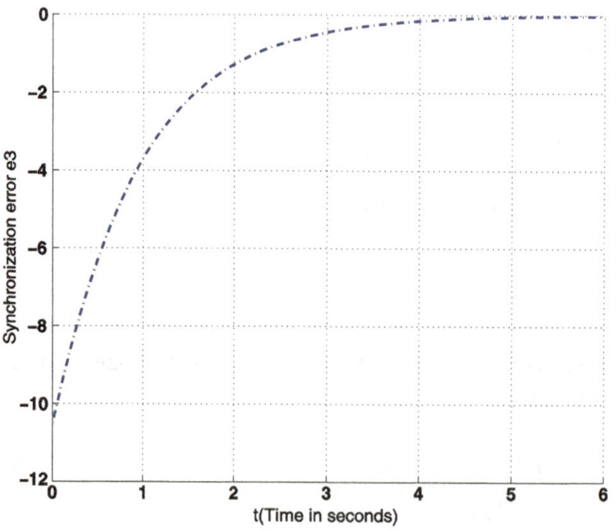

Figure 10: Time evolution of the synchronization error e_3.

7 Conclusion

In this chapter, different schemes of coexistence of some types of synchronization between fractional-order chaotic and hyperchaotic systems with different orders and dimensions have been studied. Namely, by exploiting stability theory of the linear fractional-order system, the coexistence of full state hybrid projective synchronization (FSHPS) and inverse full state hybrid projective synchronization (IFSHPS) have been proved. Additionally, by using classical Lyapunov stability theory and stability theory of linear fractional-order system, the coexistence of generalized synchronization (GS) and inverse generalized synchronization (IGS) have been rigorously proved to be achievable. Successively, based on fractional Lyapunov technique and stability theory of linear fractional-order system, different results have been derived for the coexistence of complete synchronization (CS), antisynchronization (AS) and inverse full state hybrid function projective synchronization (IFSHFPS). Finally, the effectiveness of the proposed methods has been illustrated by numerical examples.

Bibliography

[1] N. Aguila-Camacho, M. A. Duarte-Mermoud, and J. A. Gallegos, Lyapunov functions for fractional order systems, *Communications in Nonlinear Science and Numerical Simulation*, **19** (2014), 2951–2957.

[2] A. T. Azar, A. Ouannas, and S. Singh, Fractional control of co-existence of some types of chaos synchronization, in *Proceedings of the IEEE Conference ICM*, 2017.

[3] A. T. Azar, S. Vaidyanathan, and A. Ouannas, *Fractional Order Control and Synchronization of Chaotic Systems*, Springer-Verlag, Germany, 2017.

[4] S. Boccaletti, J. Kurths, G. Osipov, D. Valladares, and C. Zhou, The synchronization of chaotic systems, *Physical Report*, **366** (2002), 1–101.

[5] M. Caputo, Linear models of dissipation whose q is almost frequency independent–ii, *Geophysical Journal International*, **13** (1967), 529–539.

[6] D. Chen, R. Zhang, X. Liu, and X. Ma, Fractional order Lyapunov stability theorem and its applications in synchronization of complex dynamical networks, *Communications in Nonlinear Science and Numerical Simulation*, **19** (2014), 4105–4121.

[7] W. Deng and C. Li, The evolution of chaotic dynamics for fractional unified system, *Physical Letter A*, **372**(4) (2008), 401–407.

[8] A. Gasri and A. Ouannas, A general control method for inverse hybrid function projective synchronization of class of chaotic systems, *International Journal of Mathematical Analysis*, **9**(9) (2015), 429–436.

[9] G. Grassi, A. Ouannas, and V.-T. Pham, A general unified approach to chaos synchronization in continuous-time systems (with or without equilibrium points) as well as in discrete-time systems, *Archives of Control Sciences*, **28**(1) (2018), 5–24.

[10] R. Gorenflo and F. Mainardi, Fractional calculus: integral and differential equations of fractional order, in A. Carpinteri and F. Mainardi (eds.) *Fractals and Fractional Calculus in Continuum Mechanics*, pp. 223–276, Springer-Verlag, Wien, 2017.

[11] N. Heymans and I. Podlubny, Physical interpretation of initial conditions for fractional differential equations with Riemann–Liouville fractional derivatives, *Rheologica Acta*, **45** (2006), 765–772.
[12] C. Li and J. Yan, The synchronization of three fractional differential systems, *Chaos, Solitons and Fractals*, **22**(3) (2007), 751–757.
[13] Y. Li, Y. Chen, and I. Podlubny, Stability of fractional-order nonlinear dynamic systems: Lyapunov direct method and generalized Mittag-Leffler stability, *Computers & Mathematics with Applications*, **59** (2010), 1810–1821.
[14] C. Liu, T. Liu, L. Liu, and K. Liu, A new chaotic attractor, *Chaos, Solitons and Fractals*, **22** (2004), 1031–1038.
[15] D. Matignon, Stability results for fractional differential equations with applications to control processing, in *Proceedings of the IMACS–SMC*, pp. 963–968, 1996.
[16] P. Muthukumar, P. Balasubramaniam, and K. Ratnavelu, Synchronization of a novel fractional order stretch-twist-fold (STF) flow chaotic system and its application to a new authenticated encryption scheme (AES), *Nonlinear Dynamics*, **77** (2014), 1547–1559.
[17] Z. Odibat, Adaptive feedback control and synchronization of non-identical chaotic fractional order systems, *Nonlinear Dynamics*, **60**(4) (2010), 479–487.
[18] E. Ott, C. Grebogi, and J. Yorke, Controlling chaos, *Physical Review Letters*, **64** (1990), 1196–1199.
[19] A. Ouannas, Co-existence of complete synchronization and anti-synchronization in a class of discrete rational chaotic systems, *Far East Journal of Dynamical Systems*, **23**(1–2) (2013), 41–48.
[20] A. Ouannas, A new Q-S synchronization scheme for discrete chaotic systems, *Far East Journal of Applied Mathematics*, **84**(2) (2013), 89–94.
[21] A. Ouannas, Chaos synchronization approach for coupled of arbitrary 3-D quadratic dynamical systems in discrete-time, *Far East Journal of Applied Mathematics*, **86**(3) (2014), 225–232.
[22] A. Ouannas, A new chaos synchronization criterion for discrete dynamical systems, *Applied Mathematical Sciences*, **8**(41) (2014), 2025–2034.
[23] A. Ouannas, Nonlinear control method for chaos synchronization of arbitrary 2D quadratic dynamical systems in discrete-time, *International Journal of Mathematical Analysis*, **8**(53) (2014), 2611–2617.
[24] A. Ouannas, On full state hybrid projective synchronization of general discrete chaotic systems, *Journal of Nonlinear Dynamic*, **2014** (2014), 983293.
[25] A. Ouannas, Some synchronization criteria for n-dimensional chaotic systems in discrete-time, *Journal of Advanced Research in Applied Mathematics*, **6**(4) (2014), 1–10.
[26] A. Ouannas, Synchronization and inverse synchronization of different dimensional discrete chaotic systems via scaling matrix, *International Journal of Chaos, Control, Modelling and Simulation*, **3**(4) (2014), 1–12.
[27] A. Ouannas, A synchronization criterion for a class of sinusoidal chaotic maps via linear controller, *International Journal of Contemporary Mathematical Sciences*, **9**(14) (2014), 677–683.
[28] A. Ouannas, A new generalized-type of synchronization for discrete chaotic dynamical systems, *Journal of Computational and Nonlinear Dynamics*, **10**(6) (2015), 061019.
[29] A. Ouannas, A new synchronization scheme for general 3D quadratic chaotic systems in discrete-time, *Nonlinear Dynamics and Systems Theory*, **15**(2) (2015), 163–170.
[30] A. Ouannas, Synchronization criterion for a class of n-dimensional discrete chaotic systems, *Journal of Advanced Research in Dynamical and Control Systems*, **7**(1) (2015), 82–89.
[31] A. Ouannas, Co-existence of various synchronization-types in hyperchaotic maps, *Nonlinear Dynamics and Systems Theory*, **16**(3) (2016), 312–321.

[32] A. Ouannas, On inverse generalized synchronization of continuous chaotic dynamical systems, *International Journal of Applied and Computational Mathematics*, **2**(1) (2016), 1–11.

[33] A. Ouannas, S. Abdelmalek, and S. Bendoukha, Coexistence of some chaos synchronization types in fractional-order differential equations, *Electronic Journal of Differential Equations*, **2017**(128) (2017), 1–15.

[34] A. Ouannas and R. Abu-Saris, On matrix projective synchronization and inverse matrix projective synchronization for different and identical dimensional discrete-time chaotic systems, *Journal of Chaos*, **2015** (2015), 4912520.

[35] A. Ouannas and R. Abu-Saris, A robust control method for q-s synchronization between different dimensional integer-order and fractional-order chaotic systems, *Journal of Control Science and Engineering*, **2015** (2015), 703753.

[36] A. Ouannas and M. M. Al-sawalha, On $\lambda - \varphi$ generalized synchronization of chaotic dynamical systems in continuous-time, *The European Physical Journal Special Topics*, **225**(1) (2016), 187–196.

[37] A. Ouannas and M. M. Al-sawalha, Synchronization between different dimensional chaotic systems using two scaling matrices, *Optik—International Journal for Light and Electron Optics*, **127** (2016), 959–963.

[38] A. Ouannas, M. M. Al-sawalha, and T. Ziar, Fractional chaos synchronization schemes for different dimensional systems with non-identical fractional-orders via two scaling matrices, *Optik—International Journal for Light and Electron Optics*, **127**(20) (2016), 8410–8418.

[39] A. Ouannas and I. Aljazaery, A new method to generate a discrete chaotic dynamical systems using synchronization technique, *Far East Journal of Dynamical Systems*, **24**(1–2) (2014), 15–24.

[40] A. Ouannas, A. T. Azar, and R. Abu-Saris, A new type of hybrid synchronization between arbitrary hyperchaotic maps, *Intenational Journal of Learning Machine and Cybernitic*, **8**(6) (2017), 1887–1894.

[41] A. Ouannas, A. T. Azar, and A. G. Radwan, On inverse problem of generalized synchronization between different dimensional integer-order and fractional-order chaotic systems, in *Proceedings of the IEEE International Conference on Microelectronics (ICM)*, pp. 193–196, 2016.

[42] A. Ouannas, A. T. Azar, and S. Vaidyanathan, A new fractional hybrid chaos synchronization, *International Journal of Modelling, Identification and Control*, **27**(4) (2017), 314–323.

[43] A. Ouannas, A. T. Azar, and S. Vaidyanathan, New hybrid synchronization schemes based on coexistence of various types of synchronization between master-slave hyperchaotic systems, *International Journal of Computer Applications in Technology*, **55**(2) (2017), 112–120.

[44] A. Ouannas, A. T. Azar, and S. Vaidyanathan, On a simple approach for Q-S synchronization of chaotic dynamical systems in continuous-time, *International Journal of Computing Science and Mathematics*, **8**(1) (2017), 20–27.

[45] A. Ouannas, A. T. Azar, and S. Vaidyanathan, A robust method for new fractional hybrid chaos synchronization, *Mathematical Methods in the Applied Sciences*, **40** (2017), 1804–1812.

[46] A. Ouannas, A. T. Azar, and T. Ziar, On inverse full state hybrid function projective synchronization for continuous-time chaotic dynamical systems with arbitrary dimensions, *Differential Equations and Dynamical Systems*, (2017), 1–14.

[47] A. Ouannas and G. Grassi, Inverse full state hybrid projective synchronization for chaotic maps with different dimensions, *Chinese Physics B*, **25**(9) (2016), 090503.

[48] A. Ouannas and G. Grassi, A new approach to study co-existence of some synchronization types between chaotic maps with different dimensions, *Nonlinear Dynamics*, **86**(2) (2016), 1319–1328.

[49] A. Ouannas, G. Grassi, T. Ziar, and Z. Odibat, On a function projective synchronization scheme between non-identical fractional-order chaotic (hyperchaotic) systems with different

dimensions and orders, *Optik—International Journal for Light and Electron Optics*, **136** (2017), 513–523.
[50] A. Ouannas and A. Karouma, Different generalized synchronization schemes between integer-order and fractional-order chaotic systems with different dimensions, *Differential Equations and Dynamical Systems*, **26**(1–3) (2018), 125–137.
[51] A. Ouannas and E. E. Mahmoud, Inverse matrix projective synchronization for discrete chaotic systems with different dimensions, *Journal of Computational Intelligence and Electronic Systems*, **3**(3) (2014), 188–192.
[52] A. Ouannas and Z. Odibat, Generalized synchronization of different dimensional chaotic dynamical systems in discrete-time, *Nonlinear Dynamics*, **81**(1) (2015), 765–771.
[53] A. Ouannas, Z. Odibat, and T. Hayat, Fractional analysis of co-existence of some types of chaos synchronization, *Chaos, Solitons and Fractals*, **105** (2017), 215–223.
[54] A. Ouannas, Z. Odibat, and N. Shawagfeh, A new Q–S synchronization result for discrete chaotic systems, *Differential Equations and Dynamical Systems*, (2016), 1–10.
[55] A. Ouannas, Z. Odibat, and N. Shawagfeh, Universal chaos synchronization control laws for general quadratic discrete systems, *Applied Mathematical Modeling*, **45** (2017), 636–641.
[56] A. Ouannas, X. Wang, V.-T. Pham, G. Grassi, and T. Ziar, Co-existence of some synchronization types between non-identical commensurate and incommensurate fractional-order chaotic systems with different dimensions, *Advances in Difference Equations*, **2018** (2018), 35.
[57] A. Ouannas, X. Wang, V.-T. Pham, and T. Ziar, Dynamic analysis of complex synchronization schemes between integer order and fractional order chaotic systems with different dimensions, *Complexity*, **2017** (2017), 4948392.
[58] A. Ouannas, O. Zehrour, and Z. Laadjal, Nonlinear methods to control synchronization between fractional-order and integer-order chaotic systems, *Nonlinear Studies*, **25**(1) (2018), 91–106.
[59] L. Pecora and T. Carrol, Synchronization in chaotic systems, *Physical Review Letters*, **64** (1990), 821–824.
[60] V.-T. Pham, A. Ouannas, C. Volos, and T. Kapitaniak, A simple fractional-order chaotic system without equilibrium and its synchronization, *AEÜ. International Journal of Electronics and Communications*, **86** (2018), 67–79.
[61] I. Podlubny, *Fractional Differential Equations*, Academic Press, 1999.
[62] K. Pyragas, Continuous control of chaos by self-controlling feedback, *Physical Letters A*, **170** (1992), 421–428.
[63] S. S. Samko, A. A. Klibas, and O. I. Marichev, *Fractional Integrals and Derivatives: Theory and Applications*, Gordan and Breach, 1993.
[64] M.-J. Wang and X.-Y. Wang, Dynamic analysis of the fractional order Newton–Leipnik system, *Acta Physica Sinica*, **59**(3) (2010), 01583.
[65] X. Wang, A. Ouannas, V.-T. Pham, and H. R. Abdolmohammadid, A fractional-order form of a system with stable equilibria and its synchronization, *Advances in Difference Equations*, **2018** (2018), 20.
[66] X.-Y. Wang and J.-M. Song, Synchronization of the fractional order hyperchaos Lorenz systems with activation feedback control, *Communications in Nonlinear Science and Numerical Simulation*, **14**(8) (2009), 3351–3357.
[67] P. Zhou, L. J. Wei, and X. F. Cheng, A novel fractional-order hyperchaotic system and its synchronization, *Chinese Physical B*, **18** (2009), 2674.
[68] H. Zhu, S. Zhou, and J. Zhang, Chaos and synchronization of the fractional-order Chua's system, *Chaos, Solitons and Fractals*, **39**(4) (2009), 1595–1603.

Index

https://doi.org/10.1515/9783110571745-019